T0417720

Supercritical Fluid Nanotechnology

Supercritical Fluid Nanotechnology

Advances and Applications in Composites and Hybrid Nanomaterials

edited by

Concepción Domingo

Pascale Subra-Paternault

Published by

Pan Stanford Publishing Pte. Ltd.
Penthouse Level, Suntec Tower 3
8 Temasek Boulevard
Singapore 038988

Email: editorial@panstanford.com
Web: www.panstanford.com

British Library Cataloguing-in-Publication Data
A catalogue record for this book is available from the British Library.

Supercritical Fluid Nanotechnology: Advances and Applications in Composites and Hybrid Nanomaterials

ISBN 978-981-4613-40-8 (Hardcover)
ISBN 978-981-4613-41-5 (eBook)

Printed in the USA

Contents

Preface

Nanotechnology development is directly linked to long-term energy and environment sustainability. However, many new nanomaterials require new commercial production techniques. In this respect, more and more industries are recognizing compressed and supercritical CO_2 as a powerful green and safe technology for nanomaterial design and manufacturing. Supercritical CO_2 technology has made a transition over the past 25 years from a laboratory curiosity to a large-scale commercial reality for materials processing, with very diverse applications, such as pharmaceuticals, nutraceuticals, polymers, and textiles. Moreover, the use of recycled CO_2 in industries instead of more pollutant solvents would mitigate the CO_2 detrimental effect on climate change.

This book illustrates the basis of currently important supercritical CO_2 processing techniques, as well as the main laboratory and industrial applications. The chapters in this book provide tutorial accounts of topical areas to better understand the capacity of this environmentally friendly technology for creating and manipulating nanoscale materials for the next generation of products and technologies.

C. Domingo
P. Subra-Patternault

Chapter 1

Sustainable Processing and Nanomanufacturing

Concepción Domingo
Materials Science Institute of Barcelona (CSIC),
Campus de la UAB, 08193 Bellaterra, Spain
conchi@icmab.es

Nanotechnology development is directly linked to the long-term energy and environment sustainability. Hence, the critical aspects of sustainable nanotechnology, such as life cycle assessment, green synthesis, green energy, industrial partnerships, environmental and biological fate, and the overall sustainability of engineered nanomaterials, must always be addressed for designing a development. Many green nanomaterials require new commercial production techniques. In this respect, more and more industries are recognizing $scCO_2$ technology as a promising green technology for nanomaterial manufacturing, since increased environmental awareness had led to restrictions on previously used toxic solvents. This chapter covers the transition of $scCO_2$ technology over the past 25 years from a laboratory curiosity to a commercial reality for

Supercritical Fluid Nanotechnology: Advances and Applications in Composites and Hybrid Nanomaterials
Edited by Concepción Domingo and Pascale Subra-Paternault

ISBN 978-981-4613-40-8 (Hardcover), 978-981-4613-41-5 (eBook)
www.panstanford.com

nanomaterials processing, with applications not only in high-added value products, such as pharmaceuticals, nutraceuticals, foods and flavors, polymers, and chemicals, but also to mass commodity products, such as textiles and concrete. Chapter 2 addresses the fundamentals of fluids at supercritical conditions, while Chapter 3 treats the critical properties of confined fluids into nanometric pores.

1.1 Nanotechnology and Nanoproducts

Nanotechnology is commonly defined as the design, control, and structuring of matter of sizes less than 100 nm, at least in one dimension, to create materials with fundamentally new properties and functions [1]. The meaning of nanotechnology varies from field to field of science and it is widely used as a "catch all" description for anything very small. Nanotechnology encompasses two main approaches, (i) the top-down, in which large structures are reduced in size to the nanoscale, while maintaining their original properties (e.g., miniaturization in the domain of electronics) or deconstructed from larger structures into their smaller, composite parts, and (ii) the bottom-up, in which materials are engineered from atoms or molecular components through a process of either forced or self-assembly.

Global world market predictions show that by 2015 nanoproducts will reach a 10% of the total industrial output of materials, representing about $2.5 trillion business and more than 1 million workers involved in R&D, production, and related activities [2]. However, further than figures themselves, more impressing seems to be the rapid evolution of nanobased applications and their expansion to new technological areas. Nanoelectronics and nanoenergy have currently the highest visibility in nanotechnology, but the greatest short-term business opportunities lie in the materials for the medicine sector, mainly due to the already attained progress in manufacturing nanoparticles that can be now exploited to prepare useful nanoproducts [3, 4]. Nanostructured hybrid materials are an especial class of composite systems comprising organic, inorganic, or biological components distributed at the nanometric scale. The synergistic combination of at least two of these components in a single material at the nanosize level provides novel properties for the development of multifunctional materials [5]. There are two key approaches for the creation of composite structures at the nanometer scale, self-aggregation and dispersion. Nanostructured self-aggregated hybrid composites are materials with spatially well-defined domains for each component and with control of their mutual arrangement at the nanolevel. On the contrary, the combination by dispersion of small nanoparticles (fillers) within soft continuous matter, particularly polymers, or a second particulate phase allows the easy preparation of hybrid materials with improved properties but with a disordered nanostructure. The fascinating characteristics of these unique nanocomposites enable a wide range of applications in the fields of energy, biomedicine, optoelectronics, etc. [6]. Moreover, to effectively explore the remarkable properties of nanoparticles and to manipulate them to form nanostructured hybrid composites, one essential step is the surface modification or functionalization of the nanoparticles. The reason is that dissimilar phases in the composite are often incompatible due to low interfacial interactions [7]. The lack of interfacial interaction, coupling or bonding between the components could lead to the preparation of hybrid materials with nonisotropic properties and relatively poor mechanical behavior that limited their applications [8]. This fact becomes especially relevant for nanometric components, which have a large surface-area-to-volume ratio.

1.2 Nanomanufacturing

In their widest sense, nanomanufacturing has been used by industries for decades (semiconductors) and, in some cases, considerably longer (chemicals). However, developments over the past 20 years in the tools used to characterize materials (microscopes) have led to an increased understanding of the behavior and properties of matter at very small size scales. Currently, nanomanufacturing has developed to a stage that allows the large-scale production of different-tailored single-component nanosized entities, ranging from inorganic nanoparticles to carbon nanotubes [9]. At this point, a major challenge is to demonstrate the feasibility of up-scaling the fabrication of complex nanostructured products or devices. Current manufacturing bottom-up methods of nanostructured materials have severe limitations for mass production. On one side, the rapid condensation vapor-related physical routes lead to products with low contamination levels, but they are not easily scaled up at a reasonable cost. On the other hand, liquid-related chemical bulk approaches provide large quantities of nanosized entities at a low cost but with reduced purity. Furthermore, with size reduction to nanoscale, classical solvent approaches may be destructive for the production of complex nanostructures, since the liquid solvent itself can damage the extremely reactive surfaces that it is helping to create. This is due to undesired liquid viscosity and surface tension effects. Besides, nanoentities are extremely difficult to manipulate in liquid solvents and to integrate them into final products whilst avoiding agglomeration, degradation, or contamination. Finally, the efficient dispersion of nanometric fillers in the bulk of diverse matrices through their surface modification is a process technically needed. Consequently, and to effectively move nanoproducts commercialization forward during the next decades, new up-scaling approaches are required and, preferentially, minimization of the use of hazardous materials such as toxic volatile organic solvents [10].

In general, there are two alternatives to conventional solvent technology. The first option is to use processes not involving solvents. This includes top-down approaches (e.g., wet milling and high-pressure homogenization) and bottom-up physical techniques (e.g., vapor condensation and freeze drying), which are all difficult to apply to the processing of thermally labile compounds. On the list of damaging chemicals, solvents rank highly because they are often used in huge amounts and because they are volatile liquids that are difficult to contain. Thus, the second option is to replace toxic organic solvents by intrinsic benign solvents, applying the principles of green chemistry. Water is the obvious choice for research in green chemistry next to the development of environmentally benign new solvents, so-called neoteric solvents. Water is rarely seen as the solvent of choice in which to carry out synthetic chemistry, simply because many substances are not soluble in water. The term "neoteric solvents" refers principally to ionic liquids and supercritical fluids that have remarkable new properties and that offer a potential for industrial application [11]. These neoteric solvents are characterized by physical and chemical properties that can be finely tuned for a range of applications by varying the chemical constituents in the case of ionic liquids and by varying physical parameters in the case of supercritical fluids.

1.3 Green Technology

Green technology encourages innovation and promotes the creation of products that are both environmentally and economically sustainable. Nowadays, although most of the current products and goods are obtained using (nongreen) conventional methodologies, the relevance of the green counterpart is increasing rapidly. Green industry aims at designing products and processes that eliminate or reduce significantly the use and generation of hazardous substances and prevent negative environmental and health impacts. Hence, green technology, at its core, is about sustainability and efficiency and this means smarter use of fewer resources—raw materials, volatile organic solvents, energy, equipment, facilities, and time—as well as less waste and, finally, low cost.

There are fewer areas of more importance in green chemistry than that of solvents [12]. A solvent is defined as any substance that dissolves another substance so that the resulting mixture is a homogeneous solution. Solvents are used in manufacturing as a medium for chemical transformations, as well as in separation and extraction procedures, together with cleaning solutions. With solvents being of extremely high volume and very broad breadth of applicability, their potential for negative impact on human health and the environment is very large. In nanomanufacturing, when using liquid-related chemical bulk processes, the choice of solvent must be done in an early stage of industrial development and it is usually not questioned unless dramatic problems are encountered during up-scaling. Hence, environmentally friendly solvents must be considered in the designed processing pathway already at a lab scale. Major advances are needed to reduce the use of organic solvents (methanol, toluene, xylene, methyl ethyl ketone, methyl chloride, etc.), which account for 27% of the total toxic-release chemicals [13]. Some are recognized as carcinogens (e.g., benzene, carbon tetrachloride, trichloroethylene), while others have reproductive hazards (e.g., 2-ethoxyethanol, 2-methoxyethanol, methyl chloride) or are neurotoxics (e.g., *n*-hexane, tetrachloroethylene, toluene). Millions of workers around the world are exposed to those organic solvents that are used in a large variety of chemical industries, such as paints, varnishes, lacquers, adhesives, glues, and degreasing/cleaning agents, and in the production of dyes, polymers, plastics, textiles, printing inks, agricultural products, and cosmetics or pharmaceuticals. Common hazards include not only inherent toxicity, but also flammability, explosivity, ozone depletion, and global warming potential. When selecting a solvent for one application, often the key parameters for performance criteria are solute solubility, polarity, viscosity, and volatility. However, green chemistry suggests that reduced hazard and sustainability are equal performance criteria that need to be matched in the selection of a solvent.

The immediate environmental solution on problems related to managing waste or reducing pollution relies strongly on end-of-pipe solutions, such as waste treatment. EU policies include in the measures to reduce greenhouse gas emissions technological solutions in the chemical industry that incorporate both prevention and minimization of pollution [14]. Nowadays, a priority objective is to promote the shift from old polluting technologies to new clean technologies, incorporating environmental considerations into their design from the beginning to minimize the use of persistent organic solvents and preventing pollution from arising. Using the principles of green chemistry, new solvents or solvent systems, which reduce the intrinsic hazards associated with traditional solvents, are being developed. In some cases, new substances are being designed, while in other cases some of the best known and characterized compounds are finding new applications as solvents. Some of the leading areas of work in alternative green solvents comprise (i) aqueous solvents, (ii) supercritical fluids, (iii) ionic liquids, (iv) immobilized solvents, (v) solvent-less conditions, (vi) reduced hazard organic solvents, and (vii) fluorinated solvents [15, 16]. Water in aqueous solvents is one of the most innocuous substances on the planet; however, the correct use of water global resources and water pollution control must be taken into account. Ionic liquids or room-temperature molten salts are solvents with a negligible vapor pressure and, thus, reduced hazards [17]. Immobilized solvents or solvent molecules tethered to a polymeric backbone follow the same logic as the ionic liquids. The complete elimination of those solvent can obviously be the ultimate in eradicating the hazards associated with solvent usage. However, this is an area where the life cycle assessment must be considered, since, usually, a rigorous and material intensive separation processes is needed downstream. In a different approach, the investigation and application of various next-generation, less toxic, solvents is being pursued. The application of fluorous systems has advantages for certain types of processes, but the environmental impacts and their persistence must be first evaluated. An extremely important area of green solvent research and development is the area of supercritical and compressed fluids, particularly the area of supercritical carbon dioxide ($scCO_2$) technology [18–21]. In this case, an innocuous, extremely well-characterized substance is used as the solvent system. What is more, the US Environmental Protection Agency has performed a scrutiny over existing 25 next-generation environmental technologies [22] focused on the redesign, at the molecular and nanoscale levels, of manufacturing processes with the aim of reducing the use of hazardous materials. The report covered a wide range of near-term application technologies in varying economic sectors and at differing stages of development and commercialization. Compressed CO_2 fluid technology has been chosen as the top case 1 by considering the range of current development from early research to a full deployment in profitable businesses.

1.4 Supercritical CO_2 Fluid Technology

1.4.1 Physicochemical Characteristics of Supercritical Fluids

The critical point is defined by the temperature (T_c) and pressure (P_c) coordinates, above which no physical distinction exists between the liquid and the gas phases (Fig. 1.1a). By using the unique properties of $scCO_2$ (Fig. 1.1b), which are intermediate between those of liquids (high density that allows dissolving several compounds of low vapor pressure) and gases (low viscosity and null surface tension that enhance mass transfer properties), this fluid can be used to design innovative processes taking advantage of both above-mentioned physical and chemical nanofabrication bottom-up approaches [23]. In the supercritical region, the physical properties values can be varied continuously by modifying the pressure and/or the temperature, which allows the tailoring of fluids with the characteristics required specifically for each application. The solvent capacity of a supercritical fluid is related with its density and, then, depends on the pressure and the temperature (Fig. 1.1c). The supercritical region is highly compressible. In the area close to the critical point, small fluctuations in pressure lead to a large variation in density and, as a consequence, it is in this area where most of the synthesis and materials processing are carried out. In the compressed region, the dependence between variables can be calculated using cubic

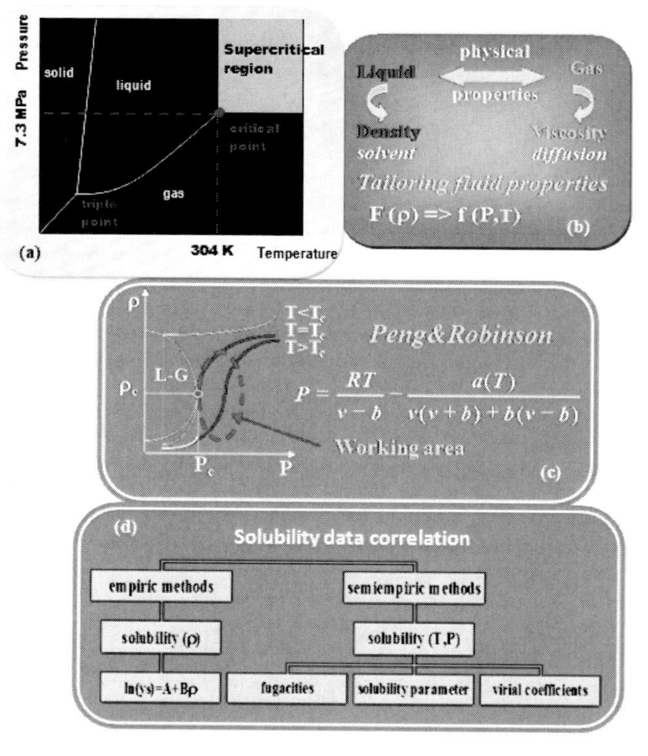

Figure 1.1 CO_2 pressure–temperature behavior: (a) phase diagram, (b) physicochemical properties, (c) schematic density diagram and cubic equation of state, and (d) approximations for solubility data correlation.

equations of state, like Peng–Robinson [24–27] in Fig. 1.1c. The measurement of solubility values at high pressure and temperature is expensive and time consuming. Hence, several empirical and semi-empirical approximations have been developed to correlate data [28, 29] (Fig. 1.1d). Using this approximations, solubility diagrams as a function of density or pressure and temperature (Fig. 1.1c) can be build, needing only a reduced amount of experimental data.

1.4.2 Historical Perspective of $scCO_2$ Technology

When Prof. Andrews of Queens College [30] (U.K.) determined the critical point of CO_2 in 1875 and later when, at a meeting of the Royal Society (London) in 1879, Hannay and Hogarth reported that "compressed gases have a pressure-dependent dissolving power" [31], they could probably not envisage that huge supercritical plants will be built in the 1970s and 1980s in Germany, decaffeinating coffee with $scCO_2$ in a process that eliminates regulatory concerns about solvent residues, environmental concerns about ozone depletion or hydrocarbon emissions, and worker safety concerns, while simultaneously producing a product with superior taste and at a cost competitive with the standard organic solvent-based processes. Starting in the 1960s, many research groups, primarily in Europe, and then later in the U.S., examined $scCO_2$ technology for developing advanced extraction processes [32].

Extraction was emphasized from botanical substrates, such as hops, spices, and herbs, and by the 1980 there were several large supercritical fluid extraction plants in operation in Germany, the U.K., and the U.S. It was not until the 1980s that supercritical fluid technology started to be developed for nonextractive industrial applications. In this decade and during the 1990s, the production of complex materials using supercritical fluids was explored, the formation of ultrafine particles being foremost among them. Even though during this time supercritical fluid technology showed a huge progress related to materials production on a laboratory scale, created innovations did not easily reached industrial implementation. Reasons have been identified as a lack of awareness of the potentials of CO_2 processing as industrial instead of academic technology, the complexities encountered in the replacement of current mature technology by still insecure technology, insufficient controlled evidence that CO_2 processes lead to value-added products, and the use of pressure equipment together with the general impression of the high initial capital cost need for supercritical fluid processes relative to other technologies. However, the 21st century is witness to many supercritical fluid advanced processes that have been developed to exploit the properties of $scCO_2$, since first the pharmacy industry and then others have finally recognized the technical, regulatory, and market attributes of supercritical fluids. Reasons are related to a reduced loss of production due to minimization of thermal and chemical degradation, health motivations in regard of solvent use and residual traces side reactions, and environmental concepts linked to the need of replacing organic solvents and minimizing waste disposal. Currently, the challenges of introducing the CO_2 fluid technology into manufacturing lines producing single solid micro- and nanoparticles have been overcome and several scaled-up plants are currently in operation [33]. Industrial applications for compressed supercritical technology have been mostly applied to natural products and food/cosmetics/pharmaceutical manufacturing lines, where scaled-up plants are currently in operation with extremely low waste factors. Large particle size reduction plants have been built and are operated for food applications, especially for precipitation derived processes, providing a basis for scale-up and cost estimation for the industrial production of nanoproducts. Although $scCO_2$ technology is originally European (Germany), its development for specific applications, including pharmaceuticals, has been more rapidly implemented in the U.S. Moreover, emerging countries (Brazil-Russia-India-China [BRIC]), principally China and India, are beginning to establish themselves as serious global players in $scCO_2$ extraction technology implementation. Still, the most active companies in CO_2 patent activities are German firms (BASF and Degussa/Evonik) followed by DuPont and Exxon Mobil from the U.S. Next, more than 10 Japanese companies go behind, from Hitachi to Mitsubishi.

1.4.3 $scCO_2$ Processing Advantages

CO_2 has an extended range of characteristics that facilitates materials processing:

(i) A low critical temperature (304.25 K), which allows processing of thermally labile materials.

(ii) A relatively low critical pressure (7.29 MPa), which allows for use of working standard pressures in the order of 10–25 MPa. These pressures are in the low limit of high-pressure systems and can be up-scaled in a straightforward way.

(iii) A large degree of tunability of solvent power at supercritical conditions through simple pressure changes, a characteristic used for process design.

(iv) A low dielectric constant and no dipole moment but a substantial quadrupole moment ($-1.4\ 10^{-34}\ J^{0.5}\ m^{2.5}$), which confers solubility to organic compounds of relatively low polarity (e.g., esters, ethers, lactones, epoxides, hydrocarbons, aromatics). A large solubility enhancement in $scCO_2$ is reported for fluorinated compounds and silicon atom–containing molecules. Moreover, the addition of a small amount (5–10 wt%) of a cosolvent (e.g., acetone, ethanol) to $scCO_2$ is used to increase the solubility of polar molecules in the resulting mixture. Recently, the solubility of a large number of organic compounds in $scCO_2$ has been compiled in a review book [34].

(v) A high diffusivity at supercritical conditions, which facilitates mass transfer, reducing processing times.

(vi) A gas-like viscosity and null surface tension at supercritical conditions, converting this fluid in a nondamaging ideal solvent for nanostructures and allowing the internal functionalization of nanoporous materials not possible using conventional solvents.

(vii) An additional degree of freedom related to the density of compressed CO_2, which allows the technology to carry out simultaneous control of the composition and the (nano) structure and the design of one-stage processes. Reactions and processes can be carried out homogeneously in the solution bulk and heterogeneously on specific surfaces or confined into microreactors (microemulsions, micro/mesoporous zeolites).

(viii) A gas state at ambient conditions; hence, produced materials can be isolated by simple depressurization, resulting in dry compounds and reducing reaction/postprocessing steps, which eliminate the cost of filtration (particularly complicated for nanoparticulated matter) and any energy-intensive drying procedure.

(ix) Low cost, no flammability, and negligible toxicity.

1.4.4 Green Chemistry and $scCO_2$ Technology

The 12 principles of green chemistry [35] provide a road map for engineers to implement green chemistry and can be applied to processes designed using $scCO_2$:

1. Design chemical processing to prevent waste (leaving no waste to treat): (i) $scCO_2$ fluid technology avoids or minimizes the use of persistent organic solvents, (ii) the engineered (nano) structures prepared using $scCO_2$ do not need any additional final washing and drying steps, and (iii) processes are designed to be closed-loop (after end-product fabrication the CO_2 is recycled back to the pressure vessel). It is considered a zero-waste technology.
2. Design safer products (preserving efficacy of function while reducing toxicity): (i) Health hazards are minimized by reducing the organic solvent concentration in the fabricated products, thus reducing risks of adverse side effects, while improving the conservation period of manufactured products due to the lack or residual solvent traces, and (ii) CO_2 is considered by the Food and Drug Administration (FDA) as having "Generally Recognized as Safe" (GRAS) status.
3. Design less hazardous processes (no toxicity to humans/ environment): (i) CO_2 has no flammability and (ii) negligible toxicity.
4. Use renewable raw materials and feedstock (rather than depleting): (i) Used CO_2 is a by-product of other processes and the technology does not generate new CO_2, (ii) shifting from organic solvent use must, at the end, decrease the total CO_2 output, and (iii) more efficient use of current high-cost organics and active agent resources is performed by decreasing the rate of degradation when stored by eliminating solvent inclusions.
5. Use catalysts (accelerating the reactions): The use of catalysts is a common practice in CO_2 technology, for example, catalytic hydrogenation [36], photoreduction of CO_2 [37], or immobilization of organometallic catalysts [38].
6. Avoid chemical derivatives (blocking or protecting groups): (i) $scCO_2$ is mostly a nonreactive molecule and the protection of functional groups from solvent attack is usually not necessary.

7. Maximize the incorporation of materials used in the process into the final product: (i) Product recovery in single-stage $scCO_2$ processes is generally high (ca. 80%–90%), offering significantly higher yields than conventional multistage processing, and (ii) the used compressed fluid is easily recycled and the remaining high-value raw materials are simply recovered by lowering the pressure of the CO_2.
8. Use safer solvents (innocuous): CO_2 is an environment-friendly nontoxic solvent.
9. Increase energy efficiency (energy and economic impacts should be minimized): The energy usage is reduced in the designed one-stage processes due to the minimization of postprocessing steps and in the generated volume of waste for disposal, recycling, or destruction.
10. Design chemicals and products to degrade after use: (i) Amorphous polymers plasticize and swell by absorbing $scCO_2$, becoming viscous liquids without the need of elevated temperatures, and (ii) newly designed (nano)material approaches have a tendency to shift to the use of polymer-based materials versus metal-based materials, since polymers can be more easily recycled or even biodegraded.
11. Perform in-process monitoring and control (analyze in real time to prevent pollution): (i) In-process chemical monitoring instruments for high-pressure equipment is being developed, and (ii) high-pressure reactors are supplied with sapphire windows.
12. Minimize the potential for accidents (releases to the environment, explosions): (i) CO_2 is a friendly solvent that improves working conditions when replacing organic solvents by avoiding workers' exposure to hazardous, flammable, and toxic materials, and (ii) high-pressure risks are limited by taking correct established engineering measures.

Companies' mission statements often emphasize environmentally responsible care; investments in clean technologies therefore fit in general policies. Moreover, industries constantly have to comply with the newest regulations regarding the environment. In the longer run, an intrinsic clean process will eliminate others that have damaging effects on the environment and health. Next to this, there is a growing tendency, for both consumers and investors to award products or companies that operate applying clean technology. The CO_2 used is a waste product of other chemical industries. Therefore, the application of $scCO_2$ as a solvent does not increase the anthropogenic greenhouse gas emission, as it does not generate any additional CO_2. Moreover, a net reduction of CO_2 emissions to the atmosphere is expected by using CO_2 in chemical processes that produce valuable carbon containing products. In general, the use of recycled CO_2 in industries will mitigate the CO_2 detrimental effect on climate change. These advantages could easily offset the initial high investment cost due to the pressure equipment needed and safety requirements.

1.5 Supercritical CO_2 Applications in Sustainability and Nanoprocessing

The concept of sustainability is not just about resource use, efficiency of utilization, and conservation but also contains strong social, economic, and cultural attributes. Today, it is often taken for granted that a cost–benefit analysis is needed to ensure better regulation and to avoid inefficiency in public and private investment. However, once upon a time, protection of human health and the natural environment could not always require such a severe economic analysis [39]. Supercritical fluid technology has several inherent processing advantages that could lead to cost/benefit gain, respecting the environment at the same time. Supercritical fluid nanotechnology can be designed for the decentralization of chemical processes (just-enough production), which would promote a sustainable society, with less dependence on goods transport. This means to achieve factors such as reduction in equipment cost, environmental compatibility, safety and handling through compact apparatus, and simplified conversion steps. High pressure and fast processes are effective for reducing the size of production apparatus. Even more, changing toward chemical systems that can be converted to products in a single stage gives a cost–benefit analysis of the technology favorable in many applications.

Commensurate with the goal of environmentally benign processing is the use of supercritical CO_2 for nonpolar to moderately polar solutes, and compressed water between its boiling and critical points for more polar solutes and reactants. Ethanol, a naturally derivable sustainable solvent, is suggested as a preferred cosolvent to be paired with carbon dioxide. This simple, renewable compressed fluids platform has many advantages and it can be used to achieve multiple results, particularly when used sequentially or in tandem with multiple-unit processes. In material-manufacturing processes, supercritical fluid technology remains as an attractive alternative to the use of organic solvents, which are widely recognized as pollutant or toxins. The advantage of the compressed fluid–processing route lies in its applicability in various industrial sectors, with the ability of producing primary (organic, inorganic) or complex (composite, hybrids) nanoparticles. The advantage of $scCO_2$ processing is that it cuts across many industries, from pharmaceutical to textile and from catalysis to concrete. The production of simple and complex functional high-added-value products for the food, biomedical, and pharmaceutical industries is the most developed supercritical sector. However, during the last years, technology transfer to other industries without the food/biomedical special requirements, for instance, high-tech industries (catalysis, sensors) and commodity or consumer product sectors (plastics, packaging, construction, specialty paper), is taking place speedily. $scCO_2$ technology is used for the production of high-performance existing and new products with unique characteristics in regard of composition (purity), size (micro or nanoscale), and architecture (particles, films, fibers, foams). Importantly, some classical boundaries between types of materials related to processing methods can be broken down using $scCO_2$ technology, used for the processing of a large number of products: inorganic, organic, and polymeric. $scCO_2$ technology holds the exceptional promise of creating new markets for the emerging manufacturing industries of nanoparticles, nanodevices, polymers, surface modification, and functionalization. The viability of the application of any of the developed supercritical processes must be subjected to a detailed case-by-case study.

Acknowledgments

This work was partially financed by the Generalitat de Catalunya with project Nassos 2014SGR377 and by the the Spanish Ministerio de Economía y Competitividad through the research project Superfactory CTQ2014-56324.

References

1. Nalwa, H. S. (2004). *Encyclopedia of Nanoscience and Nanotechnology* (American Scientific, USA).
2. RNCOS (2006). *The World Nanotechnology Market* (Delhi, India).
3. Nalwa, H. S. (2003). *The Nanotechnology Opportunity Report* (Cientifica, London, UK).
4. AIRA/Nanotech IT (2007). *Roadmaps at 2015 on Nanotechnology Application in the Sectors of Materials, Health & Medical Systems, Energy* (European Commission, EU).
5. Nalwa, H. S. (2002). *Nanostructured Materials and Nanotechnology* (Elsevier, The Netherlands).
6. Ramsden, J. (2014). *Applied Nanotechnology: The Conversion of Research Results to Products*, 2nd Ed. (Elsevier, The Netherlands).
7. Hakim, L. F., King, D. M., Zhou, Y., Gump, C. J., George, S. M., and Weimer, A. W. (2007). Nanoparticle coating for advanced optical, mechanical and rheological properties, *Adv. Funct. Mater.*, **17**, pp. 3175–3181.
8. Gómez-Romero, P., and Sanchez, C. (2006). *Functional hybrid materials*, 6th Ed. (Wiley-VCH, Germany).
9. Accenture FTF Report (2006). *Nanotechnology: The Industrial Revolution of the 21st Century* (Fundación de la Innovación Bankinter, Spain).
10. Borisenko, V. E., and Ossicini, S. (2004). *What Is What in the Nanoworld: A Handbook on Nanoscience and Nanotechnology*, 2nd Ed. (Wiley-VCH, Germany).
11. Seddon, K. R. (1996). Room-temperature ionic liquids-neoteric solvents for clean catalysis, *Kinet. Catal.*, **37**, pp. 693–697.
12. Abraham, M. A., and Moens, L. (2002). *Clean Solvents*, Vol. 819 (ACS Symposium Series, Washington DC, USA).
13. US Environmental Protection Agency (2013). *Toxics Release Inventory Chemical List for Reporting Year 2012*.
14. European Commission (2000). *European Climate Change Programme*, COM/2000/0088.

15. Yang, Z., and Pan, W. (2005). Ionic liquids: green solvents for nonaqueous biocatalysis-review article, *Enzyme Microb. Tech.*, **37**, pp. 19–28.

16. Nalawade, S. P., Picchioni, F., and Janssen, L. P. B. M. (2006). Supercritical carbon dioxide as a green solvent for processing polymer melts: processing aspects and applications-review article, *Prog. Polym. Sci.*, **31**, pp. 19–43.

17. Chen, S. L., Chua, G. L., Ji, S. J., and Loh, T. P. (2007) *Ionic Liquids in Organic Synthesis,* Vol. 950 (ACS Symposium Series, Washington DC, USA).

18. McHugh, M. A., and Krukonis, V. J. (1994). *Supercritical Fluid Extraction: Principles and Practice* (Butterworth-Heinemann, Boston, USA).

19. Sun, Y. (2002). *Supercritical Fluid Technology in Materials Science and Engineering* (Marcel Dekker, New York, USA).

20. Beckman, E. J. (2004). Supercritical and near-critical CO_2 in green chemical synthesis and processing, *J. Supercrit. Fluids*, **28**, pp. 121–191.

21. Machida, H., Takesue, M., and Smith Jr., R. L. (2011). Green chemical processes with supercritical fluids: properties, materials, separations and energy, *J. Supercrit. Fluids,* **60**, pp. 2–15.

22. US Environmental Protection Agency (2003). *Next Generation Environmental Technologies* (RAND Report, USA).

23. Eckert, C. A., Knutson, B. L., and Debenedetti, P. G. (1996). Supercritical fluids as solvents for chemical and materials processing, *Nature*, **383**, pp. 313–318.

24. Peng, D. Y., and Robinson, D. B. (1976). A new two-constant equation of state, *Ind. Eng. Chem. Fund.*, **15**, pp. 59–64.

25. Stryjek, R., and Vera, J. H. (1986). PRSV: an improved Peng-Robinson equation of state for pure compounds and mixtures, *Can. J. Chem. Eng.*, **64**, pp. 323–333.

26. Chrastil, I. (1982). Solubility of solids in supercritical gases, *J. Phys. Chem.*, **86**, pp. 3016–3021.

27. Lucien, F. P., and Foster, N. R. (2000). Solubilities of solid mixtures in supercritical carbon cioxide: a review, *J. Supercrit. Fluids*, **17**, pp. 111–114.

28. Goldman, S., Gray, C. G., Li, W., Tomberli, B., and Joslin, C. G. (1996). Predicting solubilities in supercritical fluids, *J. Phys. Chem.*, **100**, pp. 7246–7249.

29. Jouyban, A., Chan, H. K., and Foster, N. R. (2002). Mathematical representation of solute solubility in supercritical carbon dioxide using empirical expressions, *J. Supercrit. Fluids*, **24**, pp. 19–35.

30. Andrews, T. (1875). The Bakerian lecture: on the gaseous state of matter, *Proc. R. Soc. Lond.*, **24**, p. 455.

31. Hannay, J. B., and Hogarth, J. (1879). On the solubility of solids in gases, *Proc. R. Soc. Lond.*, **29**, p. 324.

32. Schutz, E. (2007). Supercritical fluids and applications: a patent review, *Chem. Eng. Technol.*, **30**, pp. 685–688.

33. www.nektar.com (EU), www.ferro.com (USA), and www.lavipharm.gr (EU-USA).

34. Gupta, R. B., and Shim, J. J. (2006) *Solubility in Supercritical Carbon Dioxide* (CRC Press, USA).

35. Anastas, P., and Warner, J. (1998). *Green Chemistry: Theory and Practice* (Oxford University Press, New York, USA).

36. Stephenson, P., Licence, P., Rossb, S. K., and Poliakoff, M. (2004). Continuous catalytic asymmetric hydrogenation in supercritical CO_2, *Green Chem.*, **6**, pp. 521–523.

37. Voyame, P., Toghill, K. E., Méndez, M. A., and Girault, H. H. (2013). Photoreduction of CO_2 using $[Ru(bpy)_2(CO)L]^{n+}$ catalysts in biphasic solution/supercritical CO_2 systems, *Inorg. Chem.*, **19**, pp. 10949–10957.

38. Leitner, W. (2004). Recent advances in catalyst immobilization using supercritical carbon dioxide, *Pure Appl. Chem.*, **76**, pp. 635–644.

39. Heinzerling, L., and Massey, R. I. (2007). Wrong in retrospect: cost-benefit analysis of past successes, in *Frontiers in Ecological Economic Theory and Application*, eds. Erickson, J. D., and Gowdy, J. M., (Cheltenham, UK).

Chapter 2

Fundamentals of Supercritical Fluids and the Role of Modeling

Lourdes F. Vega
MATGAS Research Center and Carburos Metálicos, Air Products Group, Campus de la UAB, 08193 Bellaterra, Spain
vegal@matgas.org

Supercritical fluid technology has allowed significant improvements in several industrial processes and there are still many opportunities for further applications in new emergent areas. To optimize the conditions at which supercritical fluid processes take place, a precise characterization of this region is needed; however, the particular behavior and singularities observed in the properties of compounds and their mixtures in the near-critical region makes the study of this region a challenge from any modeling and experimental approaches. This chapter focuses on fundamentals of the supercritical fluids. It deals the singularities found in the critical region of fluids as a consequence of the inhomogeneities in the density (and composition for the case of mixtures) and how to take them into account in a modeling approach. The theory is illustrated with the incorporation of a crossover term, based on renormalization theory,

Supercritical Fluid Nanotechnology: Advances and Applications in Composites and Hybrid Nanomaterials
Edited by Concepción Domingo and Pascale Subra-Paternault

ISBN 978-981-4613-40-8 (Hardcover), 978-981-4613-41-5 (eBook)
www.panstanford.com

into a molecular-based equation of state (soft-SAFT) and applied to some experimental systems.

2.1 Introduction: The Near-Critical Region of Fluids

2.1.1 What Is a Supercritical Fluid?

Before we get into the details of the singularities of the critical point, let's start with some definitions. A phase is a homogeneous region of matter in which there is no spatial variation in average density, energy, composition, or other macroscopic properties. The coexistence of phases in thermodynamic equilibrium with one another in a system consisting of two or more phases is called phase equilibrium. The simplest examples of phase equilibrium are the equilibrium of a liquid and its saturated vapor, such as liquid water and its vapor, and the equilibrium of a liquid in equilibrium with its solid phase, such as liquid water and ice at the melting point of ice. The temperature at which a phase transition occurs—for example, a boiling point or a melting point—changes if the pressure changes. The temperature change that results from an infinitesimal change in pressure is given by the Clapeyron equation.

Graphs that represent the interrelation of the various thermodynamic variables at phase equilibrium are called phase transition curves or surfaces; a set of such curves or surfaces is known as a phase diagram. A phase transition curve may either intersect two other phase transition curves at a triple point or terminate at a critical point. At the critical point the densities of the liquid and gas phases become equal and the distinction between them disappears, resulting in a single phase, called supercritical. A supercritical fluid is then any substance at a temperature and pressure above its critical point. Close to the critical point, small changes in pressure or temperature result in large changes in density. In addition, there is no surface tension in a supercritical fluid, as there is no liquid-/gas-phase boundary. By changing the pressure and temperature of the fluid, the properties can be "tuned" to be more liquid- or more gas-like, allowing, as a consequence, increasing the solubility or the reactivity of specific compounds in the supercritical phase.

Figure 2.1 shows a schematic of the pressure–temperature phase diagram of a substance where the critical point is shown, together with the solid, liquid, and gas phases, and the lines corresponding to two phases in equilibrium. The proximity of the molecules (related to the density of the fluid) is also sketched to illustrate the changes in density, depending on pressure and temperature.

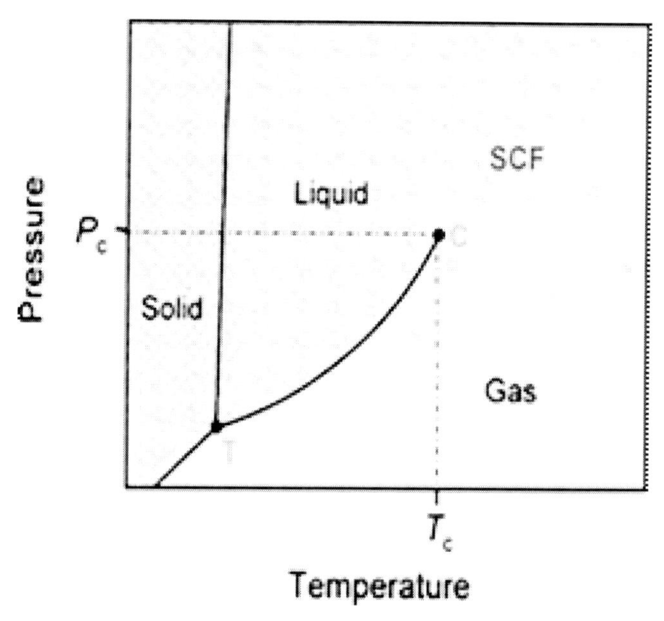

Figure 2.1 Pressure–temperature phase diagram of a fluid, showing the phase equilibria lines, the triple point, T, (where solid, liquid, and gas coincide), and the critical point, C, (where liquid and gas coincide). The supercritical fluid (SCF) region is found above the critical point.

The critical point was discovered in 1822 by Baron Charles Cagniard de la Tour in his cannon barrel experiments. He identified the critical temperature by listening to discontinuities in the sound of a rolling flint ball in sealed cannon filled with fluids at various temperatures. Above a certain temperature, the densities of the liquid and gas phases become equal and the distinction between them disappears, resulting in a single supercritical fluid phase. Figure 2.2 shows pictures of the phase transition from vapor–liquid equilibria to the critical region taken at different pressures–temperatures at equilibrium [1]. The critical temperature and pressure of some common substances are provided in Table 2.1.

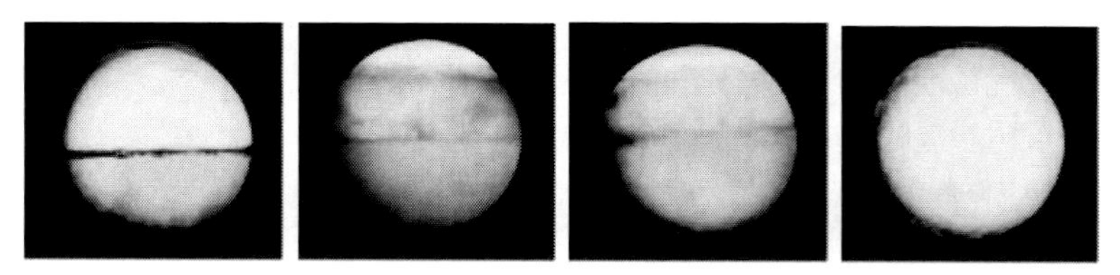

Figure 2.2 Pictures of pure CO_2 at different temperatures–pressures along equilibrium, from vapor–liquid up to the critical point [1].

Table 2.1 Critical points of common substances

Substance	Critical temperature (K)	Critical pressure (atm)
Hydrogen (H)	33.3	12.8
Nitrogen (N)	126	33.5
Argon (Ar)	151	48.5
Methane (CH_4)	191	45.8
Ethane (C_2H_6)	305	48.2
Carbon dioxide (CO_2)	305	72.9
Ammonia (NH_3)	406	112
Water (H_2O)	647	218

The properties of a supercritical fluid are those between a gas and a liquid (see Table 2.2 for viscosity, density, and diffusivity) and, as stated, can be tuned by changing the pressure or the temperature.

Table 2.2 Comparison of average values of gases, supercritical fluids, and liquids

	Density (kg/m^3)	Viscosity (μPa·s)	Diffusivity (mm^2/s)
Gases	1	10	1–10
Supercritical fluids	100–1000	50–100	0.01–0.1
Liquids	1000	500–1000	0.001

2.1.2 Molecular Singular Behavior at the Near-Critical Region

It is known that the reliable estimation of properties of fluid has a direct influence on the cost of equipment of the process operations. The increasing number of industrial applications near the critical conditions has encouraged researchers to seek for an accurate description of the thermodynamic behavior of pure fluids and their mixtures in the vicinity of the critical point. However, the complex molecular behavior of any compound at these conditions makes it a difficult task.

In the near-critical region, density and concentration fluctuations caused by long-range correlations among all the molecules lead to singularities of the pure compound properties (and mixtures) at the critical point. The value of some properties (e.g., speed of sound, isochoric heat capacity, thermal compressibility, etc.) change very rapidly with a small modification of the operating conditions. For instance, when approaching the critical point of CO_2 from the two-phase region with a density equal to the critical, the thermal conductivity tends to infinity as the temperature approaches the critical temperature. This means that unless an accurate method to describe the region is in place, small deviations in the calculations may lead to large errors for the final operating conditions of the process. As an example, Fig. 2.3 shows the behavior of the isobaric heat capacity of methanol as a function of the density for three different temperatures close to the critical point [2].

From the mathematical point of view, the approach to the critical point leads to a nonanalytic asymptotic behavior different that the one observed far away from the critical region. This is usually known as the crossover from classical to nonclassical behavior. From a microscopic point of view, the classical theory of critical points corresponds to a mean-field approximation, which neglects local inhomogeneities fluctuations in density. It is well known that the nonclassical critical behavior of the thermodynamic properties is a consequence of the long-range fluctuations of an order parameter, the density for the case of pure fluids. The spatial extend of the density fluctuations diverge at the critical point and become larger than any molecular scale in the critical region. Only theories which account for these density fluctuations can provide the correct approach to the critical point. In fact, most equations of state (EoSs), accurate to describe the vapor–liquid equilibria, cannot reproduce this change of behavior in the near-critical region because they are based in mean-field theories that do not take into account the inherent fluctuations in this region. The problem is also shared with more refined molecular-based EoSs, they also fail in the representation of the near-critical properties.

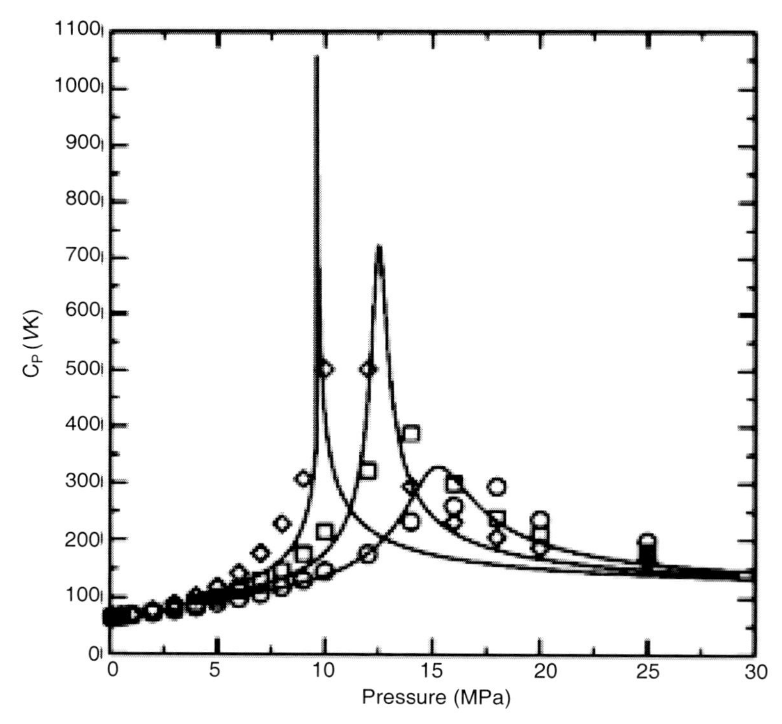

Figure 2.3 Total isobaric heat capacity of methanol at three near-critical temperatures, T = 533.15, 553.15, and 573.15 K. Symbols represent the experimental data obtained from Vargaftik et al. [3], while the lines represent soft-SAFT calculations of Llovell and Vega [2].

Hence, the rigorous estimation of vapor–liquid equilibria far from and close to the critical region requires a specific treatment taking into account the long-range fluctuations as one approaches the critical point. One of the most successful treatments comes from the theory of scaling of the Nobel Laureate Kenneth Wilson, based on how fundamental properties and forces of a system vary depending on the scale over which they are measured. He devised a strategy for calculating how phase transitions occur by considering each scale separately and then abstracting the connection between contiguous ones, in a novel appreciation of renormalization group (RG) theory [4].

On the basis of that work, some authors have developed different treatments to couple the RG theory to several EoSs, being able to reproduce the whole phase envelope with the same degree of accuracy (see, for instance, Refs. [5, 6]). These treatments are usually called *crossover theories* as they describe the crossover behavior from classical to nonanalytical as the critical region is approached. The accuracy of these so-called crossover equations lays on the accuracy of the equation to which the nonclassical behavior is coupled to, since the equation reduces to the original equation far from the critical point. Figure 2.4 illustrates the phase diagram of CO_2 obtained with a classical mean-field equation and the same one with a specific crossover treatment [6]. Note that the dashed line (mean-field equation) accurately describes the behavior of the fluid far from the critical temperature (below 290 K), but it provides an overestimation of the critical temperature. On the contrary, the equation with the crossover treatment is able to accurately describe the whole phase envelope, far from and close to the critical point.

In addition to theory, the thermodynamic properties of supercritical fluids, including thermal properties, can also be calculated by molecular simulations, and several works are available in the literature. For instance, Colina et al. [7] performed Monte Carlo simulations to calculate the volume expansivity, isothermal compressibility, isobaric heat capacity, Joule–Thomson coefficient and speed of sound for carbon dioxide in the supercritical region, using the fluctuation method based on Monte Carlo simulations in the isothermal–isobaric ensemble. However, simulations are time consuming, especially when dealing with the critical region, as taking into account long-range fluctuations requires large simulation boxes and long simulation times, making this approach of no practical use for engineering purposes with the speed of the computers and algorithms available today.

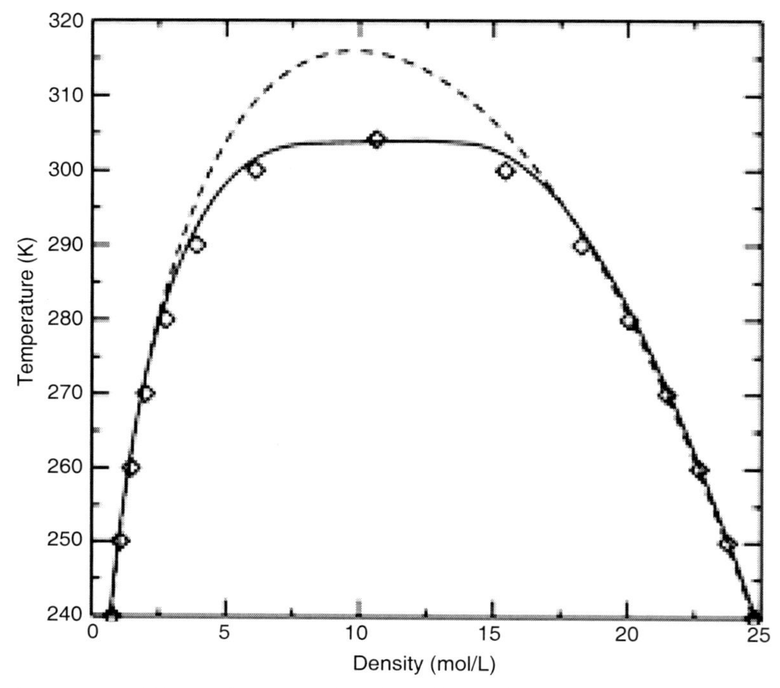

Figure 2.4 Phase diagram of CO_2 where the experimental data is represented by symbols. Dashed line: Calculation of the phase enveloping with a mean-field equation. Full line: Calculation of the phase diagram with the same equation adding a crossover term [6]. See text for details.

2.2 Incorporating Nonclassical Behavior in the Near-Critical Region: Crossover Soft-SAFT EOS

In this section we summarize the mathematical treatment required to include a crossover term, based on renormalization theory, to a molecular-based EoS, coming from the Statistical Associating Fluid Theory (SAFT, the so-called soft-SAFT equation [8–10]. It should be mentioned that the same procedure can be used in a classical cubic EoS (see, for instance, the work by Llovell et al. [11]).

2.2.1 The SAFT Approach and the Soft-SAFT Equation of State

SAFT stands for statistical associating fluid theory, as the approach is based on Wertheim's first-order thermodynamics perturbation theory (TPT1) of associating fluids [12–14]. The original equation was proposed almost 25 years ago [15–17] and it has been used since then for several researchers in both academia and industrial environments. SAFT has been especially successful in some engineering applications for which other classical EoSs fail. The success of the equation in its different versions is proved by the amount of published works since its development [18].

The key of the success of SAFT-based equations is their solid statistical–mechanics basis, which lets a physical interpretation of the system. It provides a framework in which the effects of molecular shape and interactions on the thermodynamic properties can be separated and quantified. Besides, its parameters are few in number, with physical meaning and transferable, which makes SAFT a powerful tool for engineering predictions.

SAFT equations are usually written in terms of the residual Helmholtz energy, where each term in the equation represents different microscopic contributions to the total free energy of the fluid. For associating chain systems the equation is written as

$$a^{\mathrm{res}} = a^{\mathrm{ref}} + a^{\mathrm{chain}} + a^{\mathrm{assoc}} + a^{\mathrm{polar}} \tag{2.1}$$

where a^{res} is the residual Helmholtz free energy density of the system. The superscripts *ref, chain, assoc,* and *polar* refer to the contributions from the monomer, the formation of the chain, the associating sites, and a polar contribution, respectively. Most SAFT equations differ in the reference term [8, 15–20], keeping formally identical the chain and the association term, both obtained from Wertheim's theory [12–14].

Soft-SAFT [8–10] uses as the reference term a Lennard–Jones (LJ) spherical fluid, which accounts both for the repulsive and attractive interactions of the monomers forming the chain. The free energy and derived thermodynamics of a mixture of LJ fluids are obtained through the accurate equation of Johnson et al. [21]. The chain contribution for a LJ fluid of tangent spherical segments, obtained through Wertheim's theory, in terms of the chain length m and the pair correlation function g_{LJ} of LJ monomers, evaluated at the bond length σ is

$$a^{\mathrm{chain}} = \rho\ k_{\mathrm{B}}T\sum_{i} x_i(1-m_i)\ln g_{\mathrm{LJ}} \tag{2.2}$$

where ρ is the molecular density of the fluid, T is the temperature, m is the chain length, k_{B} the Boltzmann constant, and g_{LJ} is the radial distribution function of a fluid of LJ spheres at density $\rho_m = m\rho$. We use the function fitted to computer simulation data for g_{LJ} as a function of density and temperature.

The association term, within the first-order Wertheim's perturbation theory for associating fluids, is expressed as the sum of contributions of all associating sites of component i:

$$a^{\mathrm{assoc}} = \rho k_{\mathrm{B}}T\sum_{i} x_i \sum_{\alpha}\left(\ln X_i^{\alpha} - \frac{X_i^{\alpha}}{2}\right) + \frac{M_i}{2} \tag{2.3}$$

with M_i being the number of associating sites of component i and X_i^{α} the mole fraction of molecules of component i nonbonded at site α, which accounts for the contributions of all the associating sites in each species. Figure 2.5 illustrates the molecular model of propane.

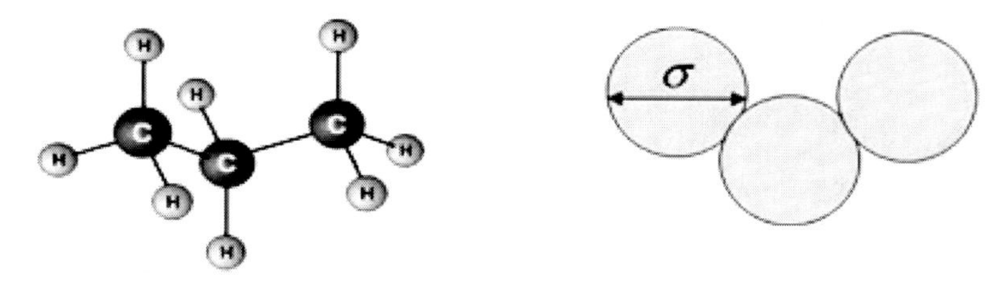

Figure 2.5 Representation of a propane molecule. Left: Chemical structure. Right: Soft-SAFT model of propane, a chain of LJ segments.

For the case of mixtures of different compounds each term in the equation should be expressed in terms of composition. Since the chain and association term in SAFT are already applicable to mixtures, the extension needs to be performed just for the reference term; extended Lorentz–Berthelot combining rules are used, together with van der Waals one-fluid theory [6].

2.2.2 The Crossover Soft-SAFT Equation

As stated, a term taking into account the long-range fluctuations in the near-critical region should be added to the equation to accurately describe the critical region. Llovell and Vega have followed the approach of incorporating White's global RG theory [22, 23] into soft-SAFT. White's approach consists of a set of recursion relations where the contribution of increasingly longer-wavelength density fluctuations up to the correlation length is successively taken into account in the free-energy density. In this way, properties approach the asymptotic behavior in the critical region, and they exhibit a crossover between the classical and the universal scaling behavior in the near-critical region.

In our work [5] we followed the implementation of White's global RG method, as done by Prausnitz and collaborators [24–26]. Since the method has been widely explained in the literature, only a brief summary of the most relevant details is presented here. To include the long-wavelength fluctuations into the free energy, the interaction potential is divided into a reference contribution, mainly due to the repulsive interactions, and a perturbative contribution, mainly due to the attractive interactions. The RG theory [22, 23] is only applied to the attractive part, since it is considered that the other term contributes mostly with density fluctuations of very short wavelengths. The effect of the density fluctuations due to the attractive part of the potential is then divided into short-wavelength and long-wavelength contributions. It is assumed that contributions from fluctuations of wavelengths less than a certain cutoff length L can be accurately evaluated by a mean-field theory. The effect of this short wavelength can be calculated using the soft-SAFT equation or any other mean-field theory. However, the choice of the mean-field theory is of relevance for the overall behavior of the crossover equation. The RG term corrects the approach to the critical region but it does not improve the performance of the underlying original equation far from the critical point. The strengths and limitations of the original equation will always be there. This is why different crossover equations have been developed and continue under development nowadays.

Following the RG approach, the contribution of the long-wavelength density fluctuations is taken into account through the phase–space cell approximation. In a recursive manner, the

Helmholtz free energy per volume of a system at density ρ can be described as [5, 22, 23]

$$a_n(\rho) = a_{n-1}(\rho) + da_n(\rho) \tag{2.4}$$

where a is the Helmholtz free energy density and da_n the term where long-wavelength fluctuations are accounted for in the following way:

$$da_n(\rho) = -K_n \cdot \ln\frac{\Omega_n^s(\rho)}{\Omega_n^l(\rho)}, \qquad 0 \le \rho \le \frac{\rho_{\max}}{2} \tag{2.5a}$$

$$da_n(\rho) = 0, \qquad \frac{\rho_{\max}}{2} \le \rho \le \rho_{\max} \tag{2.5b}$$

where Ω^s and Ω^l represent the density fluctuations for the short-range and the long-range attraction respectively, and K_n is a coefficient:

$$K_n = \frac{k_B T}{2^{3n} \cdot L^3} \tag{2.6}$$

where T is the temperature and L the cutoff length. Expressions for calculating each one of the terms can be found in Refs. [5, 6], including the extension to mixtures.

The process of successively taking into account the interactions between the different molecules (and hence density fluctuations) in a recursive manner is graphically illustrated in Fig. 2.6.

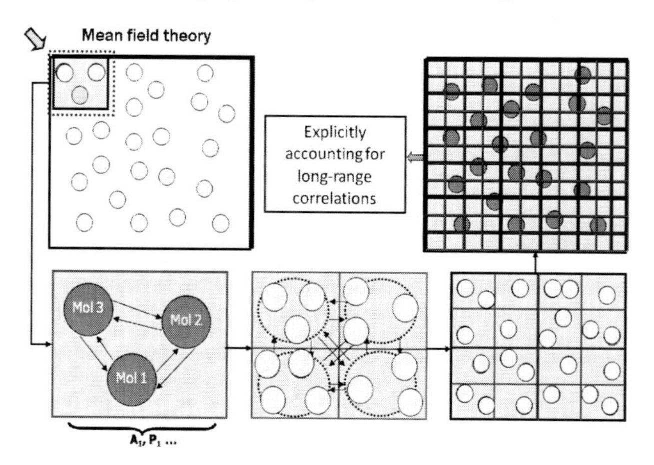

Figure 2.6 Schematic representation of the phase-cell space approximation of White to take into account the fluctuations near the critical region.

2.2.3 Calculation of Second-Order Thermodynamic Derivative Properties

A complete analysis of the performance of any EoS should include the description of second-order thermodynamic derivative properties. These properties are obtained by direct derivation from the Helmholtz energy and the pressure [27]:

$$C_{\mathrm{v}} = -T\left(\frac{\partial^2 A}{\partial T^2}\right)_{\mathrm{v}} \tag{2.7}$$

$$k_{\mathrm{T}}^{-1} = 2P + \rho^3\left(\frac{\partial^2 A}{\partial \rho^2}\right)_{\mathrm{T}} = \rho\left(\frac{\partial P}{\partial \rho}\right)_{\mathrm{T}} \tag{2.8}$$

$$\alpha = k_{\mathrm{T}}\rho^2\left(\frac{\partial^2 A}{\partial P \partial T}\right)_{\mathrm{v}} = k_{\mathrm{T}}\left(\frac{\partial P}{\partial T}\right)_{\mathrm{v}} \tag{2.9}$$

$$\mu = T\left(\frac{\partial P}{\partial T}\right)_{\mathrm{v}} - \rho\left(\frac{\partial P}{\partial \rho}\right)_{\mathrm{T}} \tag{2.10}$$

$$C_{\mathrm{p}} = C_{\mathrm{v}} + \frac{T\alpha^2}{\kappa_{\mathrm{T}}\rho} \tag{2.11}$$

$$\omega = \sqrt{\frac{C_{\mathrm{p}}}{C_{\mathrm{v}}}\left(\frac{\partial P}{\partial \rho}\right)_{\mathrm{T}}} \tag{2.12}$$

with C_{v} being the isochoric heat capacity, κ_{T} the reduced bulk modulus, μ the Joule–Thomson coefficient, α the thermal expansion coefficient, C_{p} the isobaric heat capacity, and ω the speed of sound.

As shown in Eqs. 2.7–2.12 these properties can be calculated by direct derivation of a thermodynamic potential function, however, it has been shown that they are more sensitive to errors than the original function; in fact, an accurate description of the singular behavior second order derivative properties show near the critical region become a challenge for any EoS. In addition, as most of these properties experience great changes in their values with respect to small changes in the operating conditions near the critical point, it is of interest to see how accurately they are obtained as predicted from the crossover EoS.

2.2.4 Phase Equilibria and Critical Line Calculations

For phase equilibria calculations, thermal stability and mechanical stability are satisfied by imposing the equality of chemical potentials of each component in the coexisting phases at fixed temperature and pressure. Because SAFT is formulated as an explicit function of temperature, chemical potentials and the pressure are equated at a given temperature, as follows:

$$P^{\mathrm{I}}(T, \rho^{\mathrm{I}}, x^{\mathrm{I}}) = P^{\mathrm{II}}(T, \rho^{\mathrm{II}}, x^{\mathrm{II}}) \tag{2.13}$$

$$\mu_i^{I}(T, \rho^{I}, x^{I}) = \mu_i^{II}(T, \rho^{II}, x^{II}) \tag{2.14}$$

The critical point is reached in a pure fluid when the densities of the vapor and the liquid phase become identical at a given pressure and temperature.

In the case of mixtures, given the additional degree of freedom provided by the composition, instead of a critical point, a critical line is obtained. These critical lines of mixtures are calculated by numerically solving the necessary conditions [28], which involve second and third derivatives of the Gibbs free energy with respect to the molar volume (pure systems) or the composition (in mixtures) as well as the extra condition to assure stability for critical points. In the SAFT case these derivatives are rewritten in terms of the Helmholtz free energy with respect to volume and composition at constant temperature:

$$A_{2x}A_{2v} - A_{Vx}^2 = 0 \tag{2.15}$$

$$A_{3x} - 3A_{V2x}\left(\frac{A_{Vx}}{A_{2V}}\right) + 3A_{2Vx}\left(\frac{A_{Vx}}{A_{2V}}\right)^2 - A_{3V}\left(\frac{A_{Vx}}{A_{2V}}\right)^3 = 0 \tag{2.16}$$

where the notation $AnVmx = (\delta^{n+m}A/\delta V^n \delta x^m)$ is used for the derivatives of the Helmholtz free energy.

2.3 Application to Pure Fluids

As an illustration of the importance of adding the right terms of an EoS to accurately describe the near-critical region we present next the application of the crossover soft-SAFT equation to selected compounds and mixtures. The performance of the equation is checked versus experimental data for several thermodynamic properties of relevance in supercritical processes.

To apply soft-SAFT to experimental systems it is necessary to choose a molecular model to represent each compound in the mixture. Within the soft-SAFT context *n*-alkanes [9, 10] are modeled as homonuclear chainlike molecules composed by m LJ segments of equal diameter σ and the same dispersive energy ε, m being the chain length. The quadrupolar interactions present in CO_2 are taken into account by an additional term into the equation. This term involves a new molecular parameter Q that represents the quadrupolar moment of the molecule. These molecular parameters plus the crossover parameters, ϕ and L, are enough to describe all thermodynamic properties of the pure components. These parameters are usually obtained by fitting vapor–liquid equilibrium data and are provided in Table 2.3 for completeness [6, 10].

Table 2.3 Soft-SAFT molecular parameters used in this work [5, 6]

Substance	m	σ (Å)	ε/k_B (K)	ϕ	L/σ
Methane	1.000	3.741	151.1	5.50	1.04
Propane	1.776	3.831	225.8	6.75	1.16
n-Butane	2.134	3.866	240.3	7.25	1.22
n-Pentane	2.497	3.887	250.2	7.57	1.27
n-Hexane	2.832	3.920	259.8	7.84	1.33
n-Heptane	3.169	3.937	266.0	8.15	1.38
n-Octane	3.522	3.949	271.0	8.30	1.43
n-Decane	4.259	3.960	278.6	8.44	1.55
1-Propanol*	1.941	3.815	249.8	7.30	1.32
Carbon dioxide**	1.606	3.158	159.9	5.79	1.18

* K_{HB} (Å^3) = 2300, ε_{HB}/k_B (K) = 3600

** Quadrupole value: 4.4×10^{-40} cm^2

The modeling approach is first illustrated for the case of CO_2 and the *n*-alkane series. Figure 2.7 depicts the temperature–density and pressure–temperature diagrams of pure CO_2, while *n*-alkanes are presented in Fig. 2.8. It is clearly observed that original soft-SAFT accurately describes the phase envelope, except near the critical region, while crossover soft-SAFT is able to describe, with the same degree of accuracy, the phase diagram far from and close to the critical point.

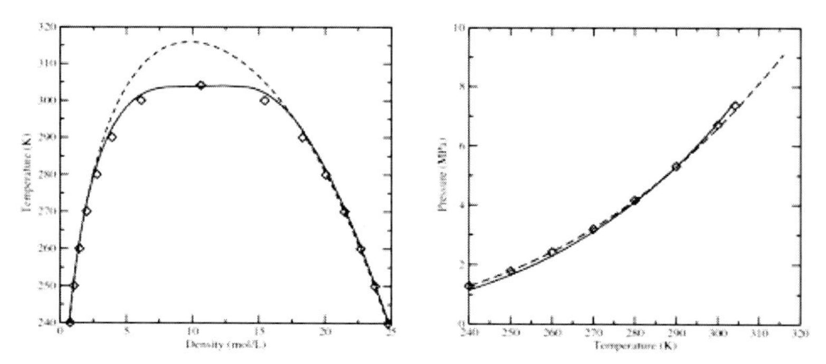

Figure 2.7 Phase diagram of carbon dioxide. Left: Temperature–density diagram Right: Pressure–temperature diagram. Symbols represent experimental data from Ref. [29], dashed lines are soft-SAFT calculations and the solid lines are the crossover soft-SAFT results.

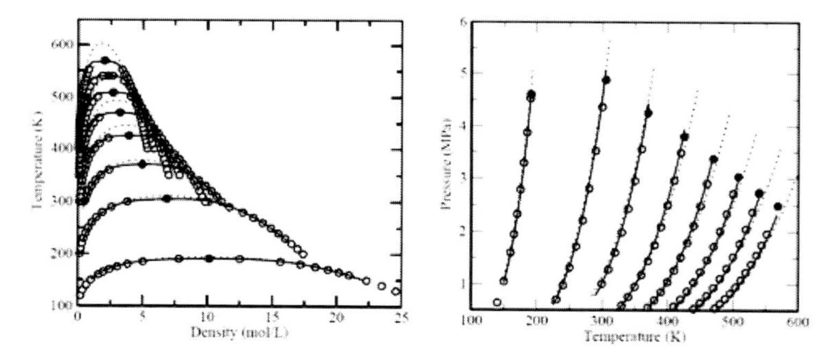

Figure 2.8 Phase diagram for the light members of the n-alkane series, from methane to n-octane. Left: Temperature–density diagram. Right: Pressure–temperature diagram. Symbols represent the experimental data taken from Ref. [29]; dashed lines, soft-SAFT predictions; and solid lines, crossover soft-SAFT predictions.

To further illustrate the performance of the equation for the near-critical region three selected supercritical isotherms and isobars of n-butane are presented in Fig. 2.8, corresponding to reduced temperatures $T_r = T/T_c = 1.05$, 1.10, and 1.15 and reduced pressures $P_r = P/P_c = 1.05$, 1.10, and 1.15. As observed, crossover soft-SAFT gives quantitative agreement with the experimental data, while the original equation overestimates the temperature and pressure in the near-critical region, giving an incorrect shape to the curve. Results shown in Fig. 2.9 represent pure predictions, as the molecular parameters were fitted from vapor–liquid equilibrium data [29].

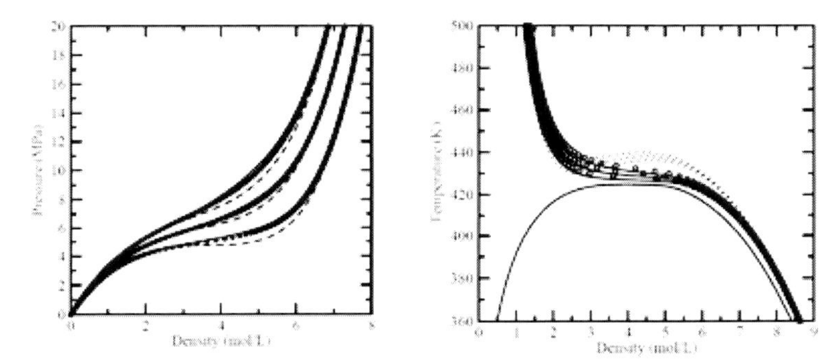

Figure 2.9 *n*-Butane. Left: Pressure–density diagrams for three temperatures above the critical point. Right: Temperature–density diagrams for three pressures above the critical point. Circles: Experimental data [29]; dashed lines: original soft-SAFT predictions; solid lines: crossover soft-SAFT predictions. See text.

2.3.1 Derivative Properties

The singular behavior of derivative properties in the vicinity of the critical point is exemplified for the case of propane in Fig. 2.10. In these calculations, there is not any additional molecular parameter fitting. As derivative properties were not included in the fitting, results obtained are pure predictions of the equation. The data used for comparison was taken from the correlated data provided in *NIST Chemistry Webbook* [29]. Figure 2.10a depicts the vapor–liquid equilibria as obtained by two versions of soft-SAFT (with and without a crossover treatment) versus a correlation to the experimental data of propane [29] from which the molecular parameters were obtained. Figure 2.10b–d shows the residual isochoric heat capacity, residual isobaric heat capacity, and isothermal compressibility of propane. The chosen reduced temperatures for comparison, where

T_c = 1.1, 1.25, and 1.5 (three T_c different values are used in this case, the experimental one, the one that is obtained from the original soft-SAFT, and the equivalent one with crossover soft-SAFT).

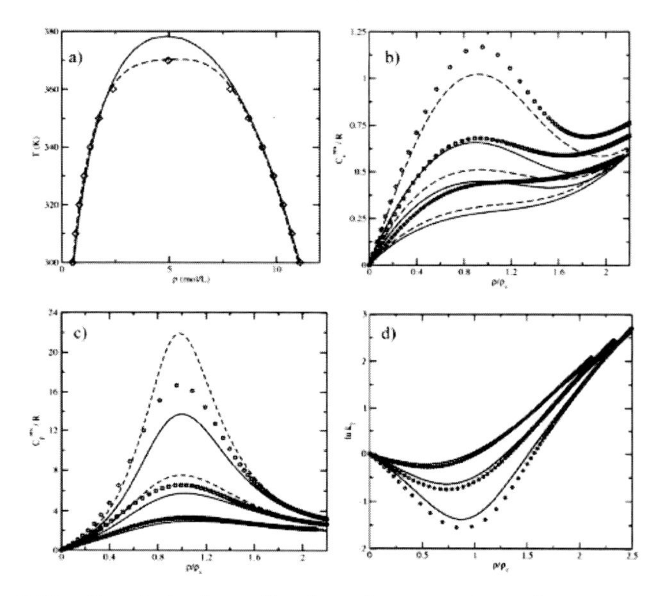

Figure 2.10 The behavior of selected properties of *n*-propane. (a) Temperature–density diagram, (b) residual isochoric heat capacity, (c) residual isobaric heat capacity, and (d) isothermal compressibility [29]. Symbols as in Fig. 2.9.

As observed in Fig. 2.10, the overall agreement is very good in all cases. The largest deviations are obtained for the residual heat capacities very close to the critical point. Soft-SAFT with parameters fitted to vapor–liquid equilibria is able to capture the extrema (maxima and minima) in very good agreement with the experimental data, except very close to the critical region.

2.4 Application to Mixtures in the Near-Critical Region

To apply the equation to mixtures isomorphism assumption is used, replacing the one-component density by the total density of the mixture in the RG approach, in the same way as Cai and Prausnitz

[26]. Details on the implementation of crossover soft-SAFT to mixtures can be found in the original references [6].

As an illustration of the crossover equation we present next the phase equilibria, critical behavior, and derivative properties of selected *n*-alkane/*n*-alkane, *n*-alkane/1-alkanol and CO_2/*n*-alkane binary mixtures because of their particular importance from both scientific and technical points of view. The goal is to show the robustness of the approach using the same set of molecular parameters fitted to vapor–liquid equilibria for all properties. Mixtures among *n*-alkanes are commonly found in several industrial processes, including those of the petrochemical industry. The behavior of *n*-alkane/1-alkanol mixtures is also of interest due to the nonideal behavior found in these mixtures, because of the presence of the hydroxyl group in the 1-alkanol molecules, while the behavior of CO_2/*n*-alkanes mixtures is of practical interest for several chemical and supercritical processes.

In addition to the three parameters already mentioned for *n*-alkanes (m, σ, and ε), 1-alkanols need two extra parameters to mimic the hydroxyl group; this is modeled by two square-well sites embedded off-center in one of the LJ segments, with volume k_{HB} and association energy $\varepsilon_{\mathrm{HB}}$. According to our model, these three (alkanes) or five (1-alkanols) molecular parameters plus the crossover parameters ϕ and L are enough to describe all thermodynamic properties of the pure components.

2.4.1 The Phase and Critical Behavior of Binary Mixtures

Figure 2.11 shows the performance of crossover soft-SAFT for the mixtures of similar light alkanes. Figure 2.11a shows the temperature composition projection for the mixture *n*-butane/*n*-octane at two different pressures, while Fig. 2.11b depicts the *PT* projections of the *PTxy* surfaces for the *n*-butane series, with mixtures between *n*-butane and *n*-pentane, *n*-hexane, *n*-heptane, and *n*-octane. As it was expected, crossover soft-SAFT predicts type I critical behavior, according to the classification of van Konynenburg and Scott [30], in quantitative agreement with experimental data for all mixtures.

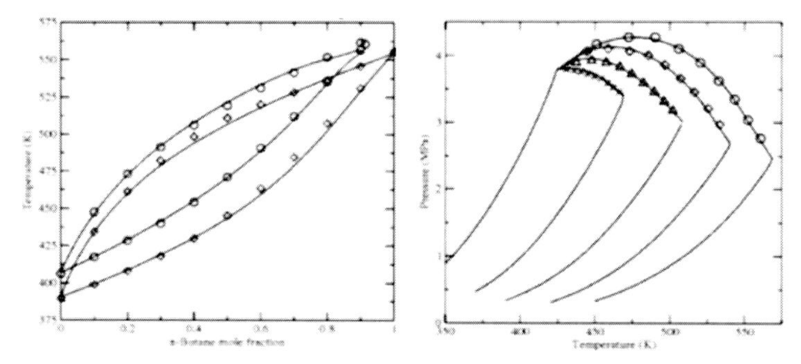

Figure 2.11 Left: Temperature–composition diagram for the binary mixture *n*-butane/*n*-octane at 2.07 and 2.76 MPa. Right: PT projection of the binary mixtures of the *n*-butane series: *n*-butane/*n*-pentane, *n*-butane/*n*-hexane, *n*-butane/*n*-heptane, and *n*-butane/*n*-octane. Symbols represent experimental data from Refs. [31, 32] and solid lines represent crossover soft-SAFT.

The mixture CO_2/ethane has been chosen as representative of a nonideal CO_2/*n*-alkane binary mixture. Figure 2.12a shows the vapor–liquid equilibria of the mixture at different temperatures (crossover soft-SAFT calculations were obtained with a binary energy parameter $\xi = 0.990$ fitted at 283K), while Fig. 2.12b shows predictions of the critical lines for this binary mixture. Note that the azeotropes are predicted at the correct pressure and mole fraction and the critical line is also in quantitative agreement with experimental data.

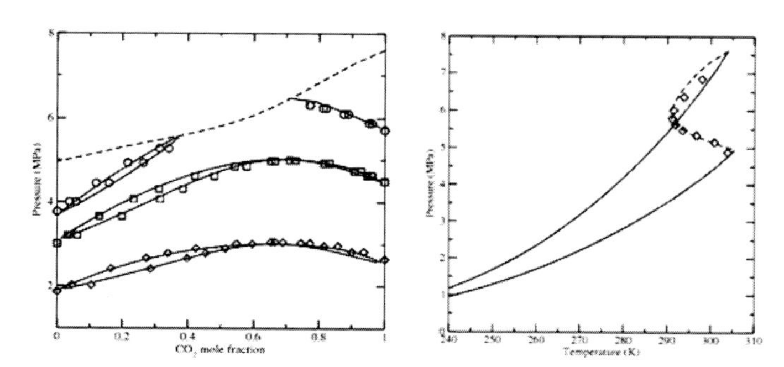

Figure 2.12 Left: PTx projection of the CO_2/ethane mixture at 263.15, 283.15, and 293.15 K. Right: PT projection of the PTx surface for the CO_2/ethane mixtures. Symbols represent experimental data [33] and solid lines represent crossover soft-SAFT predictions. The dashed line represents the critical line of the mixture.

Finally, as an example of the application of the equation to calculate second-order derivative properties of mixtures, two cases were selected, as shown in Fig. 2.13, the speed of sound of the mixture methane/*n*-butane and the isobaric heat capacity of the 1-propanol/*n*-decane binary mixture at two temperatures.

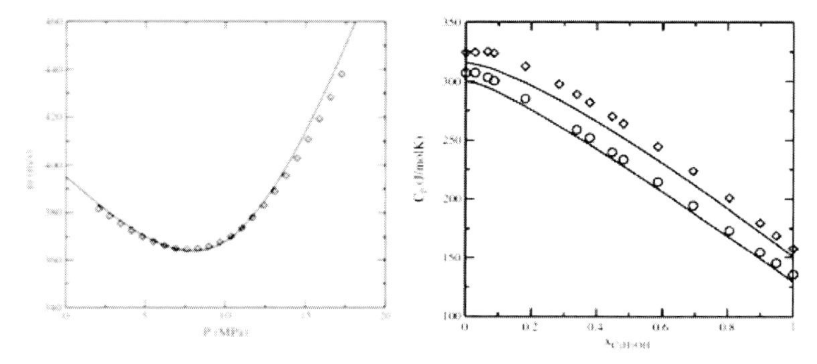

Figure 2.13 Left: Speed of sound vs. pressure of the methane/*n*-butane binary mixture at 311 K and methane composition of 0.894. Right: Isobaric heat capacity–density diagram of the 1-propanol/*n*-decane binary mixture at 280 and 318 K. Symbols represent experimental data from [34, 35], respectively, and solid lines represent crossover soft-SAFT.

2.5 Summary and Conclusions

This chapter has focused on fundamentals of the supercritical fluids, dealing with the singularities found in the critical region as a consequence of the inhomogeneities in the density (and composition for the case of mixtures). After a general introduction of what a supercritical fluid is and the main properties of them directly linked to supercritical process optimization, a detailed procedure has been explained on how to take into account the fluctuations in the critical region in a modeling approach based on renormalization theory, illustrated for the case of the crossover soft-SAFT EoS. The importance of taking into account this term to accurately describe phase equilibria, critical lines, and second-order derivative properties such as heat capacities, speed of sound, and isothermal heats capacities has been highlighted with specific examples of experimental systems. Experimental data has been compared with the original soft-SAFT equation (mean field) and with the crossover soft-SAFT (which includes the nonanalytical behavior in the near-critical region). When dealing with supercritical processes one should be aware of the proximity of the operating conditions to the critical region of the fluid, and, hence, the possible influences of the fluctuations of the density (or composition) for the given property in order to accurately design the process using the appropriate tool.

References

1. Retrieved from the University of Leeds (England) webpage http://www.chem.leeds.ac.uk/People/CMR/criticalpics.html.
2. Llovell, F., and Vega, L. F. (2006). Prediction of thermodynamic derivative properties of pure fluids through the soft-SAFT equation of state, *J. Phys. Chem. B*, **110**, pp. 11427–11437.
3. Vargaftik, N. B., Vinogradov, Y. K., and Yargin, V. S. (1996). *Handbook of Physical Properties of Fluids and Gases: Pure Substances and Mixtures* (Begell House), p. 714.
4. Wilson, K. G. (1971). Renormalization group and critical phenomena, 2. Phase space cell analysis of critical behaviour, *Phys. Rev. B*, **4**, pp. 3174–3183.
5. Llovell, F., Pàmies, J. C., and Vega, L. F. (2004). Thermodynamic properties of Lennard-Jones chain molecules: renormalization-group corrections to a modified statistical associating fluid theory, *J. Chem. Phys.*, **121**, pp. 10715–10724.
6. Llovell, F., and Vega, L. F. (2006). Global fluid phase equilibria and critical phenomena of selected mixtures using the crossover soft-SAFT equation, *J. Phys. Chem. B*, **110**, pp. 1350–1362.
7. Colina, C. M., Olivera-Fuentes, C. G., Siperstein, F. R., Lisal, M., and Gubbins, K. E. (2003). Thermal properties of supercritical carbon dioxide by Monte Carlo simulations, *Mol. Sim.*, **29**, pp. 405–412.
8. Blas, F. J., and Vega, L. F. (1997). Thermodynamic behaviour of homonuclear and heteronuclear Lennard-Jones chains with association sites from simulation and theory, *Mol. Phys.*, **92,** pp. 135–150.
9. Blas, F. J., and Vega, L. F. (1998). Prediction of binary and ternary diagrams using the statistical associating fluid theory (SAFT) equation of state, *Ind. Eng. Chem. Res.*, **37**, pp. 660–674.

10. Pàmies, J. C., and Vega, L. F. (2001). Vapor-liquid equilibria and critical behaviour of heavy n-alkanes using transferable parameters from the soft-SAFT equation of state, *Ind. Eng. Chem. Res.*, **40**, pp. 2532–2543.

11. Llovell, F., Vega, L. F., Mejía, A., and Segura, H. (2008). An accurate direct technique for parametrizing cubic equations of state: Part III. Application of a crossover treatment, *Fluid Phase Equilib.*, **264**, pp. 201–210.

12. Wertheim, M. S. (1984). Fluids with highly directional attractive forces. 2. Thermodynamic-perturbation theory and integral-equations, *J. Stat. Phys.*, **35**, pp. 35–47.

13. Wertheim, M. S. (1986). Fluids with highly directional attractive forces. 3. Multiple attraction sites, *J. Stat. Phys.*, **42**, pp. 459–476.

14. Wertheim, M. S. (1986). Fluids with highly directional attractive forces. 4. Equilibrium polymerization, *J. Stat. Phys.*, **42**, pp. 477–492.

15. Chapman, W. C., Jackson, G., and Gubbins, K. E. (1988). Phase equilibria of associating fluids. Chain molecules with multiple bonding sites, *Mol. Phys.*, **65**, pp. 1057–1079.

16. Chapman, W. C., Jackson, G., Gubbins, K. E., and Radosz, M. (1990). New reference equation of state for associating liquids, *Ind. Eng. Chem. Res.*, **29**, pp. 1709–1721.

17. Huang, S. H., and Radosz, M. (1990). Equation of state for small, large, polydisperse and associating molecules, *Ind. Eng. Chem. Res.*, **29**, pp. 2284–2294.

18. Tan, S. P., Adidharma H., and Radosz, M. (2008). Recent advances and applications of statistical associating fluid theory, *Ind. Eng. Chem. Res.*, **47**, pp. 8063–8082.

19. Gil-Villegas, A., Galindo, A., Whitehead, P. J., Mills, S. J., Jackson, G., and Burguess, A. N. (1997). Statistical associating fluid theory for chain molecules with attractive potentials of variable range, *J. Chem. Phys.*, **106**, pp. 4168–4186.

20. Gross, J., and Sadowski, G. (2001). Perturbed-chain SAFT: an equation of state based on a perturbation theory for chain molecules, *Ind. Eng. Chem. Res.*, **40,** pp. 1244–1260.

21. Johnson, J. K., Zollweg, J. A., and Gubbins, K. E. (1993). The Lennard-Jones equation of state revisited, *Mol. Phys.*, **78**, pp. 591–618.

22. White, J. A. (1992). Contribution of fluctuation to thermal properties of fluids with attractive potential force of limited range: Theory compared with pρT and CV data for argon, *Fluid Phase Equilib.*, **75**, pp. 53–64.

23. Salvino, L. W., and White, J. A. (1992). Calculation of density fluctuation contributions to thermodynamic properties of simple fluids, *J. Chem. Phys.*, **96**, pp. 4559–4568.
24. Lue, L., and Prausnitz, J. M. (1998). Renormalization group corrections to an approximate free-energy model for simple fluids near to and far from the critical region, *J. Chem. Phys.*, **108**, pp. 5529–5536.
25. Jiang, J., and Prausnitz, J. M. (1999). Equation of state for thermodynamic properties of chain fluids near-to and far-from the vapor-liquid critical region, *J. Chem. Phys.*, **111**, pp. 5964–5974.
26. Cai, J., and Prausnitz, J. M. (2004). Thermodynamics for fluid mixtures near to and far from the vapour-liquid critical point, *Fluid Phase Equilib.*, **219**, pp. 205–217.
27. Llovell, F., and Vega, L. F. (2006). Prediction of thermodynamic derivative properties of pure fluids through the soft-SAFT equation of state, *J. Phys. Chem. B*, **110**, pp. 11427–11437.
28. Sadus, J. R. (1992). *High-Pressure Phase Behaviour of Multicomponent Fluid Mixtures* (Elsevier, The Netherlands).
29. *NIST Chemistry WebBook*, http://webbook.nist.gov/chemistry.
30. Van Konynenburg, P. H., and Scott, R. L. (1980). Critical lines and phase-equilibria in binary van der Waals mixtures, *Philos. Trans. R. Soc. Lond. A*, **298**, pp. 495–540.
31. Kay, W. B., Genco, J., and Fichtner, D. A. (1974). Vapor-liquid-equilibrium relationships of binary-systems propane octane an butane octane, *J. Chem. Eng. Data*, **19**, pp. 275–280.
32. Kay, W. B., Hoffman, R. L., and Davies, O. (1975). Vapor-liquid-equilibrium relationships of binary-systems normal-butane normal-pentane and normal-butane normal-hexane, *J. Chem. Eng. Data*, **20**, pp. 333–338.
33. Fredenslund, A., and Mollerup, J. (1974). Measurement and prediction of equilibrium ratios for C_2H_6+CO_2 system *J. Chem. Soc., Faraday Trans. I*, **70**, pp. 1653–1660.
34. Plantier, F., Danesh, A., Sohrabi, M., Daridon, J. L., Gozalpour, F., and Todd, A. C. (2005). Measurements of the speed of sound for mixtures methane + butane with a particular focus on the critical state, *J. Chem. Eng. Data*, **50**, pp. 673–676.
35. Peleteiro, J., González-Salgado, D., Cerdeiriña, C. A., and Romaní, L. (2002). Isobaric heat capacities, densities, isentropic compressibilities and second-order excess derivatives for (1-propanol + *n*-decane), *J. Chem. Thermodyn.*, **34**, pp. 485–497.

Chapter 3

A Statistical Mechanical Equation of State for Predicting Critical Properties of Confined Fluids

Eldred Chimowitz[a] **and Pedro López-Aranguren**[b]
[a]*University of Rochester, School of Engineering and Applied Sciences, Rochester, New York, NY, USA*
[b]*Materials Science Institute of Barcelona (CSIC), Campus de la UAB, 08193 Bellaterra, Spain*
eldred.chimowitz@rochester.edu

The prediction of properties in porous materials is of continuing interest in the fields of chemical and materials engineering. Application areas include (i) the use of supercritical fluids to synthesize porous materials [1–4], (ii) physical adsorption of trace components from gaseous effluents, (iii) gas storage using microporous materials [5], and (iv) chemical separations using inorganic membranes [6]. Given this situation, there has been substantial effort over the years devoted to developing equations of state suitable for thermodynamic property predictions in fluids confined in porous media. However, developing tractable physically based models has remained elusive [7]. An important class of these

Supercritical Fluid Nanotechnology: Advances and Applications in Composites and Hybrid Nanomaterials
Edited by Concepción Domingo and Pascale Subra-Paternault

ISBN 978-981-4613-40-8 (Hardcover), 978-981-4613-41-5 (eBook)
www.panstanford.com

equations of state is of the mean-field type, which are of interest because they are often analytic and therefore amenable for use in process engineering calculations [8]. Two important questions arise in the context of such equations of state: (i) How accurate are they for real fluids, and (ii) can the number of adjustable parameters required for their use be kept to a minimum? The main purpose in this chapter tutorial is to describe a mean-field statistical mechanical approach for developing equations of state for predicting the properties of confined fluids, including their critical properties, which can be quite different from their bulk fluid counterparts.

3.1 The Hamiltonian for a Confined Lattice Gas

In porous systems, the host material consists of void space with complementary regions taken up by solid matrix (adsorbate) material. In the void space one will, in general, find isolated molecules, those with nearest neighbors, as well as molecules adjacent to pore surfaces. Furthermore, pore surface heterogeneities will generally be present and all of these effects should be captured in any realistic analytical theory [6]. A lattice gas with a fixed number of matrix pore-blocked sites randomly assigned throughout the structure is considered (see Fig. 3.1 for a 2D schematic) with the remaining lattice sites' void spaces that may, or may not, be occupied by fluid particles.

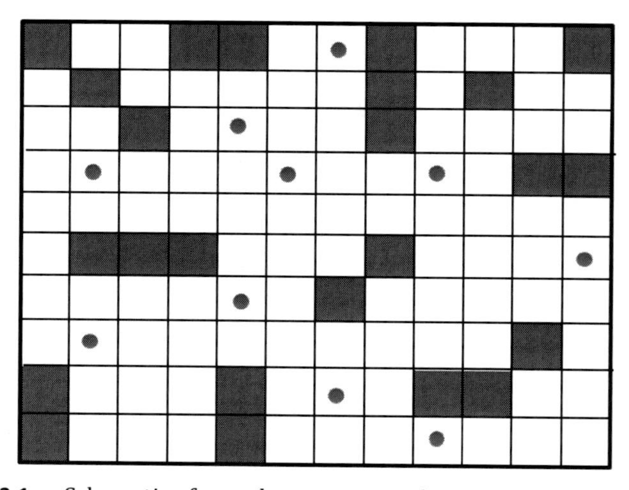

Figure 3.1 Schematic of a random porous medium.

Nearest-neighbor fluid particles interacted through a constant interaction energy, while solid matrix–fluid energetic interactions were represented by a probability density function that represents the effect of binding site energy heterogeneities on the solid surface. They considered a simple cubic lattice in d dimensions and for each site defined a quenched random variable, where d can take on the value of either 0 or 1. The value of 0, found with probability p, implies the existence of a solid matrix particle at that position in the lattice, while the value 1 designates a void space. In addition, at each void site they assigned an annealed variable, which can assume either the value 0, denoting the absence of a fluid particle, or the value 1, which represents the presence of a fluid particle; the variable is the usual density variable associated with the lattice gas partition function. For a completely random quenched structure (like aerogels) the variables are assumed to be uncorrelated. The Hamiltonian for such a system is given by the equation

$$-H_{\mathrm{LG}} = 4\Im\sum_{<ij>}\varepsilon_i\varepsilon_j n_i n_j + \Gamma\sum_{<ij>}[\varepsilon_i n_i(1-\varepsilon_j)+\varepsilon_j n_j(1-\varepsilon_i)] + \mu\sum_i \varepsilon_i n_i \tag{3.1}$$

where $4\Im$ is the coupling constant between two adjacent fluid particles, Γ the coupling constant between a fluid and a solid particle, and μ the chemical potential of the confined fluid. The summation in Eq. 3.1 (represented by the notation $<ij>$) is over all i, j pairs but the effect of the ε_i, ε_j variables is to ensure that the first term in the Hamiltonian captures fluid–fluid interactions, the second term represents fluid–solid ones, and the third term is the standard field term.

3.1.1 Mean-Field Treatment: The Confined Lattice Gas

Next, a mean-field approximation for the exact Hamiltonian given in Eq. 3.1 is described. The "exact" thermal average of molecular density at site i, which is defined as $\langle n_i \rangle \equiv \rho$, is given by the rigorous statistical mechanical equation

$$\langle n_i \rangle \equiv \rho = \frac{\sum\limits_{n_i=0,1} \cdots \sum\limits_{n_k=0,1} n_i \exp\left[\beta H_{LG}\right]}{\sum\limits_{n_i=0,1} \cdots \sum\limits_{n_k=0,1} \exp\left[\beta H_{LG}\right]} \tag{3.2}$$

where H_{LG} is given by Eq. 3.1 and $\beta \equiv 1/kT$.

In the mean-field regime, however, Eq. 3.2 can be considerably simplified. Consider a site i surrounded by z sites, each of which may be pore blocked, may be empty, or may contain another molecule. The fundamental mean-field assumption is that molecules in any of these surrounding sites must be at their respective mean densities. If amongst these z neighboring sites, N of them are pore blocked, then the probability of observing this configuration is given by the quantity Ω_N defined as

$$\Omega_N = \frac{z!}{(z-N)!N!}(1-p)^{z-N}p^N \tag{3.3}$$

with the total number of all possible configurations designated as $\Omega_{\text{total}} = \sum_{N=0}^{Z}\Omega_N = 1$. Using this approach, Eq. 3.2 can be approximated in the mean-field situation by summing over the values $n_i = 0, 1$ to get the following equations:

$$\langle n_i \rangle = \frac{1}{\Omega_{\text{total}}}\sum_{N=0}^{z}\frac{\Omega_N \exp[\beta(y_N+\mu)]}{\exp[\beta(y_N+\mu)]+1} \tag{3.4}$$

or

$$\langle n_i \rangle = \frac{1}{\Omega_{\text{total}}}\sum_{N=0}^{z}\frac{\Omega_N \exp\left[\tfrac{1}{2}\beta(y_N+\mu)\right]}{\exp\left[\tfrac{1}{2}\beta(y_N+\mu)\right]+\exp\left[-\tfrac{1}{2}\beta(y_N+\mu)\right]} \tag{3.5}$$

which can be simplified to yield the equation

$$\langle n_i \rangle = \frac{1}{2\Omega_{\text{total}}}\sum_{N=0}^{z}\Omega_N\left[1+\tanh \tfrac{1}{2}\beta(y_N+\mu)\right] \tag{3.6}$$

making use of a quantity Y_N defined as

$$y_N \equiv 4\Im\langle n_i \rangle(z-N)+N\Gamma \tag{3.7}$$

Substituting Eq. 3.3 in Eq. 3.6 yields the following result for the density of the confined fluid:

$$\rho = \frac{1}{2}\left\{\begin{array}{l} 1+\sum_{N=0}^{z}\frac{z!}{(z-N)!N!}(1-p)^{z-N}p^N \\ \tanh\left[\tfrac{1}{2}\beta(4\Im\rho(z-N)+N\Gamma+\mu)\right] \end{array}\right\} \tag{3.8}$$

which to first order in p yields the equation

$$\rho = \frac{1}{2}\left\{\begin{aligned}&1+(1-zp)\tanh\left[\beta'\left(2z\rho+\frac{\mu'}{2}\right)\right]\\&+zp\tanh\left[\beta'(2\rho(z-1)+\frac{1}{2}\Gamma'+\frac{\mu'}{2})\right]\end{aligned}\right\} \tag{3.9}$$

where the following dimensionless quantities are used:

$$\beta' \equiv \beta\Im \tag{3.10}$$

$$\mu' \equiv \frac{\mu}{\Im} \tag{3.11}$$

$$\Gamma' \equiv \frac{\Gamma}{\Im} \tag{3.12}$$

$$T' \equiv \frac{1}{\beta'} = \frac{kT}{\Im} \tag{3.13}$$

Equation 3.9 is the exact mean-field equation of state for a confined fluid in a highly porous quenched random structure.

3.1.2 The Critical Point for the Model

The conditions for the critical point in this system are defined by the following two thermodynamic equations:

$$\left(\frac{\partial\mu'}{\partial\rho}\right)_{\beta'_c} = 0 \tag{3.14}$$

and

$$\left(\frac{\partial^2\mu'}{\partial\rho^2}\right)_{\beta'_c} = 0 \tag{3.15}$$

To establish these derivatives Eq. 3.9 is rewritten more generally as

$$\rho = f(\rho, \beta', \mu') \tag{3.16}$$

which together with Eqs. 3.14 and 3.15 gives rise to the following equations at the critical point:

$$\rho_c = f(\rho_c, \beta'_c, \mu_c') \tag{3.17}$$

$$\left(\frac{\partial f}{\partial\rho}\right)_{\beta'_c, \mu'_c} = 1 \tag{3.18}$$

$$\left(\frac{\partial^2 f}{\partial \rho^2}\right)_{\beta'_c,\mu'_c} = 0 \tag{3.19}$$

3.1.3 The Low *p* Limit: For Use in Highly Porous Aerogels

The conditions given by Eqs. 3.17–3.19 provide three equations that can be solved for the critical values of T_c, μ_c, and ρ_c; however, these are implicit equations and for the low p limit an analytic approximation for Eq. 3.9 is used. This is done by defining perturbation functions χ and Φ as follows:

$$\mu' - \mu'_{p=0} \equiv \chi(\beta', \Gamma')p \tag{3.20}$$

$$\rho - \rho_{p=0} \equiv \phi(\beta', \Gamma')p \tag{3.21}$$

The functions χ and Φ are, as yet, unknown but solutions of Eq. 3.9 for $p = 0$ yield the values $\mu'_{p=0} = -2z$, $\rho_{p=0} = 1/2$. Using these equations in Eqs. 3.18 and 3.19, it is found that

$$2\phi(1-\beta' z) = 2\beta'\chi + z\tanh\left(\beta'\left(\tfrac{1}{2}\Gamma' - 1\right)\right) \tag{3.22}$$

and

$$2z^2\beta'(z\phi + \chi) + (z-1)^2 z\tanh\left(\beta'\left(\tfrac{1}{2}\Gamma' - 1\right)\right)\operatorname{sec}h^2\left(\beta'\left(\tfrac{1}{2}\Gamma' - 1\right)\right) = 0 \tag{3.23}$$

Solving both of these equations simultaneously for χ and Φ leads to the results

$$\phi = \frac{1}{2z}\tanh\left(\beta'\left(\tfrac{1}{2}\Gamma' - 1\right)\right)\left[z^2 - (z-1)^2\operatorname{sech}^2\left(\beta'\left(\tfrac{1}{2}\Gamma' - 1\right)\right)\right] \tag{3.24}$$

and

$$\chi = \frac{1}{2z\beta'}\tanh\left(\beta'\left(\tfrac{1}{2}\Gamma' - 1\right)\right). \left[z\beta'\left\{(z-1)^2\operatorname{sec}h^2\left(\beta'\left(\tfrac{1}{2}\Gamma' - 1\right)\right) - z^2\right\} - (z-1)^2\operatorname{sec}h^2\left(\beta'\left(\tfrac{1}{2}\Gamma' - 1\right)\right)\right] \tag{3.25}$$

When Eqs. 3.24 and 3.25 are combined with Eqs. 3.9 and 3.18–3.21, they yield the following results for the critical properties T_c, μ_c, and ρ_c in the confined fluid to first order in p:

$$\rho_c = \frac{1}{2} + \frac{1}{2}\tanh\left[\frac{\Gamma' - 2}{2z}\right]\left[z^2 - (z-1)^2 \sec h^2\left(\frac{\Gamma' - 2}{2z}\right)\right]p \tag{3.26}$$

$$T'_c = z\left[1 - p\left\{z - (z-1)\sec h^2\left((\Gamma' - 2)/_{2z}\right)\right\}\right] \tag{3.27}$$

$$\mu'_c = -2z^2 \tanh\left[\frac{(\Gamma' - 2)}{2z}\right]p - 2z \tag{3.28}$$

At the $p = 0$ limit, Eqs. 3.26–3.28 may be used to get the critical properties of the pure 3D lattice gas model within a mean-field approximation. These results are given by the equations

$$\rho_c(p = 0) = 1/_2 \tag{3.29}$$

$$T'_c(p = 0) = z \tag{3.30}$$

$$\mu'_c(p = 0) = -2z \tag{3.31}$$

and conform to the established results for this system.

3.2 Energy Heterogeneity in the Fluid–Solid Interaction

These results can be extended to the situation where the fluid–solid interaction term is given by a statistical distribution, that is, the surface is composed of sites with energetically heterogeneous binding energies. In this case, it is assumed that Γ' is given by a probability distribution $\eta(\Gamma')$.

The mean-field equation of state can be modified so that the average of the right hand side in Eq. 3.27 is now taken, leading to a critical temperature of the confined fluid given by the equation

$$T'_c = z\left[1 - p\{z - (z-1)\gamma\}\right] \tag{3.32}$$

where γ is the average of $\sec h^2(\Gamma'/z)$, defined by the integral

$$\gamma = \int \eta(\Gamma') \sec h^2\left(\frac{(\Gamma' - 2)}{2z}\right) d\Gamma' \tag{3.33}$$

For a narrow distribution around some average energy Γ'_0, it is derived a simple expression for γ by Taylor expansion of $\sec h^2((\Gamma' - 2)/2z)$. This yields the result

$$\gamma = \sec h^2\left(\frac{(\Gamma_0{}'-2)}{2z}\right) + \frac{1}{2}\left[\frac{d^2 \sec h^2\left(\frac{(\Gamma'-2)}{2z}\right)}{d\Gamma'^2}\right]_{\Gamma'=\Gamma'_0} \int (\Gamma'-\Gamma'_0)^2 \eta(\Gamma') d\Gamma' \quad (3.34)$$

For a narrow Gaussian distribution, with variance $\sigma << \Gamma'_0$ it is found that

$$\gamma = \sec h^2\left(\frac{(\Gamma_0{}'-2)}{2z}\right) + \frac{1}{2}\left[\frac{d^2 \sec h^2\left(\frac{(\Gamma'-2)}{2z}\right)}{d\Gamma'^2}\right]_{\Gamma'=\Gamma'_0} \sigma^2 \quad (3.35)$$

3.3 Model Predictions: Pure Fluid

Following, some of the predictions available from this theory are illustrated. Figure 3.2 shows three adsorption isotherms at various values of p calculated at a dimensionless temperature $T' = 6$ with the value of the fluid–pore surface interaction parameter $\Gamma' = 4$.

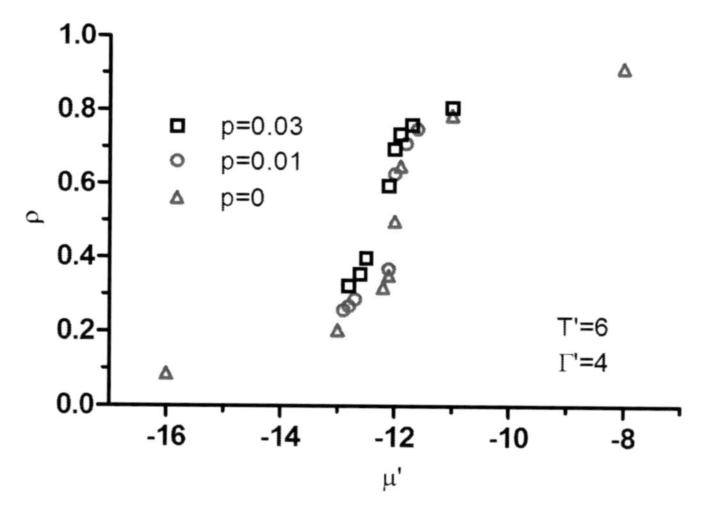

Figure 3.2 Equation-of-state predictions of adsorption isotherms of a confined fluid.

This temperature is the critical one for the zero p case as derived in Eq. 3.30 and the infinite slope of the $p = 0$ adsorption isotherm is evident at the critical chemical potential $\mu'_c(p = 0) = -2z = -12$. As p increases, the system moves further away from criticality (into a supercritical temperature regime) and the slopes of the various isotherms become finite throughout the region shown.

The variation of the system's critical temperature with the fluid–pore surface interaction parameter Γ', at various values of p, is presented in Fig. 3.3, where it can be observed that the critical temperature decreases with increasing values of p, a result that follows mathematically from Eq. 3.27.

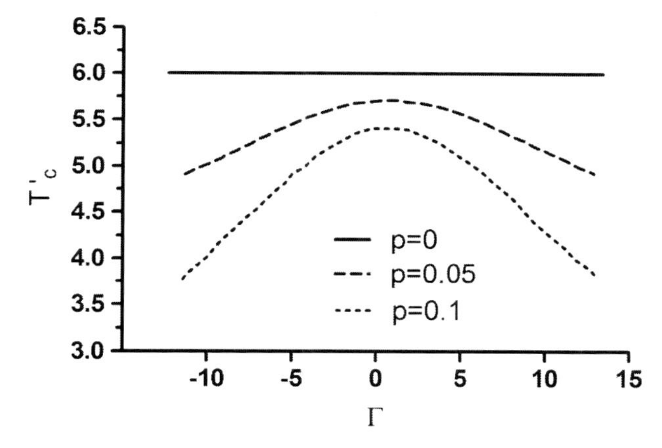

Figure 3.3 Equation-of-state predictions for the critical temperature of a confined fluid.

The effect of porosity on the *density enhancement* of the adsorbed fluid, relative to the bulk fluid in thermodynamic equilibrium with it, is also seen in the results given in Fig. 3.2. In the technologically interesting case (attractive fluid–pore interactions, that is, $\Gamma' > 0$) the adsorbed fluid density is enhanced with increasing p at a given value of both Γ' and temperature. This enhancement is particularly pronounced in the lower pressure (chemical potential) region, a result that finds use in practical applications involving light gas storage in porous media. One would also expect that increasing the fluid–pore surface interaction parameter Γ', at a given temperature and value of p, should increase the magnitude of the adsorbed fluid's density. This is the case as shown in the results presented in Fig. 3.4 for the situation $p = 0.03$ and a dimensionless temperature $T' = 6$.

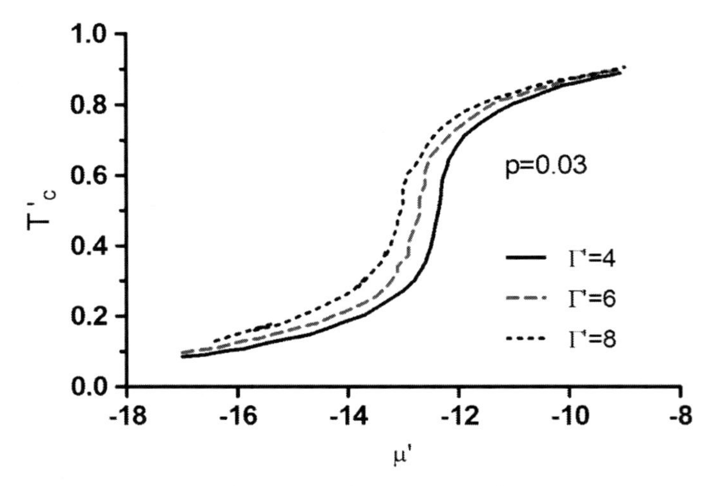

Figure 3.4 Equation-of-state predictions of adsorption isotherms of a confined fluid at various values of a surface interaction parameter.

In any real material the fluid–pore energy interaction parameter Γ' is likely to be heterogeneous. For the results shown in Fig. 3.5 for $p = 0.1$, the distribution of the energy heterogeneity was assumed to follow a Gaussian form with various values for the mean/variance ratio.

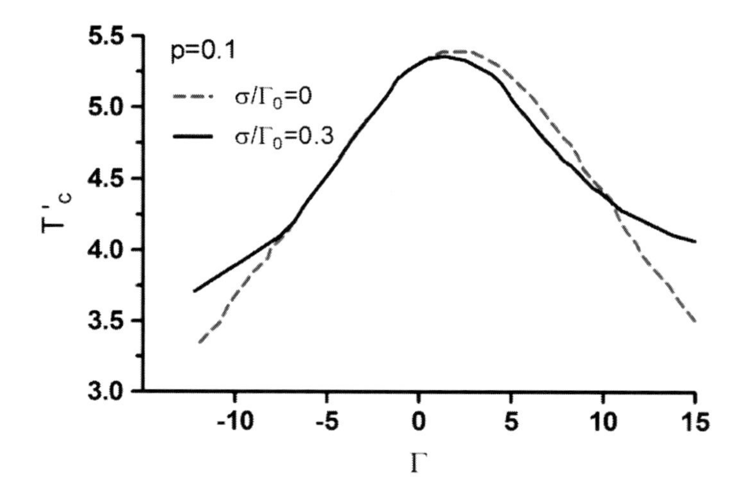

Figure 3.5 Equation-of-state predictions of the effects of surface energy heterogeneity on the critical temperature of a confined fluid.

The effect of this heterogeneity on the critical temperature of the system is to effectively widen the distribution of the $T'_c(\Gamma')$ function about Γ'_{max}.

3.3.1 Model Comparison with GCMC Computer Simulations

Some comparisons between this mean-field model and grand canonical Monte Carlo (GCMC) computer simulation [9] results are given in Figs. 3.6 and 3.7. The reducing parameters for all the thermodynamic properties in these comparisons are the corresponding critical values in the respective pure systems (i.e., $p = 0$). These are known exactly for the lattice gas and used here for the simulation results, while those for the mean-field model were found from Eqs. 3.29–3.31.

Figure 3.6 shows a comparison of adsorption isotherms of both the mean-field model and simulation results in a highly porous system ($p = 0.01$) at a slightly supercritical temperature corresponding to a reduced temperature of 1.01. In Fig. 3.7 comparisons are shown at a higher porosity and two different temperatures. The average absolute deviation between the mean-field model and simulation results at the lower (closer to critical) temperature is about 10.3%, while at the higher temperature this is 7.4%.

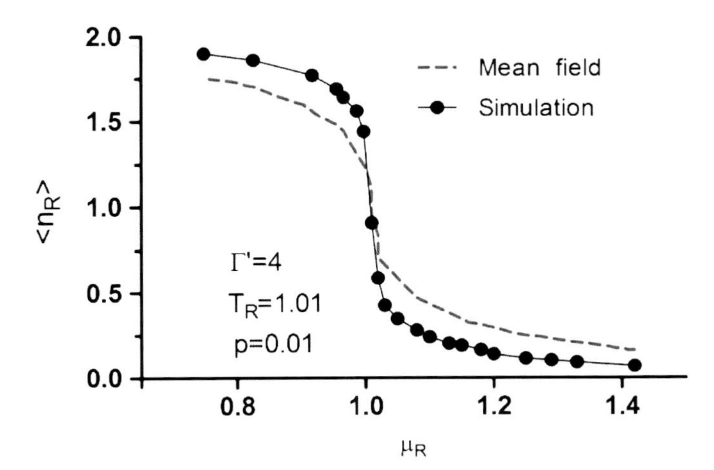

Figure 3.6 Comparison between theory and GCMS computer simulation results.

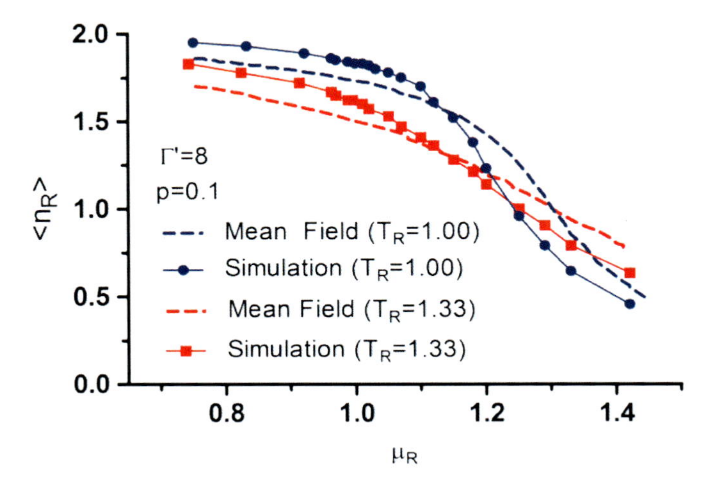

Figure 3.7 Comparison between theory and GCMS computer simulation results.

Experimental adsorption data for methane in a silica aerogel at 308 K is used with the model in Fig. 3.8.

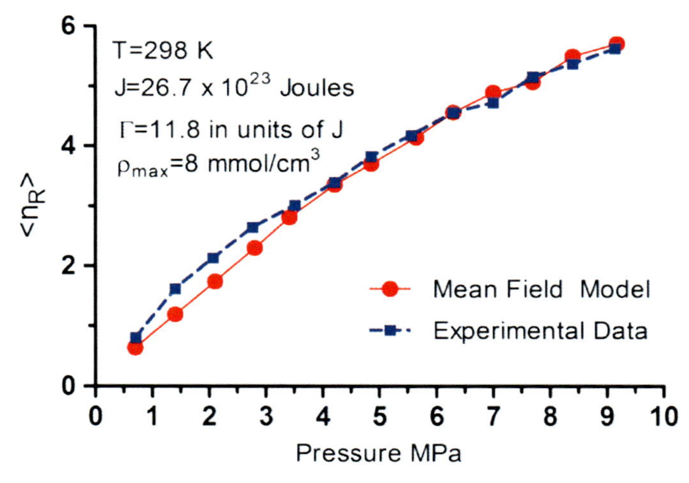

Figure 3.8 Comparison between theory and experimental data for methane adsorbed in a silica aerogel.

Such data is usually represented as pressure (chemical potential)-density adsorption isotherms and the mean-field equation of state given earlier in Eq. 3.9 provides a way of calculating the variation of density with chemical potential along the experimental isotherm, given assumed values for the model parameters and at a given value of porosity, which is often known from independent measurements. Integration of this function provides the pressure change in the system between any two chosen states 1 and 2 using the following thermodynamic identity:

$$\Delta P_{1-2} = \int_{\mu'_1}^{\mu'_2} \rho(\Gamma', \Im, \beta') d\mu' \quad (3.36)$$

This series of calculations is repeated in an attempt to optimize the values of Γ' and $\Im$ by establishing the values that provide the best fit with the data. These results are shown in Fig. 3.9 for various porosities and illustrate very good agreement between the model and the results recently obtained in supercritical carbon dioxide–porous silica aerogels at the porosity of 95% given in that paper [10, 11]. Moreover, Fig. 3.9 also shows model predictions of large pressure shifts and density enhancements at other conditions for which experimental data is unavailable.

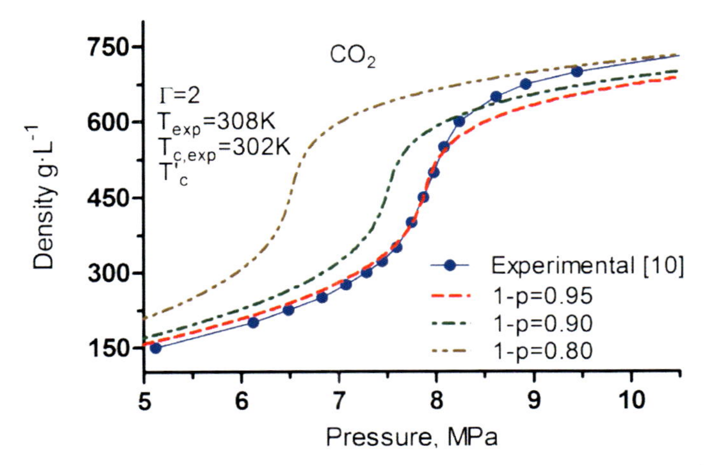

Figure 3.9 Comparison between theory and experimental data for carbon dioxide adsorbed in a silica aerogel.

On a similar performance, Fig. 3.10 shows results for pure bulk and confined C_3H_8 on the same aerogel, at 371 K (T_c = 369.7 K). The model shows in this case a good agreement with a porosity of 92%.

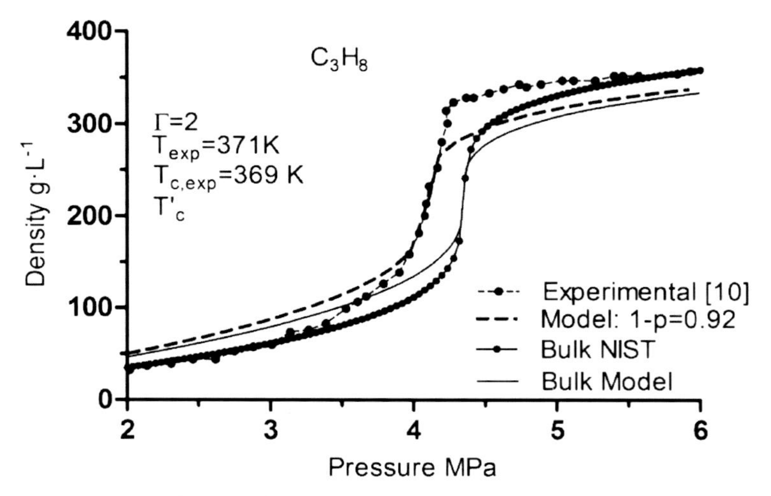

Figure 3.10 Model calculations showing the agreement for experimental adsorption data of bulk and confined C_3H_8 at 371 K in a silica aerogel.

In Fig. 3.11 the model is applied for the case of CO_2 at the critical confined temperature, that is, $T = T_c'$ from Eq. 3.27, at various porosities (80%, 90% and 95%) and compared to the bulk isotherm.

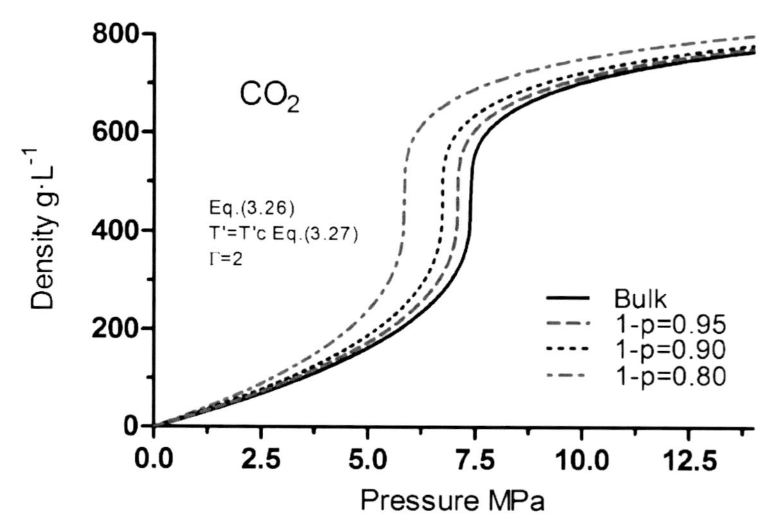

Figure 3.11 Adsorption isotherms at the critical temperature of confined CO_2 at various porosities and compared to the bulk isotherm.

3.4 Summary and Conclusions

Here's a summary of the important characteristics of the effects of confinement on the critical behavior of a confined fluid:

- Critical temperatures are *always lowered* by confinement, while densities of the confined fluid increase with *increasing porosity* at given thermodynamic conditions.
- The confined fluid density is *substantially enhanced* relative to that of the bulk fluid in equilibrium with it.
- The *critical universality class* of these systems remains an open question to the present time but it appears to fall within the random field the Ising system [12].

References

1. Domingo, C., García-Carmona, J., Llibre, J., and Rodríguez-Clemente, R. (1998). Organic-guest/microporous-host composite materials obtained by diffusion from a supercritical solution, *Adv. Mater.*, **10**, pp. 672–676.
2. López-Aranguren, P., Saurina, J., Vega, L. F., and Domingo, C. (2012). Sorption of tryalkoxysilane in low-cost porous silicates using a supercritical CO_2 method, *Microp. Mesop. Mater.*, **148**, pp. 15–24.
3. Murillo-Cremaes, N., Lopez-Periago, A. M., Saurina, J., Roig, A., and Domingo, C. (2010). A clean and effective supercritical carbon dioxide method or the host-guest synthesis and encapsulation of photoactive molecules in nanoporous matrices, *Green Chem.*, **12**, pp. 2196–2204.
4. von Behren, J., Chimowitz, E. H., and Fauchet, P. M. (1997). Critical behavior and the processing of nanoscale porous materials, *Adv. Mater.*, **9**, pp. 921–926.
5. Cracknell, R. F., Gordon, P., and Gubbins, K. E. (1993). Influence of pore geometry on the design of microporous materials for methane storage, *J. Phys. Chem.*, **97**, pp. 494–499.
6. Afrane, G., and Chimowitz, E. H. (1996). Experimental investigation of a new supercritical fluid: inorganic membrane separation process, *J. Membr. Sci.*, **116**, pp. 293–299.
7. Duong, D. D. (1998). *Adsorption Analysis: Equilibria and Kinetics* (Imperial College Press, UK).

8. Kaminsky, R. D., and Monson, P. A. (1994). A simple mean field theory of adsorption in disordered porous materials, *Chem. Eng. Sci.*, **49**, pp. 2967–2977.
9. Allen, M. P., and Tildesley, D. J. (1989). *Computer Simulation of Liquids* (Oxford Science).
10. Gruszkiewicz, M. S., Wesolowski, D. J., and Cole, D. R. (2011). Thermophysical properties of pore-confined supercritical carbon dioxide by vibrating tube densitometry, in *Proc. 36th Workshop Geotherm. Reservior Eng.*, Stanford University, California, SGP-TR-191.
11. López-Aranguren, P., Vega, L. F., Domingo, C., and Chimowitz, E. H. (2012). An equation of state for pore-confined fluids, *AIChE J.*, **58**, pp. 3597–3600.
12. DeGennes, P. G. (1984). RFIM as model for porous media, *J. Chem. Phys.*, **88**, p. 6649.

Chapter 4

General Description of Nonreactive Precipitation Methods

Concepción Domingo
Materials Science Institute of Barcelona (CSIC), Campus de la UAB, 08193 Bellaterra, Spain
conchi@icmab.es

Particle generation may be performed by expansion, for example, RESS or PGSS, or by dilution of liquids by dense gases, for example, GAS or SAS. Precipitation techniques in $scCO_2$ have already been investigated quite in detail for pure model and real substances, but interdependences between several compounds in composite products are less known. This chapter illustrates the basis of the most important $scCO_2$ precipitation techniques for the generation of powders and composites, particularly those with applications in the pharmaceutical industry. The bioavailability of pharmaceuticals presented in a solid formulation strongly depends on the size, particle size distribution, and morphology of the particles and, thus, on the used precipitation technique.

Supercritical Fluid Nanotechnology: Advances and Applications in Composites and Hybrid Nanomaterials
Edited by Concepción Domingo and Pascale Subra-Paternault

ISBN 978-981-4613-40-8 (Hardcover), 978-981-4613-41-5 (eBook)
www.panstanford.com

4.1 Particle Formation Processes with Supercritical Fluids

Several supercritical fluid (SCF) nonreactive spray precipitation processes are now under development, involving many pharmaceutical applications. The SCF is used as a precipitation medium for materials recrystallization or composite product formation. Spray processes, supported by compressed CO_2, allow for the generation of powders and composites with characteristics different from those obtained using conventional methods. Next, the different concepts used for SCF particle design are shortly described in order to clarify a rather confusing domain where many similar processes are designated by different acronyms. Process design is strongly dependent on the solubility of CO_2 of the ingredients to be precipitated [1–6].

4.1.1 Supercritical Fluid as a Solvent

Processes designed for compounds soluble in supercritical carbon dioxide ($scCO_2$) consist of dissolving the product in a compressed fluid at pressures in the order of 150–300 MPa and rapidly depressurizing the formed solution. This process takes advantage of the large variations in the solvent power of the SCF with changes in pressure. Known for long, this process is attractive due to the absence of organic solvent use; unfortunately, its application is restricted to products that present a reasonable solubility in $scCO_2$ (low-molecular-weight and low-polarity organic or organometallic compounds). Even in the most favorable cases, the production capacity of these processes is limited. This problem can be alleviated by the use of a cosolvent, but in this case the benefit of the total elimination of the organic solvent is lost.

RESS: This acronym refers to rapid expansion of supercritical solutions. The depressurization is carried out through an adequate nozzle, causing an extremely rapid nucleation of the product into a highly dispersed material.

4.1.2 Supercritical Fluid as an Antisolvent

Processes designed for compounds with negligible solubility in $scCO_2$. The SCF is not used as a solvent but as an antisolvent that causes the precipitation of a solute initially dissolved in an organic liquid. These processes typically operate at moderate pressures (5–10 MPa). The saturation of the organic solvent with CO_2 causes a decrease in the solvent power of the liquid mixture and the precipitation of the solute. The main advantage of this process over the processes in which the solute is dissolved in $scCO_2$ is versatility, as with a proper selection of the solvent, it is possible to micronize a very wide range of products. The only requisite for the application of this technology is that the solute must be soluble in an organic solvent and it must be insoluble in $scCO_2$.

GAS: This refers to batch gas antisolvent in which the compressed CO_2 is added into a continuous phase of an organic solution.

ASES: This refers to a batch aerosol solvent extraction system. The process consists of pulverizing an organic solvent solution of the solute(s) into a vessel filled by the SCF. Sometimes, it is also used to designate semicontinuous processes in which both fluids are delivered in countercurrent mode.

SAS or *PCA*: This refers to the semicontinuous supercritical antisolvent or precipitation with a compressed antisolvent, respectively. These processes are developed to overcome the limitations in the production capacity of the batch processes. In this precipitation method, CO_2 and a solution of the substance of interest are continuously fed to a precipitator. The mixing between these streams causes the precipitation of the solute due to the antisolvent effect of compressed CO_2.

SEDS: This refers to solution-enhanced dispersion by supercritical fluids. It works as a semicontinuous process in which an organic solution and a stream of $scCO_2$ are sprayed simultaneously through properly designed coaxial nozzles.

SFEE: This refers to supercritical fluid extraction of emulsions. The application of SCFs in the particle technology with emulsions appears as a natural decision to avoid the main problems of each separated technology. Emulsions techniques usually involve large quantities of organic solvents. SCFs often are not able to produce particles in the nanometer range or the products obtained present agglomeration problems. The combination of this two techniques give place to a process involving oil/water emulsions extracted with $scCO_2$.

4.1.3 Supercritical Fluid as a Solute

Processes designed for compounds with negligible solubility in $scCO_2$, but which can dissolve a significant amount of compressed fluid. Typically, the SCF is dissolved into a liquid substrate or a melted solid. They are particularly suitable to process solid lipids and polymers or to encapsulate liquids. Developed processes are well-established technologies with favorable economical conditions for commercial application. They show significant advantages over other formulation processes, including the reduced use of organic solvents, the possibility of operating at moderate temperatures in an inert atmosphere, thus avoiding oxidation or thermal degradation of the solute, as well as solute losses due to evaporation, and an enhanced control of particle size enabled by the fast precipitation kinetics.

PGSS^{TM}: This acronym refers to particles from gas-saturated solutions. It works by rapid depressurization of CO_2/substrate mixtures through a nozzle, producing the formation of solid particles due to the intense cooling caused by CO_2 expansion.

CPF^{TM}: This refers to concentrated powder form. Most of the supercritical processes are designed to produce powders from liquids that are solid at ambient temperature. This process generates powders from liquids that are not solidified during spraying. In this process, a liquid and a gas are pressurized, mixed together, and expanded into a spray tower. A powder carrier is dosed into the produced fine droplet spray and a powder with a liquid content of up to 90 wt% is formed. Particles loaded with liquids and encapsulated liquids are thus obtained.

4.2 Supercritical CO_2 Precipitation Technology Applied to Nanopharmaceuticals and Biomaterials

Scientists have dated the first medical record in ca. 2100 BC in China. The document describes the extraction of an active agent from an herb using a process of pulverization and maceration in oil or water, followed by boiling, filtering, and then adding the extract to beer for oral administration (Fig. 4.1a). This is basically the same technology that it is use today to formulate oral liquids, tablets, or capsules. Of course, the manufacturing process has been improved, with great advances in machinery and product quality control.

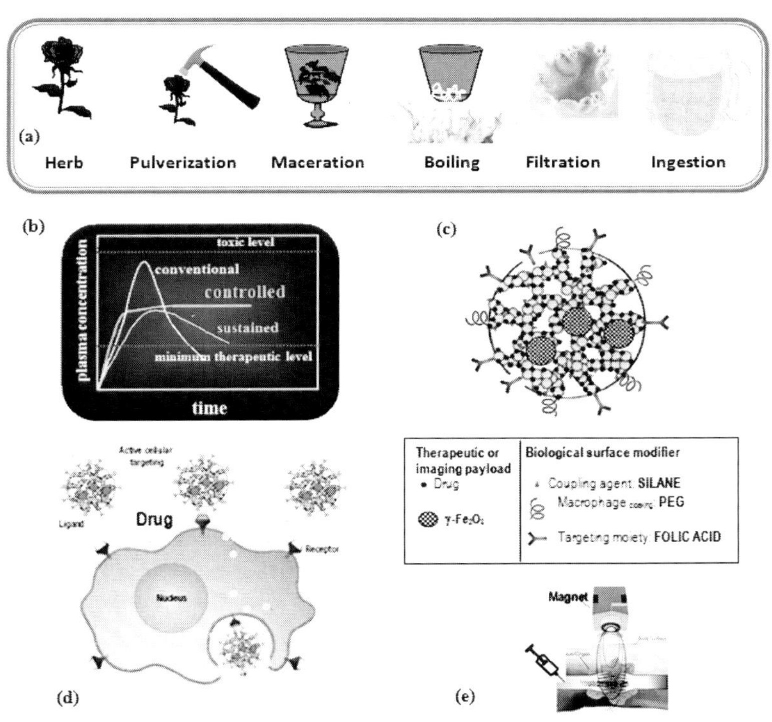

Figure 4.1 Drug delivery systems. (a) The first example of a processed drug and its administration. (b) Conventional vs. controlled or sustained drug delivery. (c) An example of a complex target drug delivery system based on porous nanoparticles of γ-Fe_2O_3@SiO_2 functionalized with the necessary moieties: The matrix is loaded with a selected drug bonded to the matrix through a coupling agent (silane) and with a label for targeting malignant cells (folic acid). Moreover, an amphiphilic coating (PEG) is necessary to minimize the attack of the immune system. (d) The nanoparticles can release their contents in close proximity to the target cells or internalize into the cell. (e) The use of magnetic nanoparticles (γ-Fe_2O_3) allows guided targeting by means of an external magnetic field and, simultaneously, hyperthermia treatment and imaging diagnosis using the γ-Fe_2O_3 as a contrast enhancer.

Moreover, the number of active principles is nowadays vast and the efficacy and safety of these products is excellent. Nevertheless, the crucial need for innovation in the pharmaceutical and biomedical industries is continuous. One important innovation route is through the development of green products and processes [7]. As the ratio of waste to useful product is very high in the pharmaceutical sector, there is a considerable need for green pharmacy innovations that are "benign by design." The emergence of nanotechnology and $scCO_2$ has had a significant impact on the drug delivery and tissue engineering sectors of the pharmaceutical industry [8, 9].

4.2.1 Drug Delivery Systems

Drug delivery systems offer several advantages compared to conventional dosage forms. These systems act as a reservoir of therapeutic agents, with specific time release profiles of the drug, thus leaving to a control of the pharmacokinetics and bioavailability (Fig. 4.1b). By using these systems with an engineered biodistribution profile, problems like insufficient drug concentration, rapid metabolization, high fluctuation of plasma levels, and negative side effects could be overcome. The control becomes complete if the system is also designed for drug targeting with specific influences on the biodistribution (Fig. 4.1c–e). Current technology for *controlled drug delivery* is mainly based on the use of micron or submicron matrixes of homopolymers or blends in which the drug is encapsulated in the form of microspheres or microparticles, dispersed, adsorbed, or chemically bonded [10]. The cytotoxicity of nanoparticles or their degradation products remain a major problem, and improvements in biocompatibility obviously are a main concern of on-going research. Moreover, up-scaling in the pharmaceutical industry from product development to marketing is not a straightforward process due to the necessary interaction with safety regulatory authorities [11, 12]. Figure 4.2 shows the scheme of the several stages involved in the full development and up-scaling of a target drug delivery system, which is similar to that of a new dug.

The application of nanotechnology to health care has led to the development of an array of novel nanoproducts for controlled drug delivery, the characteristics of which are changing the foundations of disease diagnosis, monitoring, and treatment and turning promising old and new drug discoveries into benefits for patients. The advancement in drug delivery is expected by designing nanopharmaceutical products where the in vivo fate of the active agent is determined mainly by the properties of the matrix (Fig. 4.1d,e), combining several of the properties required from an ideal drug delivery system: (i) diffusion-controlled sustained delivery and/or increased bioavailability, (ii) passive targeting based on nanometric size and the possibility of accumulation in pathological areas with compromised vasculature, (iii) active targeting by using cell-labeling molecules or external stimuli (magnetic, optical, etc.), and (iv) contrast properties for in vivo visualization and diagnosis protocols. In recent years, it has been shown that nanocarriers can penetrate through small capillaries, across numerous physiological barriers and can be taken up by cells, thus inducing efficient drug accumulation at the target site.

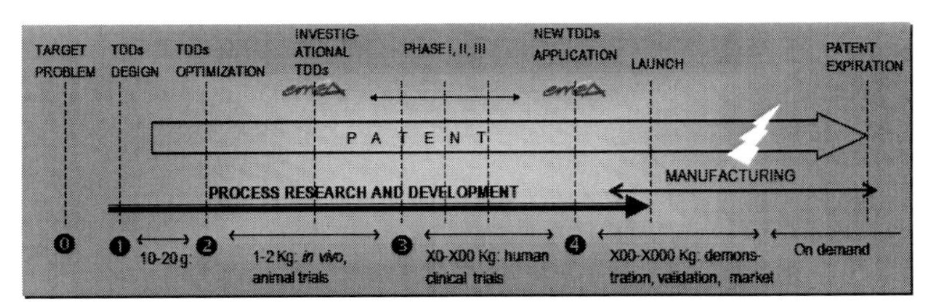

Figure 4.2 Simplified view of the several stages involved in the full development of a target drug delivery system (TDDS) for industrial take-up (not to the time scale).

Nanometric or colloidal drug delivery systems (Fig. 4.3), consisting of small simple or composite particles of 10–400 nm diameter, permit the effective administration of drugs that are highly water soluble or water insoluble or unstable in the biological environment, since the contents of the drug delivery system are effectively protected against hydrolysis and enzymatic degradation.

A broad variety of useful properties in both diagnosis and drug delivery have been reported for formulated pharmaceutical nanocarriers [13–15]. Nanocarriers can be designed for different administration routes: intravenous, intramuscular, subcutaneous, oral, nasal, or even ocular. Nanoparticles provide a unique opportunity for rapid topical and transdermal delivery, given the ability to penetrate human tissue. Nanotechnology is also opening up new opportunities in implantable release systems applied in tissue engineering and ophthalmic delivery in which the drug release can be externally controlled by using stimuli-sensitive nanocomponents. Finally, nanotechnology has the potential to revolutionize cancer diagnosis and therapy, specifically by exploiting the potential of magnetic nanoparticles and quantum dots [16] in hyperthermia and in vivo imaging. Some of those carriers have already made their way into the clinical arena (e.g., ferrofluids and polymeric particles), while others are still under preclinical development (e.g., liposomes) or in fundamental research (dendrimers and quantum dots) (Fig. 4.4).

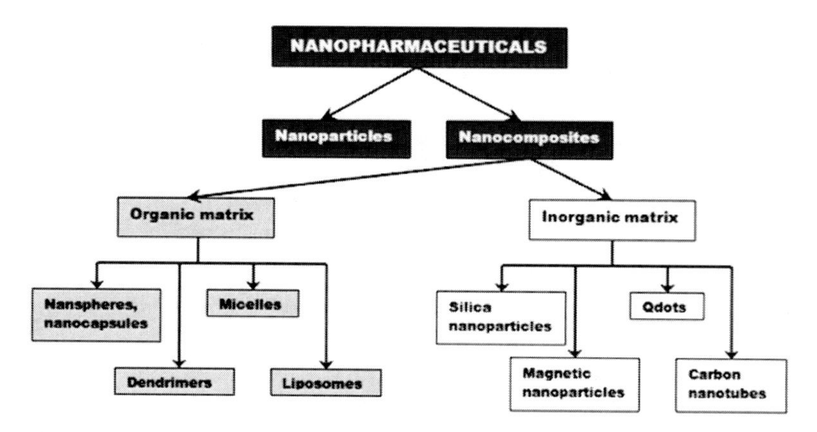

Figure 4.3 Nanopharmaceuticals for drug delivery systems.

The Würster fluid-bed process (Fig. 4.5a) is recognized by the pharmaceutical industry as the best technology for precision application of a polymeric film coating onto particulate drugs to prepare drug delivery systems of the type shown in Fig. 4.4a (nanocapsules) [17]. This technology can be used to encapsulate efficiently only micrometric powder. Following a similar idea, the spray-drying method (Fig. 4.5b) has been developed for nanopharmaceuticals [18]. The most commonly applied technique for the preparation of drug delivery systems is, however, the emulsion–solvent method [19]. The single-emulsion method (Fig. 4.5c) has been used primarily to encapsulate hydrophobic drugs dissolved in an organic solvent containing the polymeric matrix emulsified in a water phase (oil in water). Water-soluble drugs can be encapsulated by the double-emulsion water-in-oil-in-water method (Fig. 4.5d). The solvent in the emulsion is removed by either evaporation at high temperature, liophilization, or extraction. Coprecipitation, phase separation by addition of an organic antisolvent, and complex coacervation processes are also widely used methods for the manufacturing of nanostructured organic matrix-based drug delivery systems [20] (Fig. 4.4b–e). Ceramic nanoparticles, mesoporous silica, ferrofluids, and quantum dots are prepared using the sol–gel and colloidal precipitation approaches [21] (Fig. 4.4f–i).

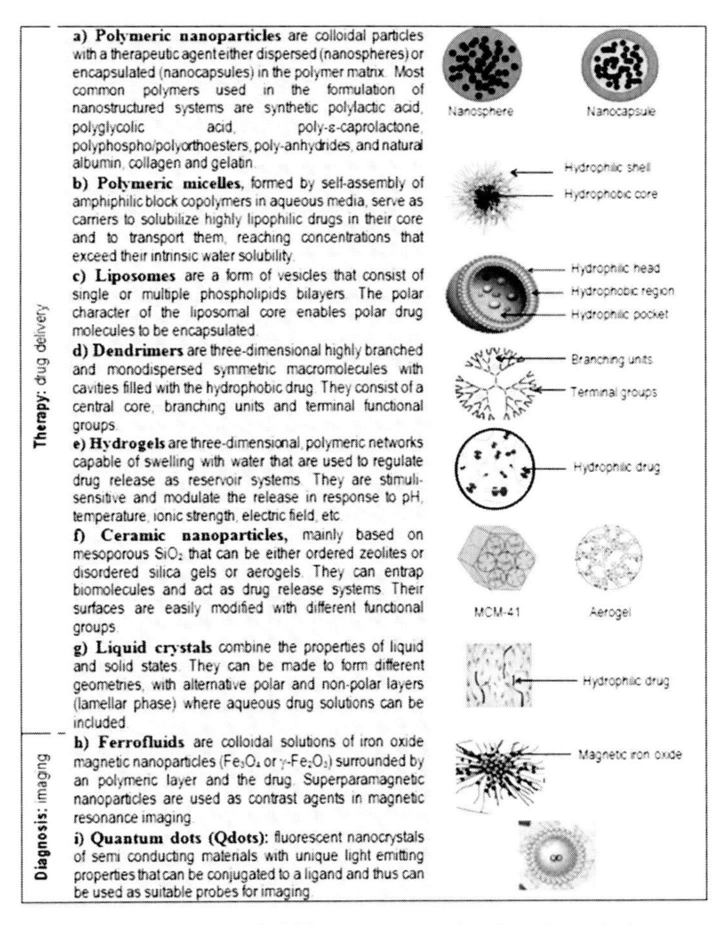

Figure 4.4 Schematics of different nanotechnology-based therapy and diagnosis systems.

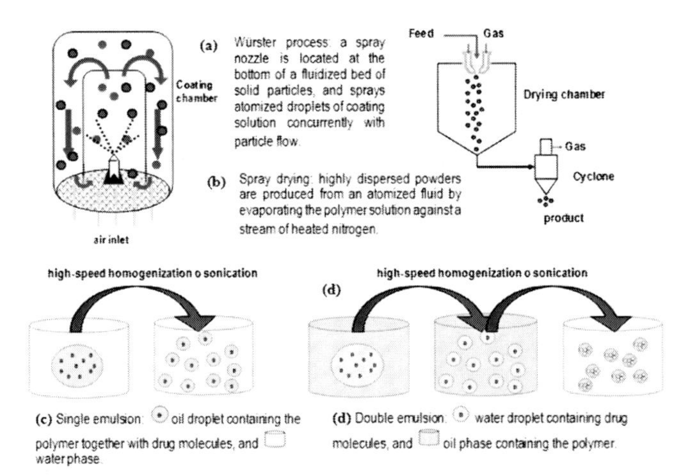

Figure 4.5 Schematic representation of techniques used for drug delivery system preparation: (a) Würster process, (b) spray drying, and (c) single- and (d) double-emulsion techniques.

With a difference, the most used green process in the production of both simple and complex pharmaceutical products is $scCO_2$ technology. The application of SCFs for the precipitation of pharmaceuticals and natural substances has attracted great attention due to the peculiar properties of these fluids. The null or little use of organic solvents, the straight preparation of dry products in confined autoclaves, and CO_2 intrinsic sterility are of particular interest to produce food, cosmetic, and pharmaceutical products and for their stabilization and formulation. Several particle formation techniques of the micro- to the nanorange, both in pure and encapsulated forms, have been developed for pharmaceutical applications using $scCO_2$ [22] (Fig. 4.6). Particle generation by spray processes using $scCO_2$ as a precipitation aid may be performed by expansion, for example, RESS [23, 24] and PGSS [25, 26] processes, or by dilution of liquids with dense gases, for example, GAS, SAS, PCA, and ASES [27] processes.

A common feature in all those particle formation techniques is that the fluid is expanded through a restriction device or nozzle in a controlled fashion. The geometry of the nozzle influences the morphology of the precipitated particles. In RESS and PGSS processes, the nozzle controls the nucleation and crystal growth by affecting the dynamics of jet expansion and the Joule–Thompson temperature drop. On the contrary, the nozzle in antisolvent processes affects particle size by controlling the initial liquid droplet diameter and the rate of solvent extraction by the SCF.

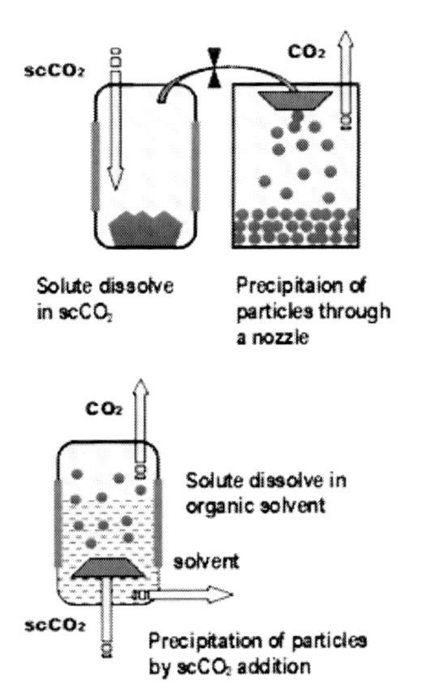

RESS (Rapid Expansion of a Supercritical Solution): $scCO_2$ is used as a solvent. The process works efficiently for compounds exhibiting significant solubility in the fluid (lipophilic compounds of relatively low molecular weight). The product is dissolved in a compressed fluid and rapidly depressurized through a nozzle resulting in the precipitation of small and monodispersed particles. Scale-up of RESS process is limited by the poor solubility of many pharmaceuticals active compounds in $scCO_2$. For drugs that are more CO_2 phobic, polar co-solvents such as acetone or ethanol are used to increase the solubility.

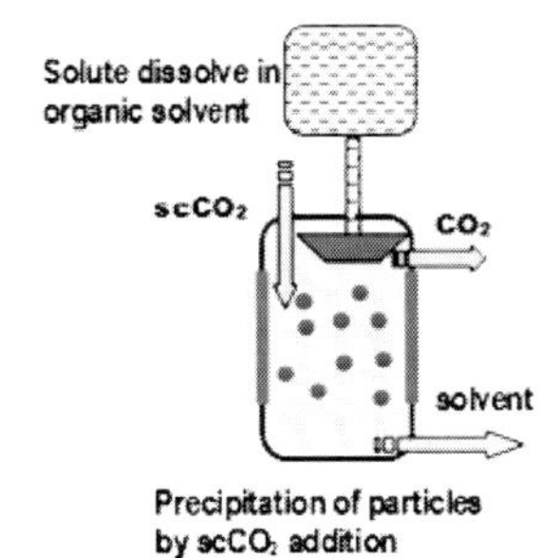

GAS/SAS (Gas/Supercritical Antisolvent)

PCA (Precipitation with a compressed antisolvent)

$scCO_2$ is used as an antisolvent to reduce the density and solvating power of an organic solvent in which the drug is contained. The antisolvent approaches provide high flexibility in product processing by choosing the liquid solvent to dissolve the treated compound, from organics or organomethalics to polymers. Many variations of the antisolvent technique exist, including the original batch GAS and SAS processes, the semicontinuous PCA, and the aerosol solvent extraction system (ASES) using a coaxial nozzle.

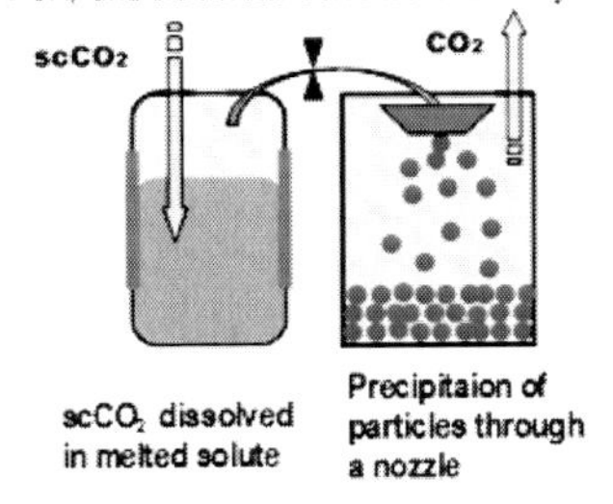

PGSS (Particles from Gas Saturated Solutions): $scCO_2$ is used as a solute dissolved in amorphous or semicrystalline high molecular weight materials (polymers and solid lipids or even some drugs). The pressurized gas diffuses into the product, lowering both its melting point and viscosity. This process produces particles by spraying the mixture via a nozzle. The process has been demonstrated on a large scale, but currently it is only applicable to produce micrometric particles.

Figure 4.6 $scCO_2$ spray precipitation techniques.

In addition to particle formation, $scCO_2$ has been used to fabricate porous polymers that can be further impregnated using $scCO_2$ as a mobile phase to obtain a drug delivery system [28–31].

After impregnation, the drug is often dispersed at a molecular level into the polymer instead of being crystallized (Fig. 4.7a). The $scCO_2$ impregnation process is also applied to modify intrinsic porous materials, such as zeolites and mesoporous silica [32–37] (Fig. 4.4f). For instance, aerogels nanoparticles can only be produced and impregnated using SCF technology. Coating or encapsulation with a thin layer of polymer using $scCO_2$ processes are powerful alternatives to the conventional chemical coating [38, 39] (Fig. 4.7b). Complex pharmaceuticals have also been precipitated in $scCO_2$/water emulsions and in water/oil emulsions extracted with $scCO_2$ [40].

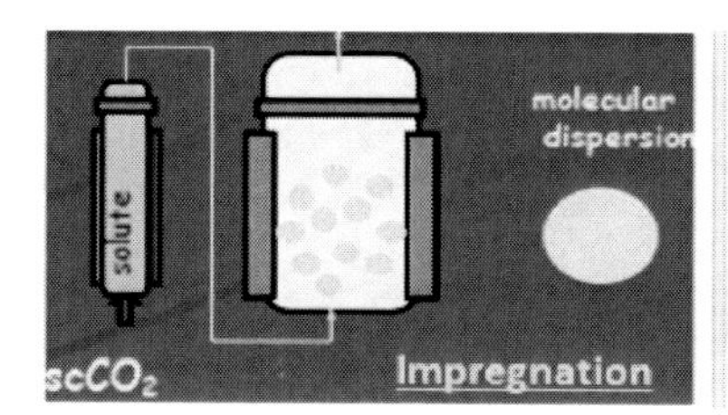

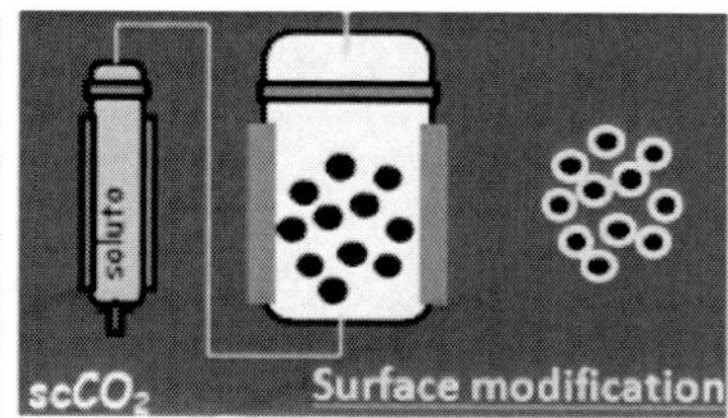

Figure 4.7 Schematic setups for (a) impregnation and (b) coating $scCO_2$ processes.

4.2.2 Scaffolds in Tissue Engineering

Tissue engineering is evolving from the use of implants that repair or replace damaged parts, to the use of controlled 3D scaffolds that induce the formation of new functional tissues either in vitro or in vivo [41]. Many polymeric scaffolds are being evaluated for a possible role in tissue engineering. Biodegradable homopolymers and copolymers of poly(lactic acid) (L-PLA) and poly(glycolic acid) (PGA) are typically chosen for these applications [42]. Poly(ε-caprolactone) (PCL) had also attracted attention as a biocompatible polymer due to the lack of toxicity and low cost, although it is a slow degradation polymer. Nonbiodegradable materials such as polymethylmethacrylate (PMMA) are being also tested in tissue engineering to achieve long-term mechanical stability after implantation as well as high hydrophobicity that enhances cells adhesion [43]. Along with the ongoing search for new materials, polymeric blends composed by a biostable polymer and a biodegradable counterpart have recently gained significant attention in tissue engineering, due to the possibility of providing them with specific advantages in morphological, degradation, and mechanical properties [44]. Not only the chemical composition but also the architecture of a scaffold plays an important role in modulating tissue growth and response behavior of cultured cells. A correct architecture of a scaffold would be the one that mimics the natural extracellular matrix that surrounds cells in the body [45]. Hence, required characteristics include high porosity and an interconnected 3D macroporous network necessary for cell proliferation [46], and a high surface area to promote cell adhesion.

Scaffolds can take forms ranging from complex monolithic 3D microcellular structures (macroporous sponges) to networks of fibers [47–49] (Fig. 4.8). Scaffolds can be produced in a variety of ways, using either conventional techniques such as solvent casting, foaming, physical separation, and freeze drying or advanced processing methods such as rapid prototyping technologies, where the scaffolds are built layer by layer from a computer-aided design model [50]. The main drawbacks of conventional scaffold production methods involve a reduced capability to control pore size and pore interconnectivity. Moreover, when using solvent approaches, the residual organic molecules left in the polymer after processing may be harmful to the transplanted cells and can inactivate biologically active growth factors. Finally, the preparation of blend systems presents a significant challenge using conventional solvent technology, as the constituent polymers are highly immiscible. Technology based on $scCO_2$ is an alternative to overcome some of the problems associated with the use of traditional organic solvents for polymers processing. $scCO_2$ technology has been already proposed for the production of a great range of biocompatible polymers with excellent control on morphology, surface properties, and purity [51–54]. Many of these applications exploit the fact that $scCO_2$ is an excellent nonsolvating porogenic agent for amorphous polymers, while being a poor solvent [55]. It is used to produce polymer foams by pressure-induced phase separation (Fig. 4.8a). Furthermore, spray processes involving $scCO_2$ have achieved considerable success in addressing polymer particles and fiber production [56–58].

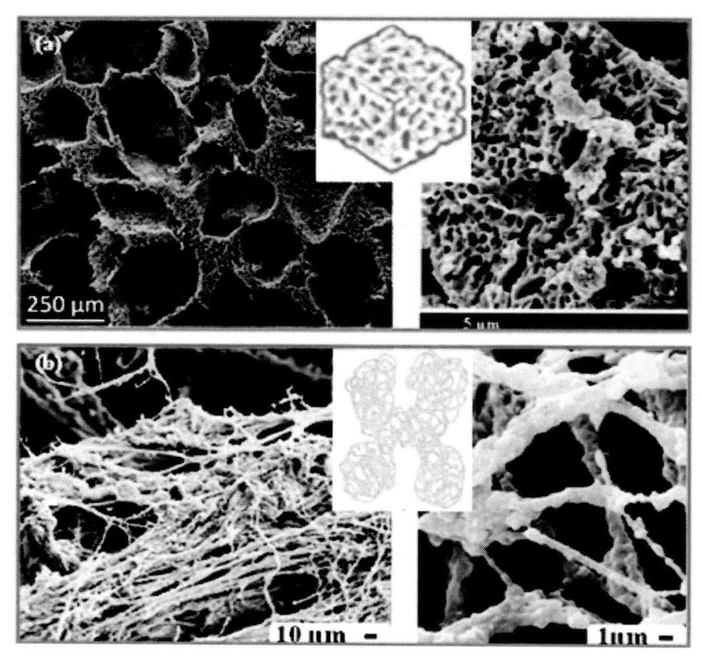

Figure 4.8 Tissue engineering scaffolds produced using $scCO_2$ technology: (a) PLA monolithic blocks obtained by foaming with $scCO_2$ and (b) PMMA/PCL fibers precipitated by a $scCO_2$ antisolvent technique.

For practical purposes, the use of solvent spray approaches, such as RESS, is limited, since $scCO_2$ can only dissolve a very small number of polymers. Alternatively, antisolvent spray processes uses common organic solvents to dissolve the polymer and, thus, a large variety of these compounds can be processed (Fig. 4.8b). Finally, the potential role of CO_2 in assisting the blending of polymers has been studied in either a reactive or a nonreactive way. In the reactive way, $scCO_2$ acts as a transport medium used to impregnate a $scCO_2$-swollen matrix with an initiator and a monomer, thus facilitating the in situ polymerization of the impregnated component [59–62]. In the nonreactive way, $scCO_2$ acts as an interfacial tension and viscosity reducing agent, which facilitates the blending of immiscible polymers [63]. Solid monolithic porous scaffolds are prepared using $scCO_2$ by simple swelling and foaming [64–67]. Fibers are prepared using the semicontinuous antisolvent SAS technique [68–70].

4.3 Conclusions and Remarks

This chapter has illustrated the basis of the most important scCO_2 precipitation techniques for the generation of powders and composites, particularly those with applications in the pharmaceutical industry. In the following chapters, a detailed description of some of the most important nonreactive scCO_2 precipitation techniques is given. First, Chapter 5 presents a review of the phase behavior and physical properties that are relevant to particle formation processes with CO_2, using as a practical example the system CO_2 plus polyethylene glycol. Next, Chapters 6, 7, and 8 describe exhaustively supercritical antisolvent approaches, the PGSS process, and the SCF extraction of emulsion precipitation techniques.

References

1. Yeob, S. D., and Kiran, E. (2005). Formation of polymer particles with supercritical fluids: a review, *J. Supercrit. Fluids*, **34**, pp. 287–308.
2. Reverchon, E., and Adamia, R. (2006). Nanomaterials and supercritical fluids review, *J. Supercrit. Fluids*, **37**, pp. 1–22.
3. Aymonier, C., Loppinet-Serani, A., Reveron, H., Garrabos, Y., and Cansell, F. (2006). Review of supercritical fluids in inorganic materials science, *J. Supercrit. Fluids*, **38**, pp. 242–251.
4. Cocero, M. J., and Martin, A. (2008). Precipitation processes with supercritical fluids: patents review, *Rec. Patent Eng.*, **2**, pp. 9–20.
5. Martín, A., Mattea, F., Varona S., and Cocero M. J. (2009). Encapsulation and co-precipitation processes with supercritical fluids: fundamentals and applications, *J. Supercrit. Fluids*, **47**, pp. 546–555.
6. Martín, A., Varona, S., Navarrete, A., and Cocero, M. J. (2010). Encapsulation and co-precipitation processes with supercritical fluids: applications with essential oils, *Op. Chem. Eng. J.*, **4**, pp. 31–41.
7. Tucker, J. L. (2006). Green chemistry, a pharmaceutical perspective, *Org. Process Res. Dev.*, **10**, pp. 315–319.
8. Lopez-Periago, A. M., Vega, A., Subra, P., Argemí, A., Saurina, J., García-González, C. A., and Domingo, C. (2008), Supercritical CO_2 processing of polymers for the production of materials with applications in tissue engineering and drug delivery, *J. Mater. Sci.*, **43**, pp. 1939–1947.

9. Domingo, C., and Saurina, J. (2012). An overview of the analytical characterization of nanostructured drug delivery systems: towards green and sustainable pharmaceuticals-a review, *Anal. Chim. Acta*, **744**, pp. 8–22.

10. Oupicky, D. (2008). Design and development strategies of polymer materials for drug and gene delivery applications, *Adv. Drug Delivery Rev.*, **60**, pp. 955–1094.

11. European Medicines Agency (2013). *The International Conference on Harmonization of Technical Requirements for Registration of Pharmaceuticals for Human Use* (ICH).

12. Nafissi, M., Ragan, J. A., and DeVries, K. M. (2002) *From Bench to Pilot Plant: Process Research in the Pharmaceutical Industry*, Vol. 817 (ACS Symposium Series, Washington DC, USA).

13. Torchilin, V.P. (2006). Multifunctional nanocarriers, *Adv. Drug. Delivery Rev.*, **58**, pp. 1532–1555.

14. Sahoo. S.K., and Labhasetwar, V. (2003). Nanotech approaches to drug delivery and imaging, *DDT*, **8**, pp. 1112–1120.

15. Peer, D., Karp, J. M., Hong, Farokhzad, S., Margalit, R., and Langer R. (2007). Nanocarriers as an emerging platform for cancer therapy, *Nat. Nanotech.*, **2**, pp. 751–760.

16. Klostranec, J. M., and Chan, W. C. W. (2006). Quantum dots in biological and biomedical research: recent progress and present challenges, *Adv. Mater.*, **18**, pp. 1953–1964.

17. Christensen, F. N., and Bertelsen, P. (1997). Qualitative description of the wurster-based fluid-bed coating process, *Drug Dev. Ind. Pharm.*, **23**, pp. 451–463.

18. Thybo, P., Hovgaard, L., Lindeløv, J. S., Brask, A., and Andersen, S. K. (2008), Scaling up the spray drying process from pilot to production scale using an atomized droplet size criterion, *Pharm. Res.*, **25**, pp. 1610–1620.

19. Antona, N., Benoita, J. P., and Saulniera, P. (2008). Design and production of nanoparticles formulated from nano-emulsion templates-a review, *J. Controlled Release*, **128**, pp. 185–199.

20. Yu, L. X. (2008). Pharmaceutical quality by design: product and process development, understanding, and control, *Pharm. Res.*, **25**, pp. 781–791.

21. Mann, S., Burkett, S. L., Davis, S. A., Fowler, C. E., Mendelson, N. H., Sims, S. D., Walsh, D., and Whilton, N.T. (1997). Sol–gel synthesis of organized matter, *Chem. Mater.*, **9**, pp. 2300–2310.

22. Jung, J., and Perrut, M. (2001). Particle design using supercritical fluids, literature and patent survey. Review, *J. Supercrit. Fluids*, **20**, pp. 179–219.

23. Domingo, C., Berends, E., and van Rosmalen, G. M. (1996). Precipitation of ultrafine benzoic acid by expansion of a supercritical CO_2 solution through a porous plate nozzle, *J. Cryst. Growth*, **166**, pp. 989–995.

24. Domingo, C., Berends, E., and van Rosmalen, G. M. (1997). Precipitation of ultrafine organic crystals from the rapid expansion of supercritical solutions over a capillary and a frit nozzle, *J. Supercrit. Fluids*, **10**, pp. 39–55.

25. García-Gonzáleza, C. A., Sampaio da Sousa, A. R., Argemí, A., López-Periago, A., Saurina, J., Duarte, C. M. M., and Domingo, C. (2009). Production of hybrid lipid-based particles loaded with inorganic nanoparticles and active compounds for prolonged topical release, *Int. J. Pharm.*, **382**, pp. 296–304.

26. García-González, C. A., Argemí, A., Sampaio da Sousa, A. R., Duarte, C. M. M., Saurina, J., and Domingo, C. (2010). Encapsulation efficiency of solid lipid hybrid particles prepared using the PGSS® technique and loaded with different polarity active agents, *J. Supercrit. Fluids*, **54**, pp. 342–347.

27. Subra-Paternault, P., Roy, C., Vrel, D., Vega-González, A., and Domingo, C. (2007). Solvent effect on tolbutamide crystallization induced by compressed CO_2 as antisolvent, *J. Cryst. Growth*, **309**, pp. 76–85.

28. Elvira, C., Fanovich, A., Fernández, M., Fraile, J., San Román, J., and Domingo, C. (2004). Evaluation of drug deliver characteristics of microspheres of PMMA/PCL-cholesterol obtained by supercritical-CO_2 impregnation and by dissolution-evaporation techniques, *J. Controlled Release*, **99**, pp. 231–240.

29. Argemí, A., López-Periago, A., Domingo, C., and Saurina, J. (2008). Spectroscopic and chromatographic characterization of triflusal delivery systems prepared by using supercritical impregnation technologies, *J. Pharm. Biomed. Anal.*, **46**, pp. 456–462.

30. Andanson, J. M., López-Periago, A., García-González, C. A., Domingo, C., and Kazarian, S. G. (2009). Spectroscopic analysis of trifusal impregnated into PMMA from supercritical CO_2 solution, *Vibrat. Spectr.*, **49**, pp. 183–189.

31. López-Periago, A., Argemí, A., Andanson, J. M., Fernández, V., García-González, C. A., Kazarian, S. G., Saurina, J., and Domingo, C. (2009). Impregnation of a biocompatible polymer aided by supercritical CO_2:

evaluation of drug stability and drug-matrix interactions, *J. Supercrit. Fluids*, **48**, pp. 56–63.

32. Domingo, C., García-Carmona, J., Llibre, J., and Rodríguez-Clemente, R. (1998). Organic-guest/microporous-host composite materials obtained by diffusion from a supercritical solution, *Adv. Mater.*, **10**, pp. 672–676.

33. Domingo, C., García-Carmona, J., Fanovich, M. A., Llibre, J., and Rodríguez-Clemente, R. (2001). Single or two-solute adsorption processes at supercritical conditions: an experimental study, *J. Supercrit. Fluids*, **21**, pp. 147–157.

34. García-Carmona, J., Fanovich, M. A., Llibre, J., and Rodríguez-Clemente, R., and Domingo, C. (2002). Processing of microporous VPI-5 molecular sieve by using supercritical CO_2: stability and adsorption properties, *Microp. Mesop. Mater.*, **54**, pp. 127–137.

35. López-Periago, A. M., Fraile, J., García-González, C., and Domingo, C. (2009). Impregnation of a triphenylpyrylium cation into zeolite cavities using supercritical CO_2, *J. Supercrit. Fluids*, **50**, pp. 305–312.

36. López-Periago, A. M., García-González, C. A., Saurina, J., and Domingo, C. (2010). Preparation of trityl cations in faujasite micropores through supercritical CO_2 impregnation, *Microp. Mesop. Mater.*, **132**, pp. 357–362.

37. Murillo-Cremaes, N., López-Periago, A. M., Saurina, J., Roig, A., and Domingo, C. (2010). A clean and effective supercritical carbon dioxide method for the host-guest synthesis and encapsulation of photoactive molecules in nanoporous matrices, *Green Chem.*, **12**, pp. 2196–2204.

38. Loste, E., Fraile, J., Fanovich, A., Woerlee, G. F., and Domingo, C. (2004). Anhydrous supercritical carbon dioxide method for the controlled silanization of inorganic nanoparticles, *Adv. Mater.*, **16**, pp. 739–744.

39. Roy, C., Vega-González, A., García-González, C. A., Tassaing, T., Domingo, C., and Subra-Paternault, P. (2010). Assesment of $scCO_2$ techniques for surface modification of micro- and nanoparticles: process design methodology based on solubility, *J. Supercrit. Fluids*, **54**, pp. 362–368.

40. Fraile, M., Martín, A., Deodato, D., Rodriguez-Rojo, S., Nogueira, I. D., Simplício, A. L.,Cocero, M. J., and Duarte, C. M. M. (2013). Production of new hybrid systems for drug delivery by PGSS (particles from gas saturated solutions) process, *J. Supercrit. Fluids*, **81**, pp. 226–235.

41. Hutmacher, D. W. (2000). Scaffolds in tissue engineering bone and cartilage, *Biomaterials*, **21**, pp. 2529–2543.

42. Vert, M., Li, M. S., Spenlehauer, G., and Guerin, P. (1992). Bioresorbability and biocompatibility of aliphatic polyesters, *J. Mater. Sci.*, **3**, pp. 432–446.

43. Downes, S., Archer, R. S., Kayser, M. V., Patel, M. P., and Braden, M. (1994). The regeneration of articular cartilage using a new polymer system. *J. Mater. Sci.: Mater. Med.*, **5**, pp. 88–95.

44. Amass, W., Amass, A., and Tighe, B. (1998). A review of biodegradable polymers: uses, current developments in the synthesis and characterization of biodegradable polyesters, blends of biodegradable polymers and recent advances in biodegradation studies, *Polym. Int.*, **47**, pp. 89–144.

45. Hentze, H. P., and Antonietti, M. (2002). Porous polymers and resins for biotechnological and biomedical applications, *Rev. Mol. Biotechnol.*, **90,** pp. 27–53.

46. Whang, K., Thomas, C. H., Healy, K. E., and Nuber, G. (1995). A novel method to fabricate bioabsorbable scaffolds, *Polymer*, **36**, pp. 837–842.

47. López-Periago, A. M., García-González, C. A., and Domingo, C. (2005). Solvent- and thermal-induced crystallization of poly-L-lactic acid in supercritical CO_2 medium, *J. Appl. Polym. Sci.*, **111**, pp. 291–300.

48. Karageorgiou, V., and Kaplan, D. (2005). Porosity of 3D biomaterial scaffolds and osteogenesis, *Biomaterials*, **26**, pp. 5474–5491.

49. Moroni, L., de Wijn, J. R., and Blitterswijk, C. A. (2006). 3D fiber-deposited scaffolds for tissue engineering: influence of pores geometry and architecture on dynamic mechanical properties, *Biomaterials*, **27**, pp. 974–985.

50. Yeong, W. Y., Chua, C. K., Leong, K. F., and Chandrasekaran, M. (2004). Rapid prototyping in tissue engineering: challenges and potential, *Trends Biotechnol.*, **22**, pp. 643–652.

51. Cooper, A. I. (2000). Synthesis and processing of polymers using supercritical carbon dioxide, *J. Mater Chem.*, **10**, pp. 207–234.

52. Cooper, A. I. (2003). Porous materials and supercritical fluids, *Adv. Mater,* **15**, pp. 1049–1059.

53. Tomasko, D. L., Li, H., Liu, D., Han, X., Wingert, M. J., Lee, L. J., and Koelling, K. W. (2003). A review of CO_2 applications in the processing of polymers, *Ind. Eng. Chem. Res.*, **42**, pp. 6431–6456.

54. Quirk, R. A., France, R. M., Shakesheff, K. M., and Howdle, S. M. (2004). Supercritical fluid technologies and tissue engineering scaffolds, *Curr. Opin. Solid State Mater. Sci*, **8**, pp. 313–321.

55. Domingo, C., Vega, A., Fanovich, M. A., Elvira, C., and Subra, P. (2003). Behavior of poly(methyl methacrylate)-based systems in supercritical CO_2 and CO_2 plus cosolvent: solubility measurements and process assessment, *J. Appl. Polym. Sci.*, **90**, pp. 3652–3659.

56. Reverchon, E., Della Porta, G., de Rosa, I., Subra., P., and Letourneur, D. (2000). Super-critical antisolvent micronization of some biopolymers, *J. Supercrit. Fluids*, **18**, pp. 239–245.

57. Sane, A., and Thies, M. C. (2007). Effect of material properties and processing conditions on RESS of poly(L-lactide), *J. Supercrit. Fluids*, **40**, pp. 134–143.

58. Reverchon, E., and Antonacci, A. (2007), Polymer microparticles production by supercritical assisted atomization, *J. Supercrit. Fluids*, **39,** pp. 444–452.

59. Woods, H. M., Silva, M. M. C. G., Nouvel, C., Shakesheff, K. M., and Howdle, S. M. (2004). Materials processing in supercritical carbon dioxide: surfactants, polymers and biomaterials, *J. Mater. Chem.*, **14**, pp. 1663–1678.

60. Mawson, S., Kanakia, S., and Johnston, K. P. (1997). Metastable polymer blends by precipitation with a compressed fluid antisolvent, *Polymer*, **38**, pp. 2957–2967.

61. Naylor, A., and Howdle, S. M. (2005). The preparation of novel blends of ultra high molecular weight polyethylene with polymethacrylate based copolymers using supercritical carbon dioxide, *J. Mater. Chem.*, **15**, pp. 5037–5042.

62. Liu, Z., Wang, J., Dai, X., Han, B., Dong, Z., Yang, G., Zhang, X., and Xu, J. (2002). Synthesis of composites of silicon rubber and polystyrene using supercritical CO_2 as a swelling agent, *J. Mater. Chem.*, **12**, pp. 2688–2691.

63. Elkovitch, M. D., Lee, L. J., and Tomasko, D. L. (2000). Supercritical fluid assisted polymer blending, *Polym. Eng. Sci.*, **40**, pp. 1850–1861.

64. Salerno, A., Netti, P. A., Di Maio, E., and Iannace, S. (2009). Engineering of foamed structures for biomedical application, *J. Cell. Plast.*, **45**, pp. 103–117.

65. Salerno, A., and Domingo, C. (2014). Making microporous nanometre-scale fibrous PLA aerogels with clean and reliable supercritical CO_2 based approaches, *Microp. Mesop. Mater.*, **184**, pp. 162–168.

66. Salerno, A., and Domingo, C. (2013). A clean and sustainable route towards the design and fabrication of biodegradable foams by means

of supercritical CO_2/ethyl lactate solid-state foaming, *RSC Adv.*, **3**, pp. 17355–17363.

67. Salerno, A., and Domingo, C. (2013). Effect of blowing agent composition and processing parameters on the low temperature foaming of poly(L-lactide/caprolactone) co-polymer by means of supercritical CO_2/ethyl lactate binary mixtures, *J. Supercrit. Fluids*, **84**, pp. 195–204.
68. Vega-González, A., Domingo, C., Elvira, C., and Subra, P. (2004). Precipitation of PMMA/PCL blends using supercritical carbon dioxide, *J. Appl. Polym. Sci.*, **91**, pp. 2422–2426.
69. Vega-González, A., Subra-Paternault, P., López-Periago, A. M., and Domingo, C. (2008). Supercritical CO_2 antisolvent precipitation of polymer networks of L-PLA, PMMA and PMMA/PCL blends for biomedical applications, *Eur. Polym. J.*, **44**, pp. 1081–1094.
70. García-González, C. A., Vega-González, A., López-Periago, A. M., Subra-Paternault, P., and Domingo, C. (2009). Composite fibrous biomaterials for tissue engineering obtained using a supercritical CO_2 antisolvent process, *Acta Biomater.*, **5**, pp. 1094–1103.

Chapter 5

Phase Equilibria, Densities, and Viscosities of Carbon Dioxide–Poly(Ethylene Glycol) Mixtures for Particle Formation Applications

Masayuki Iguchi,[a] Yoshiyuki Sato,[b] and Richard Lee Smith, Jr.[b,c]

[a]*Faculty of Food Science and Technology, University Putra Malaysia, Serdang, Selangor 43400, Malaysia*

[b]*Graduate School of Engineering, Research Center of Supercritical Fluid Technology, Tohoku University Aramaki Aza Aoba 6-6-11, Aoba-ku, Sendai 980-8579, Japan*

[c]*Graduate School of Environmental Studies, Tohoku University Aramaki Aza Aoba 6-6-11, Aoba-ku, Sendai 980-8579, Japan*

smith@scf.che.tohoku.ac.jp

In this, a review on the phase behavior and physical properties that are relevant to particle formation processes with CO_2 is given. As a practical example, the polymer poly(ethylene glycol), or PEG, with many applications in food and pharmaceutical product processing, used with supercritical CO_2 for particle formation and encapsulation is described. Vapor–liquid equilibrium (VLE) and solid–liquid equilibrium (SLE) data is necessary because particle formation processes begin by dissolution of CO_2 into the PEG solution and

Supercritical Fluid Nanotechnology: Advances and Applications in Composites and Hybrid Nanomaterials
Edited by Concepción Domingo and Pascale Subra-Paternault

ISBN 978-981-4613-40-8 (Hardcover), 978-981-4613-41-5 (eBook)
www.panstanford.com

then are formed through solidification of atomized droplets and ex-solution of CO_2. CO_2 solubility in PEG for given temperatures (333 and 353 K) is not distinctly different for molecular weights from 400 to 20,000 g/mol and varies from 3 wt% CO_2 at 2 MPa to 25 wt% CO_2 at 28 MPa. The density of PEG solutions (PEG 400 and 1600) initially decreases by about 1% when pressurized with CO_2 to around 10–15 MPa but then shows trends to increase at higher pressures. When using CO_2 to reduce the viscosity of PEG solution, the viscosity reduction for PEG solutions (PEG 200–12,000) is above 40% in the presence of 10 wt% of dissolved CO_2. The viscosity reduction ratios can be calculated with theory that uses equations of state.

5.1 Introduction

Supercritical fluid technologies have been applied to extraction, material synthesis, separation, and energy processes [1–3] and analytical methods [4]. Many researchers have focused on supercritical water and carbon dioxide (CO_2) because they are benign for the environment. Particle formation processes using supercritical water and CO_2 have been developed for producing small fine particles and elimination of residual solvents in the products, especially in applications of pharmaceutical and food processing [5–8]. Polymers in these applications are used as substrates and encapsulating materials. The polymers should be biodegradable and biocompatible for which poly(ethylene glycol) (PEG) and poly(lactic acid) (PLA) are some of the most popular choices.

PEG polymers belong to the family of poly(oxyethylene) (POE) that have molecular weights (M_w) below 20,000 g/mol. PEGs that have molecular weights greater than 20,000 g/mol are referred to as poly(ethylene oxide) [9]. The characteristic properties of PEGs are (i) high solubility in both aqueous and organic solvents, (ii) no known immunogenicity, antigenicity, and toxicity effects, and (iii) high flexibility and hydration of the polymer chain [10]. PEGs are established reference polymers for pharmaceutical and biomedical applications, and they have been used as additives in food and cosmetic products [10–12].

Supercritical CO_2 has been applied to the processing of PEG particles through rapid expansion of supercritical solutions with a nonsolvent (RESS-N) [13], particles from a gas-saturated solution (PGSS), aerosol solvent extraction system (ASES) [14], and supercritical assisted atomization (SAA) [15]. In these processes, particles are formed through solidification of atomized droplets dissolved with CO_2, so that the phase behavior of a CO_2–PEG mixture is necessary for understanding the particle formation mechanism and for developing practical applications. The size of the atomized droplets strongly depends on the fluid flow characteristics of the solutions [16] and on the fundamental physical properties of the droplet such as viscosity, η(Pa·s); surface tension, σ(J/m^2); and density, ρ(kg/m^3) [17]. These characteristics and properties affect mass transport of CO_2 into and out of the droplet and control particle size and morphology (Fig. 5.1).

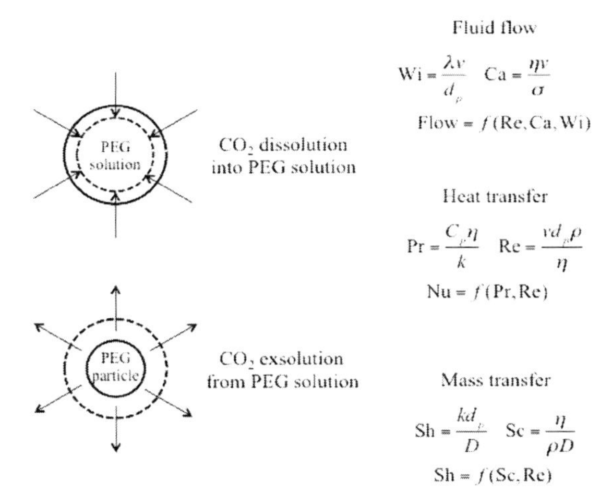

Figure 5.1 Schematic of CO_2 dissolution and CO_2 + additive ex-solution from a PEG solution. Basic fluid flow, heat transfer, and mass transfer relationships control particle formation. Fluid flow is described by Weissenberg (Wi), Capillary (Ca), and Reynolds (Re) numbers; heat transfer is described by Prandtl (Pr) and Reynolds numbers; and mass transfer is described by Schmidt (Sc) and Reynolds numbers. In the Weissenberg number, λ is the characteristic relaxation time of the solution.

This section focuses on the phase behavior that is relevant to particle formation using supercritical CO_2—vapor–liquid equilibrium (VLE) and solid–liquid equilibrium (SLE)—and two properties important for particle formation, density and viscosity of CO_2–PEG mixtures. Among the transport properties, the viscosity is particularly important because it affects size and morphology of the formed particles. Micron-sized PEG particles (≤10 μm) have been formed by the PGSS method by adding fatty acids to PEG–CO_2 solutions for viscosity reduction of the PEG solutions [18], while formed particles without adding solvents have diameters about 80–400 μm [19–21]. The SAA method has been used to produce 1–4 μm of micron particles from PEG dissolved in acetone [15]. The size of the formed particles with these methods tends to decrease when the viscosity of the PEG solutions is reduced by chemical or physical means. Both VLE and physical property data is required for understanding and modeling the viscosity [22, 23]. In this, physical properties, including VLE and SLE density and viscosity of CO_2–PEG solutions, are reviewed. Each section includes measurement and calculation methods for the property and shows how temperature, pressure, and molecular weight affect the property. By understanding the property variations, it is possible to estimate conditions for desired particle formations with molecular models.

5.2 Vapor–Liquid Equilibria of CO_2–PEG Mixtures

5.2.1 Measurement

Experimental studies reported for VLE of CO_2–PEG mixtures in the literature are shown in Table 5.1. Although a lot of VLE data for PEG of lower molecular weights below 8000 g/mol has been reported, data for PEG of higher molecular weights is relatively scarce. Among the methods to measure VLE of CO_2–PEG mixtures, analytical methods are applied to measure the composition of both phases, while other methods have been used to measure the CO_2 composition of only the liquid phase. In analytical methods, the composition of each phase can be determined from mass balance [24, 25], volume [26–28], pressure [29], refractometry [29, 30], UV spectroscopy [30], or colorimetry [30]. The analytical method can influence the reliability of the measured values of PEG when UV is used for analysis of the vapor phase or when molecular weights lower than 300 g/mol are analyzed with colorimetry [30].

Table 5.1 Experimental studies reported for vapor–liquid equilibrium of carbon dioxide–poly(ethylene glycol) mixtures*

M_w (g/mol)	Method	p range (MPa)	T range (K)	Ref.
200	Pressure decay	0.1–1	303–333	[31]
	Static-analytical	1–25	303–348	[24, 29, 30]
300	Pressure decay	0.1–1	303–333	[31]
400	Pressure decay	0.1–1	303–333	[31]
	Spectroscopic	2–20	313, 423	[32, 33]
	Synthetic	7–20	298–318	[34]
	Static-analytical	5–25	293–348	[25, 29, 30]
	Flow-analytical	1–26	313, 323	[35]
600	Static-analytical	6–27	313, 348	[29, 30]
	Flow-analytical	1–29	313, 323	[35]
1000	Flow-analytical	2–26	323	[35]
1500	Gravimetric	6–25	298, 323	[36]
	Spectroscopic	3–20	308, 328	[37]
	Flow-analytical	5–40	308, 328	[38]
	Static-analytical	2–29	323–373	[28, 39]
4000	Gravimetric	6–25	298, 323	[36]
	Static-analytical	1–29	328–373	[28, 39]
6000	Gravimetric	4–34	335–393	[26, 27]
	Static-analytical	2–33	353–393	[26, 27, 40]
8000	Gravimetric	3–34	353–393	[26]
	Static-analytical	3–33	353–393	[26, 28, 39]
12,000	Gravimetric	3–33	353–393	[26]
	Static-analytical	2–33	353–393	[26, 27]

* Data of poly(ethylene glycol) 9000 and 20,000 is available in the literature [26].

Figure 5.2 shows the VLE of CO_2–PEG 400 systems from the literature. The VLE data shown in Fig. 5.2 was measured by the pressure decay [31], spectroscopy [32], synthetic [34], static-analytical [25, 29, 30], and flow-analytical [35] methods. Values reported in the literature are in general agreement except for the liquid phase composition reported by Daneshvar et al. [35], whose deviations in the data might be due to nonhomogeneous liquid phase samples taken that overestimate the composition [39]. Foaming of PEG in the presence of CO_2 especially during pressure reduction is a systematic problem in the measurement of liquid phase compositions of PEG solutions.

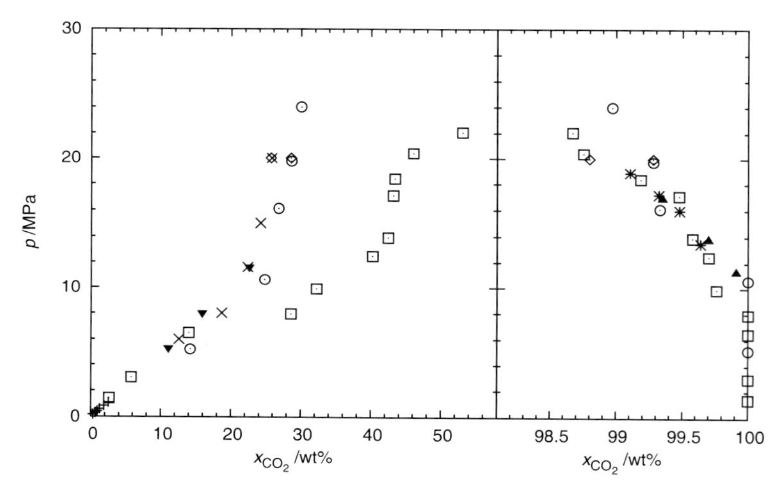

Figure 5.2 Vapor–liquid equilibrium of carbon dioxide–poly(ethylene glycol) 400 at 313 K in the literature: □ Daneshvar et al. [5, 6]; ○ Gourgouillon et al. [29, 41]; ▼ Guadagno et al. [32]; + Li et al. [31]; +, ◇ Lopes et al. [30] ; × Matsuyama et al. [13, 25]; × Vitoux et al. [33]; and ▲ Xue et al. [34].

The solubility of CO_2 in PEG 1500 at several temperatures is shown in Fig. 5.3. Values at 323 K reported by Weidner et al. [39] using static-analytical are consistent with those by Aionicesei et al. [36] using gravimetry. Values at 328 K reported by Pasquali et al. [37] using spectroscopy are much larger than those at 323 K reported by Weidner et al. [39] and Aionicesei et al. [36]. This inconsistency seems to be caused by factors other than the difference in the measured temperature (5 K), since values of PEG 400 at 313 K reported by Vitoux et al. [33] were similar to those of Weidner et al. [39] and Aionicesei et al. [36]. Vapor pressures of PEG are relatively low and decrease with increasing molecular weight [42]. Solubility data of PEG in the vapor phase of CO_2 has been reported for molecular weights ranging from 200 to 1500 g/mol [25, 29, 34, 35]. An increase in the molecular weight reduces the solubility of PEG in CO_2. For example, the solubility of PEG 1500 in CO_2 at 328 K and 40 MPa is below 0.05 wt% [38]. For higher molecular weights of PEG (>1500 g/mol), the solubility of PEG in CO_2 can be usually neglected in absence of other factors such as cosolvents and surfactants.

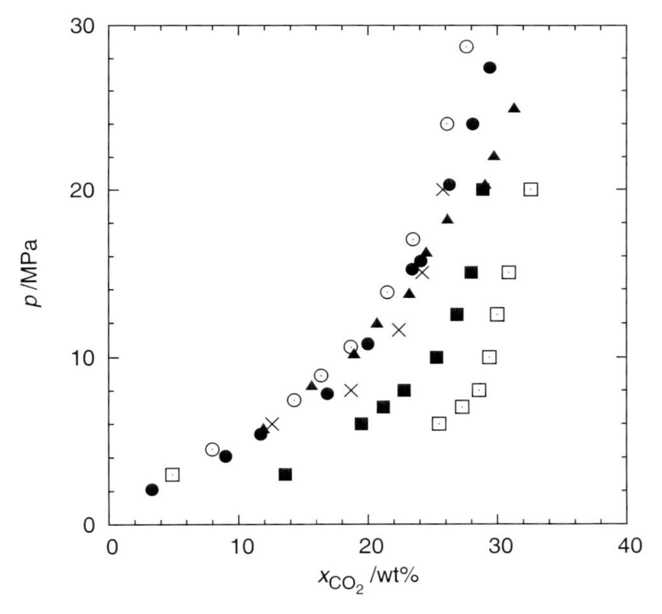

Figure 5.3 Solubility of carbon dioxide in poly(ethylene glycol) 1500 at several temperatures in the literature: ▲ Aionicesei et al. [36] at 323 K; ■, □ Pasquali et al. [37, 38] at (328 and 308) K; ○, ● Weidner et al. [21, 39] at (333 and 323) K and poly(ethylene glycol) 400, and × Vitoux et al. [33] at 313 K.

CO_2 solubilities in PEGs of different molecular weights are shown in Fig. 5.4. The solubility of CO_2 in PEG 200 is lower than that of PEG having a molecular weight above 400 g/mol (Fig. 5.4a). There are no distinct differences of CO_2 solubility in PEG for the range of molecular weights from 400 to 20,000 g/mol (Fig. 5.4).

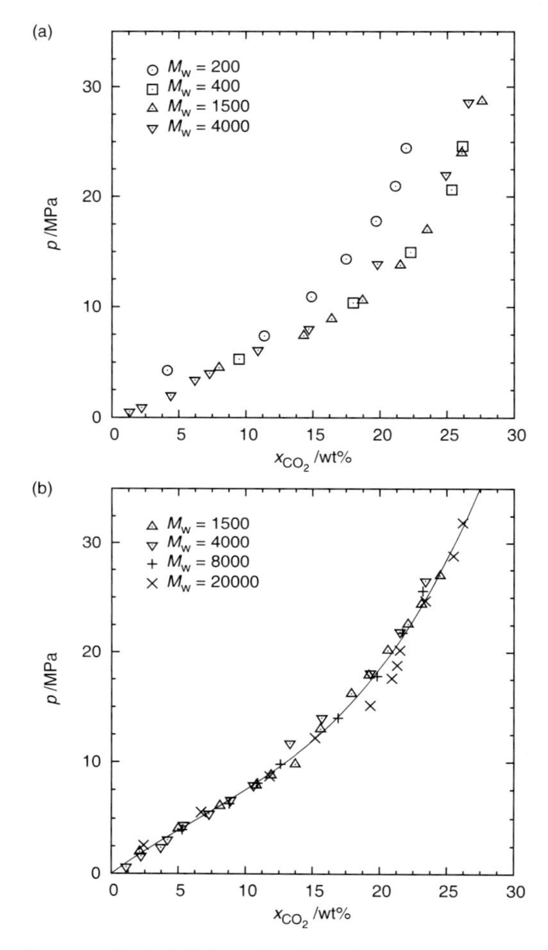

Figure 5.4 Reported solubilities of carbon dioxide in poly(ethylene glycol) of different molecular weights at different temperatures: (a) 333 K, 200 to 4000 g/mol, and (b) 353 K, 1500 to 20,000 g/mol. Data is taken from Gourgouillon et al. [29, 41] for 200 and 400 g/mol, Weidner et al. [21, 39] for 1500 g/mol, Wiesmet et al. [28] for 4000 and 8000 g/mol, and Kukova [26, 27] for 20,000 g/mol. Line shown is calculated with Eq. 5.1.

At 353 K for PEG molecular weights from 1500 to 20,000, the CO_2 solubility in PEG solution can be calculated with a simple correlation equation as follows:

$$w_{CO2} = 7.01\times10^{-5}p^3 - 2.43\times10^{-2}p^2 + 1.51p \quad (5.1)$$

where w_{CO2} is in weight percent and p is in MPa. The low solubility of CO_2 in PEG 200 can be attributed to the hydroxyl end group in the PEG chain. The end group of PEG also has some effects on the solubility of CO_2 in PEG having a molecular weight above 400 g/mol. This trend for PEG of lower molecular weight is consistent with measured partition coefficients for n-mers between vapor and liquid phases [30, 43]. An increase in system temperature slightly reduces the solubility of CO_2 in PEG, which implies that the dissolved CO_2 tends to aggregate within the polymer [24, 26–29, 31, 33, 35–37, 39]. A comparison of solubilities of gases in PEG such as CO_2, nitrogen (N_2), and propane (C_3H_8) shows that the amount of gas dissolved in PEG is in increasing order of $CO_2 > C_3H_8 > N_2$.

5.2.2 Correlation

VLE data of CO_2–PEG mixtures has been correlated with cubic equations of state (EOSs) [24], lattice-fluid (LF) EOSs [25, 29, 35, 39], and EOSs based on statistical associating fluid theory (SAFT) [28, 36]. In the calculations with EOSs, pure parameters and binary parameters for CO_2 and PEG are required. Binary parameters are typically determined by fitting VLE data. Usually, pure component parameters are determined from vapor pressures, critical constants or combined vapor pressures and liquid densities that are used at conditions of saturation. For substances of very low vapor pressure like PEG, the pure parameters are generally determined from fitting pressure-volume-temperature (PVT) data [28, 35, 44], although, as pointed out by Martin et al. [40], some pure component parameters are insensitive to PVT data and the procedure can lead to unreliable results [45]. Two approaches have been established to overcome these problems—(i) simultaneous fitting of PVT data of a pure substance and binary phase equilibrium data and (ii) extrapolation of correlated parameters on the basis of a series of low-molecular-weight substances. Martin et al. determined the pure parameters for PEG from PVT data of pure PEG and from VLE data of CO_2–PEG mixture [40]. Pedrosa et al. used the molecular weight dependence of the parameters determined from a series of ethylene glycol oligomers [46].

5.3 Solid–Liquid Equilibria of CO_2–PEG Mixtures

Weidner and coworkers have measured SLE of CO_2–PEG mixtures over a range of molecular weights from 1500 to 12,000 g/mol using a capillary method [26, 27, 39]. Pasquali et al. measured the melting temperature of PEG 1500 and 4000 at CO_2 pressures using the capillary method and reported good agreement with the results of Weidner et al. Figure 5.5 shows the SLE of CO_2–PEG mixtures for different molecular weights including that of PEO of molecular weight 35,000 g/mol.

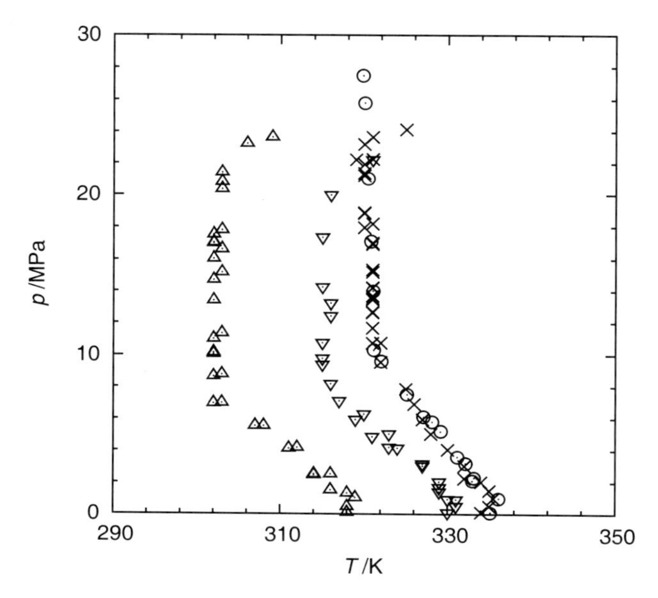

Figure 5.5 Reported solid–liquid equilibria of carbon dioxide–poly(ethylene glycol) of different molecular weights: △ 1500 g/mol; ▽ 4000 g/mol; ○ 12,000 g/mol; and × 35,000 g/mol. Data is taken from Weidner et al. [21, 39] for 1500, 4000, and 35,000 g/mol and Kukova et al. [26, 27] for 12,000 g/mol.

The melting temperature of PEGs initially increases as CO_2 pressure is applied up to 1 MPa and then it decreases. This behavior is due to the interplay between two factors, (i) the fugacity of the solid PEG phase that increases with pressure and (ii) the solubility of CO_2 in a liquid PEG phase that becomes substantial at pressures greater than 1 MPa (Figs. 5.2–5.4). In the range of pressures between 6 and 22 MPa, the melting temperature is insensitive to the pressure as a result of a balance between the increase in solid PEG fugacity with pressure and that in CO_2 solubility in liquid PEG with pressure. The melting temperature increases with pressures greater than 22 MPa, which can be attributed to the lack of CO_2 solubility in the liquid phase at the pressure and the increase of solid phase fugacity with pressure. PEG solubility in the vapor phase becomes more significant such that a solid-liquid-vapor equilibrium should terminate in an upper critical end point for which liquid–vapor phases become critical in the presence of a solid phase. The change in ratio of the melting temperature of PEG at a given CO_2 pressure to that of PEG at atmospheric pressure can be reduced to a simple curve for PEG of different molecular weights [39].

5.4 Densities and Swelling Ratios of CO_2–PEG Mixtures

5.4.1 Measurement

Table 5.2 shows experimental studies reported in the literature for volumetric properties including densities (ρ) and swelling ratios (S_W) of CO_2–PEG mixture. Volumetric data for CO_2–PEG mixtures is scarce compared to VLE data and is limited to CO_2–PEG-saturated solutions. Some volumetric data has been reported for PEG 400 [32, 33], while data for CO_2–PEG mixtures having a high molecular weight is scarce. For analysis of the reported data, the volumetric data can be compared using a density change ratio with CO_2 pressure (ρ/ρ^o) calculated as follows:

$$\frac{\rho}{\rho^o} = \frac{1}{(1 - w_{CO2}/100)(1 + Sw)} \quad (5.2)$$

where ρ^o is the density at atmospheric pressure and x_{CO2} is the weight percentage of CO_2 in the PEG solution. The swelling ratio (S_W) defines the ratio of the expanded volume with dissolved CO_2 to the initial volume.

Table 5.2 Experimental studies reported for density (rho), swelling ratio (S_W), and viscosity (η) of carbon dioxide–poly(ethylene glycol) mixtures

M_w (g/mol)	Property	Method	p range (MPa)	T range (K)	Ref.
400	ρ	Vib. wire	>0.1–25	313–348	[41]
600	ρ	Vib. tube	10–30	318	[47]
2000	ρ	Magnetic suspension balance	6–10	374	[44]
400	S_W	Near IR	2–20	313, 423	[32, 33]
1500		ATR-IR	3–20	308, 328	[37]
		Image	10–40	308, 328	[38]
200	η	Vib. wire	>0.1–25	313–348	[48]
		Dynamic light scattering	1–8	303–333	[49]
400		Vib. wire	>0.1–25	313–348	[41]
		Dynamic light scattering	1–8	303–333	[49]
600		Dynamic light scattering	1–8	303–333	[49]
6000		Torsional oscillating quartz	2–33	353–393	[26, 27]
12,000		Torsional oscillating quartz	2–32	353–393	[26, 27]

Measured densities and calculated density change ratios of CO_2–PEG-saturated solutions for some system are shown in Fig. 5.6. The densities of PEG seem to slightly decrease up to around 10 MPa and then increase in the presence of CO_2 (Fig. 5.6a). This is consistent with the melting temperature behavior described in the previous section. The density behavior of CO_2–PEG-saturated solutions is different depending on the temperature. At higher temperatures, the density does not change so much with an increase of CO_2 pressure

[41]. In the region of lower pressures (<4 MPa), the density tends to have complicated behavior, although less data exists for critical comparison.

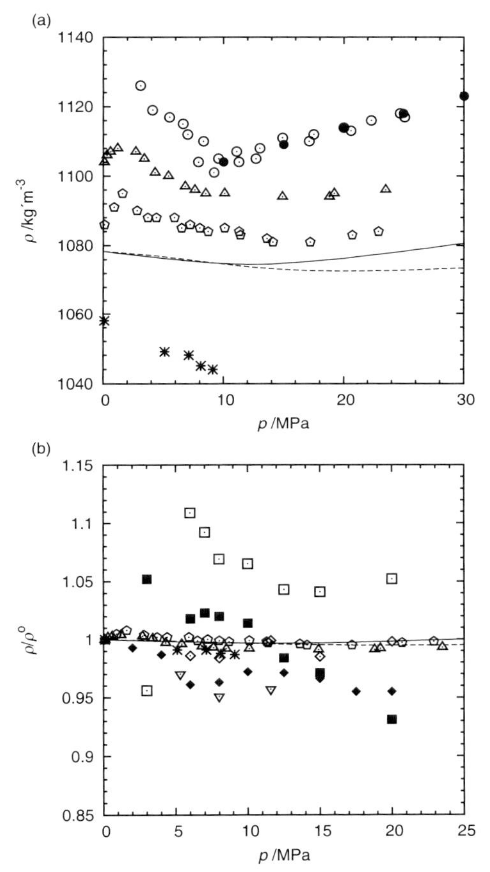

Figure 5.6 (a) Measured densities and (b) calculated density change ratios of a poly(ethylene glycol)–CO_2-saturated solution at several temperatures: ⌂, ○, △from Gourgouillon et al. [29, 41] at 313, 333, and 348 K for 400 g/mol; ●, + Harrison et al. [47] at 318 K for 600 g/mol; × Funami et al. [44] at 374 K for 2000 g/mol; ▽ Guadagno et al. [32] at 313 K for 400 g/mol; □, ■ Pasquali et al. [37, 38] at 308 and 328 K for 1500 g/mol; and ◇, ◆ Vitoux et al. [33] at 313 and 423 K for 400 g/mol. Lines represent the densities of PEG 8300 mixture estimated with equations based on a lattice fluid (---) and on statistical associating fluid theory (–), which are taken from Kasuya [50].

The swelling ratio of PEG increases with CO_2 pressure up to 10 MPa to values around 30% to 40% and then changes only slightly with increase in CO_2 pressure [32, 33, 38]. Vitoux et al. reported that swelling ratios at higher temperature (328 K) were smaller than those at low temperature (308 K) [33]. Values reported by Pasquali et al. [37] were larger than the other researchers and the temperature dependence on the swelling ratio was contrary to those reported by Vitoux et al. [33]. Thus, studies on the swelling properties of CO_2–PEG mixture are needed in the near future.

Density change ratios calculated from the measured densities do not change much with an increase in CO_2 pressure, as shown in Fig. 5.5b. For the density change ratio calculated from the swelling ratio, it changes with an increase in CO_2 pressure and is close to 1 when the CO_2 pressure is low except for the data reported by Pasquali et al. [37]. The data of Pasquali et al. might overestimate the swelling ratio, which leads to a larger solubility of CO_2 in PEG that was pointed out in Section 5.2.1.

5.4.2 Calculation

Harrion et al. correlated experimental densities of CO_2–PEG-saturated solution using a linear function of pressure [47]. Funami et al. estimated the densities of PEG with saturated CO_2 using an EOS based on an LF in which the binary parameter was determined by fitting VLE data of CO_2–PEG mixture [44]. In our laboratory, the densities of CO_2–PEG-saturated solutions were estimated using EOSs based on an LF and on SAFT [50]. Both equations give similar trends for densities (Fig 5.6), with differences among the EOSs appearing only at pressures greater than 12 MPa. At high pressures, values estimated with the LF EOSs do not change appreciably with an increase in pressure, while SAFT gives better qualitative description of the data (Fig. 5.6a). However, the EOS is able to not only estimate the density of CO_2–PEG mixtures, but it can be applied to a wide range of PEG mixtures. Volume translation of LF type of equations may be helpful in obtaining quantitative agreement of the volumetric properties.

5.5 Viscosities of CO_2–PEG Mixtures

5.5.1 Measurement

A list of experimental studies reported for viscosity of CO_2–PEG mixtures is provided in Table 5.2. The reported data is limited to PEG solutions saturated with CO_2. While some data for PEG 200 and 400 is available in the literature [41, 49], data for other molecular weights is scarce and has not been reported at moderate molecular weights ($600 < M_w < 6000$ g/mol).

A viscosity reduction ratio for PEG solutions in the presence of CO_2 ($\Delta\eta/\eta^o$) can be calculated as follows:

$$\frac{\Delta\eta}{\eta^o} = 1 - \frac{\eta}{\eta^o} \tag{5.3}$$

where η is the viscosity in presence of CO_2 and η^o is the viscosity of the PEG solution at atmospheric pressure. Figure 5.7 shows measured viscosities and calculated viscosity reduction ratios for a CO_2–PEG-saturated solution.

Regardless of the molecular weight, the viscosity of the PEG solutions decreases with an increase in CO_2 pressure and does not change much at pressures higher than 20 MPa (Fig. 5.7a). Bae reported an increase in the viscosity at pressures above 5 MPa for PEG 200 and 400 [49], which is in conflict with the results reported by Sarrada et al. [48] and Gourgouillon et al. [41]. The decrease in the viscosity of CO_2–PEG-saturated solutions can be attributed to the dissolution of CO_2 into PEG solution, and the reduction ratio for the viscosity varies with the PEG molecular weight (Fig. 5.7b). At a given CO_2 pressure, the viscosity of CO_2–PEG saturated solutions decreases with an increase in temperature. PEG–CO_2 solutions are generally assumed to behave as Newtonian fluids in the literature [26, 27, 41, 48–50]. However, shear rate dependence on the viscosity of PEG–CO_2 solutions was examined by Gourgouillon et al. who concluded that CO_2–PEG 400–saturated solutions behaved as Newtonian fluids [41]. PEG solutions of higher molecular weight behave as non-Newtonian fluids [51, 52]; therefore, viscosity measurements of PEG of higher molecular weight requires consideration of the shear rate.

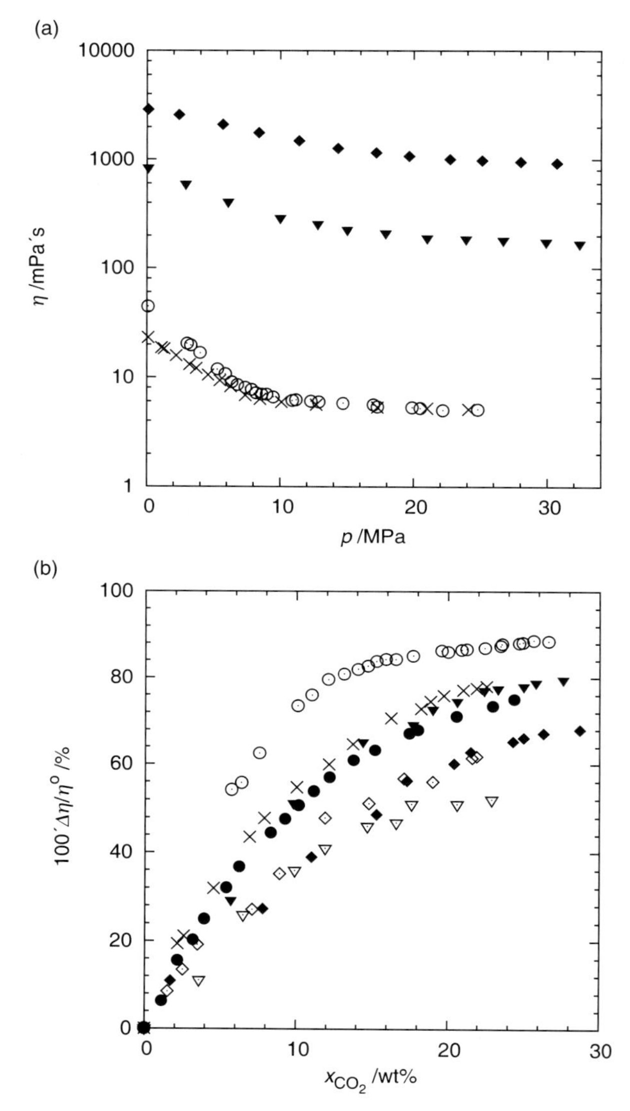

Figure 5.7 (a) Measured viscosities and (b) calculated viscosity reduction ratios for poly(ethylene glycol) saturated with carbon dioxide at several temperatures in the literature: × from Sarrade et al. [43] at 313 K for 200 g/mol; ○, ● Gourgouillon et al. [29, 41] at 313 and 348 K for 400 g/mol; ▼, ▽ Kukova et al. [26] at 353 and 393 K for 6000 g/mol; and ◆, ◇ Kukova et al. [27] at 353 and 393 K for 12,000 g/mol.

5.5.2 Calculation

Gourgouillon et al. correlated measured viscosities of PEG solutions saturated with CO_2 using an Arrhenius-type equation at constant temperature [41]. Bae used the Kelley–Bueche equation combined with the free volume for correlation of the measured viscosities of CO_2–PEG-saturated solutions [49]. Friction theory and free-volume models combined with equations of state can probably be used to provide adequate correlation of CO_2–PEG viscosities [50].

References

1. Machado, B. A. S., Pereira, C. G., Nunes, S. B., Padilha, F. F., and Umsza-Guez, M. A. (2013). Supercritical fluid extraction using CO_2: main applications and future perspectives, *Sep. Sci. Technol.*, **48**, pp. 2741–2760.
2. Machida, H., Takesue, M., and Smith, Jr., R. L. (2011). Green chemical processes with supercritical fluids: properties, materials, separations and energy, *J. Supercrit. Fluids*, **60**, pp. 2–15.
3. Smith, Jr., R. L., Inomata, H., and Peters, C. (2013). *Introduction to Supercritical Fluids, Vol. 4: A Spreadsheet-Based Approach* (Elsevier, The Netherlands).
4. Mattea, F., Martín, Á., and Cocero, M. J. (2009). Carotenoid processing with supercritical fluids, *J. Food Eng.*, **93**, pp. 255–265.
5. Deshpande, P. B., Kumar, G. A., Kumar, A. R., Shavi, G. V., Karthik, A., Reddy, M. S., and Udupa, N. (2011). Supercritical fluid technology: concepts and pharmaceutical applications, *PDA J. Pharm. Sci. Technol.*, **65**, pp. 333–344.
6. Ezhilarasi, P. N., Karthik, P., Chhanwal, N., and Anandharamakrishnan, C. (2013). Nanoencapsulation techniques for food bioactive components: a review, *Food Bioprocess Technol.*, **6**, pp. 628–647.
7. Fages, J., Lochard, H., Letourneau, J.-J., Sauceau, M., and Rodier, E. (2004). Particle generation for pharmaceutical applications using supercritical fluid technology, *Powder Technol.*, **141**, pp. 219–226.
8. Tabernero, A., Martín del Valle, E. M., and Galán, M. A. (2012). Supercritical fluids for pharmaceutical particle engineering: methods, basic fundamentals and modelling, *Chem. Eng. Proces.: Process Intensif.*, **60**, pp. 9–25.
9. Harris, J. M., (1992). Introduction to biotechnical and biomedical applications of poly(ethylene glycol), in *Poly(Ethylene Glycol)*

Chemistry: Biotechnical and Biomedical Applications, ed. Harris, J. M. (Plenum Press, New York, USA), pp. 1–13.

10. Obermeier, B., Wurm, F., Mangold, C., and Frey, H. (2011). Multifunctional Poly(ethylene glycol)s, *Angew. Chem., Int. Ed.*, **50**, pp. 7988–7997.
11. Fruijtier-Pölloth, C. (2005). Safety assessment on polyethylene glycols (PEGs) and their derivatives as used in cosmetic products, *Toxicology*, **214**, pp. 1–38.
12. Saltmarsh, M. (2013). *Essential Guide to Food Additives*, 4th Ed. (The Royal Society of Chemistry, UK).
13. Matsuyama, K., Mishima, K., Hayashi, K.-I., Ishikawa, H., Matsuyama, H., and Harada, T. (2003). Formation of microcapsules of medicines by the rapid expansion of a supercritical solution with a nonsolvent, *J. App. Polym. Sci.*, **89**, pp. 742–752.
14. Barrett, A., Dehghani, F., and Foster, N. (2008). Increasing the dissolution rate of itraconazole processed by gas antisolvent techniques using polyethylene glycol as a carrier, *Pharm. Res.*, **25**, pp. 1274–1289.
15. Liparoti, S., Adami, R., and Reverchon, E. (2012). PEG micronization by supercritical assisted atomization, operated under reduced pressure, *J. Supercrit. Fluids*, **72**, pp. 46–51.
16. McKinley, G. H. (2005). Dimensionless groups for understanding free surface flows of complex fluids, *Rheol. Bull.*, **74**, pp. 6–9.
17. Mugele, R. A. (1960). Maximum stable droplets in dispersoids, *AIChE J.*, **6**, pp. 3–8.
18. Vijayaraghavan, M., Stolnik, S., Howdle, S. M., and Illum, L. (2013). Suitability of polymer materials for production of pulmonary microparticles using a PGSS supercritical fluid technique: preparation of microparticles using PEG, fatty acids and physical or chemicals blends of PEG and fatty acids, *Int. J. Pharm.*, **441**, pp. 580–588.
19. Hao, J., J. Whitaker, M., Serhatkulu, G. M., Shakesheff, K., and Howdle, S. M. (2005). Supercritical fluid assisted melting of poly(ethylene glycol): a new solvent-free route to microparticles, *J. Mater. Chem.*, **15**, pp. 1148–1153.
20. Nalawade, S. P., Picchioni, F., and Janssen, L. P. B. M. (2007). Batch production of micron size particles from poly(ethylene glycol) using supercritical CO_2 as a processing solvent, *Chem. Eng. Sci.*, **62**, pp. 1712–1720.
21. Weidner, E., Steiner, R., and Knez, Z. (1996). Powder generation from polyethyleneglycols with compressible fluids, in *Proc. 3rd Int. Symp. High Pressure Chem. Eng.*, **12**, pp. 223–228.

22. Poling, B. E., Prausnitz, J. M., and O'Connell, J. P. (2001). *The Properties of Gases and Liquids*, 5th Ed. (McGraw-Hill, USA).

23. Viswanath, D. S., Ghosh, T. K., Prasad, D. H. L., Dutt, N. V. K., and Rani, K. Y. (2007). *Viscosity of Liquids: Theory, Estimation, Experiment, and Data* (Springer, The Netherlands).

24. Hou, M., Liang, S., Zhang, Z., Song, J., Jiang, T., and Han, B. (2007). Determination and modeling of solubility of CO_2 in PEG200 + 1-pentanol and PEG200 + 1-octanol mixtures, *Fluid Phase Equilb.*, **258**, pp. 108–114.

25. Matsuyama, K., and Mishima, K. (2006). Phase behavior of CO_2 + polyethylene glycol + ethanol at pressures up to 20 MPa, *Fluid Phase Equilb.*, **249**, pp. 173–178.

26. Kukova, E. (2003). *Phasenverhalten und transporteigenschaften binärer systeme aus hochviskosen polyethylenglykolen und kohlendioxid*, Ruhr-Universität Bochum (in Germanise).

27. Kukova, E., Petermann, M., and Weidner, E. (2004). Phasenverhalten (S-L-G) und transporteigenschaften binärer systeme aus hochviskosen polyethylenglykolen und komprimiertem kohlendioxid, *Chem. Ing. Tec.*, **76**, pp. 280–284.

28. Wiesmet, V., Weidner, E., Behme, S., Sadowski, G., and Arlt, W. (2000). Measurement and modelling of high-pressure phase equilibria in the systems polyethyleneglycol (PEG)–propane, PEG–nitrogen and PEG–carbon dioxide, *J. Supercrit. Fluids*, **17**, pp. 1–12.

29. Gourgouillon, D., and Nunes da Ponte, M. (1999). High pressure phase equilibria for poly(ethylene glycol)s + CO_2: experimental results and modelling, *Phys. Chem. Chem. Phys.*, **1**, pp. 5369–5375.

30. Lopes, J. A., Gourgouillon, D., Pereira, P. J., Ramos, A. M., and Nunes da Ponte, M. (2000). On the effect of polymer fractionation on phase equilibrium in CO_2+poly(ethylene glycol)s systems, *J. Supercrit. Fluids*, **16**, pp. 261–267.

31. Li, J., Ye, Y., Chen, L., and Qi, Z. (2012). Solubilities of CO_2 in poly(ethylene glycols) from (303.15 to 333.15) K, *J. Chem. Eng. Data*, **57**, pp. 610–616.

32. Guadagno, T., and Kazarian, S. G. (2004). High-pressure CO_2-expanded solvents: simultaneous measurement of CO_2 sorption and swelling of liquid polymers with in-situ near-IR spectroscopy, *J. Phys. Chem. B*, **108**, pp. 13995–13999.

33. Vitoux, P., Tassaing, T., Cansell, F., Marre, S., and Aymonier, C. (2009). In situ IR spectroscopy and ab initio calculations to study polymer swelling by supercritical CO_2, *J. Phys. Chem. B*, **113**, pp. 897–905.

34. Xue, Z., Zhang, J., Peng, L., Li, J., Mu, T., Han, B., and Yang, G. (2012). Nanosized poly(ethylene glycol) domains within reverse micelles formed in CO_2, *Angew. Chem., Int. Ed.*, **51**, pp. 12325–12329.

35. Daneshvar, M., Kim, S., and Gulari, E. (1990). High-pressure phase equilibria of polyethylene glycol-carbon dioxide systems, *J. Phys. Chem.*, **94**, pp. 2124–2128.

36. Aionicesei, E., Škerget, M., and Knez, Ž. (2008). Measurement and modeling of the CO_2 solubility in poly(ethylene glycol) of different molecular weights, *J. Chem. Eng. Data*, **53**, pp. 185–188.

37. Pasquali, I., Andanson, J.-M., Kazarian, S. G., and Bettini, R. (2008). Measurement of CO_2 sorption and PEG 1500 swelling by ATR-IR spectroscopy, *J. Supercrit. Fluids*, **45**, pp. 384–390.

38. Pasquali, I., Comi, L., Pucciarelli, F., and Bettini, R. (2008). Swelling, melting point reduction and solubility of PEG 1500 in supercritical CO_2, *Int. J. Pharm.*, **356**, pp. 76–81.

39. Weidner, E., Wiesmet, V., Knez, Ž., and Škerget, M. (1997). Phase equilibrium (solid-liquid-gas) in polyethyleneglycol-carbon dioxide systems, *J. Supercrit. Fluids*, **10**, pp. 139–147.

40. Martín, Á., Pham, H. M., Kilzer, A., Kareth, S., and Weidner, E. (2009). Phase equilibria of carbon dioxide + poly ethylene glycol + water mixtures at high pressure: measurements and modelling, *Fluid Phase Equilb.*, **286**, pp. 162–169.

41. Gourgouillon, D., Avelino, H. M. N. T., Fareleira, J. M. N. A., and Nunes da Ponte, M. (1998). Simultaneous viscosity and density measurement of supercritical CO_2-saturated PEG 400, *J. Supercrit. Fluids*, **13**, pp. 177–185.

42. Aschenbrenner, O., Supasitmongkol, S., Taylor, M., and Styring, P. (2009). Measurement of vapour pressures of ionic liquids and other low vapour pressure solvents, *Green Chem.*, **11**, pp. 1217–1221.

43. Daneshvar, M., and Gulari, E. (1992). Supercritical-fluid fractionation of poly(ethylene glycols), *J. Supercrit. Fluids*, **5**, pp. 143–150.

44. Funami, E., Taki, K., and Ohshima, M. (2007). Density measurement of polymer/CO_2 single-phase solution at high temperature and pressure using a gravimetric method, *J. App. Poly. Sci.*, **105**, pp. 3060–3068.

45. Gross, J., and Sadowski, G. (2002). Modeling polymer systems using the perturbed-chain statistical associating fluid theory equation of state, *Ind. Eng. Chem. Res.*, **41**, pp. 1084–1093.

46. Pedrosa, N., Vega, L. F., Coutinho, J. A. P., and Marrucho, I. M. (2007). Modeling the phase equilibria of poly(ethylene glycol) binary mixtures with soft-SAFT EoS, *Ind. Eng. Chem. Res.*, **46**, pp. 4678–4685.

47. Harrison, K. L., Johnston, K. P., and Sanchez, I. C. (1996). Effect of surfactants on the interfacial tension between supercritical carbon dioxide and polyethylene glycol, *Langmuir*, **12**, pp. 2637–2644.

48. Sarrade, S., Schrive, L., Gourgouillon, D., and Rios, G. M. (2001). Enhanced filtration of organic viscous liquids by supercritical CO_2 addition and fluidification. Application to used oil regeneration, *Sep. Purif. Technol.*, **25**, pp. 315–321.

49. Bae, Y. C. (1996). Gas plasticization effect of carbon dioxide for polymeric liquid, *Polymer*, **37**, pp. 3011–3017.

50. Kasuya, K. (2013). *Measurement and Estimation of Poly(Ethylene Glycol) Viscosities in the Presence of CO_2*, Tohoku University (in Japanese).

51. Teramoto, A., and Fujita, H. (1965). Temperature and molecular weight dependence of the melt viscosity of polyethylene oxide in bulk, *Die Makrom. Chem.*, **85**, pp. 261–272.

52. Törmälä, P., Weber, G., and Lindberg, J. J. (1978). Viscoelasticity of polyethylene oxide melts, *Rheol. Acta*, **17**, pp. 201–203.

Chapter 6

Methods for Particle Production: Antisolvent Techniques

Pascale Subra-Paternault
Laboratoire Chimie Biologie des Membranes et des Nanoobjets (CBMN), CNRS UMR 5248, Université Bordeaux 1, Institut Polytechnique de Bordeaux, Allée Geoffroy St Hilaire, Bât 14B, Pessac, 33600, France
subra@enscbp.fr

The key concept of antisolvent precipitation is the decrease of solubility in solvent–nonsolvent mixtures. Taking advantage of the poor solubility of many compounds in CO_2, the use of compressed CO_2 as antisolvent was rapidly developed for processing polymers and opened the route for composite formulation. This chapter first introduces the compressed CO_2 antisolvent concept based on the relevant phase equilibria (CO_2–solvent mixture, solubility of a species in CO_2+solvent, ternary diagram of a polymer–CO_2–solvent system) and describes the key features of the two versions of the process known as gaseous antisolvent (GAS) and supercritical antisolvent (SAS). The second part of the article sorts out examples of composites produced by GAS and SAS as coprecipitation of two components (drug–polymer powders), precipitation of one

Supercritical Fluid Nanotechnology: Advances and Applications in Composites and Hybrid Nanomaterials
Edited by Concepción Domingo and Pascale Subra-Paternault

ISBN 978-981-4613-40-8 (Hardcover), 978-981-4613-41-5 (eBook)
www.panstanford.com

compound on a slurry (polymer–inorganic spheres or fibers, drug–excipient powders) and coprecipitation on slurry (drug–polymer–inorganic ternary hybrids).

6.1 Introduction

Coating or encapsulation of particles is of great interest in pharmaceutical, agrochemical, food, and cosmetic industries since it enables the tailoring of surface or bulk properties of the material. With many molecules poorly soluble in compressed CO_2, the concept of using compressed CO_2 as antisolvent was developed in the 1995s to bypass the restriction of rapid expansion of supercritical solutions (RESS) to the few molecules soluble enough in CO_2 to be economically viable. The CO_2 antisolvent process offers hence solutions for formulating CO_2 nonsoluble compounds.

In antisolvent approach, the compound(s) to precipitate is initially dissolved in an organic solvent and a nonsolvent for the solute is further mixed with the solution. The key concept is the decrease of solubility of the compound in the formed solvent–antisolvent mixtures compared to its solubility in neat solvent. Compared to RESS, the extend of the solubility shift induced by a CO_2 antisolvent is less pronounced than the shift induced by the depressurization encountered in RESS, so, antisolvent techniques are prone to produce larger sizes of particles than RESS [1]. Precipitation by an antisolvent, compressed or not, obeys to general laws of crystallization. The control of particle size goes along with the control of the supersaturation level, that is, the knowledge of the solubility variation with the solvent–antisolvent composition and of the time and space scales at which supersaturation is obtained and crystallization develops (mixing scales of solvent and antisolvent). Compared to conventional precipitation carried out at atmospheric pressure, the interest of the compressed antisolvent resides in the rapidity at which supersaturation can be obtained due to the higher mass and thermal diffusivities in fluids compared to liquids. As postprocessing advantage, the elimination of solvent from the precipitated material is realized in situ by a flow of CO_2; by designing a one-pot process, it avoids the multiple filtration and drying steps detrimental to the product integrity and to the technician's health.

Two variants of antisolvent techniques were developed, the gaseous antisolvent (GAS) in which CO_2 is added initially as a gas to the solution, with, as a consequence, pressurization of the medium; and the supercritical antisolvent (SAS) in which the solution is directly injected in a compressed CO_2 maintained at desired pressure. The operating lines of the two variants in the pressure-composition diagram are very different, as illustrated in Fig. 6.1, so that different sizes of particles or composites can be obtained [2].

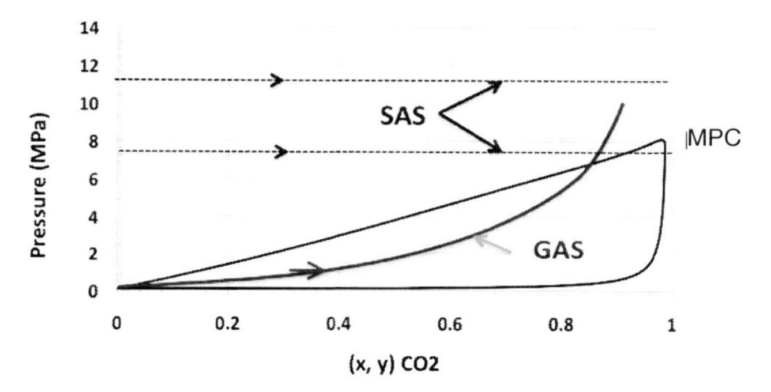

Figure 6.1 Operating lines of SAS and GAS in a solvent–CO_2 liquid–vapor phase diagram showing the evolution of the mixture composition when CO_2 is added to the solvent. In GAS, the pressure increases as well. In SAS that proceeds by injection of the solution in CO_2, the pressure is fixed.

Composites described in this chapter incorporate mostly a polymer in their formulation. Processing or coprocessing polymers by the SAS technique is not as "easy" as single processing molecules of low molecular weight. Polymer behavior in dense fluids has been reviewed by Kiran in 2009 [3]. First, CO_2 might interact strongly with polymers, especially with the amorphous ones. The great solubility of CO_2 in the polymer phase can result in plasticization and agglomeration of the final product. Secondly, polymers in solution exhibit a peculiar behavior since their chains can behave as independent entities at dilute concentration or are capable of entangling with one another when more concentrated. The increased viscosity of the polymer melt with concentration makes atomization more difficult. Combined to a precipitation in presence of CO_2 that occurs by liquid–liquid phase split, various morphologies ranging from particles to fibers can be formed, especially with SAS. Moreover, in the case of coprocessing, the polymer and the molecule might not be soluble in the same solvent, so a mixture of solvent could be necessary for the composite production.

This chapter is divided into two parts. The first section gives a general introduction to the SAS approach, enhancing the phase behaviors and the specificity of the GAS and SAS versions, whereas several examples of composites selected mostly from author's background are given in the second section. More examples can be found in reviews [4–6].

6.2 An Explanation of the Process and Approach

6.2.1 Antisolvent Effect Related to Solubility Considerations

Two types thermodynamic data have to be considered when dealing with CO_2-assisted precipitation from solution:

1. The binary CO_2–solvent diagram that indicates in what conditions the solution and CO_2 merge as a single phase fluid or mix as a two-phase system
2. The solubility of the substance to precipitate in CO_2–solvent mixtures

The miscibility of solvent and CO_2 is given by the vapor–liquid equilibrium (VLE) diagram (Fig. 6.2). The liquid and vapor phases merge at the mixture critical pressure (MCP), above which they are indiscernible. Below this pressure, two phases coexist and their compositions are given by the abscissa of the respective dew and bubble point lines. Such phases envelops are specific to a solvent–CO_2 mixture, so at a given pressure, for example, 9 MPa in the case of systems described hereby, ethanol, acetone, methylene chloride are fully miscible with CO_2, whereas dimethylsulfoxide (DMSO) of a higher MCP would split into a liquid DMSO-rich phase and a vapor CO_2-rich phase. VLE envelopes are function of temperature; hence, depending on the solvent and temperature, the respective composition of the liquid and the vapor phase in CO_2 will be different as well (see, for instance, difference between ethanol–CO_2 and acetone–CO_2).

The location of the operating pressure to that of the MCP is especially critical in SAS, since it originates two regimes for the injection. In conditions below MCP, the solution breaks up as droplets and the mutual diffusion of fluids that occurs at the droplet interface creates conditions of precipitation in a confined environment. In conditions far above the MCP, solvent and CO_2 are fully miscible so distinct droplets never form and the two fluids merge as an expanding plume in which precipitation takes place.

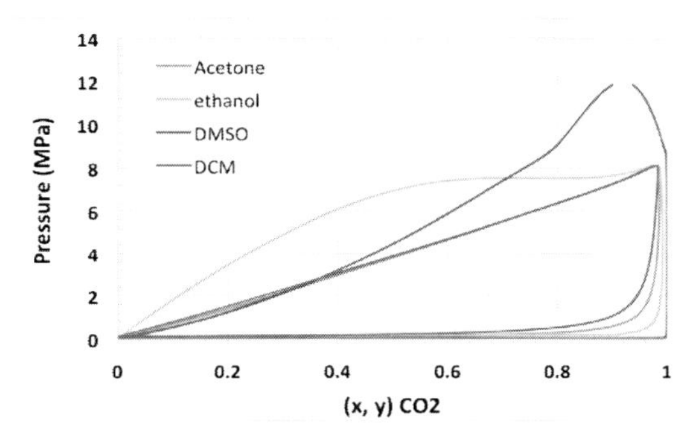

Figure 6.2 Liquid–vapor phase diagrams of CO_2–acetone, CO_2–ethanol, CO_2–methylene chloride (DCM), and CO_2–dimethylsulfoxide (DMSO) calculated at 40°C (ProPhyPlus software).

Precipitation occurs when the saturation limit of the solute is exceeded. The saturation curve of two small molecules in CO_2–solvent is given in Fig. 6.3 (more in Refs. [7–11]). The curve is specific to pressure and temperature.

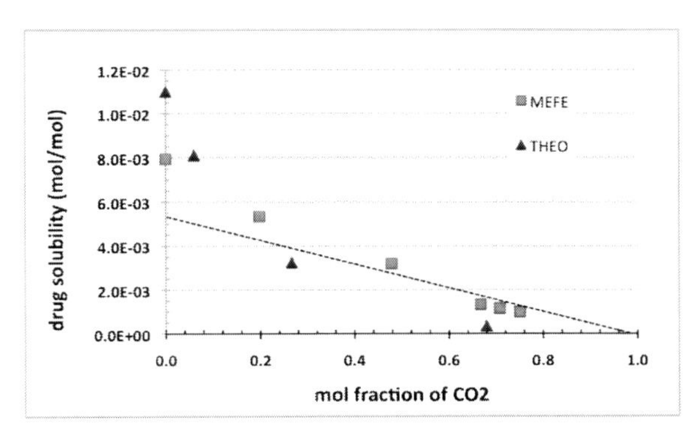

Figure 6.3 Solubility of mefenamic acid (MEFE) and theophylline (THEO) in CO_2+solvent mixtures of variable composition, at 10 MPa and 35°C/acetone for MEFE and 36°C/ethanol–methylene chloride for THEO. Operation line in dots.

The solubility decreases as the composition of the mixture increases in CO_2 hence the precipitation of these drugs by CO_2 from solution is potential and was indeed successful [7, 10]. However, the solubility decrease is not very steep, which lets anticipate low levels of supersaturation. To enable moreover the precipitation during the process, the operating line that depicts the evolution of the initial concentration with the addition of CO_2 has to cross the saturation curve. For theophylline that exhibits a sharper decrease, the crossing occurs for a CO_2 content around 20 mol%, whereas for mefenamic acid, a higher content around 60 mol% should be attained. Diluted initial solutions would require large composition of CO_2 for the crossing, and too diluted solutions can even never precipitated. Such saturation curves are thus useful to understand the influence of concentration and of the solvent on the precipitation characteristics and allows for envisaging alternative routes of precipitation like RESS with a cosolvent or the Depressurization of an Expanded Liquid Organic Solution (DELOS) process [12].

When the solute to precipitate is a polymer, the situation is more complicated since a liquid–liquid phase equilibrium generally occurs. A schematic ternary phase diagram for solvent–polymer–CO_2 (Fig. 6.4) is helpful to explain the precipitation of the polymer [13, 14].

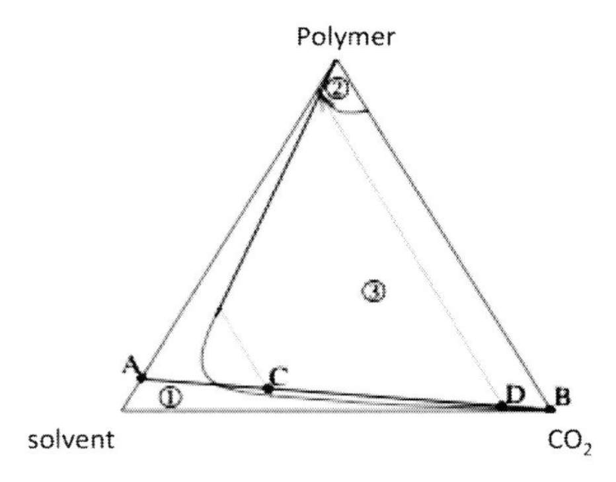

Figure 6.4 Schematic ternary diagram for polymer–solvent–antisolvent CO_2.

The specificity of polymer behavior compared to a low-molecular-weight compound is the existence of a two-phase region () where a polymer-rich and a polymer-lean phase coexist. Region ① is a single-phase region of a diluted polymer dissolved in the solvent with some

CO_2 absorbed, whereas the single-phase region ② consists mostly of a polymer with some solvent and CO_2 absorbed. The AB line depicts the operating line when CO_2 is added to the polymer solution. Starting from an initial solvent–polymer solution (A), the addition of CO_2 provokes the attainment of the solubility limit at which the splitting in two phases occurs (region ③). As the mutual diffusion of CO_2 and solvent continues (toward D), the polymer-rich phase becomes more concentrated until the further removal of solvent in B induces a phase transition to the glassy region ② where the polymer vitrifies. Eventually, if foreign particles are present in the medium, the polymer embeds the particles. The final morphology of the polymer product is related to the mass transfer pathway on the phase diagram and especially to the concentration of the initial solution with respect to the concentration C^*, which divides the dilute and semidilute regimes in binary polymer-solvent mixture [15]. At low initial polymer concentration, the polymer-rich phase nucleates and growth in a polymer-lean phase, that is, in a solvent-rich continuous phase. Moreover, polymer chains in a solvent behave as independent entities (the so-called diluted regime). At higher concentration (semidilute regime), the chains of the polymer are capable of entangling with one another and the solvent-rich phase nucleate and growth in a polymer-rich continuous phase. The change of morphology from microspheres to fibers with concentration was reported several times for SAS processing [15–20]. Operating below or above the solvent–CO_2 critical pressure was also a way to manipulate the morphology, from large particles of 50 μm below the MCP to particles around 2 μm and microfibers above it [14].

6.2.2 The Two Modes: SAS and GAS

SAS proceeds by the injection of the solution into a steady flow of CO_2 maintained at a given pressure and temperature. Variants of SAS have been introduced over years but they are basically identical by regards of the mechanism of precipitation: solution-enhanced dispersion by supercritical fluids (SEDS) [21] and supercritical antisolvent with enhanced mass transfer [SAS-EM] [22]. These variants use coflows nozzle (SEDS) or ultrasonic waves (SAS-EM) to enhance the jet breakup. To be distinguished from GAS, SAS was also called ASES, for aerosol solvent extraction system [23].

The main interrogations in SAS were about the possible existence of droplets and the scales of the mass transfer between solvent and CO_2. Droplets are clearly formed in conditions that place the operating point below the mixture critical point due to the liquid–vapor coexistence. The droplet sizes or size distribution are governed by the mechanism of jet breakup. The four well-known disintegration mechanisms of a liquid jet were simulated in subcritical environment [24, 25]. Results indicate that transitions between regimes occur in the same range of Ohnesorge and Reynolds numbers than transitions at atmospheric conditions. For instance, an increased velocity of the injection from 0.5 to 30 m/s would induce a transition from a Rayleigh breakup that forms droplets of size about twofold the nozzle diameter to an atomization regime characterized by smaller droplets below 0.15-fold the nozzle diameter. Pressure influences the distance at which the jet breaks and the regime of disintegration as well: close to the MCP, asymmetric instabilities arise from the sharp decrease of the interfacial tension, and the jet now breaks up by a first-wind-induced mechanism rather than by Rayleigh.

During the jet breakup or more probably at the droplet surface, a countercurrent transfer of CO_2 and solvent occurs. The swelling-shrinking behavior of ethanol droplets in CO_2 at conditions below MCP was computed and was found to be less than 0.1 sec [26]. The very fast saturation of droplets with CO_2, which decreases the solute solubility, and their shrinking, which increases the solution concentration, result in creating supersaturation that enables particle precipitation. Confronted to experimental results, it was found that precipitation of nicotinic acid from ethanol was controlled by the droplet size.

In conditions above the MCP, CO_2 and the solvent are fully miscible, and since interfacial tension tends to be null, distinct droplets never form, except very near the MCP due to the existence of a nonzero dynamic interfacial tension [27]. The mixing of CO_2 and solvent in those miscible conditions was first modeled by considering a pseudo droplet, that is, a droplet of a solvent-rich phase in a continuum of CO_2 ([28] and references therein). The swelling-shrinking times of various solvents at different pressures and temperatures were evaluated. Whatever the conditions, the time necessary for a spot of solvent to fully mix with the compressed environment was below 0.5 sec. The value is in the same order of magnitude than those reported for two-phase region. Computational fluid dynamics were used to simulate the mixing of the two fluids in a more realistic way [7, 29–31]. The mixing of a methylene chloride jet and ambient CO_2 is illustrated in Fig. 6.5. It is very similar to that of acetone [7] or ethanol [30].

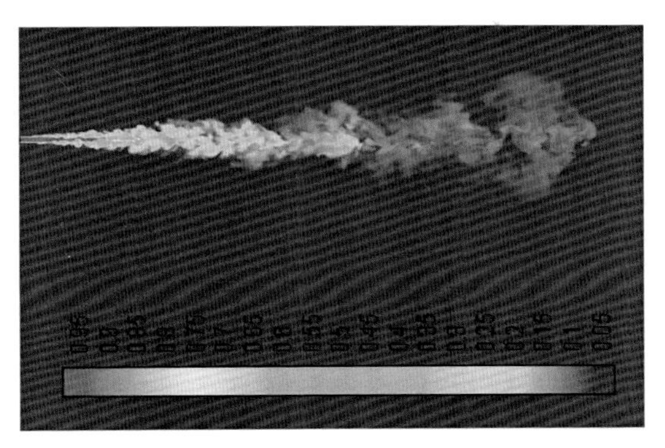

Figure 6.5 Injection of methylene chloride in CO_2 at 35°C and 13 MPa: plot of solvent mass fraction. Injection velocity: 3 m/s, capillary of 130 μm diameter.

The solution and CO_2 merge as gaseous plumes with a jet core mostly composed of the solvent at proximity of the nozzle. As the merging proceeds in time and space through convection and diffusion, the solvent content in the jet axis and in plumes decreases in favor of CO_2, which, according to the solubility behavior, will induce the solute precipitation. Simulations show that areas propitious for precipitation are localized within 1 cm from the nozzle tip. According to the Peclet dimensionless number, the mass transfer mechanism is dominated by convection rather than diffusion, which explains the relative insensitivity of the contour with pressure. The effect of MCP proximity on particle size is not uniform throughout literature examples [32]. In the case of theophylline that precipitated as large crystals around 67 μm mean size, the proximity to MCP led to smaller sizes; for paracetamol obtained as 2–3 μm particles whatever the pressure [33], the particles tend to be more spherical and agglomerated near the critical locus. Narrowest size distribution was obtained as well by Tabernero et al. [21]. It is worth recalling that although the mechanism of jet breakup or mixing times and scales need to be specified when dealing with such process, the final particle size and distribution are governed by the nucleation/growth kinetics that are dependent on pressure and temperature as well. Moreover, the presence of the solute in solution and the exothermic effect of the mixing enthalpy might produce a displacement from a single supercritical phase to a biphasic phase, which can induce a change in the precipitation regime from a gas plume to transient droplets.

Fundamentals of GAS version have been less studied. In this version, the solute is dissolved in an organic solvent and the solution is poured into a vessel. Upon the gradual addition of CO_2 to the vessel, the CO_2 diffuses into the solvent and induces a volumetric expansion that causes a reduction of the solute solubility. The volumetric expansion is macroscopic evidence that amount of CO_2 has diffused into the solution according to the liquid phase equilibrium reported in Fig. 6.2. Hence, the rate of supersaturation buildup in solution is function of the operating temperature, the pressure buildup or CO_2 introduction rate, the type of solvent and the stirring power input. The precipitation in GAS is likely to occur in a CO_2 expanded liquid phase in equilibrium with a CO_2-rich gaseous phase. Compared to SAS, the CO_2-expanded solution is not confined in droplets or in mixing plumes, a scale that could explain the larger particle sizes encountered in GAS compared to SAS [32, 33]. The analysis of mass transfer fluxes at the gas–liquid interface was carried out by Lin et al. [34] using time-dependent material balances. Calculations revealed that the flux of the solvent from the liquid to the gas phase was negligible compared to the flux of CO_2 into the liquid phase. The influence of operating conditions on mass transfer was evaluated: the larger effect came from the stirring rate that favored the dissolution of CO_2 in the liquid phase. The theoretical analysis was confronted to experimental results of paracetamol precipitation from acetone [35] that indeed showed a decrease of drug mean size with higher stirring and addition rates. Griseofulvine [12] and theophylline [32] did not show the same sensitivity to these parameters, a discrepancy that could be explained by a different dominant mechanism of crystallization [36]. Raman spectroscopy was developed as well for in situ monitoring of the sorption of CO_2 in the liquid phase, detecting the crystallization onset and for monitoring the crystallization kinetics [37]. A first burst of precipitation that produced 70% of the total mass of solids was detected at pressure around 5 MPa and 60% of CO_2 in the liquid phase, indicating that precipitation by GAS can be carried out at moderate pressure.

6.3 Applications

6.3.1 Coprecipitation: Two Species Coprecipitate by Antisolvent

SAS coprecipitation has been mostly proposed to formulate therapeutic molecules with biodegradable polymers [38]. The general goal of encapsulation is to elaborate vehicles that enable the delivery of the active molecule in a controlled way and on specific targets. Polymers or blends of polymers are then selected to comply with the diversity of foreseen release rates as faster dissolution or sustained release.

A poorly water-soluble drug (phenytoin) was coprocessed with hydrophilic poly(vinyl-pyrrolidone) to enhance the dissolution behavior of the drug [36]. As a first attempt, the GAS recrystallization of poly vinyl-pyrrolidone (PVP) alone from ethanol/acetone mixture or methylene chloride was studied. Liquid–liquid phase separation was observed prior to vitrification of the polymer-rich phase, yielding typically large, porous, interconnected polymer structures rather than discrete particles. Therefore, SAS was preferred. The main operating parameters were chosen on the basis of outcomes of experiments on both pure phenytoin and PVP, in particular, the use of an ethanol/acetone mixture (24 wt% ethanol) and operating conditions in the two-phase region (i.e., 8 MPa and 25°C) to maximize product yield. The maximum processed concentration of PVP and drug in solvent were of 2 wt%. While the pure drug had formed high-aspect-ratio crystals, the coformulations were amorphous, provided that the drug-to-polymer ratio was small enough; the product appeared as micron-size agglomerates of spherical nanoparticles that showed no recrystallization tendency after several weeks of storage. At a drug-to-polymer ratio above 1:2, both drug crystals and polymer particles were observed. As major result, it was found that solid dispersions precipitated by CO_2 markedly outperformed neat drug microcrystals, physical mixtures, as well as spray-dried composite particles in in vitro dissolution experiments. Residual solvent contents were below 1 wt% for both acetone and ethanol. Later, the drug and the polymer were coprocessed at 15 MPa and 40°C and extra experiments at 8 MPa and 25°C were performed [39]. Products that contained less than 40 wt% in drug were fully amorphous and similar to pure PVP spheres, while crystals of the drug were also present in products of higher drug contents. The same threshold was found at both operating conditions.

Contrary to previous examples, compacts of hydrocortisone and PVP prepared by SAS from ethanol solutions did not dissolve faster than systems processed by solvent evaporation or spray drying [40]. The CO_2-prepared materials exhibited microcrystals of the drug at PVP loadings between 0 and 40% and were amorphous at PVP content of 60% (i.e., at drug loading below 40%), a result that fit previous conclusions for PVP-phenytoin. In contrast, spray drying yielded amorphous solids whatever the mass fraction of PVP.

Formulations can be prepared to sustain a release. Poly(L-lactic acid) (L-PLA) and diuron herbicide were coprecipitated by SAS from methylene chloride at 35°C and 10 or 14.5 MPa [41]. Various biodegradable polymers such as ethyl cellulose, polymethylmethacrylate (PMMA), cellulose acetate, or waxes were coprecipitated as well using methylene chloride or tetrahydrofuran. With PLA, microparticles of polymer entrapping diuron was recovered only at 14.5 MPa and providing a processed concentration of 1 wt% and 0.15 wt% for polymer and herbicide, respectively. Otherwise, the two populations of large diuron needles and PLA microspheres coexisted in the product. The other coating materials did not give convincing results because of the high degree of particles agglomeration. PLA was thus the best coating substance, since the formulation was recovered as nonaggregated particles and provided a long-lasting release of the herbicide (40% of diuron released after 20 days of immersion).

PLA was selected as well to sustain the release of theophylline [2, 42]. Theophylline is moderately soluble in CO_2 and its precipitation as neat drug was effective providing the use of a two-solvents mixture, methylene chloride and ethanol [10]. When formulated with L-PLA, theophylline was found to retain its specific crystallinity. The presence of polymer slightly prevented the growth of the drug particles but was unable to reduce significantly the size down to limits acceptable for pulmonary delivery. Whatever the conditions, the products were made of two populations in which the coated plate-like drug existed besides the PLA spheroids (Fig. 6.6a). Formulations yielded to a release of the drug over 14 h with 50% of the drug released in 4 h that was longer than the 10 min of the neat theophylline. Residual solvent contents were below 100 ppm for both solvents, which is below the acceptable limits.

A mixture of methylene chloride (good solvent for the polymer) and DMSO (good solvent for the drug) was also used to produce azacytidine/PLA particles [43]. A scanning electron microscopy (SEM) picture (Fig. 6.6b) of a coprocessed drug shows a population of deformed balloon-like structures attributable to the PLA-coated drug and some elongated structures attributable to uncoated particles by reference to the elongated shape of the single-processed drug. The produced composite particles showed size roughly below 2 μm that was smaller than the 3–6 μm of the neat drug. Drug monitoring indicated a drug loading of 25 wt%, with 95% of the drug amount entrapped/coated by the polymer. Moreover, besides a small size and a sustained delivery of the drug over a period of several hours, an improved stability of the coated drug with respect to the neat azacytidine was found, thus proving the suitability of this approach for protecting unstable compounds.

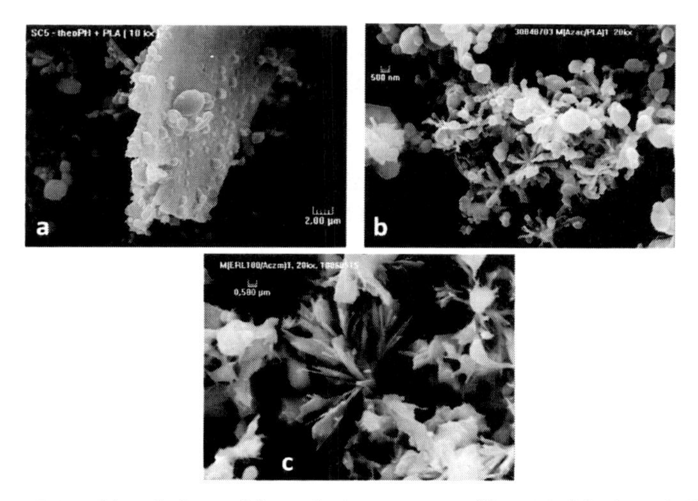

Figure 6.6 Morphology of formulations prepared by SAS. (a) Theophylline PLA, 12 and 12 mg/mL in DMSO:EtOH, 36°C, 10 MPa; scale of 2 μm; (b) azacytidine–PLA, 4 and 15 mg/mL in DMSO:DCM, 40°C, 11 MPa, scale of 0.5 μm; and (c) acetazolamide–Eudragit ERL100, 8 and 17 mg/mL in acetone, 36°C, 8.5 MPa, scale of 0.5 μm.

The two modes GAS and SAS were investigated to prepare ophthalmic drug delivery systems of acetazolamide, using Eudragits RS and RL, pure or in mixture, to vary the release behavior [44]. Eudragit® is the trade name of copolymers derived from esters of acrylic and methacrylic acid, whose properties are can be tuned through the functional groups. Although both techniques produced formulations that slowed down the drug release, SAS produced smaller and less aggregated particles than the GAS operation. As seen in Fig. 6.6c, the product was a mix of elongated particles attributable to the drug and of spherical or deformed spheres of the polymer. The composition of the polymer mixtures did not influence notably the mean size. It is worth noting that the presence of the drug allows for the precipitation of the polymer(s), whereas its processing as pure has led to a film deposition onto the vessel surface. Eudragit® polymers were used as well with mesalazine (5-ASA) that is a drug that has to be specifically delivered to the colon. The molecule has thus to survive at the acidic gastric juice and pass into the small intestine. The problem is being addressed by formulations that should prevent any release at pH 1.2 during two hours. Several polymer blends were coprocessed with 5-ASA by SAS using DMSO/acetone as a solvent: Eud ERL 100 + ERS 100, Eud ERL 100 + ethylcellulose [45], or Eud EL 100 + ES 100 (unpublished results). Products were usually recovered as fine powders in which submicron crystals of drug coexisted with more or less deformed spheres of polymer (Fig. 6.7), unless a high polymer:drug ratio of 1:5 was processed. As the processed concentration increased, the drug crystals aggregated as rounded-shape particles that emerged from the polymer coating. Although some formulations showed a noticeable sustained release of 5-ASA at pH 1.2, the encapsulation of the drug was insufficient to cut down completely the release.

Figure 6.7 Formulation of 5-ASA with Eudragit EL + ES blend by SAS at 36°C and 11 MPa from a DMSO/acetone mixture. ASA:EL:ES concentration of 10:8.7:2.3 mg/mL (left) and 19:15.5:4 mg/mL (right).

Previous particles were more or less coated by the polymer, but except for theophylline, they were rather small. It is interesting to look deeper at the coating in the case of long needles, like those obtained by processing griseofulvine by SAS. Figure 6.8 details the coating of griseofulvine by D,L-poly(lactic-co-glycolic acid) (PLGA) 50:50, a copolymer of lactic and glycolic acids. Contrary to poly(lactic acid), PLGA is amorphous, hence it plasticizes easily when contacted with CO_2. This behavior precludes its precipitation as powder when processed alone, but enables a coating when coprecipitated with a drug. Crystals of drug are clearly coated by a continuous layer of polymer that, however, shows regular expansions as balloons. The drug might have nucleated and grew in the polymer-rich phase before the final vitrification. When concentrated solutions were used, the product exhibited a different morphology, as long fibers embedded in droplets-like structures. The high concentration deeply influences the viscosity of the solution, which, in turn, modifies the breakup mechanism of the jet sprayed by the nozzle and thus generates different structures than those issued from diluted regime of viscosity.

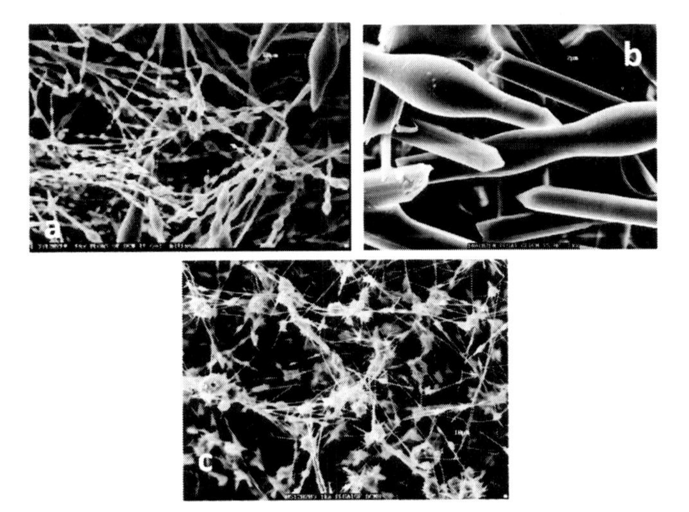

Figure 6.8 Precipitation of griseofulvine and D,L-PLGA by SAS at 40°C, 8.5 MPa, from methylene chloride: (a) GRI = PLGA = 8 mg/mL, diameter of polymer balloons below 5 μm; (b) detail of previous picture, width of needles = 10 μm; and (c) GRI = 100 mg/mL, PLGA = 50 mg/mL, "droplet" diameter = 10 μm.

Previous formulations were made with compound of low molecular weight. Encapsulation of more complex system as proteins was investigated as well [6]. Egg lysozyme was encapsulated by corn zein using a SAS process [46]. The water-soluble character of protein and carrier required the use of a water–ethanol mixture as a solvent. Precipitation of water-soluble species by a CO_2 antisolvent might seem inadequate since water and CO_2 are poorly miscible. However, water can be extracted by using a mixture of CO_2 and ethanol as in the case of processing chitosan–plasmid DNA complexes [47]. An alternative to water–ethanol mixtures is the use of DMSO to dissolve proteins, with, however, a risk of irreversible change of their secondary structure resulting of loss in functional activity [6]. Insulin–PLA or insulin–PLA–poly(ethylene glycol) (PEG) microspheres below 2 μm in size were produced by SAS using a mixture of dichloromethane (DCM) and DMSO to ensure the solubility of both polymer and protein [48–50]. The process results in high product yield, extensive organic solvent elimination, and maintenance of >80% of the insulin hypoglycemic activity. More than 80% of the processed protein amount was incorporated and finely dispersed into the polymer matrix. Morphology and biopharmaceutical properties of products were affected mostly by the PEG content in the formulation and its molecular weight. The release of insulin was of 100% over three to four months when particles were produced with a PEG 1900 content of 67% or 75%. Hence, protein-loaded particles possessing the prerequisites for pharmaceutical applications were produced.

Not only polymers but organic excipients can be processed together with a drug. For instance, the influence of mannitol and trehalose on the physicochemical characteristics of inhaled insulin powders produced by SAS was examined [51]. Results showed that when properly formulated, the process could produce particles with size and size distribution suitable for pulmonary insulin delivery that is of prime importance for improving the life quality of diabetes patients.

6.3.2 Precipitation on Slurry: Precipitation of One Species by Antisolvent in the Presence of Pre-Existing Particles

A more emerging area is the precipitation of a species in presence of pre-existing solid particles, that is, performing SAS or GAS on a slurry.

When the dissolved entity is a polymer, one expects the coating of the particles. When the dissolved species is a drug, for instance, one expects an intimate mixing between the two components that might positively influence the drug behavior.

Both GAS and SAS are potential technique for precipitation-on-slurry process, but in SAS, large solids can be incompatible with the injector diameter (risk of plugging). Moreover, it is critical to maintain a good dispersion of suspended particles to avoid agglomeration or sedimentation. This could be partly solved by introducing surfactants in the solution, with the drawbacks that they can interfere with the precipitation process or they can be incorporated within the precipitates.

SAS was selected to coat submicronic particles of silica [13, 52, 53], whereas we used GAS to coat 5–20 μm silica beads and both SAS and GAS for coating nanometric TiO_2 [54]. In their pioneer work [52], Wang et al. coprocessed hydrophobic or hydrophilic silica of 20 and 600 nm diameter with Eudragit RL100. The slurry was prepared in acetone and injected in CO_2 at 32°C and 8.3 MPa. The coating was evidenced but more as a shell encapsulating agglomerates of nanoparticles rather than coating individual particles. The mechanism was purely physical deposition of polymer on existing surface. Using PLGA, silica particles of 0.5 μm were again coated in the form of loose agglomerates [13]. Among the parameters investigated (pressure, temperature, flow rate of polymer solution, polymer-to-particle-weight ratio, polymer concentration, addition of a CO_2-soluble surfactant), the polymer ratio and the polymer concentration played a critical role in the successful encapsulation of silica with minimum agglomeration. To get a smooth particle coating with minimal agglomeration, a polymer concentration of 4 mg/mL and around 15% of that of silica was desirable. High pressure (11 MPa *vs.* 9 MPa) facilitated the agglomeration as a result of sintering because of the depressed polymer glass transition, and temperature (33°C to 42.5°C) appeared to have little effect only when settled below the glass-transition temperature.

Similar effect of polymer concentration was evidenced by Marre et al. [53], who processed silica particles of 170 and 550 nm with hydrophilic PEG or hydrophobic hydroxy-terminated polybutadiene. The thickness of the deposited layer was controlled from 3 to 30 nm by varying the polymer concentration from 0.5 to 1.25 mg/mL (silica concentration of 5 mg/mL). In the case of smaller

Aerosil particles (16–30 nm), the particles were embedded in the Eudragit RL100 polymer film that represented about 30 wt% of the product composition [55]. However, the polymer:silica ratio was much higher than previously: 8 mg/mL for silica, 4 or 8 mg/mL for polymer. The coating of nanometric particles was carried out with polymethylmethacrylate (PMMA) from methylene chloride suspensions [54] (~3 mg/mL for TiO_2 and 1, 3, and 17 mg/mL for PMMA). Agglomeration increased with the polymer content although powders were still recovered at the highest concentration. The coating of TiO_2 by hydrophobic PMMA lead to a significant change in the dispersion behavior of the material, since PMMA-coated product formed stable dispersions in toluene even after 12 h, whereas unmodified titania nanoparticles settled down completely after 2 h.

Nanometric filler–polymer composites organized in fibers network for tissue engineering were prepared as well by spraying suspensions of fillers (range of 4 mg/mL) with concentrated polymer solutions [56]. Fibers or microparticles of TiO_2 dispersed in a PMMA/poly(ε-caprolactone) matrix, and fibers of L-PLA loaded with hydroxyapatite were obtained with a content of 10–20 wt% of inorganic phase distributed throughout the composite. With GAS technique, larger particles 5–20 μm were coated by PMMA at concentrations of 12 mg/mL (unpublished results) and 3 mg/mL [54]. The SEM pictures given in Fig. 6.9a,b clearly evidence the presence of the polymer film at the beads surface. The decrease of polymer concentration down to 3 mg/mL decreases the film thickness down to 0.6 μm.

With applications in biomedical field, magnetite-lipid composites were prepared by SAS [57] as core–shell structures below 200 nm (Fig. 6.9c). PLGA–magnetite composites were prepared using a combination of emulsions and extraction of the solvent by CO_2 [58]. Known as supercritical fluid extraction of emulsion (SFEE), the technique uses, however, the antisolvent effect of CO_2 to precipitate the polymer shell, whilst the solvent, which was the dispersed phase of the emulsion, was removed. The method consists in first dispersing magnetite and PLGA in methylene chloride and mini-emulsifying them in presence of a biocompatible emulsifier, poly(vinyl alcohol) (PVA), before injecting the emulsion in a flux of CO_2 via a coaxial nozzle. The size of the composite nanoparticles was tuned from 230 nm to 140 nm by changing the concentration of PVA from 1 to 4 wt%. The morphology of the nanoparticles was of Janus type, with magnetite accumulated at the surface and usually in one cluster located on one hemisphere of the particles.

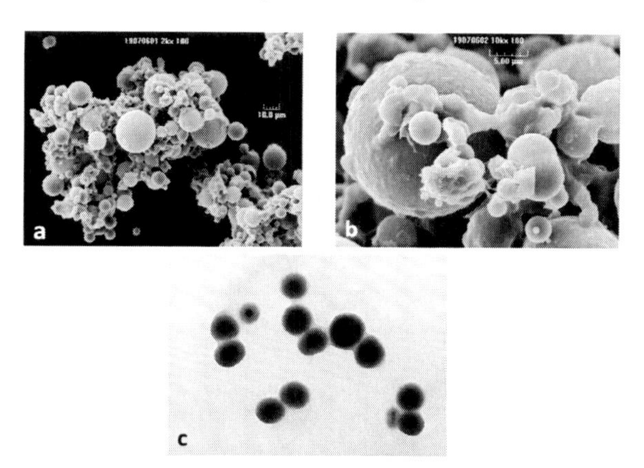

Figure 6.9 (a) Coating 5–20 μm silica beads with PMMA by GAS precipitation on slurry at 36°C, 10 MPa; silica and polymer concentration of 12 mg/mL; (b) a detail; and (c) coating 156 nm magnetic particles with lipids by SAS precipitation on slurry at 30°C, 9 MPa; oxide and lipid concentration of 3 and 15 mg/mL.

Precipitation-on-slurry can be also carried out with a nonpolymer substance in presence of pre-existing solids. The objective is to mix intimately the two species, for instance, to improve a drug characteristic or a powder flow property. Indeed, agglomeration of organic particles is very difficult to avoid since it comes from interactions between particles that, being alike, exhibit the same surface characteristics. Addition of glidants to powders, for instance, silica, is a common practice to aid the powder flowability or to achieve high dispersal efficiency in the case of pharmaceutical products. Precipitation of a drug in presence of oxide slurry was attempted by GAS [59, 60]. In the presence of oxide particles, griseofulvine particles changed from short, wide crystals to long, thin, needle-like structures with high aspect ratios [59]. In the case of naproxen [60], silica particles did not influence notably the crystallization behavior of the drug since same morphology and sizes were obtained. The silica particles adhered, however, to the crystal surface (Fig. 6.10), which can explain the improved flowability of powders.

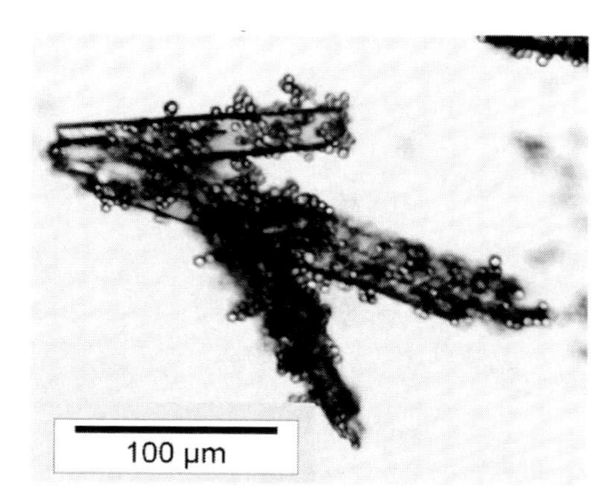

Figure 6.10 Naproxen/silica formulations prepared by GAS precipitation on slurry at 27°C, 10 MPa; acetone slurry stirred at 500 rpm, silica and naproxen concentration of 22 and 44 mg/mL.

Naproxen was also precipitated in presence of other excipients offering various sizes and potential interactions with the drug [60]. Due to the hydrophilic nature of excipients, most formulations yielded an increased drug dissolution rate. Compared to physical mixtures, the benefit of the CO_2 methodology was demonstrated in the case of mannitol excipient.

To avoid the drug recrystallization during the process, porous oxides can be first loaded at atmospheric pressure before being encapsulated by SAS or GAS to create a ternary composite. A 5-fluorouracil–SiO_2–PLA composite was thus prepared [61], immerging first 265 nm silica particles in an ethanol solution of the drug and collecting the drug-loaded SiO_2 by centrifugation. After dispersion into a 0.5% (wt/v) PLA solution in DCM, the slurry was injected through a coaxial nozzle into CO_2.

6.3.3 Coprecipitation on Slurry: Precipitation of Two Species by Antisolvent in the Presence of Pre-Existing Particles

Coprecipitation of drug and polymer on slurry was proposed to elaborate magnetic composites for targeted drug delivery. In these composites, drug and magnetite particles are encapsulated in a biodegradable polymer. The particles are injected into the blood supply of the target organ in the presence of an external magnetic field that drives the particles through the blood vessels onto the targeted site where the drug is locally delivered. Indomethacin–Fe_3O_4–PLA magnetic microparticles were successfully prepared by SAS at 12 MPa and 33°C, spraying the slurry through a coaxial nozzle [62]. The 10–30 nm Fe_3O_4 nanoparticles (oleic acid modified) were dispersed into a 0.5% (w/v) PLA solution in DCM, with a Fe_3O_4:PLA ratio of 1:2. The produced microparticles had a mean size of 901 nm that was slightly higher than the 803 nm of pure magnetite/PLA particles. Spectroscopy demonstrated that the process was a physical coating, which is favorable for drugs since there is no change in chemistry. An in vitro test showed moreover that the Fe_3O_4-PLA magnetic microparticles had no cytotoxicity and were biocompatible. Indomethacin and magnetite were coprocessed with other polymers, using classical SAS or SAS assisted by ultrasonic field (SES-EM) [63]. Polymers were PMMA, PLGA, and Eudragit RS. The magnetite was suspended in a light mineral oil using a fatty acid surfactant and mixed with a polymer and a drug dissolved in methylene chloride. Typical concentrations were of 10 mg/mL, 5 mg/mL, and 0, 1 or 2 mg/mL for the polymer, magnetite, and drug, respectively. The process was carried out at 35°C and 8.3 MPa. Regarding polymer/magnetite products, a uniform distribution of magnetite particles within the polymer matrix was evidenced. Particle sizes were in the range of 2 μm when produced by classical SAS but were decreased to the submicronic range by using SAS-EM. When the drug was coprocessed, no indomethacin was seen outside the particles, except for when the higher drug loading was used with PMMA.

In previous examples, the drug precipitated in submicron size, probably because of a very low solubility in the final CO_2–solvent mixture. We extended the concept of coprecipitation on slurry to a drug that, alone, crystallized in the form of large polyedric crystals of 250 μm mean size [64]. The drug was poorly-water soluble, so the objective of the formulation was to enhance its dissolution, by adding, therefore, silica oxide and/or water soluble polymers like PEG; Eudragits were tested as well to modulate the drug dissolution. The expectation of adding silica was that silica could act as passive spacers to deagglomerate drug particles and could impact the precipitation behavior of the drug. The slurry was processed by GAS because of the large size of silica, 5 and 60 μm. Formulations were successfully prepared as powders providing a polymer:silica ratio below 1:1. The drug grew almost unaffected by the presence of silica and polymer in the solution. Deagglomeration of drug particles was effective at low amount of polymer, otherwise the polymer, especially Eudragits, embedded both silica and drugs, giving coarse grains rather than fine powders.

Acknowledgments

P. Subra-Paternault wants to thank the coworkers who have contributed over the years to published and unpublished results, especially A. Vega-Gonzalez (ex-LIMHP) and A. Erriguible (I2M-TREFLE).

References

1. Subra, P., Berroy, P., Vega, A., and Domingo, C. (2004). Process performances and characteristics of powders produced using supercritical CO_2 as solvent and antisolvent, *Powder Technol.*, **142**, pp. 13–22.
2. Roy, C. (2008). *Generation de Particules par Procédé Assisté par CO_2 Comprimé: Cristallisation et Formulation*, PhD thesis Paris 13 University (Editions Universitaires Européennes, Germany) (in French).
3. Kiran, E. (2009). Polymer miscibility, phase separation, morphological modifications and polymorphic transformations in dense fluids, *J. Supercrit. Fluids*, **47**, pp. 466–483.
4. Kalani, M., and Yunus, R. (2011). Application of supercritical antisolvent method in drug encapsulation: a review, *Int. J. Nanomed.*, **6**, pp. 1429–1442.
5. Mishima, K. (2008). Biodegradable particle formation for drug and gene delivery using supercritical fluid and dense gas, *Adv. Drug Delivery Rev.*, **60**, pp. 411–432.
6. Davies, O., Lewis, A., Whitaker, M., Tai, H., Shakesheff, K., and Howdle, S. M. (2008). Applications of supercritical CO_2 in the fabrication of polymer systems for drug delivery and tissue engineering, *Adv. Drug Delivery Rev.*, **60**, pp. 373–387.
7. Erriguible, E., Laugier, S., Late, M., and Subra-Paternault, P. (2013). Effect of pressure and non isothermal injection on recrystallization by

CO_2 antisolvent: solubility measurements, simulation of mixing, and experiments, *J. Supercrit. Fluids*, **76**, pp. 115–125.

8. De Gioannis, B., Vega, A., and Subra, P. (2004). Antisolvent and cosolvent effect of CO_2 on the solubility of griseofulvine in acetone and ethanol solutions, *J. Supercrit. Fluids*, **29**, pp. 49–57.
9. Kikic, I., De Zordi, N., Moneghini, M., and Solinas, D. (2010). Solubility estimation of drugs in ternary systems of interest for the antisolvent precipitation processes, *J. Supercrit. Fluids*, **55**, pp. 616–622.
10. Subra, P., Laudani, C., Vega-Gonzalez, A., and Reverchon, E. (2005). Precipitation and phase behavior of theophylline in solvent-supercritical CO_2 mixtures, *J. Supercrit. Fluids*, **35**, pp. 95–105.
11. DeGioannis, B., Jestin, P., and Subra, P. (2004). Morphology and growth control of griseofulvin recrystallized by compressed carbon dioxide as antisolvent, *J. Cryst. Growth*, **262**, pp. 519–526.
12. Sala, S., Córdoba, A., Moreno-Calvo, E., Elizondo, E., Muntó, M., Rojas, P. E., Larrayoz, M. A., Ventosa, N., and Veciana, J. (2012). Crystallization of microparticulate pure polymorphs of active pharmaceutical ingredients using CO_2-expanded solvents, *Cryst. Growth Des.*, **12**, pp. 1717–1726.
13. Wang, Y., Pfeffer, R., Dave, R., and Enick R. (2005). Polymer encapsulation of fine particles by a supercritical antisolvent process, *AIChE J.*, **51**, pp. 440–455.
14. Pérez de Diego, Y., Pellikaan, H., Wubbolts, F., Witkamp, G., and Jansens, P. (2005). Operating regimes and mechanism of particle formation during the precipitation of polymers using the PCA process, *J. Supercrit. Fluids*, **35**, pp. 147–156.
15. Luna-Barcenas, G., Kanakia, S., Sanchez, I., and Johnston, K. (1995). Semicrystalline microfibrils and hollow fibres by precipitation with a compressed-fluid antisolvent, *Polymer*, **36**, pp. 3173–3182.
16. Bodmeier, R., Wang, H., Dixon, D., Mawson, S., and Johnston, K. (1995). Polymeric microspheres prepared by spraying into compressed carbon dioxide, *Pharm. Res.*, **12**, pp. 1211–1217.
17. Vega-Gonzalez, A., Domingo, C., and Subra-Paternault, P. (2006). Supercritical antisolvent process technique for the production of biopolymer scaffolds, *Proc. 8th Conf. Supercrit. Fluids Appl.*, Italy, pp. 451–456.
18. Vega Gonzalez, A., Domingo, C., Elvira, C., and Subra, P. (2004). Precipitation of PMMA/PCL blends using supercritical carbon dioxide. *J. Appl. Polym. Sci.*, **91**, pp. 2422–2426.

19. Vega-Gonzalez, A., Subra-Paternault, P., Lopez-Periago, A., Garcia-Gonzalez, C., and Domingo, C. (2008). Supercritical CO_2 antisolvent precipitation of polymer networks of L-PLA, PMMA and PMMA/PCL blends for biomedical applications, *Eur. Polym. J.*, **44**, pp. 1081–1094.

20. Lopez-Periago, A., Vega, A., Subra, P., Argemi, A., Saurina, J., Garcia-Gonzalez, C. A., and Domingo, C. (2008). Supercritical CO_2 processing of polymers for the production of materials with applications in tissue engineering and drug delivery, *J. Mater. Sci.*, **43**, pp. 1939–1947.

21. Tabernero, A., Martin del Valle, E., and Galan, M. (2012). Precipitation of tretinoin and acetaminophen with solution enhanced dispersion by supercritical fluids (SEDS). Role of phase equilibria to optimize particle diameter, *Powder Technol.*, **217**, pp. 177–188.

22. Chattopadhyay, P., and Gupta, R. (2002). Protein nanoparticles formation by supercritical antisolvent with enhanced mass transfer, *AIChE J.*, **48**, pp. 235–244.

23. Barrett, A., Dehghani, F., and Foster, N. R. (2008). Increasing the dissolution rate of itraconazole processed by gas antisolvent techniques using polyethylene glycol as a carrier, *Pharm. Res.*, **25**, pp. 1274–1289.

24. Erriguible, A., Vincent, S., and Subra-Paternault, P. (2012). Numerical investigations of liquid jet breakup in pressurized carbon dioxide: conditions of two-phase flow in supercritical antisolvent process, *J. Supercrit. Fluids*, **63**, pp. 16–24.

25. Ouchene, R., Erriguible, E., Vincent, S., and Subra-Paternault, P. (2013). Simulation of liquid solvent atomization in compressed CO_2, *Mech. Res. Commun.*, **54**, pp. 1–6.

26. Baldyga, J., Kubicki, D., Shekunov, B., and Smith, K. (2010). Mixing effects on particle formation in supercritical fluids, *Chem. Eng. Res. Des.*, **88**, pp. 1131–1141.

27. Dukhin, S., Zhu, C., Pfeffer, R., Luo, J., Chavez, F., and Shen, Y. (2003). Dynamic interfacial tension near critical point of a solvent–antisolvent mixture and laminar jet stabilization, *Physicochem. Eng. Aspects*, **229**, pp. 181–199.

28. Fadli, T., Erriguible, A., Laugier, S., and Subra-Paternault, P. (2010). Simulation of heat and mass transfer of CO_2–solvent mixtures in miscible conditions: isothermal and non-isothermal mixing, *J. Supercrit. Fluids*, **52**, pp. 193–202.

29. Martin, A., and Cocero, M. J. (2004). Numerical modeling of jet hydrodynamics, mass transfer, and crystallization kinetics in the supercritical antisolvent (SAS) process, *J. Supercrit. Fluids*, **32**, pp. 203–219.

30. Erriguible, A., Fadli, T., and Subra, P. (2013). A complete 3D simulation of a crystallization process induced by supercritical CO_2 to predict particle size, *Comp. Chem. Eng.*, **52**, pp. 1–9.

31. Sierra Pallares, J., Marchisio, D., Parra-Santo, M. T., Garcia-Serna, J., Castro, F., and Cocero, M. J. (2012). A computational fluid dynamics study of supercritical antisolvent precipitation: mixing effects on particle size, *AIChE J.*, **58**, pp. 385–398.

32. Roy, C., Vrel, D., Vega-Gonzalez, A., Jestin, P., Laugier, S., and Subra-Paternault, P. (2011). Effect of CO_2-antisolvent techniques on size distribution and crystal lattice of theophylline, *J. Supercrit. Fluids*, **57**, pp. 267–277.

33. Fusaro, F., Hanchen, M., Mazzotti, M., Muhrer, G., and Subramanian, B. (2005). Dense gas precipitation: a comparative investigation of GAS and PCA Techniques, *Ind. Chem. Eng. Res.*, **44**, pp. 1502–1509.

34. Lin, C., Murher, G., and Mazzotti, M. (2003). Vapor-liquid mass transfer during gas antisolvent recrystallization: modeling and experiments, *Ind. Eng. Chem. Res.*, **42**, pp. 2171–2182.

35. Fusaro, F., Mazzotti, M., and Murher, G. (2004). Gas antisolvent recrystallization of paracetamol from acetone using compressed carbon dioxide as antisolvent, *Cryst. Growth Des.*, **4**, pp. 881–889.

36. Muhrer, G., Meier, U., Fusaro, F., Albano, S., and Mazzotti, M. (2006). Use of compressed gas precipitation to enhance the dissolution behavior of a poorly water-soluble drug: generation of drug microparticles and drug-polymer solid dispersions, *Int. J. Pharm.*, **308**, pp. 69–83.

37. Vega-Gonzalez, A., Marteau, P., and Subra-Paternault, P. (2006). Monitoring a crystallization induced by compressed CO_2 with Raman spectroscopy, *AIChE J.*, **52**, pp. 1308–1317.

38. Tandya, A., Mammucari, R., Dehghani, F., and Foster, N. R. (2007). Dense gas processing of polymeric controlled release formulations, *Int. J. Pharm.*, **328**, pp. 1–11.

39. Kluge, J., Fusaro, F., Muhrer, G., Thakur, R., and Mazzotti, M. (2009). Rational design of drug-polymer co-formulations by CO_2 anti-solvent precipitation, *J. Supercrit. Fluids*, **48**, pp. 176–182.

40. Corrigan, O., and Crean, A. (2002). Comparative physicochemical properties of hydrocortisone-PVP composites prepared using supercritical carbon dioxide by the GAS anti-solvent recrystallization process, by coprecipitation and by spray drying, *Int. J. Pharm.*, **245**, pp. 75–82.

41. Boutin, O., Badens E., Carretier, E., and Charbit, G. (2004). Co-precipitation of a herbicide and biodegradable materials by the supercritical anti-solvent technique, *J. Supercrit. Fluids*, **31**, pp. 89–99.

42. Roy, C., Vega-González, A., and Subra-Paternault, P. (2007). Theophylline formulation by supercritical antisolvents, *Int. J. Pharm.*, **343**, pp. 79–89.

43. Argemi, A., Vega, A., Subra-Paternault, P., and Saurina, J. (2009). Characterization of azacytidine/polylactic acid particles prepared by supercritical antisolvent precipitation, *J. Pharm. Biomed. Anal.*, **50**, pp. 847–852.

44. Duarte, C., Roy, C., Vega-González, A., Duarte, C. M. M., and Subra-Paternault, P. (2007). Preparation of acetazolamide composite microparticles by supercritical antisolvent techniques, *Int. J. Pharm.*, **332**, pp. 132–139.

45. Vega-Gonzalez, A., Subra-Paternault, P., and Vrel, D. (2006). Aminosalicylate formulation using compressed CO_2, *Proc. 8th Int. Symp. Supercrit. Fluids*, Kyoto, Japan.

46. Zhong, Q. X., Jin, M. F., Davidson, P. M., and Zivanovic, S. (2009). Sustained release of lysozyme from zein microcapsules produced by a supercritical anti-solvent process, *Food Chem.*, **115**, pp. 697–700.

47. Okamoto, H., Nishida, S., Todo, H., Sakakura, Y., Iida, K., and Danjo, K. (2003). Pulmonary gene delivery by chitosan-pDNA complex powder prepared by a supercritical carbon dioxide process, *J. Pharm. Sci.*, **92**, pp. 371–380.

48. Elvassore, N., Bertucco, A., and Caliceti, P. (2001). Production of insulin-loaded poly(ethylene glycol)/poly(L-lactide) (PEG/PLA) nanoparticles by gas antisolvent techniques, *J. Pharm. Sci.*, **90**, pp. 1628–1636.

49. Elvassore, N., Bertucco, A., and Caliceti, P. (2001). Production of protein-loaded polymeric microcapsules by compressed CO_2 in a mixed solvent, *Ind. Eng. Chem. Res.*, **40**, pp. 795–800.

50. Caliceti, P., Salmaso, S., Elvassore, N., and Bertucco, A. (2004). Effective protein release from PEG/PLA nano-particles produced by compressed gas anti-solvent precipitation techniques, *J. Controlled Release*, **94**, pp. 195–205.

51. Kim, Y., Sioutas, C., and Shing, K. (2009). Influence of stabilizers on the physicochemical characteristics of inhaled insulin powders produced by supercritical antisolvent process, *Pharm. Res.*, **26**, pp. 61–71.

52. Wang, Y., Dave, R., and Pfeffer, R. I. (2004). Polymer coating/encapsulation of nanoparticles using a supercritical anti-solvent process, *J. Supercrit. Fluids*, **28**, pp. 85–99.

53. Marre, S., Cansell, F., and Aymonier, C. (2008). Tailor-made surface properties of particles with a hydrophilic or hydrophobic polymer shell mediated by supercritical CO_2, *Langmuir*, **24**, p. 252.

54. Roy, C., Vega-Gonzalez, A., Garcia-Gonzalez, C., Tassaing, T., Domingo, C., and Subra-Paternault, P. (2010). Assessment of $scCO_2$ techniques for surface modification of micro- and nano-particles: process design methodology based on solubility, *J. Supercrit. Fluids*, **54**, pp. 362–368.

55. Chong, G., Yunus, R., Abdullah, N., Choong, T., and Spotar, S. (2009). Coating and encapsulation of nanoparticles using supercritical antisolvent, *Am. J. Appl. Sci.*, **6**, pp. 1352–1358.

56. Garcia-Gonzalez, C., Vega-Gonzalez, A., Lopez-Periago, A., Subra-Paternault, P., and Domingo, C. (2009). Composite fibrous biomaterials for tissue engineering obtained using a supercritical CO_2 antisolvent process, *Acta Biomater.*, **5**, pp. 1094–1108.

57. Murillo-Cremaes, N., Subra-Paternault, P., Saurina, J., Roig, A., and Domingo, C. (2014). Compressed antisolvent process for polymer coating of drug-loaded aerogel nanoparticles and study of the release behavior, *Colloid. Polym. Sci.*, **292**, pp. 2475–2484.

58. Furlan, M., Kluge, J., Mazzotti, M., and Lattuada, M. (2010). Preparation of biocompatible magnetite-PLGA composite nanoparticles using supercritical fluid extraction of emulsions, *J. Supercrit. Fluids*, **54**, p. 348.

59. Thakur, R., Hudgins, A., Goncalves, E., and Muhrer, G. (2009). Particle size and bulk powder flow control by supercritical sntisolvent precipitation, *Ind. Eng. Chem. Res.*, **48**, pp. 5302–5309.

60. Subra-Paternault, P., Gueroult, P., Larrouture, D., Massip, S., and Marchivie, M. (2014). Preparation of naproxen-excipient formulations by CO_2 precipitation on a slurry, *Powder Technol.*, **255**, pp. 80–88.

61. Chen, A. Z., Li, Y., Chen, D., and Hu, J. Y. (2009). Development of core–shell microcapsules by a novel supercritical CO_2 process, *J. Mater. Sci.: Mater. Med.*, **20**, pp. 751–758.

62. Chen, A. Z., Kang, Y. Q., Pu, X. M., Yin, G. F., Li, Y., and Hu, J. Y. (2009). Development of Fe_3O_4-poly(L-lactide) magnetic microparticles in supercritical CO_2, *J. Col. Interface Sci.*, **330**, pp. 317–322.

63. Chattopadhyay, P., and Gupta, R. (2002). Supercritical CO_2 based production of magnetically responsive micro- and nanoparticles for drug targeting, *Ind. Eng. Chem. Res.*, **41**, p. 6049.

64. Subra-Paternault, P., Vrel, D., and Roy, C. (2012). Coprecipitation on slurry to prepare drug-oxide-polymer formulations by supercritical antisolvent, *J. Supercrit. Fluids*, **63**, pp. 69–80.

Chapter 7

Development of Hybrid Structured Particles Prepared through the PGSS® Process

Vanessa S. S. Gonçalves and Catarina M. M. Duarte
[a]*Instituto de Biologia Experimental e Tecnológica, Apartado 12, 2781-901 Oeiras, Portugal*
[b]*Instituto de Tecnologia Química e Biológica, Universidade Nova de Lisboa, Avenida da Republica, 2780-157 Oeiras, Portugal*
cduarte@itqb.unl.pt

Supercritical fluid (SCF) technology has shown to be a viable option with relevant advantages for the production of particulate systems. Alternatively to conventional methods, single-step and solvent-free SCF techniques have been optimized to process materials under mild operating conditions. Namely, particle formation through the particles from gas-saturated solutions (PGSS®) technique has been widely explored for the production of particles with application in numerous fields, like pharmaceutical, cosmetic, and food industries. Moreover, the development of hybrid and structured particles with singular features has gained particular interest in the last decade. For the pharmaceutical industry, hybrid formulations are

Supercritical Fluid Nanotechnology: Advances and Applications in Composites and Hybrid Nanomaterials
Edited by Concepción Domingo and Pascale Subra-Paternault

ISBN 978-981-4613-40-8 (Hardcover), 978-981-4613-41-5 (eBook)
www.panstanford.com

considered an attractive alternative to polymer-based and/or lipid-based delivery systems alone, which still present limitations related with inefficient pharmacokinetic, bioavailability, or cellular uptake of active compounds. In this chapter the preparation of polymer–polymer, polymer–lipid, and lipid–lipid hybrid delivery systems, through the use of the PGSS process is reviewed. The analytical tools needed for their characterization are briefly presented.

7.1 Hybrid Structured Particles as Delivery Systems of Active Compounds

The development of new drugs alone is not sufficient to ensure progress in drug therapy. The development of suitable carrier systems is fundamental to overcome barriers to drug's usefulness. During the last years, pharmaceutical technology has progressed towards the production of more complex drug delivery structures in order to fulfill four key design aspects: stability, encapsulation efficiency, controlled release, and biocompatibility. Therefore, instead of using just one carrier, hybrid particles composed of two or more classes of materials, such as polymers, lipids, silica, and metal among others, are being developed, so as to combine several functionalities in a single delivery system [1–4]. The most common strategy is to combine the enhanced cellular uptake features of highly biocompatible lipids with the structural integrity and improved stability in biological fluids of biodegradable polymeric matrices in order to overcome the weakness of a material using the strength of another. These robust structured systems are capable to incorporate drugs with distinct polarities and have already been applied as vaccine adjuvant for cancer targeting and delivery of nucleic acids [1, 3]. Although solid lipid particles have been considered as an attractive delivery system, it is well known that they have some limitations such as low drug loading capacity and drug leakage during storage mostly due to their almost perfect lipid crystal matrix [5, 6]. The most common strategy to overcome this bottleneck is to prepare hybrid structures, namely, incorporate liquid lipids into the solid matrix of lipid particles to enable the formation of imperfections in the crystal lattice structure. The resulting delivery systems, referred to as structured lipid carriers, present higher drug entrapment capacity and have been widely used for pharmaceutical and cosmetic applications (Fig. 7.1) [7, 8].

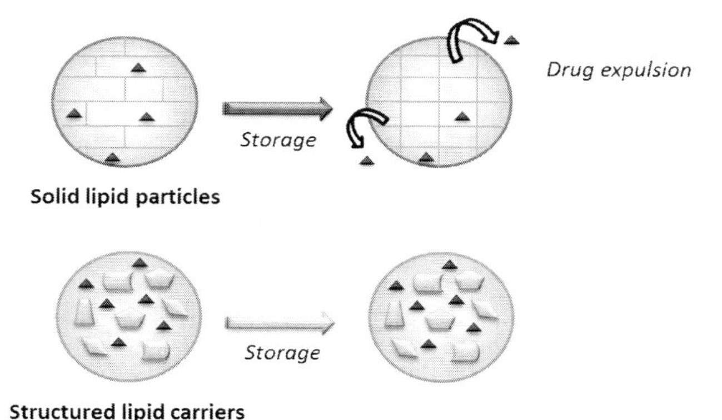

Figure 7.1 Formation of a perfect crystal lattice in solid lipid particles and a crystal structure with imperfections in structured lipid carriers.

The methods to produce hybrid particles can be divided into two major approaches, namely, the top-down and the bottom-up techniques [3, 9]. The top-down technique consists of starting with a bulk material and through the reduction of its size obtaining the hybrid particles, whereas the bottom-up method starts with very small structural units to form larger structures. In both approaches, there are still many drawbacks to overcome, in particular the difficulties in controlling the morphology, structure and size, the difficulty and cost of the purification step, and the use of organic solvents [10].

7.2 The Particles from the Gas-Saturated Solution Technique

The PGSS technique was patented by Weidner et al. [11] and it is considered one of the most attractive CO_2-based micronization processes, because, as it do not relies on the solvent strength of CO_2, it employs relatively low operating pressures and can totally eliminate the need for organic solvents [12]. The major advantages of this process, in comparison with other supercritical-based precipitation methods, is the low intake of carbon dioxide, the possibility to process thermolabile substances, and also the fact that the compound to be micronized do not need to be soluble in CO_2, PGSS processes already being in operation at a large scale [10,

13–16]. A schematic diagram of a typical PGSS process is presented in Fig. 7.2.

Figure 7.2 The particles from gas-saturated solution (PGSS) technique.

Briefly, the process consists in dissolving the compressed gas into the molten material in a stirred high pressure reactor until saturation is reached. The gas-saturated solution formed, which can typically contain between 5–50 wt% of the compressed gas, is then expanded through a nozzle to ambient pressure, causing the release of CO_2 with large cooling effect due to the energy consumption, ultimately leading to the precipitation of the compounds [9, 15, 17, 18].

The morphology, size, and apparent density of the produced particles may depend on several parameters such as the structure and viscosity of the compounds to be precipitated, the operating conditions, and even the geometry of the equipment used to perform the PGSS process [19–21]. Generally, the average particle size decreases when using high pressures, and thus high carbon dioxide content, low temperature values, and small nozzle diameters, the particles being more spherical as higher temperatures are used [13, 17, 22–24]. Thus, the prior thermodynamic knowledge of the solubility of CO_2 in the molten materials, as well as the determination of the solid-liquid transition of a compound in the presence of carbon dioxide, is of extremely importance for the development of the experimental plan [19, 25].

The PGSS process has already been applied for the micronization of polymers, fats, waxes, resins, natural products, and active pharmaceutical ingredients [15, 19, 25]. This technique is especially suited for processing polymers and fats in which CO_2 has a large solubility and moreover it has a melting point depression effect allowing to spray matrices that under classical conditions can hardly be sprayed or can even not be sprayed at all [15]. The extent of melting point depression experimented by each substance depends on the amount of CO_2 that solubilizes into the substance and it is caused by molecular interactions between dissolved CO_2 and the substance of interest [26]. Determination of solid-liquid transitions in pressurized systems is essentially as it gives information on the pressure needed to melt the substance to be micronized and form a liquid phase at a given temperature [27, 28].

The first reported PGSS application was for the generation of powders from poly(ethylene glycol)s (PEGs) [29]. PEG is a widely used hydrophilic polymer due to its biocompatibility and nontoxicity; it is used as a carrier material in the development of pharmaceutical and cosmetic formulations and has been employed by different authors as a model substance in order to obtain a better knowledge of this technique. The main limitation pointed to the PGSS process is that the solute has to be melted, which can be problematic for heat-sensitive materials [30]. Nevertheless, it is still possible to produce particles from compounds that are not melted during the PGSS process, like the case of some APIs, by incorporating these compounds in liquefied polymers or lipids, and further atomize this suspension [19, 31]. This was the strategy used by Critical Pharmaceuticals Ltd. (Nottingham, UK) for the development of a single shot tetanus vaccine [32]. To improve the PGSS method, several variants appeared as specific requirements for the production of particles [25].

7.3 Production of Hybrid Structured Particles through PGSS

PGSS is also a suitable technique for the production of well-controlled hybrid composite particles [33]. The reduction of the melting and glass transition temperatures, due to the dissolution of CO_2 in the compounds, makes possible the mixing of sensitive materials at low temperatures without the occurrence of unwanted reactions. Moreover, reducing the viscosity and surface tension enables the blend and the atomization of mixed compounds previously immiscible with one another, the generation of particles with singular properties being possible. It could be even possible to take advantage of the plasticizer effect of some carriers on other compounds to enhance the effect of CO_2 on the physical properties of the mixture. This strategy is already used by Critical Pharmaceuticals Ltd. (Nottingham, UK) with the so-called CriticalMix process. The developed formulations comprise mixtures of polymers that are able to provide a variety of release profiles. Despite the emergence of some variants of this process, PGSS remains a method capable of generating particles with morphologies and compositions that have never been achieved before in just one step of production.

The preparation of hybrid particles through PGSS started in 2003 and since then, more than 20 publications regarding this topic has been published, herein presented in Fig. 7.3 [34–57].

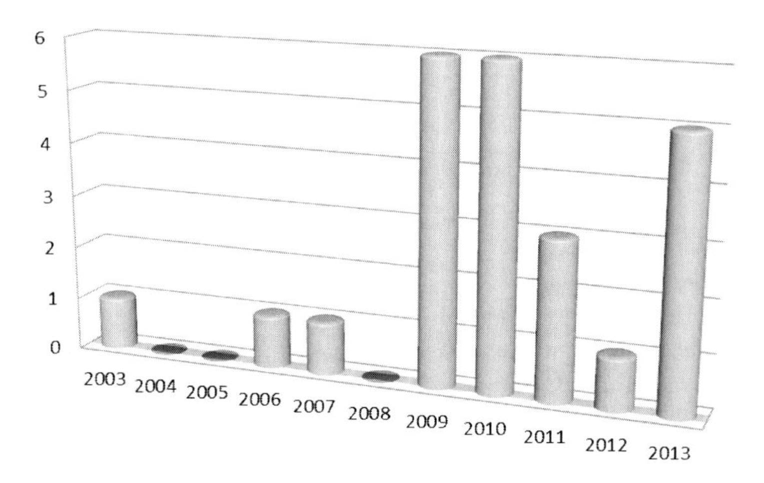

Figure 7.3 Number of publications regarding the production of hybrid particles through PGSS. Keywords used here are PGSS, particles from gas-saturated solutions or saturated solutions; hybrid or structured; and particles or drug delivery systems. Google Academics, Science Direct, and Web of Knowledge were the main reference sources.

There were an increased number of publications concerning the production of hybrid particles from saturated solutions between 2009 and 2010. During 2011 and 2012 the number of publications was reduced, but it increased again throughout 2013. By analyzing the composition of the hybrid particles produced to date and published in the literature, it is possible to see that the majority is composed by a mixture of lipids or by a mixture of lipids and polymers (Fig. 7.4).

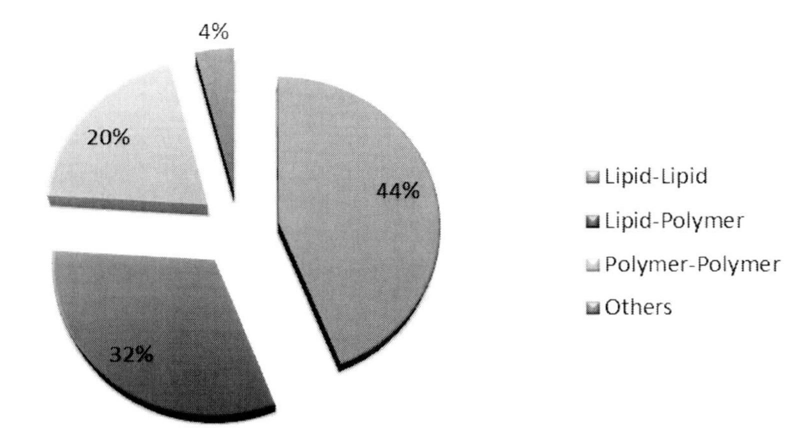

Figure 7.4 Types of hybrid particles produced through PGSS thus far reported in the literature. Keywords used here are PGSS, particles from gas-saturated solutions or saturated solutions; hybrid or structured; lipid or polymer; and particles or drug delivery systems. Google Academics, Science Direct, and Web of Knowledge were the main reference sources.

This tendency is in agreement with what has been stated above, that is, the production of structured lipid carriers (composed of solid and liquid lipids) and lipid–polymer particles are two major strategies in pharmaceutical technology to develop more robust, stable, and efficient drug delivery systems.

7.3.1 Lipid–Lipid System

A large amount of lipid–lipid systems have been produced by PGSS (Table 7.1). Elvassore and coauthors [34] started to develop hybrid lipid particles in 2003, by producing a carrier composed of triestearin and phospatidylcholine. The authors were able to produce particles of 22 μm using low temperature values. After 10 years, São Pedro et al. [35, 36] used the same carrier system with the purpose of encapsulating curcumin, a natural curcuminoid with biomedical applications, through the variant gas-assisted melting atomization (GAMA) developed by Salmaso and coauthors [45], where air streams are present in the atomizer and in the precipitation vessel, to facilitate the atomization process and prevent the formation of agglomerated particles. The presence of phosphatidylcholine not only prevented the formation of a perfect crystal lattice of tristearin, thus enhancing the drug load capacity of this carrier, but also caused a depression in its melting point. The authors verified that using dimethylsulfoxide (DMSO), with the purpose of improving the dispersion of the active compound in the lipid matrix, causes the aggregation of the particles. Furthermore, the encapsulation efficiency of curcumin decreased by increasing the amount of this bioactive compound in the initial mixture, probably due to a phase separation in the mixing vessel and the resulting presence of curcumin residues inside this chamber at the end of the experiment. Nevertheless, its structure was preserved during the PGSS process.

Table 7.1 Hybrid lipid particles produced through PGSS

Carrier	Active compound	Ref.
Triestearin and phospatidylcholine	Curcumin	[34–36]
Gelucire 50/13™ and Precirol™ ATO 5	*trans*-Chalcone	[37–38]
Ceramide, fatty acids, cholesterol	–	[39]
Glyceryl monostearate and Cutina™ HR	TiO_2 and caffeine, glutathione or ketoprofen	[40–42]
Triestearin and phosphatidylcholine	Magnetic nanoparticles	[43]
Peceol™ and Gelucire 43/01™, or Gelucire 50/13™ or Geleol™	–	[44]

Choosing hybrid materials as drug delivery systems is an increasingly used strategy for the enhancement of drug's solubility. By using Gelucire 50/13™ (stearoyl polyoxyl-32 glycerides) and Precirol™ ATO 5 (glyceryl distearate) as carriers, with a mass ratio of 1:1, Sampaio de Sousa and coauthors [37, 38] were able to increase the solubility of *trans*-chalcone, both in gastric and in intestinal fluid. The particles obtained with this new lipid formulation had a slightly broad size distribution, with an average size of 6.7 μm. Since both carrier matrices have glycerides in their composition, the authors had difficulty distinguishing them through attenuated total reflection–Fourier transform infrared (ATR-FTIR) spectroscopy, due their overlapping infrared (IR) spectra. This demonstrates the challenge in characterizing drug delivery systems composed by similar matrices. Nevertheless, ATR-FTIR analyses showed an effective mixing between the carriers and the active compound.

Semenzato and coauthors [39, 58] developed lipid nanoparticles composed of epidermal lipids with great potential to act as cosmetic ingredient in dermatological products. The authors precipitated binary mixtures of ceramide, cholesterol, and fatty acids through PGSS, producing bigger particles while working at higher temperature, the effect of pressure being almost irrelevant. Interestingly, the lipid nature had also an effect in the final size of the particles, that is, larger particles were produced when ceramide and cholesterol, which have a high molecular weight, were present in major quantities. The solid lipid nanoparticles were further incorporated into cream–gel systems, revealing that they were appropriate for emulsion formulation through a cold process.

García-González and coauthors [40, 41] developed a structured matrix composed of two lipids at equal mass ratio, glyceryl monostearate and Cutina™ HR (hydrogenated castor oil, waxy triglyceride), with supercritical silanized nanoparticulate titanium dioxide (TiO_2), suitable to deliver both hydrophilic and hydrophobic drugs. The aim of the authors was to produce hybrid particles that could be used not only as sunscreens with UV-radiation protection properties free of organic absorbers, but also as lipid formulations for the treatment of skin diseases. The purpose of using a mixture of lipids to produce a matrix was again to achieve a crystal structure with more imperfections, thus, with increased drug load capacity. Moreover, the use of lipid compounds reduces the risk of TiO_2 nanoparticles toxicity. The authors verified that the hydrophobic drug used in this study, ketoprofen, was encapsulated with a higher efficiency into the lipophilic particles in comparison to the hydrophilic drugs used, namely, caffeine or glutathione. The low encapsulation efficiency of caffeine was not only due to its high solubility in supercritical carbon dioxide ($scCO_2$) and further precipitation inside the mixing vessel throughout depressurization but also due to its low solubility in the lipid carriers used. To increase the dissolution of caffeine in the lipophilic matrix, the authors added water to the initial mixture, thus allowing the formation of an emulsion. The precipitation of the emulsion has led to the formation of particles with higher amount of active compound and the presence of water led to the formation of the nonstoichiometric caffeine hydrate, which was quickly dissolved during the release profile assay. The strategy to add water to the system in order to increase the solubility of hydrophilic molecules in lipophilic compounds could not be used for glutathione, since this drug rapidly oxidizes in aqueous solutions. Despite the high solubility of ketoprofen in $scCO_2$, the drug was encapsulated with high efficiency due to its hydrophobic nature and high compatibility with the lipid matrices. This hybrid system was further studied by Argemí and coauthors [42] concluding that the carrier provided a sustained release of ketoprofen for at least 24 hours, being suitable for topical administration of active compounds, since it avoids the skin irritation that sometimes occurs with burst releases. Moreover, the hybrid structure prevented the degradation of ketoprofen by light.

Magnetic nanoparticles (MNPs) are promising carriers for biomedical applications, not only as contrast agents, but also as drug delivery systems, mostly due to the possibility of moving them to the target site through the use of an external magnetic field. However, it is necessary to encapsulate these particles in biocompatible materials to enable them to be administrated into the human body. Vezzù and coauthors [43] successfully encapsulated MNPs into mixtures of triestearin and phosphatidylcholine by the variant GAMA, increasing their biocompatibility for further diagnostic and therapeutic applications. The nanoparticles produced had bi- and trimodal size distributions, yielding smaller particles when the mixture of lipids was used, probably due to the surfactant effect of phosphatidylcholine. The authors have also verified that the encapsulation efficiency of the MNPs was independent of the type of lipids used, increasing with the concentration of magnetite in the initial mixture. The lipid particles produced containing magnetite presented themselves as a viable alternative to polymeric matrices.

In 2013, Gonçalves and coauthors [44] have produced for the first time structured lipid particles through the GAMA variant. With the purpose of obtaining lipid particles with mucoadhesive properties, the authors blended glyceryl monooleate (Peceol^{TM}), a mucoadhesive liquid lipid, with three different types of glycerolipid matrices. The hydrophilic-lipophilic balance (HLB) of the solid matrices proved to be a key parameter to obtain homogeneous mixture with Peceol™. The authors concluded that the more different the HLB of the solid lipids compared with the HLB of glyceryl monooleate, the more difficult was the mixture of the compounds and hence the hardest to produce fine and handlable particles.

7.3.2 Lipid–Polymer System

The PGSS technique can also be applied for the production of protein-loaded hybrid particles, since it does not need the use of organic solvents and therefore maintains the structure and functions of these compounds. During preliminary studies, Salmaso and coauthors [45] were able to encapsulate up to 80% of insulin in a mixture of lipid carriers, namely, tristearin and phosphatidylcholine, with and without the incorporation of PEG 5000, through the use of the variant GAMA. The authors used DMSO to obtain homogeneous protein/carrier mixtures and thus were able to prevent protein precipitation and segregation during the mixing step. Should be noticed that DMSO is highly miscible with CO_2 and, consequently, it was removed from the final particles during the depressurization. The carrier composition was found to play a significant role in the protein release profile, as the formulation with PEG showed a burst and faster release of insulin, while the formulation without the polymer showed a slowly release. Also, the formulation containing the hydrophilic polymer was more stable to sedimentation in aqueous media, since PEG prevented lipid particle aggregation. In any case the activity of the protein was maintained. Further studies were performed by the authors [46] in order to investigate the pharmacokinetic and pharmacodynamic in vivo performance of the drug delivery systems developed by using insulin and recombinant human growth hormone (rh-GH) as model proteins. Once again, the combination of tristearin, phosphatidylcholine and PEG 5000, in the presence of DMSO and $scCO_2$, enabled a low melting homogenous dispersion. Small amounts of PEG were used so as to avoid the burst release of proteins which could compromise the therapeutic purpose of the drug delivery system. As to improve the atomization step and consequently obtain a homogeneous distribution of particle size, the authors not only slightly decreased the temperature of the process in comparison to the preliminary studies, increasing the CO_2 dissolved in the lipid mixture, but also equipped the plant with a peristaltic pump that supports the mixture's atomization. However, the encapsulation efficiency of the proteins was unexpectedly low. The authors explained this outcome owing to partial precipitation of the compounds in the mixing vessel or to interactions between the proteins and the lipid compounds of the mixture. Nevertheless, the proteins formulated in particles were released in a typical diffusive mechanism for four days with preservation of their biological structure and activity. Furthermore, during experiments performed using appropriate mouse models, the authors verified that the hybrid formulation enhanced the oral bioavailability of the proteins due to their protection from enzymatic degradation in the stomach.

Some bioactive compounds need to be functionalized with PEG molecules in order to increase their stability and hydrophilic properties. Vezzù and coauthors [47] used the same hybrid carrier previously described, composed by triestearin and phosphatidylcholine, to incorporate the model protein ribonuclease A (RNAse) functionalized with PEG 5000, so as to develop a drug delivery system with improved therapeutic performance. The authors concluded that the temperature of the process had influence not only in the final size of the particles produced but also in the yield of the method, that is, both parameters increased with increasing temperature values. Furthermore, pressure increase during the PGSS experiments led to the formation of smaller particles up to a certain pressure value; thereafter the particle size was increased when using higher pressures. The addition of DMSO to the mixture of compounds, so as to facilitate the dispersion of the protein in the carrier matrix, led also to the precipitation of smaller particles. However, while using high amounts of DMSO, the authors observed the formation of large aggregates due to the incomplete solvent extraction by CO_2. Despite the fact that the product yield obtained with PEGylated RNAse was very low, its encapsulation efficiency was considerably higher than the encapsulation of the native protein. Moreover, the enzymatic activity of functionalized RNAse was strongly preserved, leading the authors to conclude that the PEGylation prevented the tight interaction between the protein and the carrier compounds, preserving the protein activity. Lastly, the hybrid formulation developed allowed a slow release of PEGylated protein. Nunes et al. [48] explored the supercritical fluid (SCF) precipitation technology to prepare lipid–polymer hybrid drug delivery systems. The authors developed structured spherical particles comprising PEG 4000 and Gelucire 43/01 (hard fat), with different mass ratios. Differential scanning calorimetry (DSC) and transmission electron microscopy (TEM) analyses suggested a possible organization of the materials with the formation of a core–shell structure. Rodriguez-Rojo and coauthors [49] proceeded with the characterization of these particles by confocal microscopy using a hydrophilic fluorescent dye, concluding that their shell was constituted by PEG 4000. Moreover, the authors produced flufenamic acid–loaded particles, whose release profile also suggested the presence of the hydrophilic polymer in the shell. Nunes et al. [48] also developed a hybrid system in order to increase the bioavailability of quercetin, an essential micronutrient present in vegetables and fruits, which has low oral bioavailability. By using glyceryl monostearate blended with hydroxypropyl-β-cyclodextrin, it was possible to produce porous particles with increased quercetin intestinal transport, and thus, with high bioavailability. Almeida et al. [50] reported that this new drug delivery system was capable not only to increase the solubility of quercetin in the intestinal fluid, but also to enhance the antioxidant activity of the flavonoid.

With the purpose of producing pulmonary microparticles, Vijayaraghavan et al. [51, 59] have studied whether it would be possible to process PEGs with fatty acids through PGSS. The authors studied the thermodynamic behavior of free PEGs, with different molecular weights, mixed with stearic, palmitic or myristic acid. They verified that fatty acids with low melting temperature (T_m), namely, palmitic and myristic acid, were capable to act as plasticizing agents and to reduce the T_m of PEG, making these blends suitable to be processed by PGSS. By analyzing the precipitated particles, it was possible to conclude that myristic acid produced the smaller particles with the best control of particle size when mixed with PEG, resulting in a higher process yield in comparison with other fatty acids. The authors believe that this fact is due to the influence of myristic acid on the T_m and viscosity of the polymer blend, being more evident when using a more viscous PEG (high molecular weight). It is easier to break up liquefied carriers into smaller particles when using materials with lower viscosity. The morphology of these hybrid particles, instead of being spherical like in the case of processing PEG alone [60], was irregularly shaped, this event being attributed to different solidification rates during the atomization process.

Fraile et al. [52] produced systems composed of pluronic L64 with the glycerolipid, gelucire 43/01™ or glyceryl monostearate, through PGSS in order to obtain delivery systems with more controlled release of drugs. In both cases, it was added some water to the initial mixture in an attempt to get a rearrangement of the carrier materials; however this was not accomplished and only larger aggregates were formed. Table 7.2 gives a list of produced lipid–polymer compounds by PGSS.

Table 7.2 Hybrid lipid–polymer particles produced through PGSS

Carrier	Active compound	Ref.
Tristearin, phosphatidylcholine, and PEG 5000	Insulin, r-hGH, RNAse	[45–47]
Gelucire 43/01™ and PEG 4000	Flufenamic acid	[48, 49]
Glyceryl monostearate and hydroxypropyl-β-cyclodextrin	Quercetin	[48, 50]
Fatty acids and PEGs	–	[51]
Pluronic L64 and Gelucire 43/01™ or glyceryl monostearate	Ibuprofen	[52]

7.3.3 Polymer–Polymer System

Most of the newly discovered drug candidates are poorly water soluble, and therefore, it is important to improve their dissolution rate in order to obtain a high bioavailability of drugs. Brion et al. [53] developed a formulation through PGSS on the basis of a mixture of hydrophilic polymers in order to promote the solubility and dissolution rate of a new poorly water-soluble active compound. The presence of a nonionic surfactant, poloxamer 407, in combination with PEG 400 and PEG 4000 allowed a more effective wetting of the particles. Moreover, the PGSS process produced smaller particles with high porosity, which also increases the diffusion of the active compound into the dissolution medium. The authors have also studied the effect of several operating parameters on the precipitation of the hydrophilic solid dispersion, concluding that the main significant parameters were the temperature and pressure. They have also concluded that the drug loading influenced the particle size of the final formulation, with larger particles being produced when using a higher drug load. Since the active compound was not soluble in the polymeric mixture, its presence resulted in an increase in viscosity of the system and thus the production of larger particles.

Jordan et al. [54] developed microparticles capable to confer sustained release of a protein for subcutaneous injections. The human growth hormone (hGH), which is a 22 kDa protein, was processed in the dry state with the biocompatible polymers poly(lactic-co-glycolic acid) (PLGA) and poly(lactic acid) (PLA) through PGSS, with an encapsulation efficiency of near 100%. After testing several polymer ratios with different emulsifiers, the elected formulation was composed by PLGA:PLA with a wt/wt ratio of 90:10 and poloxamer 407, capable to produce 100 μm rounded particles with few pores. The structure and function of the protein was preserved, since the mixture of polymers used is capable of being plasticized at temperatures just above 32 °C, thus avoiding the degradation of the drug. Microparticles showed an initial in vitro burst of around 35% and a sustained release for more than 14 days, revealing that some of the protein was positioned close to the surface being available to readily diffuse out of the polymeric particles. It is important to mention that hGH-loaded particles of PLGA or PLA alone did not show the features necessary for the subcutaneous administration of this protein, such as size and drug release properties. Thereafter, this long acting formulation of hGH established preclinical proof of concept in a nonhuman primate's pharmacokinetic and pharmacodynamic studies. Currently, this vaccine is being developed at Critical Pharmaceuticals Ltd. (Nottingham, UK) as a once every two week injection and has already completed preclinical development.

Not only drugs can be present at solid state throughout the PGSS process, as Casettari and coauthors [55] have proved by developing a new biodegradable and mucoadhesive formulation composed by PLGA, mPEG and chitosan. Both polymers were plasticized in $scCO_2$ and, thus, were used as the liquefied matrix where the polysaccharide was mixed and dispersed as a dry powder. Therefore, this suspension was further precipitated by PGSS, producing microparticles (<100 μm) under very mild conditions. Like already seen in other studies, the presence of mPEG led to an increase in particle size with the formation of more rounded and smooth particles. The presence of chitosan on the surface of the particles conferred mucoadhesive properties to this formulation, which was confirmed through an in vitro assay.

To process highly viscous polymers, like PLA, into small particles, it is often necessary to use high temperatures or volatile organic solvents to facilitate their handling. Kelly et al. [56] reported an alternative method to overcome this problem. By using a $scCO_2$ assisted mixing of PLA with PEG 6000, a polymer with lower viscosity, they developed a new formulation with properties of both carriers. Lowering the PLA viscosity by the addition of PEG, allied to the plasticizing effect of CO_2 in the polymers matrix, allowed the processability of these carriers at temperature as low as 40°C by PGSS. The fine white particles obtained were bigger, more spherical and smoother as the content of PEG increased. To explain this, the authors hypothesized that, since the solubility of CO_2 is lower in PEG than in PLA, the incorporation of PEG in the polymeric mixture would lead to less gas loss during depressurization, resulting in smoother and more spherical particles.

Fraile et al. [52] showed that PGSS is a well suited technique to formulate water-insoluble drugs with distinct features by simply varying the carrier materials. By formulating Ibuprofen with pluronic F127 and L64, it was possible to increase by fivefold the dissolution of the drug in simulated gastric fluid, in comparison with the pure unprocessed drug. Apparently, these polymers are capable to form micelles in aqueous solutions and thus stabilize the drug in their hydrophobic cavities leading to increased drug solubility. Table 7.3 gives a list of produced hybrid polymeric compounds by PGSS.

Table 7.3 Hybrid polymeric particles produced through PGSS

Carrier	Active compound	Ref.
Poloxamer 407, PEG 400, and PEG 4000	YNS3107	[53]
PLGA and PLA	hGH	[54]
PLGA, mPEG, and chitosan	–	[55]
PLA and PEG 6000	–	[56]
Pluronic F127 and L64	Ibuprofen	[52]

7.3.4 Other Applications

Nanocomposites produced through PGSS may have several applications beyond the delivery of active compounds, as Pollak and coauthors [57] demonstrated by developing microparticles from polybutylenterephthalate (PBT) with dispersed zinc oxide or bentonite nanoparticles. The obtained composites could be further used in laser sintering and coating processes, due to their improved features, such as better scratch and impact resistance or higher absorption of laser light. Instead of using a heated vessel, the authors used a single screw extruder with a dispersing zone improving the mixing of the compounds. By using carbon dioxide as an auxiliary gas, it was possible to combine the mixing step of the polymer and additive nanoparticles with the micronization process in one single plant. The authors verified that by choosing carefully the spraying conditions, namely, using high temperature and lowest pressure values, spherical polymer particles were formed. Nevertheless, the optimal operating conditions for the micronization of pure PBT in respect to morphology and particle size were not necessarily the same, even while processing particles with a small amount of bentonite.

The increasing number of publications regarding the production of hybrid particles through PGSS evidences that it is a promising method for the preparation of these singular structures. PGSS could be implemented in several industries due to its high batch-to-batch reproducibility and high purity and sterility of compounds produced, being feasible even at an economical level. This environmentally friendly method has already a simple and reproducible scale-up process, being in compliance with the regulatory demands of the current good manufacturing practices (GMPs) for the production of particles from single carriers [61]. Nevertheless, it is still necessary the acquirement of thermodynamic and mass transfer knowledge of ternary and multinary systems, that is, a detailed understanding of all the process parameters that affect the characteristics of the final hybrid particles produced. With a critical evaluation of these aspects, PGSS could become a predictable, consistent, and widespread method for the production of structured particles at the industrial level.

7.4 Characterization of Hybrid Structured Particles

The complexity of hybrid structures lead to analytical challenges for their characterization, the use of multidisciplinary studies being essential for the evaluation of the physicochemical, textural, size, and morphological features of the particles [62]. Several methods for the characterization of structured particles are summarized in Table 7.4.

Table 7.4 Methods applied for the characterization of structured particles

Parameter	Method of characterization
Size	SEM, TEM, AFM, DLS, LD, NTA, MFI, flow cytometry
Morphology	SEM, TEM, AFM, FIB, CLSM
Surface charge	LDE
Textural characterization	AFM, BET method, mercury porosimetry
Thermal behavior	DSC, TGA, XRD
Composition	DSC, XRD, Raman, NIR, FTIR, NMR, XPS, EDS, ToF-SIMS

Scanning electron microscopy (SEM); transmission electron microscopy (TEM); atomic force microscopy (AFM); dynamic light scattering (DLS); laser diffraction (LD); nanoparticle tracking analysis (NTA); microflow imaging (MFI); focused ion beam (FIB); confocal laser scanning microscopy (CLSM); laser Doppler electrophoresis (LDE); Brunauer–Emmett–Teller (BET); differential scanning calorimetry (DSC); thermogravimetric analysis (TGA); X-ray diffraction (XRD); near-infrared (NIR); Fourier transform infrared (FTIR); nuclear magnetic resonance (NMR); X-ray photoelectron spectroscopy (XPS); energy-dispersive X-ray spectrometry (EDS); time-of-flight secondary ion mass spectrometry (ToF-SIMS)

7.4.1 Size, Morphology, and Surface Charge

Particle size and morphology are two of the most important characteristics that affect the general performance and stability of drug delivery systems. Advanced microscopic techniques like atomic force microscopy (AFM), scanning electron microscopy (SEM), and TEM are widely used to analyze size and morphology of particles. By using the AFM technique, it is also possible to acquire three-dimensional surface profiles with spatial resolution up to 0.1 μm. The microscope uses a cantilever with a tip, acting as a probe, at its end to scan the entire specimen surface. Since no vacuum is applied during the analysis, it is possible to directly examine samples containing solvents [63]. SEM is capable to provide an accurate assessment of the three-dimensional morphology, size, and surface of particles with a spatial resolution of 5–10 nm. However, since it has limited throughput, it is challenging to obtain particle size distribution. Moreover, it is necessary to previously coat the samples with electrically conducting elements, such as gold or palladium, which can damage the structure of fragile particles. Nevertheless, SEM remains the technique of choice for the evaluation of the shape of particles and is the best analytical method for the characterization of nonspherical particles [62, 64]. Scanning electron microscopes could be equipped with a focused ion beam (FIB) that offers the possibility of dissecting and probing the interior of particles to verify their porosity [65]. Another electron microscopy, TEM, offers a two-dimensional picture of the particles, since a beam of electrons is transmitted through the sample. Although the resolution of this technique is better than of SEM, the sample preparation could be a complex procedure. Besides the determination of size and morphology, TEM is also a useful technique to differentiate the core–shell structure of a particle by staining the sample with uranyl acetate, osmium tetraoxide, or phosphotungstic acid for better imaging contrast. This procedure allows the observation of the lipid layer which is negatively stained [4]. Another strategy that can be applied to verify the core–shell structure of a hybrid particle is by using conventional fluorescence microscopy allied with confocal laser scanning microscopy (CLSM) and staining the components of the particles with distinct fluorescent dyes [66].

Dynamic light scattering (DLS), also known as photon correlation spectroscopy (PCS), is a reliable technique and the most used to analyze particle size and particle size distribution. DLS measures the fluctuation of the intensity of the scattered light caused by the Brownian motion of particles. The size that is measured is the hydrodynamic diameter, and so is related to a sphere with the same translational diffusion coefficient as the sample. Since DLS assumes spherical shape for particles, additional care must be taken when measuring the size of nonspherical samples due to the lower precision of the technique in this case. Besides that, DLS is not a suitable technique for multimodal particle size distribution and has a limit of size detection around 10 μm [64, 67, 68]. Laser diffraction (LD) is an alternative technique that covers a broad size range (40 nm to 2000 μm), which is based on the dependence of diffraction angle on the particle radius [67]. Another alternative for measuring particle size is the nanoparticle tracking analysis (NTA) method, which is based on fluorescence microscopy and automatic imaging analysis. Although NTA provides high resolution when analyzing multimodal samples, it requires sample dilution in order to prevent an observation field very crowded [68, 69]. A recent study showed that microflow imaging (MFI) and flow cytometry are also additional methods for determination of the nonspherical particle size [64].

When particles are in contact with a solvent there are two liquid layers surrounding them, the weakly bound outer layer and the strongly bound inner layer. The zeta potential is measured at the boundary of the outer layer by laser Doppler electrophoresis (LDE), and is a parameter that greatly influences the stability of particles. A stable particle has a zeta potential more positive than +30 mV or more negative than −30 mV due to electrostatic repulsion. However, care must be taken while performing this analysis, since the zeta potential value is greatly influenced by the ionic strength and pH of the medium and may change depending on the solvent used [67, 68].

7.4.2 Textural Characterization

The surface area of a particle is another important feature of solid materials applied for pharmaceutical purposes, since it influences the load and dissolution rate of active compounds. AFM is a technique applied to determine the surface area and textural characteristics of particles that could be complemented by physical gas adsorption methods. In the Brunauer–Emmett–Teller (BET) method, nitrogen is adsorbed at low temperatures on the surface of particles and then desorbed at room temperature. The isotherms of adsorption and desorption not only can provide the surface area but also provide information regarding the porosity of the materials, such as pore size and pore size distribution. Mercury porosimetry is another alternative method to get information regarding porosity, pore volume, pore size distribution, and density [62].

7.4.3 Thermal Behavior

To evaluate the stability and thermal transitions of particles, calorimetric and thermogravimetric techniques are usually applied. DSC is one of the most used techniques to monitor thermochemical events, giving important insight about melting, recrystallization, sublimation, glass and polymorphic transitions, or even information about decomposition of the materials. In this technique, both sample and reference are at the same temperature, the heat exchanges occurring during structural alterations of materials being measured. DSC is the technique most used to determine the glass transition temperature of noncrystalline and semicrystalline polymers, as well as the melting point of crystalline materials. The determination of the melting point of particles is very useful to estimate the purity, degree of crystallinity, and particle size of the samples analyzed. Especially, in the case of lipids, the degree of crystallinity and the modification of its lattice structure influence not only the amount of active compound that is incorporated, but also its release rate [70]. However, the cause of a thermal event is not revealed directly by DSC, such knowledge could be achieved with complementary methods like X-ray diffraction (XRD) or thermogravimetric analysis (TGA) [62, 67, 71]. TGA is a technique that measures the weight loss of a sample as a function of temperature, being particularly suitable for the quantitative determination of the volatile content of a solid material, as well as for the evaluation of its thermal stability. It is also possible to identify the volatile compound by coupling the TGA apparatus with a gas-phase IR analyzer [62].

7.4.4 Composition

Since hybrid particles are constituted by two or more compounds, it is important to study the final composition of the particles, as well as understand the way the materials are arranged. This could be performed by performing DSC, XRD, vibrational spectroscopy such as Raman, near-infrared (NIR), or FTIR, and nuclear magnetic resonance (NMR) [2, 62, 67, 72]. If the particles are intended for the encapsulation of active compounds, these techniques could be also useful to study the drug-matrix interaction and their spatial distribution.

Elemental analysis of the surface of the particles could be important in some cases, and could be performed either by X-ray photoelectron spectroscopy (XPS), by coupling a scanning electron microscope with energy-dispersive X-ray spectrometer, or by performing time-of-flight an secondary ion mass spectrometry [2, 55, 65].

References

1. Cheow, W. S., and Hadinoto, K. (2011). Factors affecting drug encapsulation and stability of lipid-polymer hybrid nanoparticles, *Colloids Surf., B*, **85**, pp. 214–2520.

2. Mandal, B., Bhattacharjee, H., Mittal, N., Sah, H., Balabathula, P., Thoma, L., and Wood, G. C. (2013). Core-shell-type lipid-polymer hybrid nanoparticles as a drug delivery platform, *Nanomedicine*, **9**, pp. 474–491.

3. Morales, C. S., Valencia, P. M., Thakkar, A. B., Swanson, E., and Langer, R. (2013). Recent developments in multifunctional hybrid nanoparticles: opportunities and challenges in cancer therapy, *Front. Biosci.*, **15**, pp. 529–545.

4. Zhang, L., Chan, J. M., Gu, F. X., Rhee, J. W., Wang, A. Z., Radovic-Moreno, A. F., Alexis, F., Langer, R., and Farokhzad, O. C. (2008). Self-assembled lipid–polymer hybrid nanoparticles: a robust drug delivery platform, *ACS Nano*, **2,** pp. 1696–1702.
5. Müller, R. H, Mäder, K., and Gohla, S. (2000). Solid lipid nanoparticles (SLN) for controlled drug delivery: a review of the state of the art, *Eur. J. Pharm. Biopharm.*, **50**, pp. 161–177.
6. Müller, R. H., Radtke, M., and Wissing, S. (2002). Solid lipid nanoparticles (SLN) and nanostructured lipid carriers (NLC) in cosmetic and dermatological preparations, *Adv. Drug Delivery Rev.*, **54**, pp. 131–155.
7. Pardeike, J., Hommoss, A., and Müller, R. H. (2009). Lipid nanoparticles (SLN, NLC) in cosmetic and pharmaceutical dermal products, *Int. J. Pharm.*, **366**, pp. 170–184.
8. Muchow, M., Maincent, P., and Muller, R. H. (2008). Lipid nanoparticles with a solid matrix (SLN, NLC, LDC) for oral drug delivery, *Drug Dev. Ind. Pharm.*, **34**, pp. 1394–1405.
9. Sanli, D., Bozbag, S. E., and Erkey, C. (2011). Synthesis of nanostructured materials using supercritical CO_2: part I. Physical transformations, *J. Mater. Sci.*, **47**, pp. 2995–3025.
10. Elizondo, E., Veciana, J., and Ventosa, N. (2012). Nanostructuring molecular materials as particles and vesicles for drug delivery, using compressed and supercritical fluids, *Nanomedicine*, **7**, pp. 1391–1408.
11. Weidner, E., Knez, Z., and Novak, Z. A. (1994). Process and equipment for production and fractionation of fine particles from gas saturated solutions, World Patent WO 95/21688.
12. Foster, N., Mammucari, R., and Dehghani, F. (2003). Processing pharmaceutical compounds using dense gas technology, *Ind. Eng. Chem. Res.*, **42**, pp. 6476–6493.
13. Knez, Ž. (2003). High pressure micronisation of polymers, *Proc. Sixth Int. Symp. Supercrit. Fluids*, pp. 1865–1870.
14. Brunner, G. (2010). Applications of supercritical fluids, *Annu. Rev. Chem. Biomol. Eng.*, **1**, pp. 321–342.
15. Weidner, E. (2009). High pressure micronization for food applications, *J. Supercrit. Fluids*, **47**, pp. 556–565.
16. Jung, J., and Perrut, M. (2001). Particle design using supercritical fluids: literature and patent survey, *J. Supercrit. Fluids*, **20**, pp. 179–219.
17. Tabernero, A., Martín del Valle, E. M., and Galán, M. (2012). Supercritical fluids for pharmaceutical particle engineering: methods, basic fundamentals and modelling, *Chem. Eng. Process Intensif.*, **60**, pp. 9–25.

18. Yeo, S. D., and Kiran, E. (2005). Formation of polymer particles with supercritical fluids: a review, *J. Supercrit. Fluids*, **34**, pp. 287–308.
19. Knez, Ž., Markočič, E., Novak, Z., and Hrnčič, M. K. (2011). Processing polymeric biomaterials using supercritical CO_2, *Chem. Ing. Tech.*, **83**, pp. 1371–1380.
20. Pollak, S., Kareth, S., Kilzer, A., and Petermann, M. (2011). Thermal analysis of the droplet solidification in the PGSS-process, *J. Supercrit. Fluids*, **56**, pp. 299–303.
21. Li, J., Matos, H. A., and de Azevedo, E. G. (2004). Two-phase homogeneous model for particle formation from gas-saturated solution processes, *J. Supercrit. Fluids*, **32**, pp. 275–286.
22. Strumendo, M., Bertucco, A., and Elvassore, N. (2007). Modeling of particle formation processes using gas saturated solution atomization, *J. Supercrit. Fluids*, **41**, pp. 115–125.
23. Kappler, P., Leiner, W., Petermann, M., and Weidner, E. (2003). Size and morphology of particles generated by spraying polymer-melts with carbon dioxide, *6th Int. Symp. Supercrit. Fluids*, PMd34.
24. Elvassore, N., Flaibani, M., Bertucco, A., and Caliceti, P. (2003). Thermodynamic analysis of micronization processes from gas-saturated solution, *Ind. Eng. Chem. Res.*, **42**, pp. 5924–5930.
25. Nunes, A. V. M., and Duarte, C. M. M. (2011). Dense CO_2 as a solute, co-solute or co-solvent in particle formation processes: a review, *Materials*, **4**, pp. 2017–2041.
26. Pasquali, I., and Bettini, R. (2008). Are pharmaceutics really going supercritical?, *Int. J. Pharm.*, **364**, pp. 176–187.
27. Knez, Z., and Weidner, E. (2003). Particles formation and particle design using supercritical fluids, *Curr. Opin. Solid. State Mater. Sci.*, **7**, pp. 353–361.
28. Knez, Ž., Škerget, M., and Mandžuka, Z. (2010). Determination of S–L phase transitions under gas pressure, *J. Supercrit. Fluids*, **55**, pp. 648–652.
29. Weidner, E., Steiner, R., and Knez, Ž. (1996). Powder generation from polyethyleneglycols with compressible fluids, *High Press. Chem. Eng.*, **12**, pp. 223–228.
30. Martín, A., and Cocero, M. J. (2008). Micronization processes with supercritical fluids: fundamentals and mechanisms, *Adv. Drug Delivery Rev.*, **60**, pp. 339–350.
31. Fages, J., Lochard, H., Letourneau, J. J., Sauceau, M., and Rodier, E. (2004). Particle generation for pharmaceutical applications using supercritical fluid technology, *Powder Technol.*, **141**, pp. 219–226.

32. Baxendale, A., van Hooff, P., Durrant, L. G., Spendlove, I., Howdle, S. M., Woods, H. M., Whitaker, M. J., Davies, O. R., Naylor, A., Lewis, A. L., and Illum, L. (2011). Single shot tetanus vaccine manufactured by a supercritical fluid encapsulation technology, *Int. J. Pharm.*, **413**, pp. 147–154.

33. Weidner, E., Petermann, M., and Knez, Z. (2003). Multifunctional composites by high-pressure spray processes, *Curr. Opin. Solid State Mater. Sci.*, **7**, pp. 385–390.

34. Elvassore, N., Flaibani, M., Vezzù, K., Bertucco, A., Caliceti, P., Semenzato, A., and Salmaso, S. (2003). Lipid system micronization for pharmaceutical applications by PGSS techniques, *6th Int. Symp. Supercrit. Fluids*, PMd27.

35. São Pedro, A., Dalla Villa, S., Caliceti, P., Salmaso, S., Albuquerque, E., and Bertucco, A. (2013). Curcumin-loaded lipid particles produced by supercritical fluid technology, *6th Int. Symp. High Press. Process Technol.*, Belgrade, Serbia, pp. 1–7.

36. São Pedro, A., Dalla Villa, S., Caliceti, P., Salmaso, S., Elvassore, N., Serena, E., Fialho, R., Vieira de Melo, S., Bertucco, A., and Cabral-Albuquerque, E. (2013). Solid lipid nanoparticles entrapping curcumin by supercritical fluid technology, *XXI Int. Conf. Bioencapsulation*, Berlin, Germany, pp. 252–253.

37. Sampaio de Sousa, A. R, Silva, R., Tay, F. H., Simplício, A. L., Kazarian, S. G., and Duarte, C. M. M. (2009). Solubility enhancement of trans-chalcone using lipid carriers and supercritical CO_2 processing, *J. Supercrit. Fluids*, **48**, pp. 120–125.

38. Duarte, C. M. M., de Sousa, A. R. S., Kazarian, S. G., and Luísa, A. (2007). Functional particles prepared by PGSS®, *11th Eur. Meet. Supercrit. Fluids*, OC_M_22.

39. Semenzato, A., Amabile, G., Vezzù, K., Caliceti, P., and Bertucco, A. (2006). Compressed fluid based process for development of cosmetic products. *AIChe Annu. Meet.*

40. García-González, C. A., Sampaio da Sousa, A. R., Argemí, A., López Periago, A., Saurina, J., Duarte, C. M. M., and Domingo, C. (2009). Production of hybrid lipid-based particles loaded with inorganic nanoparticles and active compounds for prolonged topical release, *Int. J. Pharm.*, **382**, pp. 296–304.

41. García-González, C. A., Argemí, A., Sampaio de Sousa, A. R., Duarte, C. M. M., Saurina, J., and Domingo, C. (2010). Encapsulation efficiency of solid lipid hybrid particles prepared using the PGSS® technique and loaded with different polarity active agents, *J. Supercrit. Fluids*, **54**, pp. 342–347.

42. Argemí, A., Domingo, C., Sampaio de Sousa, A. R., Duarte, C. M. M., García-González, C. A., and Saurina, J. (2011). Characterization of new topical ketoprofen formulations prepared by drug entrapment in solid lipid matrices, *J. Pharm. Sci.*, **100**, pp. 4783–4789.

43. Vezzù, K., Campolmi, C., and Bertucco, A. (2009). Production of lipid microparticles magnetically active by a supercritical fluid-based process, *Int. J. Chem. Eng.*, **2009**, pp. 1–9.

44. Gonçalves, V. S. S., Matias, A. A., Nogueira, I. D., and Duarte, C. M. M. (2013). Production of mucoadhesive hybrid lipid-based drug delivery systems containing glyceryl monooleate through supercritical fluid precipitation technology, *6th Int. Symp. High Press. Process Technol.*, pp. 1–8.

45. Salmaso, S., Elvassore, N., Bertucco, A., and Caliceti, P., (2009). Production of solid lipid submicron particles for protein delivery using a novel supercritical gas-assisted melting atomization process, *J. Pharm. Sci.*, **98**, pp. 640–650.

46. Salmaso, S., Bersani, S., Elvassore, N., Bertucco, A., and Caliceti, P. (2009). Biopharmaceutical characterisation of insulin and recombinant human growth hormone loaded lipid submicron particles produced by supercritical gas micro-atomisation, *Int. J. Pharm.*, **379**, pp. 51–58.

47. Vezzù, K., Borin, D., Bertucco, A., Bersani, S., Salmaso, S., and Caliceti, P. (2010). Production of lipid microparticles containing bioactive molecules functionalized with PEG, *J. Supercrit. Fluids*, **54**, pp. 328–334.

48. Nunes, A. V., Rodriguez-Rojo, S., Almeida, A. P., Matias A. A., Rego, D., Simplicio, A. L, Bronze, M. R., Cocero, M. J., and Duarte, C. M. M. (2010). Supercritical fluids strategies to produce hybrid structures for drug delivery, *J. Controlled Release*, **148,** pp. e11–e12.

49. Rodriguez-Rojo, S., Rego, D., Nunes, A. V. M., Nogueira, I. D., and Cocero, M. J. (2010). Supercritical Fluids (SCF) strategies to produce double- walled particles for drug delivery applications, *12th Eur. Meet. Supercrit. Fluids*, Graz, Austria, p. 75.

50. Almeida, A. P. C, Nogueira, I. D., Matias, A. A., and Duarte, C. M. M. (2011). Development of novel hybrid delivery systems using SCF technology for enhanced bioavailability of antioxidant compounds, *1st Iber. Meet. Nat. Bioact. Entrapment Food Ind.*, p. 23.

51. Vijayaraghavan, M., Stolnik, S., Howdle, S. M., and Illum, L. (2013). Suitability of polymer materials for production of pulmonary microparticles using a PGSS supercritical fluid technique: preparation of microparticles using PEG, fatty acids and physical or chemicals blends of PEG and fatty acids, *Int. J. Pharm.*, **441**, pp. 580–588.

52. Fraile, M., Martín, Á., Deodato, D., Rodriguez-Rojo, S., Nogueira, I. D., Simplício, A. L., Cocero, M. J., and Duarte, C. M. M. (2013). Production of new hybrid systems for drug delivery by PGSS (particles from gas saturated solutions) process, *J. Supercrit. Fluids*, **81**, pp. 226–235.

53. Brion, M., Jaspart, S., Perrone, L., Piel, G., and Evrard, B. (2009). The supercritical micronization of solid dispersions by particles from gas saturated solutions using experimental design, *J. Supercrit. Fluids*, **51**, pp. 50–56.

54. Jordan, F., Naylor, A., Kelly, C. A., Howdle, S. M., Lewis, A., and Illum, L. (2010). Sustained release hGH microsphere formulation produced by a novel supercritical fluid technology: in vivo studies, *J. Controlled Release*, **141**, pp. 153–160.

55. Casettari, L., Castagnino, E., Stolnik, S., Lewis, A., Howdle, S. M., and Illum, L. (2011). Surface characterisation of bioadhesive PLGA/chitosan microparticles produced by supercritical fluid technology, *Pharm. Res.*, **28**, pp. 1668–1682.

56. Kelly, C. A., Naylor, A., Illum, L., Shakesheff, K. M., and Howdle, S. M. (2012). Supercritical CO_2: a clean and low temperature approach to blending PDLLA and PEG, *Adv. Funct. Mater.*, **22**, pp. 1684–1691.

57. Pollak, S., Petermann, M., Kareth, S., and Kilzer, A. (2010). Manufacturing of pulverised nanocomposites: dosing and dispersion of additives by the use of supercritical carbon dioxide, *J. Supercrit. Fluids*, **53**, pp. 137–141.

58. Vezzù, K., Bertucco, A., Lucien, F. P. (2008). Solid-liquid equilibrium of binary and ternary mixtures of lipids under higher pressure of CO_2 for PGSS process development, *11th Eur. Meet. Supercrit. Fluids*, Barcelona, pp. 1–6.

59. Vijayaraghavan, M., Stolnik, S., Howdle, S. M., and Illum, L. (2012). Suitability of polymer materials for production of pulmonary microparticles using a PGSS supercritical fluid technique: thermodynamic behaviour of fatty acids, PEGs and PEG-fatty acids, *Int. J. Pharm.*, **438**, pp. 225–231.

60. Hao, J. J., Whitaker, M., Serhatkulu, G. M., Shakesheff, K., and Howdle, S. M. (2005). Supercritical fluid assisted melting of poly(ethylene glycol): a new solvent-free route to microparticles, *J. Mater. Chem.*, **15**, pp. 1148–1153.

61. Perrut, M., Deschamps, F., Jung, J., Leboeuf, F., and France, S. F. C. (2005). SCF Particle generation in compliance with GMP, *10th Eur. Meet. Supercrit. Fluids*, Mo17.

62. Domingo, C., and Saurina, J. (2012). An overview of the analytical characterization of nanostructured drug delivery systems: towards

green and sustainable pharmaceuticals; a review, *Anal. Chim. Acta*, **744**, pp. 8–22.

63. Sitterberg, J., Ozcetin, A., Ehrhardt, C., and Bakowsky, U. (2010). Utilising atomic force microscopy for the characterisation of nanoscale drug delivery systems, *Eur. J. Pharm. Biopharm.*, **74**, pp. 2–13.
64. Mathaes, R., Winter, G., Engert, J., and Besheer, A. (2013). Application of different analytical methods for the characterization of non-spherical micro- and nanoparticles, *Int. J. Pharm.*, **453**, pp. 620–629.
65. Heng, D., Tang, P., Cairney, J. M., Chan, H. K., Cutler, D. J., Salama, R., and Yun, J. (2007). Focused-ion-beam milling: a novel approach to probing the interior of particles used for inhalation aerosols, *Pharm. Res.*, **24**, pp. 1608–1617.
66. Troutier, A. L., Delair, T., Pichot, C., and Ladavière, C. (2005). Physicochemical and interfacial investigation of lipid/polymer particle assemblies, *Langmuir*, **21**, pp. 1305–1313.
67. Parhi, R., and Suresh, P. (2012). Preparation and characterization of solid lipid nanoparticles: a review, *Curr. Drug Discovery Technol.*, **9**, pp. 2–16.
68. Cho, E. J., Holback, H., Liu, K. C., Abouelmagd, S. A., Park, J., and Yeo, Y. (2013). Nanoparticle characterization: state of the art, challenges, and emerging technologies, *Mol. Pharm.*, **10**, pp. 2093–2110.
69. Saveyn, H., de Baets, B., Thas, O., Hole, P., Smith, J., and van der Meeren, P. (2010). Accurate particle size distribution determination by nanoparticle tracking analysis based on 2-D Brownian dynamics simulation, *J. Col. Interface Sci.*, **352**, pp. 593–600.
70. Bunjes, H., Westesen, K., and Koch, M. H. J. (1996). Crystallization tendency and polymorphic transitions in triglyceride nanoparticles. *Int. J. Pharm.*, **129**, pp. 159–173.
71. Bunjes H, Unruh T (2007). Characterization of lipid nanoparticles by differential scanning calorimetry, X-ray and neutron scattering, *Adv. Drug Delivery Rev.*, **59**, pp. 379–402.
72. Tandon, P., Förster, G., Neubert, R., and Wartewig, S. (2000). Phase transitions in oleic acid as studied by X-ray diffraction and FT-Raman spectroscopy, *J. Mol. Struct.*, **524**, pp. 201–215.

Chapter 8

Preparation of Water-Soluble Formulations of Hydrophobic Active Compounds by Emulsion Template Processes

Ángel Martín,[a] Esther de Paz,[a] Facundo Mattea,[b] and María José Cocero[a]

[a]*Department of Chemical Engineering and Environmental Technology, University of Valladolid, Spain*

[b]*Faculty of Chemistry, National University of Córdoba, CONICET, Argentina*

mjcocero@iq.uva.es

The use of many active compounds of interest for nutraceutical and pharmaceutical applications is restricted by low solubility of these compounds in water, which contributes to low bioavailability, and makes it difficult to add these compounds to beverages or other water-based products. Thus, there is great interest in the development of water-soluble formulations of these compounds such as micellar solutions or other colloidal dispersions. Emulsion techniques have been particularly useful for the production of these formulations, since the emulsion can provide a template for

Supercritical Fluid Nanotechnology: Advances and Applications in Composites and Hybrid Nanomaterials
Edited by Concepción Domingo and Pascale Subra-Paternault

ISBN 978-981-4613-40-8 (Hardcover), 978-981-4613-41-5 (eBook)
www.panstanford.com

the morphology of the final product. This chapter presents the main techniques used for this purpose, starting with conventional emulsion evaporation and solvent displacement methods, and presenting recent developments, including novel emulsification techniques such as ultrasound emulsification and high-pressure homogenization, in the intensification of the process by increase in throughput and reduction of processing time through the processing of pressurized emulsions, and the application of supercritical fluids for extraction and antisolvent precipitation from emulsions. Finally, the application of several of these techniques for the preparation of water-soluble β-carotene formulations for application as natural colorant is presented as a case study.

8.1 Water-Soluble Formulations of Hydrophobic Active Compounds

Many natural active compounds are characterized by high hydrophobicity and, consequently, low solubility in water. This property can be an important limitation for the development of applications based on these compounds. In pharmaceutical applications, low solubility in water is generally associated with a low bioavailability of the compound, which can reduce its therapeutical activity and makes it difficult to dose the compound. On the other hand, many food products have an aqueous base in which hydrophobic active compounds cannot be dissolved. Therefore, there is a great interest in the development of formulations that can enhance the stability and solubility in water of hydrophobic active compounds of interest for food and pharmaceutical applications. Different methods for the development of such formulations are based on emulsions. Emulsions can provide a template for the final morphology of the product and enable an enhanced control over its size by controlling the size of the disperse phase droplets in the emulsion. Emulsion-template techniques are suitable for the production of different types of formulations (Fig. 8.1), including:

- Simple oil-in-water (o/w) emulsions, which can be suitable final formulations without further processing when the base product already presents a certain fat content (e.g., butter or milk)

- Micro- or nanosuspensions of a pure compound, which can be obtained with a controlled particle size distribution by elimination of the organic phase from an o/w emulsion with controlled droplet size
- Micellar solutions, formed by association of surfactant molecules in aqueous environment that create an inner hydrophobic cavity that can host water-insoluble compounds
- Liposomal solutions, formed by self-assembly of phospholipids in water, which can encapsulate both hydrophilic compounds in its inner aqueous cavity, as well as hydrophobic compounds in the lipid bilayer
- Hybrid systems, formed by a combination of hydrophilic and hydrophobic carrier materials, with the objective of stabilizing hydrophobic compounds in aqueous environments at the same time that a controlled delivery of these compounds is achieved

This chapter presents the main emulsion techniques used to produce these types of formulations, starting with a brief presentation of the main conventional techniques and focusing on recently developed processes based on high-pressure and supercritical fluids, presenting the results obtained in the formulation of β-carotene as a case study.

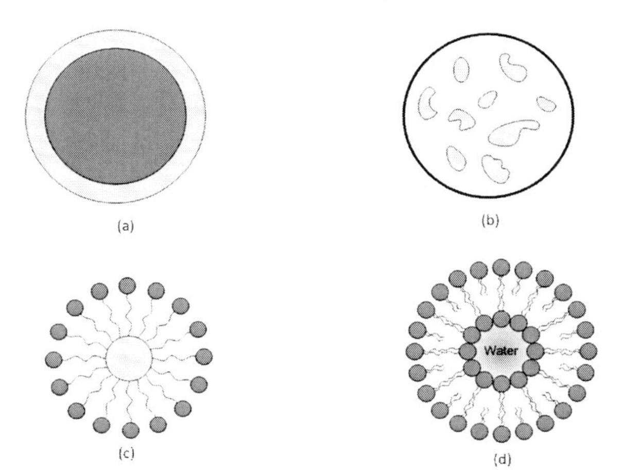

Figure 8.1 Different types of water-soluble formulations of hydrophobic compounds: (a) microcapsules, (b) microcomposites, (c) micelles, and (d) liposomes.

8.2 Emulsion Evaporation and Solvent Displacement Methods

Emulsion evaporation processes have been widely used to produce micro-/nanoparticles and composites of controlled size distribution. Figure 8.2 presents a schematic diagram of this process. It consists of forming an o/w emulsion, in which the active compound and carrier materials usually are dissolved in the disperse phase of the emulsion. Afterward, the organic solvent or oil is removed from the disperse phase by evaporation, maintaining agitation in order to ensure the preservation of the emulsion structure during the process. As a result, an aqueous suspension of particles is obtained.

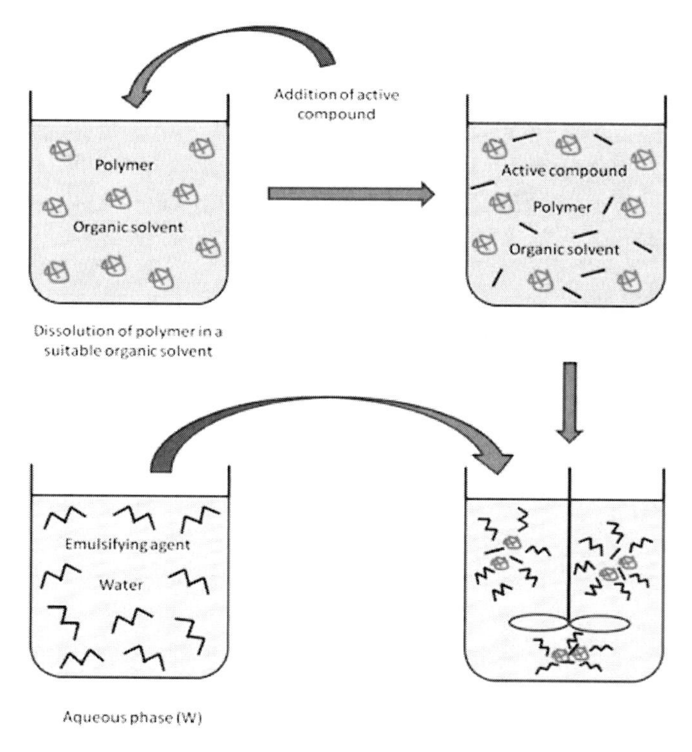

Figure 8.2 Schematic diagram of the emulsion evaporation process.

One of the main applications of the emulsion evaporation process is the production of polymer micro-/nanocapsules, using biodegradable polymers such as poly(lactic acid) (PLA) and poly(lactic-co-glycolic acid) (PLGA) [1]. Different process parameters, such as emulsification conditions, the organic–water ratio, solvents and carriers used, and the solvent evaporation rate, have an influence on the characteristics of the particles produced.

Firstly, any parameter that modifies the droplet size in the emulsion has a direct influence on the particle size, as a reduction of the initial emulsion droplet size leads to a reduction of the final particle size. Thus, increasing the energy provided for emulsification, for example, increasing the stirring rate, generally allows reducing the emulsion droplet size and therefore the final particle size. An increase in the surfactant concentration that allows stabilizing the emulsion and reducing the droplet size also allows reducing the final particle size. An increase of the organic–water ratio usually leads to an increase of particle size due to increased droplet coalescence in the emulsion. The concentration of the carrier and active compounds in the emulsion droplets can also have a direct impact on the emulsification and particle size through the variation of the viscosity. If the disperse phase is made more viscous, for example, increasing the concentration of polymer carrier in this phase, the emulsification is hindered and larger particle sizes are obtained [2]. In addition to emulsion parameters, the concentration of solute in the organic phase also has a direct influence on particle size, with larger particle sizes when the concentration is increased. The rate of removal of the organic solvent from the disperse phase has a strong influence on the morphology of the final particles and on the efficiency of encapsulation of the active compound in the polymer carrier. If solvent removal is slow, a considerable drop in the encapsulation efficiency can occur due to the partitioning of the active compound between the organic and aqueous phases. The crystallinity of particles can also change depending on the rate of elimination of the solvent. If this rate is too slow, the precipitation of polymer is also slow, and isolated crystalline particles of polymer and active compound can be obtained, which are detrimental for the efficiency of encapsulation. Accelerating solvent removal and therefore polymer precipitation can enable producing amorphous microcapsules of polymer, which are a more favorable morphology for a high efficiency of encapsulation. To remove the solvent, it must become solubilized in the aqueous phase at the droplet/water interface, then it must diffuse through water, and finally it must evaporate at the water/gas interface. Therefore, the rate of solvent removal firstly depends on the physical properties of the solvent itself, and particularly on the balance of partition coefficients between the organic/aqueous phase and between the aqueous/gas phase, as well as on the diffusion coefficient of the organic solvent in water [3]. The removal rate also depends on the method of solvent removal. Comparing the removal of solvent by evaporation at ambient pressure and at reduced pressure, the faster removal of solvent achieved at reduced pressures facilitates the production of amorphous polymer capsules, thus increasing the encapsulation efficiency [4].

Another commonly used technique for the production of micro-/nanoparticles from emulsions is the solvent displacement method. A schematic diagram of this process is presented in Fig. 8.3. The process is a combination of emulsion-template and antisolvent precipitation processes that relies on the use of a partially water-miscible organic solvent such as ethyl acetate, ethyl formate, or propylene carbonate, in which the active compound and carrier materials are dissolved.

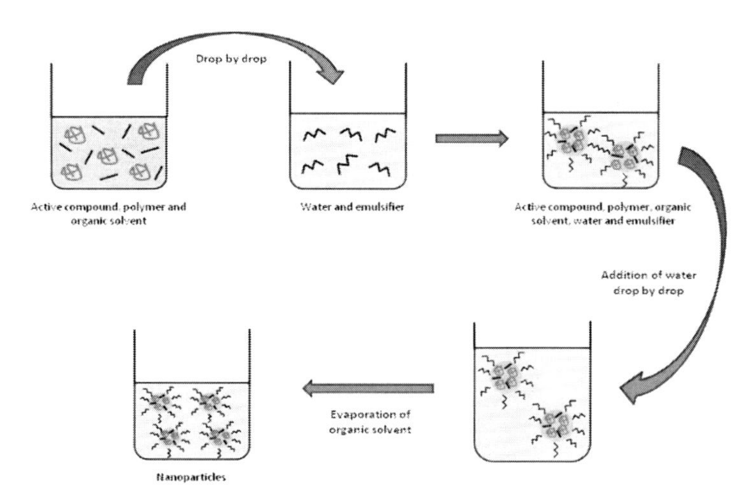

Figure 8.3 Schematic diagram of the solvent displacement process.

Upon mixing of this solution with water, a transient emulsion is formed, which serves as template for the production of particles. Once that a sufficiently high proportion of water is added, complete dissolution of the organic solvent in the aqueous phase is achieved and precipitation occurs. Finally the organic solvent is removed by vacuum evaporation, and an aqueous suspension of particles is obtained, stabilized by the surfactants added to form the emulsion [5, 6]. In addition to the parameters that influence the size of the initial emulsion (including organic–water ratio, surfactant type, and composition and emulsification method), the rate of addition of water is another key parameter, since it controls solvent displacement and antisolvent precipitation mechanisms [6].

The polymer carrier can be formed by an in situ polymerization reaction, instead of using a preformed polymer. The starting point for this process is an emulsion of monomer droplets stabilized by a surfactant. Afterward, the polymerization is started by adding an initiator in the continuous phase [7] or by alternative methods such as application of ultrasounds [8], UV light [9], or enzymes [10]. Depending on the type of emulsification method and polymer used, the need for an additional organic solvent for the disperse phase of the emulsion can be eliminated, thus allowing one to directly produce the polymeric particles without any additional solvent removal step [11]. However, the possible degradation of the active compound due to undesired interactions with initiation agents must be considered [7].

Besides simple microparticle or microcomposite formulations, the solvent evaporation and solvent displacement methods can be used to develop more complex formulations [7]. An example is the production of solid lipid micro-/nanoparticles, constituted by an active compound encapsulated in a lipid matrix that is solid at physiological temperatures [12]. This type of formulation is particularly interesting for the formulation of hydrophobic active compounds, since the particles can be stable for long periods, and the lipid matrix can form liposomal structures in aqueous environments, which can facilitate the delivery of hydrophobic compounds through cellular membranes or gastrointestinal barriers [13]. However, if the active compound shows low solubility in the lipid, the formulations can become destabilized by expelling crystalline particles of the active compound after a polymorphic transition [14]. The production of lipid–drug conjugates [15] or of hybrid hydrophilic-hydrophobic carrier systems can contribute to solve these problems. Lipid particles can generally be obtained by a solvent displacement method, or by a simple cooling method consisting of melting the lipid phase, mixing and emulsifying with an aqueous surfactant solution, and finally solidifying the lipid by cooling the emulsion down to room temperature [7]. Solvent evaporation and solvent displacement methods can also be used to generate microcapsules, consisting of an outer polymeric shell that confines a reservoir of an aqueous or an oil solution of the active compound [7, 16]. Such microcapsules can be generated both by solidification of a preformed polymer, or by an in situ polymerization reaction. The dissolution of the active compound in the inner solvent reservoir can contribute to facilitate its delivery, while the outer shell can provide protection against degradation processes and avoid burst release effects.

Finally, it must be noted that emulsion evaporation and solvent displacement methods based on simple o/w emulsions in general can only be used for the formulation of hydrophobic compounds. In the case of hydrophilic compounds, in an o/w emulsion a rapid partitioning of the compound to the aqueous phase would be observed, resulting in a very low encapsulation efficiency and a loss of control over particle size and morphology through the emulsion template. To circumvent this problem, oil-in-oil emulsions can be used [17], but this approach is limited by the relatively low stability of most such emulsions compared to o/w systems. Alternatively, double emulsions of the type water-in-oil-in-water (w/o/w) can be used, in which the active compound can be dissolved in the inner aqueous phase [18], or other complex multiple emulsion systems (w/o/o, w/o/o/o, . . .) can be considered [19].

8.3 Novel Emulsification Techniques

While macroemulsions can be easily formed by stirring a mixture of water, oil, and surfactant in appropriate proportions, more complex emulsification processes are required in order to prepare emulsions with restricted droplet sizes suitable for application as templates for production of micro- and nanoparticles according to the methods described in the previous section. From a thermodynamic point of view, the emulsification process must provide the energy required to form and stabilize the interface between emulsion droplets and the continuous phase. Depending on the method used to supply this energy, emulsification processes are usually classified as high energy or low energy methods [7]. High-energy methods are the most frequently used emulsification techniques both in industry and academia. Among them, rotor/stator devices are particularly widespread. These devices simply rely on creating a high velocity flow through orifices of restricted size, which promote droplet breakup, thus forming the emulsion. Commercial equipment at a wide range of scales are provided by different companies, including, for example, Ultra-Turrax® and Omny-mixer® apparatuses. The main drawback of these techniques is a relatively poor control over the dispersion of droplet sizes, as well as a very low efficiency in the use of energy, which is mostly dissipated by viscous friction [20]. Moreover, the considerable amount of heat produced by friction poses a problem when volatile solvents are used to prepare the emulsion, making it necessary to use complex refrigeration systems in order to avoid the loss of these solvents by evaporation within the emulsification equipment. High-pressure homogenizers can be considered as an evolution of rotor-stator devices. They are based on the application of high pressures to the fluid in order to force it to flow through microchannels, achieving velocities that can be of the order of 300 m/s [7]. This flow creates a combination of shear, impact, and cavitation forces that promote emulsion formation. Application of ultrasounds is another high-energy emulsification method that is very popular at the laboratory scale due to its simplicity, efficiency, and low cost, while industrial implementation still is limited. Although the mechanisms of emulsion formation by application of ultrasounds still are not entirely understood, it is thought that a basic aspect is the formation and cavitation of gas bubbles [21]. A limitation of ultrasound equipment in the large scale is the relatively small range of action of ultrasound probes, which makes it necessary to combine them with stirring devices in order to achieve a good homogeneity of the product.

Among the low energy methods, the solvent displacement method has been long used for the production of micro- and nanoemulsions [7]. It consists of adding a water-soluble solvent to the oil phase of the emulsion. Upon mixing with water, this solvent rapidly diffuses into the aqueous phase, promoting the formation of an emulsion under specific conditions [22, 23]. The mechanism of emulsion formation is the generation of interfacial turbulence related to surface tension gradients due to the diffusion of solutes between the two phases, as well as an spontaneous interfacial expansion process due to the generation of regions of local supersaturation of the surfactant associated with concentration gradients of solvents [24]. The phase inversion temperature (PIT) method is another low-energy emulsification procedure, which is based on the variation of the partition coefficients of polyethoxylated surfactants between aqueous and organic phases as a function of temperature [25, 26]. This allows producing a emulsion phase inversion as the affinity of the surfactant between phases is varied as a result of a gradual temperature variation, for example, from an o/w to a water-in-oil (w/o) emulsion. More interestingly, within the transition region, where the surfactant shows similar affinity for the two solvents, extremely low interfacial tensions are observed, thus allowing one to create micro- and nanoemulsions. Typical methods to implement this procedure are a generation of a phase transition by rapid cooling [27] or by a sudden dilution in water or oil [28]. It must be noted that besides being a low-energy method, the PIT procedure is organic solvent free, a highly interesting feature for food and pharmaceutical applications.

8.4 Process Intensification by Precipitation from Pressurized Emulsions

Although the methods described in the previous sections have been highly successful for the development of formulations of hydrophobic compounds, they still present some limitations. Of particular relevance for industrial applications are long processing times required by most emulsion template techniques, frequently also involving the application of high temperatures, the difficulty of implementing a continuous process, and the intensive use of organic solvents, which result in low productivity as well as in possibilities of product contamination or degradation along the process. Therefore there is a considerable interest for the development of process intensification strategies that allow reducing processing times and limiting the use of high temperatures and organic solvents. The basic idea behind the high-pressure emulsion techniques is to accelerate the kinetics of mixing and heat transfer processes, approaching them to the kinetics of the particle formation processes. With this, a higher homogeneity of the conditions in which particles are formed can be achieved, which results in a better control of the properties of the final product. Moreover, with the intensification of the mixing and heat transfer processes a reduction of the particle size of the active compound may be achieved, entering into the nanometer range (<100 nm) [29].

Figure 8.4 presents a flow diagram of the high-pressure emulsion process. The process operates with an organic solvent at high temperature (typically, 80°C–150°C) in order to increase the solubility of the active compound into the organic solvent and make the process suitable for active compounds with low solubility (or, optionally, to replace the organic solvent by other that may have more desirable properties, for example, toxicity, but cannot be normally used due to low solubility of the active compound). Moderate pressures (in the order of 5 MPa) are applied in order to keep the solvent in the liquid state. Since many active compounds are thermolabile, heating of the organic solution is achieved by direct contact between a suspension of particles of the active compound in cold organic solvent with another stream of preheated organic solvent. Immediately afterward, the hot organic solution is mixed with a third stream of cold water, which contains the surfactant and carrier materials, as well as other possible additives. By using a suitable mixer, an o/w emulsion is formed in this step, providing the template for the formation of particles. Furthermore, the emulsion is cooled down, reducing the exposure of the product to high temperatures and the possible degradation of the active compound. Finally, particles of the active compound are formed in this step by a thermal and antisolvent effect (drastic reduction of solubility by reduction of temperature and mixing with water antisolvent) rather than by a solvent evaporation method, thus providing the aforementioned intensification of the particle formation process. With this, a suspension of active compound particles in the o/w medium is obtained, which must be further processed in order to remove the organic solvent and, optionally, water, thus precipitating the surfactant and carrier materials and forming dry microcomposites. It must be mentioned that although some active compound particles are formed during these last solvent evaporation processing steps, as in a normal solvent evaporation method, the vast majority of the product precipitates before as a consequence of the thermal effect, due to the drastic increase of solubility that can be achieved with the increased temperature, which may be of several orders of magnitude.

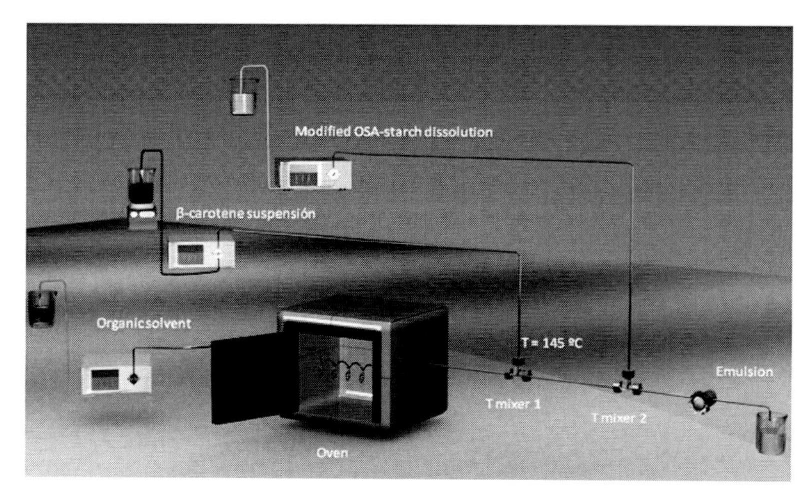

Figure 8.4 Schematic diagram of the process for precipitation from pressurized emulsions using a T-mixer.

8.5 Supercritical Fluid Processing of Emulsions: Supercritical Extraction of Emulsions and Antisolvent Precipitation from an Emulsion

As described in previous chapters of this book, processes based on supercritical carbon dioxide offer considerable advantages for the processing of natural and pharmaceutical compounds, due to the tunable properties of carbon dioxide and the possibilities to operate at mild temperatures and in an inert environment. A wide range of particle formation and formulation processes have been developed in which carbon dioxide acts as solvent, antisolvent, solute, or propellant, among other functions [30].

A new promising technology to produce nanometer particles of natural and pharmaceutical substances is the use of supercritical fluids in combination with nanoemulsions, which presents advantages over these two separated technologies. This process, denominated supercritical fluid extraction of emulsions (SFEE) [31] consists of extracting the organic solvent from the droplets of an o/w emulsion using supercritical carbon dioxide. Figure 8.5 presents a schematic diagram of the process concept. In a first approach, the technique bears resemblance with the conventional solvent evaporation technique, substituting the removal of the solvent through evaporation by an extraction with carbon dioxide. The use of supercritical carbon dioxide can contribute to eliminate some of the limitations of the solvent evaporation technique, as it allows avoiding the exposition of the product to high temperatures during prolonged times. Moreover, since organic solvents are highly soluble or even completely miscible in supercritical fluids, by supercritical extraction the elimination of residual concentrations of organic solvent in the product can be facilitated. Depending on the type of carrier materials used, different formulations can be obtained ranging from micro-/nanoparticles of pure compounds, nanocapsules with polymeric carrier materials, or micellar and liposomal solutions in water. Furthermore, as it will be discussed below, a fully continuous implementation of this process is possible.

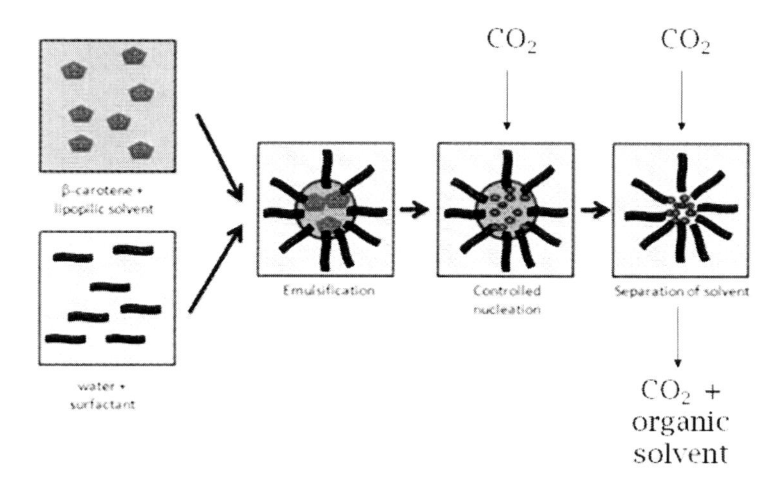

Figure 8.5 Supercritical fluid extraction of emulsions: process concept.

Figure 8.6 presents the evolution of an organic droplet of solvent in water during exposition to CO_2 in the SFEE process [32]. As the emulsion is put into contact with CO_2, a two-sided diffusion process takes place: the diffusion of CO_2 from the gas phase through water into the droplet, and the diffusion of organic solvent in the opposite direction. Due to the favorable transport properties of CO_2, its diffusion is faster than that of the organic solvent, and therefore droplet swelling is observed in the first stages of the process. If the extraction is performed at pressures below the mixture critical point, droplet swelling continues until the concentration of CO_2 approaches the saturation concentration at the operating pressure. As the diffusion of CO_2 is slowed down due to the reduction of the concentration gradient driving force, the diffusion of solvent out of the droplet eventually becomes predominant and droplet shrinking is observed. At pressures above the mixture critical point, where CO_2 and organic solvent are completely miscible, droplet swelling continues until the droplet is destabilized out of the emulsion due to the increase of concentration of CO_2.

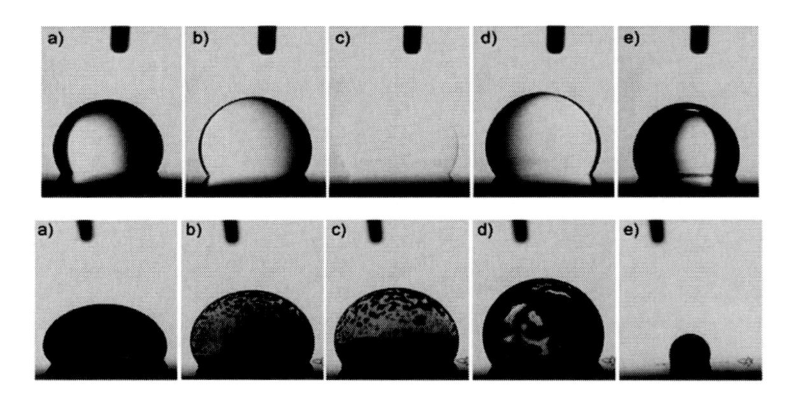

Figure 8.6 Evolution of an organic solvent drop during the supercritical extraction of emulsions. Adapted from Ref. [32].

As it can also be observed in Fig. 8.6, if a solute is dissolved in the organic solvent droplets, its precipitation occurs in the first moments of exposition to CO_2, due to the dissolution of CO_2 into the organic solvent droplets. Indeed, each emulsion droplet behaves as a miniature gas antisolvent precipitator. Since particle formation is confined inside the droplets, particle size is restricted by emulsion droplet size, thus enabling to achieve a better control over particle size distribution than with conventional gas antisolvent processes. Furthermore, as in other emulsion template techniques, the particles obtained can be stabilized by addition of other compounds to the emulsion, including for example, the encapsulation in a polymeric carrier material dissolved in the organic phase, or the stabilization of the particles in the aqueous environment without sedimentation or agglomeration by addition of a surfactant to the aqueous phase. Compared with conventional emulsion evaporation techniques, particle formation occurs quickly and simultaneously in the first moments of the process, thus enabling to obtain a product with more controlled and narrower particle size distribution.

Although particle formation occurs quickly in the first fractions of second of the process by the fast saturation of emulsion droplets with CO_2, longer processing times are required for a complete removal of organic solvent from the product, due to the slower diffusion of organic solvent out of the emulsion through the aqueous phase. This effect is illustrated in Fig. 8.7 that shows modeling results regarding the evolution of system composition during the supercritical extraction of an emulsion sprayed into CO_2 [33]. As shown in this figure, high CO_2 concentrations in the organic disperse phase, sufficient for an antisolvent precipitation of active compounds, are achieved in fractions of second. However, longer processing times, in the range of minutes, are required to eliminate the organic solvent from the liquid phase. The extraction time required to remove the organic solvent from the emulsion strongly depends on the miscibility of the organic solvent in the aqueous phase. Water-immiscible solvents like dichloromethane can be easily removed in extraction times of a few minutes, because the extraction is facilitated by the high solubility of the pure organic solvent in CO_2, which can even reach complete miscibility at pressures above the mixture critical point (usually, pressures in the range of 80–100 bar at temperatures of 40°C–60°C). However, in the case of solvents with partial water miscibility such as ethyl acetate, the extraction of the solvent is hindered by a very slow removal of the organic solvent dissolved in the aqueous phase, because, even if the pure organic solvent again can be completely miscible with CO_2 at moderate pressures, the partition coefficient of the organic solvent between aqueous and CO_2 phases is, in general, unfavorable for the extraction, making it necessary to employ long extraction times that can reach the range of hours in the case of batch extraction systems.

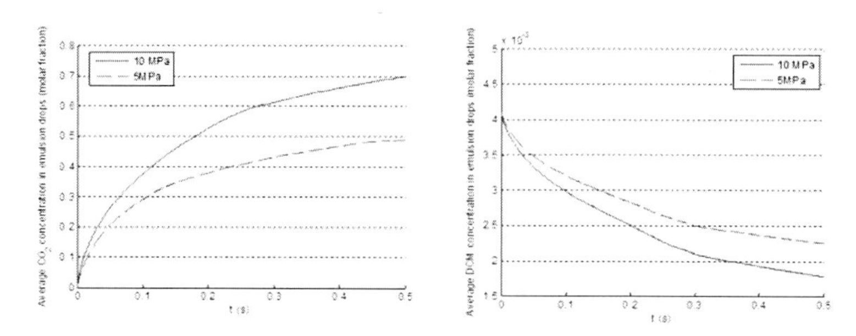

Figure 8.7 Evolution of the concentration of CO_2 and organic solvent during the supercritical extraction of an emulsion sprayed into CO_2, modeling result. Adapted from Ref. [33].

As mentioned before, an advantage of SFEE processes is the possibility to implement the process in different layouts, ranging from small batch systems to fully continuous processes. Figure 8.8 presents schematic diagrams of some possibilities. The first diagram corresponds to a simple batch system, and consists of an extraction vessel which is loaded with emulsion at the beginning of the experiment, and a system to pass supercritical CO_2 through this vessel. This CO_2 flow system can simply consist of a reservoir filled with pressurized CO_2 at the beginning of the experiment, a temperature control system, and an isobaric recirculation pump, thus resulting in a simple system that requires little supervision during the extraction and that can be suitable for the in situ production of small amounts of product. The second diagram corresponds to the extraction by atomization of the emulsion through a nozzle into an extraction/precipitation vessel filled with CO_2. This layout can be implemented using existing commercial equipment for the supercritical antisolvent (SAS) precipitation, described in previous chapters, with minor modifications. Furthermore, the process can be operated either in a semicontinuous way, if the aqueous suspension obtained as a product is accumulated inside the extraction/precipitation vessel until the end of the experiment, or it can be made fully continuous if the aqueous suspension is continuously retrieved from the vessel through a depressurization valve. The third layout corresponds to a continuous SFEE process based on the contact of CO_2 and emulsion by cocurrent or countercurrent flow in a column, designed to improve the mass transfer between phases and to provide the required extraction time for a complete removal of the organic solvent.

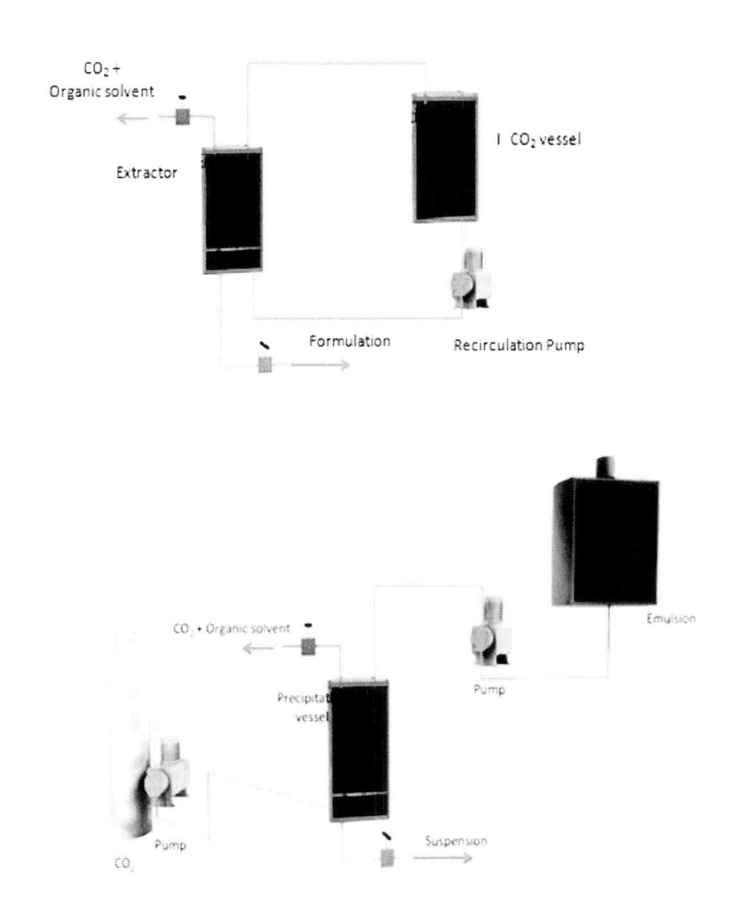

Figure 8.8 Schematic flow diagrams of processes for the supercritical fluid extraction of emulsions: batch system and continuous system with extraction from a spray or in a packed column.

Published reports on the SFEE process still are scarce, since it is a relatively recent technology. Shekunov et al. developed the original patent of the process, and described the application to produce micro- and nanoparticles both of water-insoluble pure compounds such as cholesterol acetate, griseofulvin and megestrol acetate [31], as well as microcapsules of active compounds in PLA/PLGA and Eudragit RS polymers [34]. These authors obtained particle sizes ranging from 100 to 1000 nm, observing that emulsion droplet size, organic solvent content, and drug solution concentration were the main parameters that controlled particle size. More recently, they also studied the formation of solid lipid nanoparticles for pulmonary delivery, reaching drug loadings of up to 20% [35].

The group of Della Porta and Reverchon has focused on the production of polymeric micro-/nanoparticles. They studied the encapsulation of piroxicam in PLGA, obtaining particles of 1–3 mm with narrow particle size distributions and residual solvent concentration below 40 ppm after 30 min of contact in a batch system [36]. They also developed a continuous system based on a packed column, and characterized the flow parameters of the system [37]. More recently, they proposed a supercritical-assisted emulsion diffusion method, based on a combination of a conventional emulsion-diffusion method using benzyl alcohol as solvent for the production of solid lipid particles, and a supercritical CO_2 extraction for the removal of this solvent from the final product [38].

Kluge et al. have also contributed to the development of SFEE processes, employing a semicontinuous emulsion atomization system for the encapsulation in PLGA of lyzosyme as a model hydrophylic compound, using double w/o/w emulsions [39] and ketoprofen as a model hydrophobic compound [40]. They obtained highly regular spherical microparticles, with sizes closely related to the initial emulsion droplet size distribution and dependent on the polymer concentration on these droplets and encapsulation efficiencies near 50%. They applied the same technique for the preparation of biocompatible magnetite–PLGA composites, which have applications as contrast agents in magnetic resonance imaging [41]. These authors also completed a fundamental study of the process employing phenantrene as model compound and observed two different crystallization mechanisms, primary nucleation of particles inside single emulsion droplets and growth by aggregation of emulsion droplets with existing particles, in a process bearing resemblance with the crystallization by an oiling-out mechanism [42]. They also employed SFEE as a technique for the preparation of polymer nanoparticles for measurements of active compound solubility in the polymer matrix [43].

A few other groups have also presented reports on SFEE applications. Mayo et al. produced pDNA loaded PLGA nanoparticles for gene delivery, obtaining high loading efficiency (>98%) and low residual solvent concentration (<50 ppm) [44]. Lin et al. reported the encapsulation of ibuprofen as a model low-water solubility drug in PLGA and characterized the drug release profiles [45]. Finally, Luther and Braeuer applied a microfluidic system to analyze flow patterns in the SFEE process by shadowgraphy techniques [46].

8.6 Case Study: Precipitation and Encapsulation of *β*-Carotene by Emulsion Techniques

In this section, the results obtained by the authors for the development of water-soluble and water-insoluble formulations of a carotenoide (*β*-carotene) with different techniques are presented as a case study. Carotenoids are valuable natural additives for food and nutraceutical products. One of the most common, abundant, and used carotenoid is *β*-carotene. Besides its excellent colorant properties, *β*-carotene is a precursor of retinol and retinoic acid, which have an important role in human health as vitamin A precursor and as cellular regulatory signal, respectively. Since animals and humans cannot produce carotenoids in their organisms, they need to acquire them from food. Being strong antioxidants, carotenoids are prone to degradation by the action of light, oxygen or moderate temperatures. Moreover, they are highly hydrophobic compounds that show very low solubility in water, as well as low bioavailability. Therefore, to improve their dispersability in water and coloring strength potential and also to increase their bioavailability during gastrointestinal passage, carotenoid crystals must be formulated [47]. In particular, for the use of carotenoids as natural colorants in food products, it is important to obtain an appropriate colour intensity of the formulation, which depends on the properties of the particles, including a restricted particle size and controlled crystallinity [29].

As shown in the literature review presented in Table 8.1, many authors have applied different techniques in order to develop suitable formulations of *β*-carotene. The formulation methods employed include the formation of o/w micro- and nanoemulsions

Table 8.1 Literature review of formulation of β-carotene

Ref.	F.m.	Stabilizer/ surfactant	Oil phase/solvent	Process technique	PS/EE
[48]	(o/w) nanoemulsions	β-lactoglobulin Tween 20	Orange oil	High-pressure homogeniza-tion (HPH)	PS: 158 nm
[49]		β-lactoglobulin or Tween 20 + water soluble: EDTA/ascorbic acid Addition oil-soluble: Vitamin E acetate or Coenzyme Q10	Corn oil		PS: 180 nm
[50]		Tween 20	Corn oil (long-chain triglyceride) Miglyol 812 (medium-chain triglyceride) Orange oil		PS: 140–170 nm

Ref.	F.m.	Stabilizer/ surfactant	Oil phase/solvent	Process technique	PS/EE
[51, 52]		Tween 20 Decaglycerol monolaurate (DML) Octenyl succinate st. Whey protein isolate Blend of Tween 20 & whey protein isolate	Sunflower oil		PS: 115–300 nm
				Microfluidiza-tion	PS: 120–370 nm
[53]		OSA-modified starches: HI-CAP 100, CAPSUL, CAPSUL-TA	Medium-chain triacylglycerol	HPH	PS: 150 nm
[54]		Tween 20 Tween 40 Tween 60 Tween 80	Medium-chain triglyceride		PS: 132–184 nm
[55]		Tween 20		HPH (response surface methodology)	PS: 120–170 nm

(*Continued*)

Table 8.1 (*Continued*)

Ref.	F.m.	Stabilizer/ surfactant	Oil phase/solvent	Process technique	PS/EE
[56]		Tween 20	Hexane	High-energy emulsification-evaporation	PS: 9–280 nm
[57]	Nanodispersions	Tween 20	Hexane	Emulsification (HPH)-evaporation	PS: 100 nm
[58]		Polyglycerol esters of fatty acids			PS: 85–132 nm
[59]		Protein			PS: 17 nm
[47]		Poly(D,L-lactic acid) Poly(D,L-lactic coglycolic acid)	Acetone	Solvent displacement method	PS: 100 nm
[5]		Protein			PS: 171 nm
[60]		Sodium caseinate Tween 20 Decaglycerol monolaurate Sucrose fatty acid ester			PS: 30–206 nm
[61]	Dispersions	Decaglycerol monolaurate	Soybean oil	HPH or combining emulsification-evaporation	PS: 45 nm–18.3 µm
[62]		Tween 80 or poly(lactic acid)	Chloroform	Conventional homogenization	PS: 150 nm

Ref.	F.m.	Stabilizer/ surfactant	Oil phase/solvent	Process technique	PS/EE
[63]		Modified starch	Ethyl acetate	HPH	PS: 400 nm EE: 80%
[64]	SCF precipitation		Ethyl acetate, dichloromethane	GAS	PS: < 1μm
[65]			Tetrahydrofuran	SAS	PS: 100 μm
[66]			Dichloromethane	SEDS-PA	PS: 0.4–5 μm
[67]				SEDS	PS: 3.2–96.8 μm
[68]				SEDS-EM	PS: 20–205 nm
[69]	SCF coprecipitation	Poly(ethylene glycol((PEG)	Dichloromethane	SAS	
[70]		Poly(3-hydroxybutirate-co-hydroxyvalerate) (PHBV)		SEDS	PS: 3.8–246.8 μm EE: 80%
[71]					EE: 55%
[72]		PEG		SEDS-PA	PS: 6 μm EE: 50%
[73]	SCF coformulation	OSA surfactant Blend of Tween 20 and Span 20	Dichloromethane	SFEE	PS: 400 nm
[74]		OSA-modified starch			PS: 345 nm
[75]		Tween 80	Acetone	SAILA	PS: 50–150 nm

F.m.: formulation method; PS: particle size; EE: encapsulation efficiency.

[48–56], the formation of particle microdispersions by application of conventional solvent evaporation [57–59], solvent displacement [5, 47, 60], homogenization techniques [61], and high-pressure emulsion techniques [63], and also the application of different precipitation and coprecipitation methods based on supercritical carbon dioxide, including the supercritical antisolvent precipitation [64–72] and SFEE [73, 74]. A large variety of surfactant and carrier materials have been employed, ranging from synthetic Tween surfactants, to biocompatible polymers such as poly(ethylene glycol) (PEG), poly(lactic acid) (PLA), or poly(3-hydroxybutirate-co-hydroxyvalerate) (PHBV), starches modified with the *n*-octenyl succinic anhydride (OSA) group, and natural carriers such as phospholipids or whey.

8.6.1 Formulation by Conventional Emulsification and Solvent Evaporation Techniques

Micellar formulations of β-carotene in OSA–starch surfactants were prepared according to the solvent evaporation method described in Section 2 [76]. An ultrasound emulsifier (UP400S Ultrasonic Processor, Hielscher, Germany) or a high-shear rotor-stator machine (IKA Ultra-Turrax® LABOR PILOT 2000/4) were used to prepare emulsions. A Rotavapor (BÜCHI 011-BÜCHI 461 Water Bath) was used to eliminate the organic solvent in emulsion evaporation experiments. A dissolution of β-carotene in ethyl acetate (β-carotene concentration: 5 g/L) and a dissolution of OSA–starch in de-ionized water (starch concentration: 36 g/L) were prepared. Both dissolutions were heated to 55°C under stirring to ensure complete dissolution of the compounds. The two dissolutions were mixed, using different proportions of organic and aqueous solutions resulting in organic/water volume ratios ranging from 0.275 mL organic/mL water to 0.73 mL/mL. Then, the mixture was emulsified using the ultrasound equipment. Different parameters of the emulsification were varied, including the apparatus used (ultrasound or high-shear Ultra-Turrax®), and in ultrasound emulsification experiments, the time emulsification, the ultrasound pulse amplitude and the ultrasound duty cycle (i.e., the fraction of time that ultrasound is being generated). After the emulsification, the organic solvent was eliminated by vacuum evaporation at 60°C, thus producing a suspension of β-carotene particles in water.

As presented in Fig. 8.9, the encapsulation efficiency increased and the micellar particle size decreased when the time of application of ultrasounds was increased from 6 min to 20 min. When ultrasounds were applied for more than 14 min, approximately constant values of encapsulation efficiency of 12%–13%, and micellar particle size of 1–5 μm, were obtained, or even a slight increase in particle size was observed with application times longer than 20 min, maybe due to a partial destabilization of the emulsion due to an excessively long experiment. These results indicate that a complete emulsification was achieved after application of ultrasounds during 20 min, and a longer application of ultrasounds did not improve the emulsification and the final results.

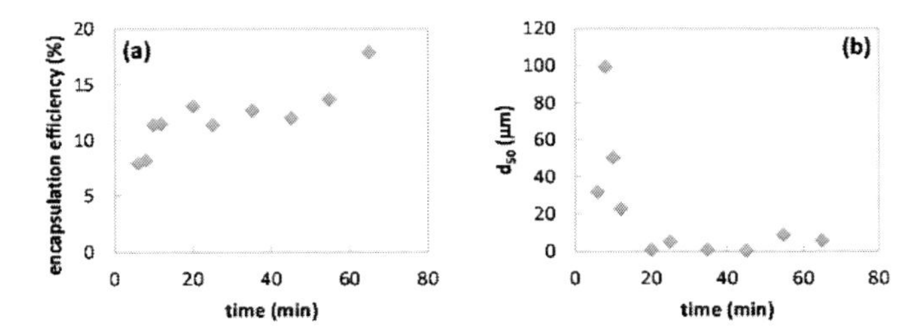

Figure 8.9 Influence of the time of application of ultrasounds on (a) the encapsulation efficiency and (b) the micellar particle size.

Another important parameter is the ratio between organic and aqueous phases in the initial emulsion. As shown in Fig. 8.10, when the ratio was reduced from $r_{\text{organic–water}}$ = 0.73 mL/mL to $r_{\text{organic–water}}$ = 0.275 mL/mL, the encapsulation efficiency was increased from 18% to 24%, and the micellar particle size was considerably reduced from 6 μm to 140 nm. These variations of results can be correlated with the reduction of emulsion droplet size and the increase of emulsion stability when the fraction of organic solvent in the emulsion is reduced and the amount of surfactant is increased.

Figure 8.10 also presents a comparison of results obtained by ultrasound emulsification and high-shear emulsification. As shown in this figure, in experiments carried out by high-shear emulsification the encapsulation efficiency remained approximately constant, ranging between 4.6% and 8.0%. This value was clearly lower than the results achieved by ultrasound emulsification. On the other hand, micelar particle sizes obtained with both techniques were similar. These results probably are a consequence of the more aggressive conditions during high-shear emulsification compared to ultrasound emulsification.

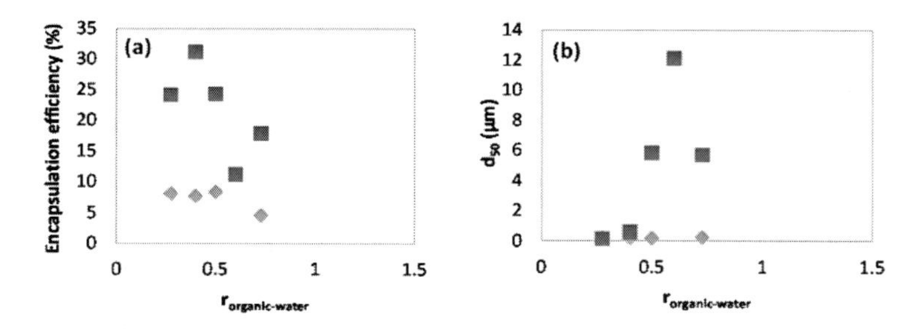

Figure 8.10 Influence of the ratio between volume of the organic phase and volume of the aqueous phase organic–water on (a) the encapsulation efficiency and (b) the micellar particle size, obtained by high-shear emulsification (◆) and by ultrasound emulsification (■).

8.6.2 Formulation by Precipitation from Pressurized Organic Solvent-on-Water Emulsions

With the high-pressure emulsion technique previously discussed in Section 4, a study of the formulation of β-carotene using a modified n-octenyl succinate starch refined from waxy maize as carrier material was carried out [63]. Formulations were prepared with a process based in the formation of an organic-in-water emulsion with pressurized fluids. As previously discussed, the aim in the conception of this process is to improve the formulation over the conventional emulsion evaporation process, accelerating the mass transfer kinetics to the time scales of the precipitation processes. Ethyl acetate was chosen as organic solvent because it is a generally recognized as safe (GRAS) solvent with low toxicity.

The experimental setup is shown in Fig. 8.4. The experimental apparatus consists of three small storages at ambient pressure, corresponding to the feed of pure organic solvent (ethyl acetate), β-carotene suspension in the same organic solvent, and the aqueous solution of the modified OSA–starch. The stream of the organic solvent is preheated in order to reach the specified operation temperature after mixing with the β-carotene suspension (typically 145°C). All streams are pressurized with the pumps in order to keep them in the liquid phase at this temperature. The suspension of β-carotene is pumped at ambient temperature. Then, it is mixed with the hot organic solvent stream in a T-mixer; at this point the β-carotene is completely dissolved because the solubility increases when temperature. Shortly afterward, the β-carotene solution is mixed with the cold aqueous solution of surfactant using another T-mixer in order to reduce the contact time of β-carotene particles with the hot organic solvent and to avoid the isomerization and degradation of the product. The contact of the hot solution of β-carotene with the cold aqueous solution in the second T-mixer causes the emulsification of the organic solvent and the precipitation of β-carotene by a combined antisolvent and cooling effect. Then, the emulsion is collected and the organic solvent is removed by vacuum evaporation in order to produce a suspension of β-carotene nanoparticles in water stabilized with the surfactant.

In this research, the influence of the main process parameters was studied: the concentration of modified OSA–starch and the organic–water ratio. The effect of the concentration of modified OSA–starch dissolution was carried out varying this concentration from 37 g/L to 367 g/L. In Fig. 8.11, the influence of the concentration of surfactant on the percentage of encapsulated β-carotene and micellar particle size is presented.

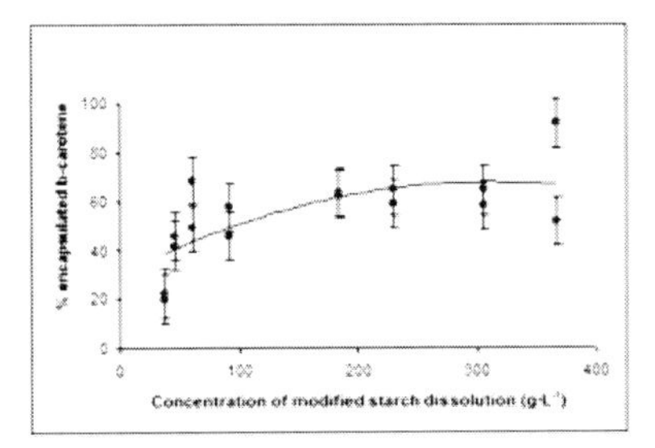

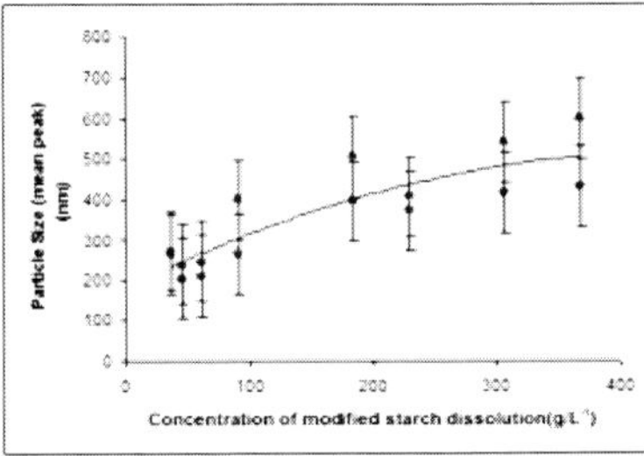

Figure 8.11 Variation of the micellar particle size and encapsulation efficiency with the concentration of an OSA–starch surfactant in the formulation of β-carotene by precipitation from pressurized o/w emulsions.

The results show that the percentage of encapsulated β-carotene is higher when the concentration of modified-starch is increased.

With regard to the micellar particle size, the main sizes obtained ranged from 200 nm to 600 nm; it is higher when the concentration of surfactant is increased. Although an increase in the micellar particle size is in general disadvantageous for the stability of the suspension, it must be taken into account that the use of high concentrations of starch allows to encapsulate a higher percentage of β-carotene and to obtain a better emulsion stability. As for the effect of the organic–water ratio, it was carried out varying this ratio from 0.6 to 1.3. It is necessary to emphasize that this organic–water ratio has a strong influence on the micellar particle size. When this ratio is increased, the micellar particle size increases as well. On the other hand, the encapsulation efficiency does not show a clear variation, achieving percentages of encapsulated β-carotene of 70%–80%. The best results were obtained with low ratios, in the range of 0.65 and 0.73. The obtained suspension can be further processed by spray-drying or similar techniques in order to obtain a dry product. As shown in Fig. 8.12, a homogeneous suspension of β-carotene in water can be easily obtained by rehydration of this formulation.

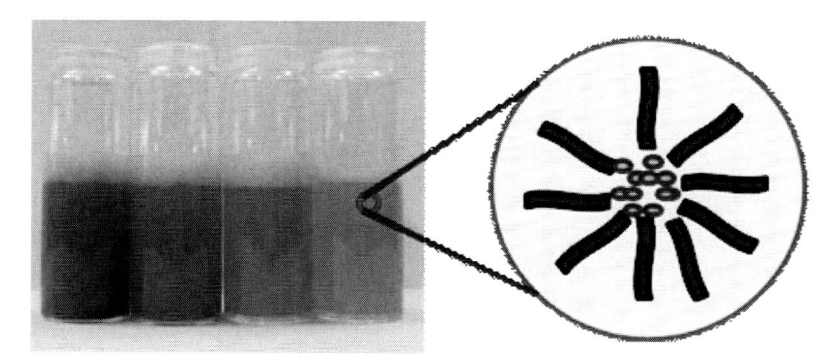

Figure 8.12 Dispersions of β-carotene formulations in water.

To better assess the mechanisms of particle formation by the pressurized emulsion method, additional experiments were carried out using ethanol instead of ethyl acetate as organic solvent [77]. Compared to ethyl acetate, ethanol is completely water miscible, and therefore using this solvent particle formation is only governed by an antisolvent mechanism, as the emulsion template is not formed with this solvent. As shown in Fig. 8.13, the trends of variation of particle size and encapsulation efficiency with the organic–water ratio obtained with both solvents are equivalent: particle size increases at higher organic–water ratios, while approximately constant encapsulation efficiencies are obtained. However, considerable differences between the values of particle size and efficiency obtained with the two solvents can be observed. In particular, the encapsulation efficiencies obtained using ethanol as organic solvent (30%–40%) are clearly lower than those achieved using ethyl acetate and the same OSA–starch (70%–80%). This can be a consequence of the lack of the emulsion template in experiments carried out with ethanol. Regarding the particle size, slightly smaller particle sizes were obtained in successful experiments with ethanol (150–200 nm) than in experiments with ethyl acetate (400–500 nm). This result indicates that the formation of an emulsion in experiments with ethyl acetate does not contribute to control particle size by restricting particle growth within emulsion droplets. On the contrary, particle size appears to be controlled by the antisolvent precipitation process or by the size of surfactant micelles in the final aqueous dispersion.

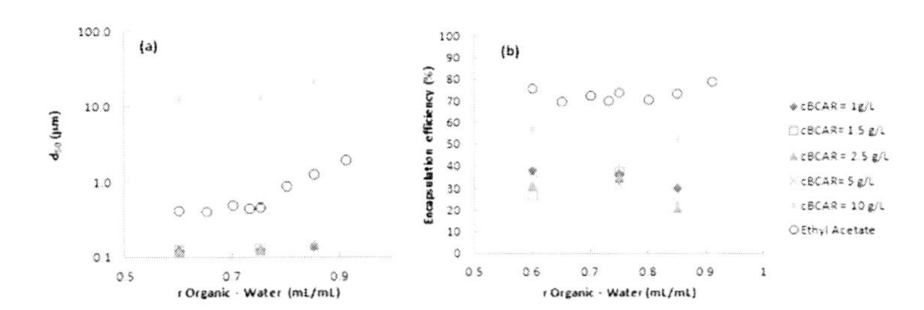

Figure 8.13 Variation of (a) micellar particle size and (b) encapsulation efficiency with the organic–water ratio in experiments using ethyl acetate or ethanol as solvents.

8.6.3 Formulation by Supercritical Fluid Extraction of Emulsions

With this technique, the formulation of β-carotene, using modified starch and a blend of Tween 20 and Span 20 as surfactants, was carried out by Mattea et al. [73]. This work was later extended to the formulation of lycopene by Santos et al. [74]. In a first series of experiments of formulation of β-carotene by batch SFEE extraction, a clear relationship between the initial emulsion droplet size and the final particle size was observed. As shown in Fig. 8.14, the relationship between the emulsion size and the droplet size is clear and limitations in the size were related to the possibility of obtaining smaller droplets sizes during the emulsion formation.

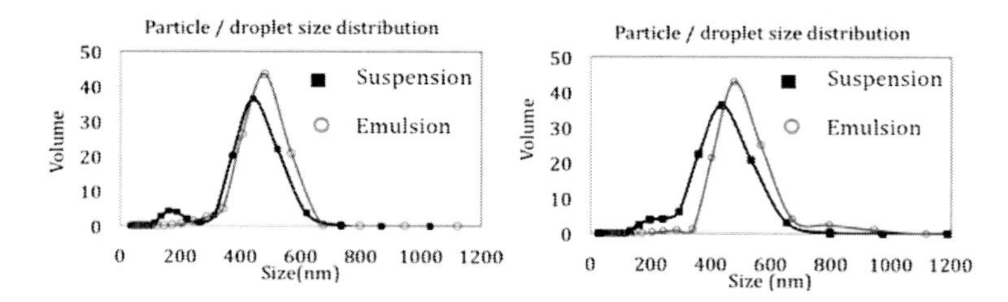

Figure 8.14 Size distribution of the starting emulsion and produced suspension for a SAS process at 323 K, 4 kg of CO_2/h: (a) 8 MPa and (b) 13 MPa.

The evolution of the concentration of dichloromethane during an experiment can be observed in Fig. 8.15. The concentration of dichloromethane obtained in the experiments was lowered from the initial 15%–20% v/v to a final value inferior of 10 ppm. The desired concentration of the organic solvent is directly related to the further application of the nanosuspension of carotenoids. If the obtained suspension-like product is going to be dried by other methods like spray drying or freeze drying, the remaining organic solvent will be removed in those processes.

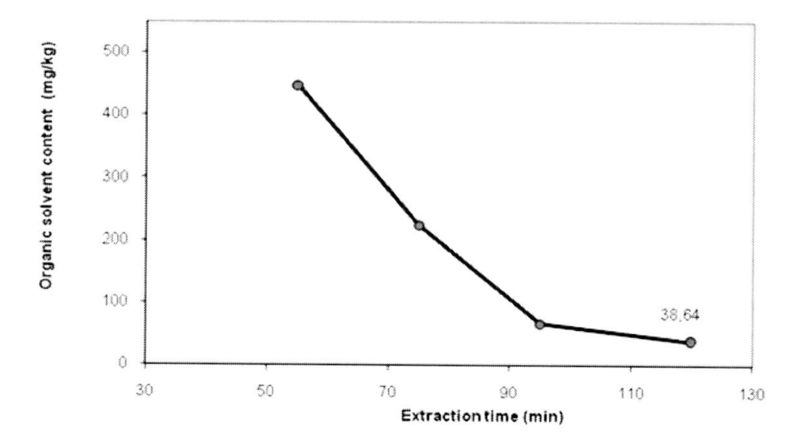

Figure 8.15 DCM content of the produced suspensions at different time intervals during a GAS process.

Figure 8.16 presents SEM and ESEM images of particles produced by semicontinuous SFEE from a spray, followed by lyophilization of the obtained aqueous suspension. In the first image the modified starch based product is exposed, and it is possible to identify large needle type particles formed during the lyophilization together with small spherical particles formed during the antisolvent precipitation process. The size of the particles obtained during the GAS and SAS process can be observed in the optical microscopy (OM) and SEM images of the particles obtained with Tween 20 and Span 20, (b) and (c), respectively, which are distributed in a continuous matrix composed by the surfactant mixture.

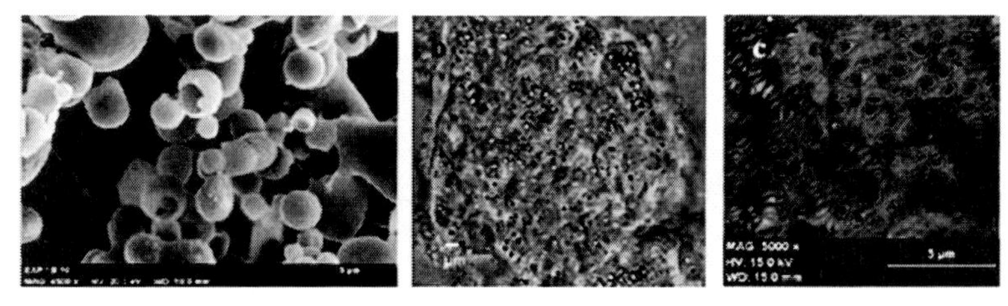

Figure 8.16 Microscopic images of the particles produced during a SAS process: (a) SEM image of the freeze-dried product from an emulsion prepared with OSA. (b) OM and (c) SEM image of the freeze-dried product from an emulsion prepared with Tween 20 and Span 20.

The content of DCM of the samples obtained in these experiments was below 10 ppm after the SAS process and negligible after the lyophilization process. The obtained particle size distributions for the processes at different operative conditions were very similar, and always below or similar to the original droplet size distribution, showing narrow distributions in the submicrometric range, as presented in Fig. 8.17. Furthermore, as presented in Fig. 8.18, high encapsulation efficiencies, ranging from 50% to 70%, were obtained. Differential scanning calorimetry (DSC) assays of lyophilized or spray-dried aqueous suspensions did not show peaks corresponding to crystalline β-carotene, indicating that all β-carotene was dispersed as an amorphous phase in the surfactant matrix, which is a positive characteristic for an enhanced bioavailability of the active compound.

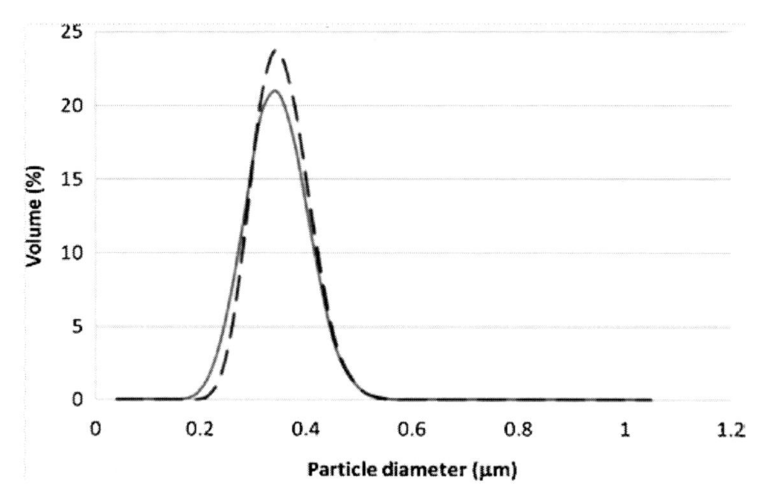

Figure 8.17 Particle size distributions obtained by SFEE processing of β-carotene (continuous line) and lycopene (dashed line) with OSA–starch surfactants Adapted from Ref. [74].

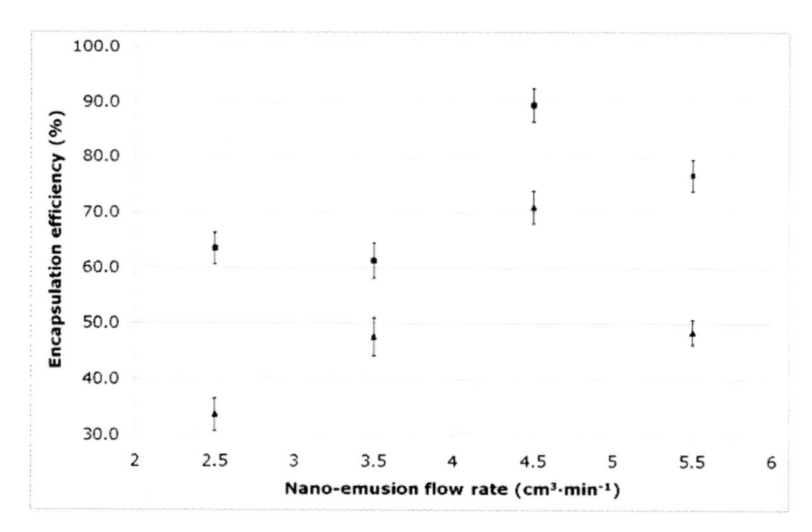

Figure 8.18 Encapsulation efficiency obtained by SFEE processing of β-carotene (■) and lycopene (▲) with OSA–starch surfactants. Adapted from Ref. [74].

The encapsulation of β-carotene in polycaprolactones by SFEE processing has also been studied. Compared to OSA–starches, polycaprolactones are hydrophobic polymers that show slow degradation in water, enabling a slow release of the entrapped β-carotene. Particles were produced by semicontinuous SFEE processing of atomized solutions, dissolving both β-carotene and the polymer in the disperse dichloromethane phase of the emulsion. Three polycaprolactones with different molecular weights (MWs) were tested: CAPA 2403D (MW: 4000 g/mol), CAPA 6100 (MW: 10,000 g/mol), and CAPA 6250 (MW: 25,000 g/mol). As shown in Fig. 8.19, the particle size distributions of the emulsion droplet and of the suspension of β-carotene with the three PCLs were similar, obtaining a mean diameter of about 300 nm. Furthermore, Fig. 8.20 presents SEM micrographs of the particles obtained. As can be observed, polycaprolactone particles exhibit in all cases a very similar morphology and particle size, independently of which polycaprolactone was used, observing spheres with a size in the order of 200 nm with a rough surface, in agreement with the results of particle size distribution measurements previously reported.

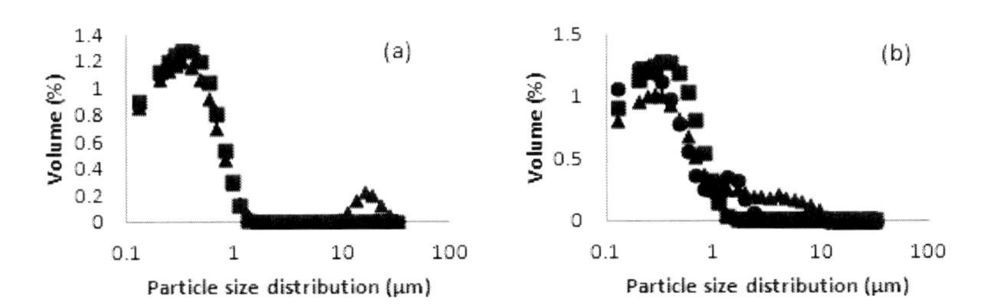

Figure 8.19 Influence of the molecular weight of PCLs on the particle size distribution of (a) emulsions and (b) suspensions (after SFEE process): (•) CAPA 2403D, (■) CAPA 6100, and (▲) CAPA 6250.

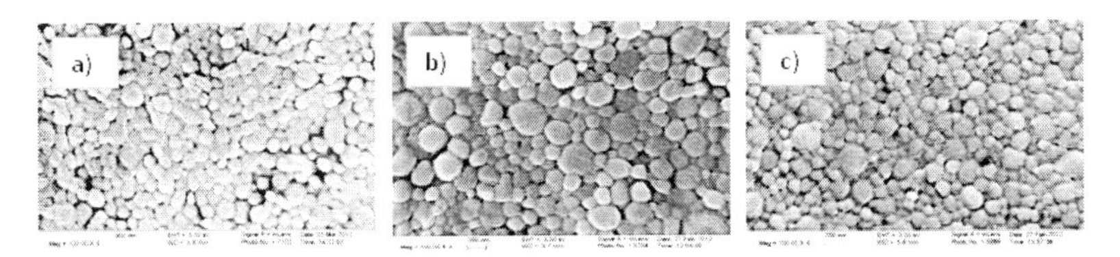

Figure 8.20 SEM micrographs of particles processed by SFEE: (a) CAPA 2403D- β-carotene, (b) CAPA 6100-β-carotene, and (c) CAPA 6250-β-carotene.

8.6.4 Comparison of Results Obtained with Different Techniques

Figure 8.21 presents a global comparison of the results obtained with the different techniques reported in this section. Comparing the results obtained by conventional solvent evaporation and by precipitation from pressurized emulsions, the most notable differences are observed in the encapsulation efficiency. Much lower encapsulation efficiencies are yielded by emulsion evaporation processes, especially when the emulsion is formed by conventional shear stress methods. This result can be related to a partial destabilization of the emulsion during the long processing times that are required by emulsion evaporation processes, which are reduced to seconds with the pressurized emulsion method used in this work. On the contrary, smaller differences are observed in the micellar particle sizes or even smaller sizes are obtained by ambient pressure emulsion evaporation techniques than in experiments with ethyl acetate reported in this work, which may indicate a more effective emulsion formation by the ultrasound or high-shear methods employed in emulsion evaporation techniques than by turbulent mixing as in experiments of precipitation from pressurized emulsions.

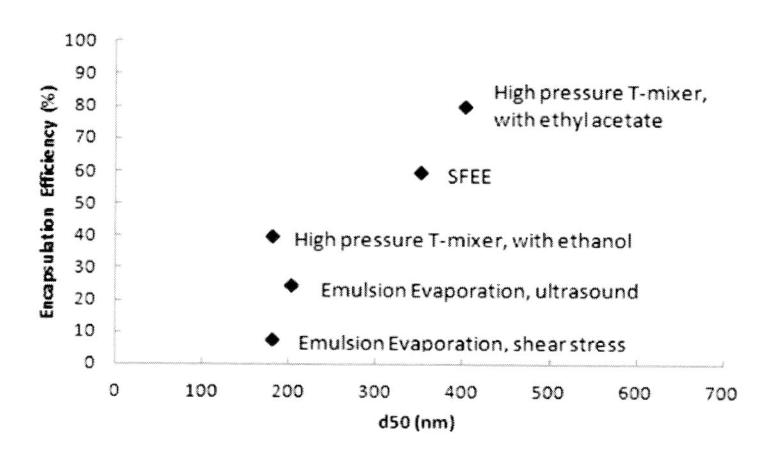

Figure 8.21 Comparison between results obtained by emulsion evaporation, precipitation from pressurized emulsions, and supercritical fluid extraction of emulsions.

8.7 Conclusions

With respect to the results obtained by supercritical extraction of emulsions, compared to experiments of precipitation from pressurized emulsions, similar micellar particle sizes and slightly lower encapsulation efficiencies were achieved. As a major difference between the morphology of product obtained by these techniques, as previously reported no crystalline β-carotene particles were observed in DSC assays of SFEE-processed formulations, while some crystalline β-carotene was detected in formulations produced by precipitation from pressurized emulsions. This indicates that SFEE processing yielded a homogeneous dispersion of amorphous β-carotene in the carrier, which is a favorable property for an enhanced bioavailability of this compound. Furthermore, SFEE processing enabled producing solvent-free products in a single step with comparatively short processing times, while solvent removal by vacuum evaporation was required in experiments of precipitation from pressurized emulsions.

References

1. O'Donnell, P. B., and McGinity J. W. (1997). Preparation of microspheres by the solvent evaporation technique, *Adv. Drug Delivery Rev.*, **28**, pp. 25–42.
2. Sansdrap, P., and Moes, A. J. (1993). Influence of manufacturing parameters on the size characteristics and the release profiles of nifedipine from poly(DL-lactide-co-glycolide) microspheres, *Int. J. Pharm.*, **98**, pp. 157–164.
3. Bodmeier, R., and McGinity, J. W. (1988). Solvent selection in the preparation of poly (DL-lactide) microspheres prepared by the solvent evaporation method, *Int. J. Pharm.*, **43**, pp. 179–186.
4. Izumikawa, S., Yoshioka, S., Aso, Y., and Takeda, Y. (1991). Preparation of poly(L-lactide) microspheres of different crystalline morphology and effect of crystalline morphology on drug release rate, *J. Controlled Release*, **15**, pp. 133–140.
5. Chu, B. S., Ichikawa S., Kanafusa, S., and Nakajima, M. (2007). Preparation and characterization of β-carotene nanodispersions prepared by solvent displacement technique, *J. Agr. Food Chem.*, **55**, pp. 6754–6760.

6. Trotta, M., Gallarte, M., Pattarino, F., and Morel, S. (2001). Emulsions containing partially water-miscinle solvents for the preparation of drug nanosuspensions, *J. Controlled Release*, **76**, pp. 119–128.
7. Anton, N., Benoit, J. P., and Saulnier, P. (2008). Design and production of nanoparticles formulated from nano-emulsion templates-a review, *J. Controlled Release*, **128**, pp. 185–199.
8. Craparo, E. F., Cavallaro, G., Bondi, M. L., Mandracchia, D., and Giammona, G. (2006). Pegylated nanoparticles based on polyaspartamide. Preparation, physico-chemical characterization and intracellular uptake, *Biomacromolecules*, **7**, pp. 3083–3092.
9. Bradley, M. A., Prescott, S. W., Schoonbrood, H. A. S., Landfester, K., and Grieser, F. (2005). Miniemulsion copolymerization of methyl methacrylate and butyl acrylate by ultrasonic initiation, *Macromolecules*, **38**, pp. 6346–6351.
10. Qi, G., Jones C. W., and Shork, F. J. (2006). Enzyme-initiated miniemulsion polymerization, *Biomacromolecules*, **7**, pp. 2927–2930.
11. Antonietti, M., and Landfester, K. (2002). Polyreactions in miniemulsions, *Prog. Polym. Sci.*, **27**, pp. 689–757.
12. Chaturvedi, S. P., and Kumar, V. (2012). Production techniques of lipid nanoparticles: a review, *Res. J. Pharm. Bio. Chem. Sci.*, **3**, pp. 525–541.
13. Freitas, C., and Muller, R. H. (1999). Correlation between long-term stability of solid lipid nanoparticles (SLN) and crystallinity of the lipid phase, *Eur. J. Pharm. Biopharm.*, **42**, pp. 125–132.
14. Samad, A., Sultana, Y., and Aquil, M. (2007). Liposomal drug delivery systems: an update review, *Curr. Drug Delivery*, **4**, pp. 297–305.
15. Olbrich, C., Gessner, A., Kayser, O., and Muller, R. H. (2002). Lipid-drug conjugate (LDC) nanoparticles as novel carrier system for the hydrophulic antitrypanosomal drug diminazenediaceturate, *J. Drug Targeting*, **10**, pp. 387–396.
16. Park, J. H., Ye, M., Yeo, Y., Lee, W. K., Paul, C., and Park, K. (2006). Reservoir-type microcapsules prepared by the solvent exchange method: effect of formulation parameters on microencapsulation of lysozyme, *Mol. Pharm.*, **3**, pp. 135–143.
17. Sturesson, C., Carlfors, J., Edsman, K., and Andersson, M. (1993). Preparation of biodegradable poly(lactic-co-glycolic) acid microspheres and their in vitro release of timolol maleate, *Int. J. Pharm.*, **89**, pp. 235–244.
18. Herrmann, J., and Bodmeier, R. (1995). Somatostatin containing biodegradable microspheres prepared by a modified solvent

evaporation method based on w/o/w multiple emulsions, *Int. J. Pharm.*, **126**, pp. 129–138.

19. Iwata, M., and McGinity, J. W. (1992). Preparation of multi-phase micro-spheres of poly(D,L-lactic acid) and poly (D,L-lactic-co-glycolic acid) containing a w/o emulsion by a multiple emulsion solvent evaporation technique, *J. Microencap.*, **9**, pp. 201–214.
20. Walstra, P. (1993). Principles of emulsion formation, *Chem. Eng. Sci.*, **48**, pp. 333–349.
21. Abismail, B., Canselier, J. P., Wilhelm, A. M., Delmas, H., and Gourdon, C. (1999). Emulsification by ultrasound: drop size distribution and stability, *Ultrason. Sonochem.*, **6**, pp. 75–83.
22. Bouchemal, K., Briancon, S., Perrier, E., and Fessi, H. (2004). Nano-emulsion formulation using spontaneous emulsification: solvent, oil and surfactant optimization, *Int. J. Pharm.*, **280**, pp. 241–251.
23. Ganachaud, F., and Katz, J. L. (2005). Nanoparticles and nanocapsules created using the ouzo effect: spontaneous emulsification as an alternative to ultrasonic and high-shear devies, *Chem. Phys. Chem.*, **6**, pp. 209–216.
24. Ruschak, K. J., and Miller, C. A. (1972). Spontaneous emulsification in ternary systems with mass transfer, *Ind. Eng. Chem. Fund.*, **11**, pp. 534–540.
25. Shinoda, K., and Saito, H. (1968). The effect of temperature on the phase equilibria and the types of dispersions of the ternary system composed of water, cyclohexane and nonionic surfactant, *J. Col. Interface Sci.*, **26**, pp. 70–74.
26. Shinoda, K., and Saito, H. (1969). The stability of O/W type emulsions as a function of temperature and the HLB of emulsifiers: the emulsification by PIT-method, *J. Col. Interface Sci.*, **30**, pp. 258–263.
27. Izquierdo, P., Esquena, J., Tdros, T. F., Dederen, J. C., Feng, J., García-Delma M. J., Azemar, N., and Solans, C. (2004). Phase behavior and nano-emulsion formation by the phase inversion temperature method, *Langmuir*, **20**, pp. 6594–6598.
28. Pons, R., Carrera, I., Caelles, J., Rouch, J., and Panizza, P. (2006). Formation and properties of miniemulsions formed by microemulsions dilution, *Adv. Col. Interface Sci.*, **106**, pp. 129–146.
29. Horn, D., and Rieger, J. (2001). Organic nanoparticles in the aqueous phase: theory, experiment and use, *Angew. Chem., Int. Ed.*, **40**, pp. 4330–4361.

30. Martín, Á., and Cocero, M. J. (2008). Micronization processes with supercritical fluid: fundamentals and mechanisms, *Adv. Drug Delivery Rev.*, **60**, pp. 339–350.

31. Shekunov, B. Y., Chattopadhyay, P., Seitzinger, J., and Huff, R. (2006). Nanoparticles of poor water-soluble drugs prepared by supercritical fluid extraction of emulsions, *Pharm. Res.*, **23**, pp. 196–204.

32. Mattea, F., Martín, A., Schulz, C., Jaeger, P., Eggers, R., and Cocero, M. J. (2010). Behavior of an organic solvent drop during the supercritical extraction of emulsions, *AIChE J.*, **56**, pp. 1184–1195.

33. Mattea, F., Martín, A., Matías-Gago, A., and Cocero, M. J. (2009). Supercritical antisolvent precipitation from an emulsion: β-carotene nanoparticle formation, *J. Supercrit. Fluids*, **51**, pp. 238–247.

34. Chattopadhyay, P., Huff, R., and Shekunov, B. Y. (2006). Drug encapsulation using supercritical fluid extraction of emulsions, *J. Pharm. Sci.*, **95**, pp. 667–679.

35. Chattopadhyay, P., Shekunov, B. Y., Yim, D., Cipolla, D., Boyd, B., and Farr, S. (2007). Production of solid lipid nanoparticle suspensions using supercritical fluid extraction of emulsions (SFEE) for pulmonary delivery using the AERx system, *Adv. Drug Delivery Rev.*, **59**, pp. 444–453.

36. Della Porta, G., and Reverchon, E. (2008). Nanostructured microspheres produced by supercritical fluid extraction of emulsions, *Biotech. Bioeng.*, **100**, pp. 1020–1033.

37. Falco, N., Reverchon, E., and Della Prota, G. (2012). Continuous supercritical emulsions extraction: packed tower characterization and application to poly(lactic-co-glycolic acid) + insulin microspheres production, *Ind. Eng. Chem. Res.*, **51**, pp. 8616–8623.

38. Campardelli, R., Cherain, M., Perfetti, C., Iorio, C., Scognamiglio, M., Reverchon, E., and Della Porta, G. (2013). Lipid nanoparticles production by supercritical fluid assisted emulsion-diffusion, *J. Supercrit. Fluids*, **82**, pp. 34–40.

39. Kluge, J., Fusaro, F., Casas, N., Mazzotti, M., and Muhrer, G. (2009). Production of PLGA micro and nanocomposites by supercritical fluid extraction of emulsions: I. Encapsulation of lysozyme, *J. Supercrit. Fluids*, **50**, pp. 327–335.

40. Kluge, J., Fusaro, F., Mazzotti, M., and Muhrer, G. (2009). Production of PLGA micro and nanocomposites by supercritical fluid extraction of emulsions: II. Encapsulation of ketoprofen, *J. Supercrit. Fluids*, **50**, pp. 336–343.

41. Furlan, M., Kluge, J., Mazzotti, M., and Lattuada, M. (2010). Preparation of biocompatible magnetite-PLGA composite nanoparticles using supercritical fluid extraction of emulsions, *J. Supercrit. Fluids*, **54**, pp. 348–356.

42. Kluge, J., Joss, L., Viereck, S., and Mazzotti, M. (2012). Emulsion crystallization of phenanthrene by supercritical fluid extraction of emulsions, *Chem. Eng. Sci.*, **77**, pp. 249–258.

43. Kluge, J., Mazzotti, M., and Muhrer, G. (2010). Solubility of ketoprofen in colloidal PLGA, *Int. J. Pharm.*, **399**, pp. 163–172.

44. Mayo, A. S., Ambati, B. K., and Kompella, U. B. (2010). Gene delivery nanoparticles fabricated by supercritical fluid extraction of emulsions, *Int. J. Pharm.*, **387**, pp. 278–285.

45. Lin, C. S., Xu, J. J., Ng, K. M., Wibowo C., and Luo, K. Q. (2013). Encapsulation of a low aqueous solubility substance in a biodegradable polymer using supercritical fluid extraction of emulsion, *Ind. Eng. Chem. Res.*, **52**, pp. 134–141.

46. Luther, S. K., and Braeuer, A. (2012). High-pressure microfluidics for the investigation into multi-phase systems using supercritical fluid extraction of emulsions (SFEE), *J. Supercrit. Fluids*, **65**, pp. 78–86.

47. Ribeiro, H. S., Chu, B.S., Ichikawa, S., and Nakajima, M. (2008). Preparation of nanodispersions containing β-carotene by solvent displacement method, *Food Hydrocol.*, **22**, pp. 12–17.

48. Qian, C., Decker, E.A., Xiao, H., and McClements, D. J. (2012). Nanoemulsion delivery systems: influence of carrier oil on β-carotene bioaccessibility, *Food Chem.*, **135**, pp. 1440–1447.

49. Qian, C., Decker, E. A., Xiao, H., and McClements, D. J. (2012). Physical and chemical stability of β-carotene-enriched nanoemulsions: influence of pH, ionic strength, temperature, and emulsifier type, *Food Chem.*, **132**, pp. 1221–1229.

50. Qian, C., Decker, E. A., Xiao, H., and McClements, D. J. (2012). Inhibition of β-carotene degradation in oil-in-water nanoemulsions: influence of oil-soluble and water-soluble antioxidants, *Food Chem.*, **135**, pp. 1036–1043.

51. Mao, L. K., Xu, D. X., Yang, F., Gao, Y. X., and Zhao, J. (2009). Effects of small and large molecule emulsifiers on the characteristics of beta-carotene nanoemulsions prepared by high pressure homogenization, *Food Tech. Biotech.*, **47**, pp. 336–342.

52. Mao, L. K., Yang, J., Xu, D. X., Yuan, F., and Gao, Y. X. (2010). Effects of homogenization models and emulsifiers on the physicochemical properties of β-carotene nanoemulsions, *J. Disp. Sci. Tech.*, **31**, pp. 986–993.

53. Liang, R., Shoemaker, C. F., Yang, X., Zhong, F., and Huang, Q. (2013). Stability and bioaccessibility of β-carotene in nanoemulsions stabilized by modified starches, *J. Agric. Food Chem.*, **61**, pp. 1249–1257.

54. Yuan, Y., Gao, Y., Zhao, J., and Mao, L. (2008). Characterization and stability evaluation of c prepared by high pressure homogenization under various emulsifying conditions, *Food Res. Int.*, **41**, pp. 61–68.

55. Yuan, Y., Gao, Y., Mao, L., and Zhao, J. (2008). Optimisation of conditions for the preparation of β-carotene nanoemulsions using response surface methodoly, *Food Chem.*, **107**, pp. 1300–1306.

56. Silva, H. D., Cerqueira, M. A., Souza, B. W. S., Ribeiro, C., Avides, M. C., Quintas, M. A. C., Coimbra, J. R. S., Carneiro-da-Cunha, M.G., and Vicente, A. A. (2011). Nanoemulsions of β-carotene using a high-energy emulsification-evaporation technique, *J. Food Eng.*, **102**, pp. 130–135.

57. Tan, C. P., and Nakajima, M. (2005). β-carotene nanodispersions: preparation, characterization and stability evaluation, *Food Chem.*, **92**, pp. 661–671.

58. Tan, C. P., and Nakajima, M. (2005). Effect of polyglycerol esters of fatty acids on physicochemical properties and stability of β-carotene nanodispersions prepared by emulsification/evaporation method, *J. Sci. Food Agric.*, **85**, pp. 121–126.

59. Chu, B. S., Ichikawa, S., Kanafusa, S., and Nakajima, M. (2007). Preparation of protein-stabilized β-carotene nanodispersions by emulsification-evaporation method, *J. Am. Oil Chem. Soc.*, **84**, pp. 1053–1062.

60. Yin, L. J., Chu, B. S., Kobayashi, I., and Nakajima, M. (2009). Performance of selected emulsifiers and their combinations in the preparation of β-carotene nanodispersions, *Food Hydrocol.*, **23**, pp. 1617–1622.

61. Wang, P., Liu, H. J., Mei, X. Y., Nakajima, M., and Yin, L. J. (2012). Preliminary study into the factors modulating β-carotene micelle formation in dispersions using an in vitro digestion model, *Food Hydrocol.*, **26**, pp. 427–433.

62. Cao-Hoang, L., Fougère, R., and Waché, Y. (2011). Increase in stability and change in supramolecular structure of β-carotene through encapsulation into poly lactic acid nano particles, *Food Chem.*, **124**, pp. 42–49.

63. De Paz, E., Martín, A., Estrella, A., Rodríguez Rojo, S., Matías, A. A., Duarte, C. M. M., and Cocero, M. J. (2012). Formulation of β-carotene by precipitation from pressurized ethyl acetate on water emulsions for application as natural colorant, *Food Hydrocol.*, **26**, pp. 17–27.

64. Cocero, M. J., and Ferrero, S. (2002). Crystallization of β-carotene by a GAS process in batch. Effect of operating conditions, *J. Supercrit Fluids,* **22**, pp. 237–245.

65. Cardoso, M. A., Antunes, S., Van Keulen, F., Ferreira, B. S., Geraldes, A., Cabral, J., and Palabra, A. (2009). Supercritical antisolvent micronisation of synthetic all-trans- β-carotene with tetrahydrofuran as solvent and carbon dioxide as antisolvent. *J. Chem. Tech. Biotech.*, **84**, pp. 215–222.

66. He, W. Z., Suo, Q. L., Hong, H. L., Li, G. M., Zhao, X. H., Li, C. P., and Shan, A. (2006). Supercritical antisolvent micronization of natural carotene by the SEDS process through prefilming atomization, *Ind. Eng. Chem. Res.,* **45**, pp. 2108–2115.

67. Franceschi, E., De Cesaro, A. M., Ferreira, S. R. S., and Oliveira, J. V. (2009). Precipitation of β-carotene microparticles from SEDS technique using supercritical CO_2, *J. Food Eng.*, **95**, pp. 656–663.

68. Yin, H. Y., Hemingway, M., Xia, F., Li, S. N., and Zhao, Y. P. (2011). Production of β-carotene nanoparticles by the solution enhanced dispersion with enhanced mass transfer by ultrasound in supercritical CO_2 (SEDS-EM), *Ind. Eng. Chem. Res.*, **50**, pp. 13475–13484.

69. Martín, A., Mattea, F., Gutiérrez, L., Miguel, F., and Cocero, M. J. (2007). Co-precipitation of carotenoids and bio-polymers with the supercritical anti-solvent process, *J. Supercrit. Fluids*, **41**, pp. 138–147.

70. Franceschi, E., De Cesaro, A. M., Feiten, M., Ferreira, S. R. S., Dariva, C. U., Kunita, M. H., Rubira, A. F., Muniz, E. C., Corazza, M. L., and Oliveira, J. V. (2008). Precipitation of β-carotene and PHBV and co-precipitation from SEDS technique using supercritical CO_2, *J. Supercrit. Fluids*, **47**, pp. 259–269.

71. Priamo, W. L., Cezaro, A. M., Ferreira, S. R. S., and Oliveira, J. V. (2010). Precipitation and encapsulation of β-carotene in PHBV using carbon dioxide as anti-solvent, *J. Supercrit. Fluids*, **54**, pp. 103–109.

72. He, W., Suo, Q., Hong, H., Shan, A., Li, C., Huang, Y., Li, Y., and Zhu, M. (2007). Production of natural carotene-dispersed polymer microparticles by SEDS-PA co-precipitation, *J. Mater. Sci.*, **42**, pp. 3495–3499.

73. Mattea, F., Martín, Á., Matías-Gago, A., and Cocero, M. J. (2009). Supercritical antisolvent precipitation from an emulsion: β-carotene nanoparticle formation, *J. Supercrit. Fluids*, **51**, pp. 238–247.

74. Santos, D. T., Martín, Á., Meireles, M. A. A., and Cocero, M. J. (2012). Production of stabilized sub-micrometric particles of carotenoids using supercritical fluid extraction of emulsions, *J. Supercrit. Fluids*, **61**, pp. 167–174.

75. Campardelli, R., Adami, R., and Reverchon, E. (2012). Preparation of stable aqueous nanodispersions of β-carotene by supercritical assisted injection in a liquid antisolvent, *20th International Congress of Chemical and Process Engineering CHISA 2012, 25–29 August 2012,* Prague, Czech Republic, *Proc. Eng.*, **42**, pp. 1493–1501.

76. De Paz, E., Martín, Á., and Cocero, M. J. (2013). Formulation of β-carotene: comparison between precipitation from a pressurized emulsion and ultrasound emulsification, *6th International Symposium on High Pressure Process Technology,* Belgrade, Serbia.

77. De Paz, E., Martín, A., and Cocero, M. J. (2011). Formulation of β-carotene by precipitation from pressurized organic solvent-on-water emulsions: comparison between ethanol and ethyl acetate, *1st Iberian Meeting on Natural Bioactives Entrapment for the Food Industry: Challenges and Perspectives, from Nanotechnology to Bioavailability (IMBEFI1),* Lisbon, Portugal.

Chapter 9

Strategies for $scCO_2$ Technology

Concepción Domingo
Materials Science Institute of Barcelona (CSIC),
Campus de la UAB, 08193 Bellaterra, Spain
conchi@icmab.es

The physicochemical properties of $scCO_2$ offer a widespread set of alternatives for novel processing protocols and strategies. Strategies of the use of $scCO_2$ for material-processing purposes could be classified into four main groups, as depicted in Fig. 9.1.

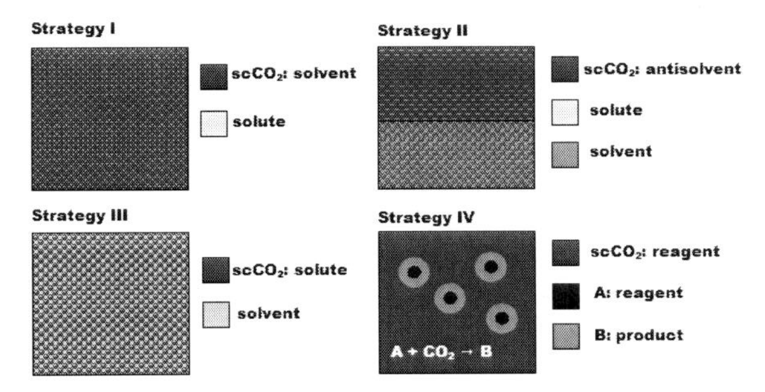

Figure 9.1 Strategies for material processing using supercritical fluid technology: (I) $scCO_2$ as a solvent, (II) $scCO_2$ as an antisolvent, (III) $scCO_2$ as a solute, and (IV) $scCO_2$ as a reagent.

Supercritical Fluid Nanotechnology: Advances and Applications in Composites and Hybrid Nanomaterials
Edited by Concepción Domingo and Pascale Subra-Paternault

ISBN 978-981-4613-40-8 (Hardcover), 978-981-4613-41-5 (eBook)
www.panstanford.com

9.1 Strategy I: Use of $scCO_2$ as a Solvent

Extraction of natural compounds, particle formation processes (e.g., rapid expansion of a supercritical solution, RESS), polymer impregnation, cleaning of impurities, solvent removal, and the reaction medium (e.g., in Diels–Alder, silanization, hydrogenation, hydroformilation, oxidation, polymerization, and enzimatic reactions) are among the most promising applications exploiting the solvent power of supercritical carbon dioxide ($scCO_2$) [1–6]. CO_2 can provide not only environmental advantages but also chemical advantages when applied strategically [7]:

(i) CO_2 cannot be oxidized; in essence, CO_2 is the result of complete oxidation of organic compounds and it is, therefore, particularly useful as a solvent in oxidation reactions,

(ii) CO_2 is benign and hence cross-contamination during extraction is not significative.

(iii) CO_2 is an aprotic solvent that can be employed without penalty in cases where labile protons could interfere with the reaction.

(iv) CO_2 is generally immune to free radicals and does not support chain transfer to solvent during free-radically-initiated polymerization; thus, it is an ideal solvent for use in such polymerizations, despite the fact that it is typically a poor solvent for high-molecular-weight polymers.

(v) CO_2 is miscible with gases in all proportions above 304 K.

(vi) CO_2 exhibits solvent properties with a variety of low-molecular-weight organic liquids, as well as with many common fluorous (perfluorinated) solvents.

(vii) CO_2 exhibits a liquid viscosity only 1/10 that of water and, hence, the Reynolds number for flowing CO_2 is approximately 10 times that for conventional fluids at comparable fluid velocity, improving convective heat transfer.

Obviously, CO_2 also exhibits some inherent disadvantages as a solvent:

- CO_2 exhibits a low dielectric constant.
- CO_2 is a Lewis acid and reacts with strong bases (amines, phosphines, alkyl anions).

- CO_2 can be hydrogenated in the presence of noble metal catalysts to produce CO.
- Compressed CO_2 produces low pH (2–3) upon contact with water.

Chapter 10 addresses two examples of organic reactive processes carried out in $scCO_2$, the synthesis of organic macrocycles [8] and the ship-in-a-bottle formation of trityl and piryl cations [9, 10]. Chapter 11 addresses the reactive precipitation of inorganic calcium carbonate in supercritical media.

9.2 Strategy II: Use of $scCO_2$ as an Antisolvent

The lack of solubility of certain solutes in $scCO_2$ may be exploited for the processing of these materials by using the supercritical fluid as an antisolvent. $scCO_2$ must be partially or totally miscible with the liquid solvent and a nonsolvent for the solute. Under these conditions, a simultaneous two-way mass transfer, caused by the rapid diffusion of CO_2 into the liquid solution and the solvent into the CO_2 phase, takes place. The diffusion of the $scCO_2$ antisolvent into the liquid solvent reduces the solvent power of the liquid, thus causing solute precipitation. The antisolvent processes were described in detail in Chapter 6.

9.3 Strategy III: Use of $scCO_2$ as a Solute

Technologies based on the use of $scCO_2$ as a solute and, in particular the particles from gas-saturated solutions (PGSS®) process, have emerged as alternative one-step methods to obtain solvent-free polymer or lipid particles at low processing temperatures. Active substance-loaded particles and particles with encapsulated liquid are also processed by the PGSS® process. The technique consists in dissolving high concentrations of $scCO_2$ in the bulk of a melted solid and subsequent quick expansion through a nozzle, causing complete evaporation of the gas and solidification of the liquid into fine particles. The PGSS® process operates at lower pressures than other $scCO_2$-asssisted particle formation processes (e.g., RESS) and no organic cosolvents are needed for processing. Moreover, it can be operated in continuous mode, giving excellent process yields. However, the control of particle size and particle size distribution of the obtained powder has to be improved. The PGSS process was described in detail in Chapter 7.

Moreover, $scCO_2$ is used as a solute when dissolved in amorphous polymers to act as a porogenic agent. Pores are formed when the $scCO_2$ solubilized in the polymeric matrix is released upon depressurization. The foaming of biopolymers using $scCO_2$ is being investigated with the aim of preparing scaffolds for tissue engineering [11–16]. These techniques will be described in detail in Chapters 12–15 of Part IV of this book.

9.4 Strategy IV: Use of $scCO_2$ as a Reagent

Much research is focused on the use of CO_2 as a green feedstock for the sustainable synthesis of chemicals [17–25]. To date, however, the economics of such processes are not very promising, since in most of the applications CO_2 technology competes with the existing routes/plants using the highly reactive and effective, but toxic, carbon monoxide (CO). Nonetheless, the high concentration of CO_2 when $scCO_2$ is used may accelerate some of these reactions and make the use of $scCO_2$ more effective than CO. Moreover, the high solubility of many gases in $scCO_2$ (e.g., H_2) is a chemical advantage for using $scCO_2$ as a reagent (e.g., in the hydrogenation of CO_2 to make formic acid and in the synthesis of alkyl formates or alkyl and dialkyl formamides using CO_2 and H_2 as reagents) [26]. The use of $scCO_2$ as a carbonation agent may also reduce the mass transfer limitations in diffusion-controlled reactions, such as in the synthesis of inorganic carbonates [27]. Finally, the simultaneous use of $scCO_2$ as both a reaction medium and a reactant is being prospected, since it is regarded as a smart reaction strategy for synthesis [28]. This approach is being evaluated for the synthesis of organic chemicals, such as ureas, Schiff base macrocycles, carbamates, and isocyanate carbodiimides [29, 30].

In Chapter 11, rhombohedral calcite with a very low degree of agglomeration, high specific surface area, and sizes going from a few microns to nanocrystals is obtained under supercritical conditions [31]. By using compressed CO_2, the reactor size, necessary for a desired production rate, is decreased. Moreover, the use of ultrasounds coupled with $scCO_2$ treatment largely improves carbonation process kinetics [32, 33]. The $scCO_2$ carbonation process is also applied to the in situ precipitation of calcite inside the pores of cellulose paper [34] or Portland cement, thus increasing the density and reducing the water permeability and pH of the material, which enhances the durability in certain applications [35, 36]. Cement carbonation not only generates a high-value-added product but also could help to CO_2 capture and storage [37], and therefore, it is regarded as a sustainable process. Calcium-based CO_2 solid sorbents for CO_2 capture reveal a different efficiency of the carbonation/calcination cycle according to their origin, natural or synthetic, and the synthesis method.

References

1. Beckman, E. J. (2004). Supercritical and near-critical CO_2 in green chemical synthesis and processing, *J. Supercrit. Fluids*, **28**, pp. 121–191.
2. Domingo, C., Berends, E., and van Rosmalen, G. M. (1997). Precipitation of ultrafine organic crystals from the rapid expansion of supercritical solutions over a capillary and a frit nozzle, *J. Supercrit. Fluids*, **10**, pp. 39–55.
3. Roig, A., Mata, I., Molins, E., Miravitlles, C., Torras, J., and Llibre, J. (1998). Silica aerogels by supercritical extraction, *J. Eur. Ceramic Soc.*, **18**, pp. 1141–1143.
4. Loste, E., Fraile, J., Fanovich, M. A., Woerlee, G. F., and Domingo, C. (2004). Anhydrous supercritical carbon dioxide method for the controlled silanization of inorganic nanoparticles, *Adv. Mater.*, **16**, pp. 739–744.
5. Martínez, J. L. (2007). *Supercritical Fluid Extraction of Nutraceuticals and Bioactive Compounds* (CRC Press, Taylor and Francis, Boca Raton, USA).
6. Santana, A., Larrayoz, M. A., Ramírez, E., Nistal, J., and Recasens, F. (2007). Sunflower oil hydrogenation on Pd in supercritical solvents: kinetics and selectivities, *J. Supercrit. Fluids*, **41**, pp. 391–403.
7. Beckman, E. J. (2004). Supercritical and near-critical CO_2 in green chemical synthesis and processing, *J. Supercrit. Fluids*, **28**, pp. 121–191.

8. Lopez-Periago, A. M., Garcıa-Gonzalez, C. A., and Domingo, C. (2010). Towards the synthesis of Schiff base macrocycles under supercritical CO_2 conditions, *Chem. Commun.*, **46**, pp. 4315–4317.
9. López-Periago, A.M., Fraile, J. García-González, C. A., and Domingo, C. (2009). Impregnation of a triphenylpyrylium cation into zeolite cavities using supercritical CO_2, *J. Supercrit. Fluids*, **50**, pp. 305–312.
10. López-Periago, A.M. García-González, C. A., Saurina, J., and Domingo, C. (2010). Preparation of trityl cations in faujasite micropores through supercritical CO_2 impregnation, *Microp. Mesop. Mater*, **132**, pp. 357–362.
11. Woods, H. M., Silva, M. M. C. G., Nouvel, C., Shakesheff, K. M., and Howdle, S. M. (2004). Materials processing in supercritical carbon dioxide: surfactants, polymers and biomaterials, *Mater. Chem.*, **14**, pp. 1663–1678.
12. Marr, R., and Gamse, T. (2000). Use of supercritical fluids for different processes including new developments: a review, *Chem. Eng. Proc.*, **39**, pp. 19–28.
13. Kim, S. H., Jung, Y., and Kim, S.H. (2013). A biocompatible tissue scaffold produced by supercritical fluid processing for cartilage tissue engineering, *Tissue Eng. Part C Methods*, **19**, pp. 181–188.
14. Montjovent, M. O., Mathieu, L., Hinz, B., Applegate, L. L., Bourban, P. E., Zambelli, P. Y., Månson, J. A., and Pioletti, D. P. (2005). Biocompatibility of bioresorbable poly(L-lactic acid) composite scaffolds obtained by supercritical gas foaming with human fetal bone cells, *Tissue Eng.*, **11**, pp. 1640–1649.
15. Duarte, A. R. C., Mano, J. F., and Reis, R. L. (2009). Preparation of starch-based scaffolds for tissue engineering by supercritical immersion precipitation, *J. Supercrit. Fluids*, **49**, pp. 279–285.
16. Salerno, A., Zeppetelli, S., di Maio, E., Iannace, S., and Netti, P. A. (2011). Design of bimodal PCL and PCL-HA nanocomposite scaffolds by two step depressurization during solid-state supercritical CO_2 foaming, *Macromol. Rapid Commun.*, **3**, pp. 1150–1156.
17. Behr, A. (1988). Carbon dioxide as an alternative C1 synthetic unit: activation by transition-metal complexes, *Angew. Chem., Int. Ed.*, **27**, pp. 661–678.
18. Vieville, C., Yoo, J. W., Pelet, S., and Mouloungui, Z. (1998). Synthesis of glycerol carbonate by direct carbonation of glycerol in supercritical CO_2 in the presence of zeolites and ion exchange resin, *Catal. Lett.*, **56**, pp. 245–247.

19. Shaikh, A. A. G., and Sivaram, S. (1996). Organic carbonates, *Chem. Rev.*, **96**, pp. 951–976.

20. Stephenson, P., Licence, P., Ross, S. K., and Poliakoff., M. (2004). Continuous catalytic asymmetric hydrogenation in supercritical CO_2, *Green Chem.*, **6**, pp. 521–523.

21. Pomelli, C.S., Tomasi, J., and Solà, M. (1998). Theoretical study on the thermodynamics of the elimination of formic acid in the last step of the hydrogenation of CO_2 catalyzed by rhodium complexes in the gas phase and supercritical CO_2, *Organometallics*, **17**, pp. 3164–3168.

22. Ikariya, T., Jessop, P. G. Hsiao, Y., and Noyori, R. (1998). *Method for Producing Formic Acid or Its Derivatives*, US Patent No. 5,763,662.

23. Gibson, D.H. (1996). The organometallic chemistry of carbon dioxide, *Chem. Rev.*, **96**, pp. 1063–2095.

24. Reetz, M. T., Könen, W., and Strack, T. (1993). Supercritical carbon dioxide as a reaction medium and reaction partner, *Chim. Int. J. Chem.*, **47**, pp. 493–499.

25. Jessop, P. G., Hsiao, Y., Ikariya, T., and Noyori, R. (1996). Homogeneous catalysis in supercritical fluids: hydrogenation of supercritical carbon dioxide to formic acid, alkyl formates, and formamides, *J. Am. Chem. Soc.*, **118**, pp. 344–355.

26. Jessop, P.G. (2006). Homogeneous catalysis using supercritical fluids: recent trends and systems studied, *J. Supercrit. Fluids*, **38**, pp. 211–231.

27. Domingo, C., Loste, E., Gómez-Morales, J., García-Carmona, J., and Fraile, J. (2006). Calcite precipitation by a high-pressure CO_2 carbonation route, *J. Supercrit. Fluids*, **36**, pp. 202–215.

28. Jessop, P. G., and Leitner, W. (1999). *Chemical Synthesis Using Supercritical Fluids* (VCH/Wiley, Weinheim, Germany).

29. Scondo, A., Dumarcay-Charbonnier, F., Marsura, A., and Barth, D. (2009). Supercritical CO_2 phosphine imide reaction on peracetylated β-cyclodextrins, *J. Supercrit. Fluids*, **48**, pp. 41–47.

30. López-Aranguren, P., Fraile, J., Vega, L. F., and Domingo, C. (2014). Regenerable solid CO_2 sorbents prepared by supercritical grafting of aminoalkoxysilane into low-cost mesoporous silica, *J. Supercrit. Fluids*, **85**, pp. 68–80.

31. Domingo, C., García-Carmona, J., Loste, E., Fanovich, A., Fraile, J., and Gómez-Morales J. (2004). Control of calcium carbonate morphology precipitation in compressed and supercritical carbon dioxide media, *J. Cryst. Growth*, **271**, pp. 268–273.

32. Lopez-Periago, A. M., Pacciani, R., Garcia-Gonzalez, C. A., Vega, L.F., and Domingo, C. (2010). A breakthrough technique for the preparation of high-yield precipitated calcium carbonate, *J. Supercrit. Fluids*, **52**, pp. 298–305.

33. Lopez-Periago, A. M., Pacciani, R., Vega, L. F., and Domingo, C. (2011). Monitoring the effect of mineral precursor, fluid phase CO_2-H_2O composition and stirring on $CaCO_3$ crystallization in a supercritical-ultrasound carbonation process, *Cryst. Growth Des.*, **11**, pp. 5324–5332.

34. Domingo, C., Loste, E., Gómez, J., García, J., and Fraile, J. (2006). Calcite precipitation by a high-pressure CO_2 carbonation route, *J. Supercrit. Fluids*, **36**, pp. 202–215.

35. García-González, C. A., Hidalgo, A., Andrade, C. Alonso, M. C., Fraile, J., López-Periago, A. M., and Domingo, C. (2006) Modification of composition and microstructure of Portland cement pastes as a result of natural and supercritical carbonation procedures, *Ind. Eng. Chem. Res.*, **45**, pp. 4985–4992.

36. Huertas, J., Hidalgo, A., Rozalén, M. L., Pellicione, S., Domingo, C., García-González, C. A., Andrade, C., and Alonso, C. (2009). Interaction of bentonite with supercritically carbonated concrete, *Appl. Clay Sci.*, **42**, pp. 488–496.

37. López-Periago, A. M., Fraile, J., López-Aranguren, P., Vega, L. F., and Domingo, C. (2013). CO_2 capture efficiency and carbonation/calcination kinetics of micro and nanosized particles of supercritically precipitated calcium carbonate, *Chem. Eng. J.*, **226**, pp. 357–366.

Chapter 10

Innovations in Organic Synthesis in $scCO_2$: The Schiff Base Reaction and a Ship-in-a-Bottle Approach for the Preparation of Hybrid Materials

Ana M. Lopez-Periago, Nerea Murillo-Cremaes, and Concepción Domingo
Materials Science Institute of Barcelona (CSIAC), Campus de la UAB, 08193 Bellaterra, Spain
amlopez@icmab.es

Synthesis of many specialty chemicals involves use of organic solvents. The objective of this chapter is to describe the feasibility of carrying out organic reactions in supercritical fluids. Supercritical fluids have properties that could make them nearly the ideal media for conducting synthetic reactions. This chapter, after a revision of chemical reactions in $scCO_2$, addresses the synthesis of organic complex molecules using this fluid as a solvent. The first section revises the formation of macrocycles in $scCO_2$, where Schiff-based (C=N) and polycyclic ring compounds are prepared through a cyclocondensation reaction between amines and aromatic aldehydes

Supercritical Fluid Nanotechnology: Advances and Applications in Composites and Hybrid Nanomaterials
Edited by Concepción Domingo and Pascale Subra-Paternault

ISBN 978-981-4613-40-8 (Hardcover), 978-981-4613-41-5 (eBook)
www.panstanford.com

in the absence of any template and catalyst or cosolvent. The second section explores the preparation of hybrid materials through a ship-in-a-bottle approach in $scCO_2$. These hybrid materials are composed of medium-sized organic cationic molecules synthetized into zeolite cages.

10.1 Introduction to Chemical Reactions in $scCO_2$

The blooming of the study of the chemical reactions carried out using supercritical carbon dioxide ($scCO_2$) as a solvent started in the early 1980s, when the potential of this solvent started to inspire scientist into the research of different methods of chemical synthesis. Later in the 1990s, the concern of carrying out chemical reactions in a greener way increased the interest in developing new approaches where all, the chemical reagents, solvents, and by-products, are environmentally friendly or, at least, less harmful [1]. This interest was reflected by McHungh and Krukonis [2] who described the advantages of carrying out the reactions in supercritical fluids, relating the increased reactions rates and selectivities resulting from the rapid diffusion and the low solvation of the reacting species. $scCO_2$ has unique properties that makes it an excellent medium in which conduct chemical reactions. Despite the potential of $scCO_2$ as a solvent medium to carry out organic synthesis, there are still some disadvantages that slow down the development of new synthetic strategies. Among these disadvantages are found the low solubility of high-molecular-weight molecules in $scCO_2$ and the difficulty of working with polar compounds in the absence of cosolvents.

Supercritical fluids are becoming increasingly important in industry, partly in response to the adverse environmental impact of solvent use and disposal. As a reaction medium, the attractive physical and toxicological inertness properties of $scCO_2$ have made it superior to conventional organic solvents in a large number of synthetic transformations [3]. Research is focused on reactions in which the outcome either cannot be obtained using traditional organic solvents or is influenced to a great extent by the unique properties of $scCO_2$. Two main categories of organic reactions in $scCO_2$ have been developed—(i) those in which $scCO_2$ is used as a solvent/reaction medium [4] and (ii) those in which $scCO_2$ takes part in the reaction as a reactant or simultaneously as a reagent and a solvent [5].

10.1.1 Organic Reactions in an $scCO_2$ Medium

In 1999, a special issue of *Chemical Reviews* was dedicated to organic reactions in supercritical media [6]. From them, the various aspects of reactions performed in $scCO_2$, used as a solvent, have been extensively reviewed [7–13]. $scCO_2$ is used as a medium in the following chemical reactions: enzyme-catalyzed [14], polymerization [15–17], radical [18], cycloaddition [19], and transition metal–catalyzed reactions [20, 21].

10.1.1.1 Transition metal–catalyzed reactions

Researchers in organometallic catalysis started to contemplate a broad use of CO_2 as a solvent in the mid-1990s, especially after sufficiently "CO_2-philic" organometallic catalysts became generally available [22, 23]. Organic reactions catalyzed by transition metals in $scCO_2$ include oxidation, reduction, carbonylation, radical addition, C–C coupling, and cycloaddition reactions [24]. All these reactions can be performed in $scCO_2$ with equal or even better results than in conventional organic solvents, owing to the high diffusion and low viscosity of $scCO_2$. Selective examples using Pd, Rh, Ru, Ir, Pt, Ni, Co, Cu, Ag, and Au are reviewed in [25]. Particularly, palladium-catalyzed C–C coupling reactions are finding increasing importance for industrial synthesis of fine chemicals, including biologically active compounds. The most studied chemical transformations in $scCO_2$ involving the C–C bond formation include the Diels–Alder reaction, the thermal cracking of hydrocarbons, the hydroformylation reactions, and Suzuki coupling [26]. Another well-described type of reaction is the hydrogenation of fats [27]. In $scCO_2$, such coupling processes occur at high rates and with excellent selectivity [28–30]. In oxidation reactions, the advantage of the process is related with the fact that oxygen and CO_2 are completely miscible under supercritical conditions. Also, CO_2 cannot be oxidized further during catalytic oxidation. Hence, oxidation reactions in $scCO_2$ eliminate side products formation due to oxidation of organic solvents. The addition of suitable amounts of cosolvents in $scCO_2$ is necessary in most reactions, because it not only improves the solubility of transition metal salts in $scCO_2$ but also regulates the ratio of products and raises the selectivity as well.

10.1.1.2 Polymerization reactions

The first homogeneous free-radical polymerization in $scCO_2$ was reported in 1992 [31]. $scCO_2$ is an ideal solvent for the polymerization of fluoropolymers and silicon-based polymers, which display limited solubility in organic solvents [32]. Moreover, although most commercial polymers are not soluble in $scCO_2$, they can be synthesized in biphasic dispersion and emulsion polymerization modes [33, 34]. The molecular weights and polymer properties prepared in $scCO_2$ are similar to those obtained by analogous polymerization methods in organic solvents. Therefore, the incentives of using $scCO_2$ as a solvent lie not in the polymerization reaction but in the decreased cost of polymer processing [35]. Polymers synthesized in $scCO_2$ can be isolated simply by depressurization of the reaction vessel. Moreover, due to the increased plasticity of polymers in $scCO_2$, the residual monomer and catalysts are easily removed from the polymer matrix.

10.1.1.3 Enzyme-catalyzed reactions

The stability and activity of enzymes exposed to CO_2 under high pressure depend on enzyme species, water content in the solution, and the pressure and temperature in the reaction system, as well as the depressurization step [36, 37]. The three-dimensional structure of enzymes may be significantly altered under extreme conditions, causing their denaturation and consequent loss of their activity. If the conditions are less adverse, the protein structure may largely be retained [38]. The initial applications of $scCO_2$ treatments involving enzymes were designed as an alternative to thermal microbial inactivation, which can be safely used in foods and bioactive materials at relatively low temperatures [39]. This method has received increasing attention since $scCO_2$ was shown to be effective for the treatment of *Escherichia coli* [40]. It was stated that enzyme inactivation with $scCO_2$ could be predominantly attributed to the pH-lowering effect during treatment [41]. Further, the finding that some enzymes, such as lipases, several phosphatases, dehydrogenases, oxidases, amylases, and others, are well suited for enzymatic reactions in $scCO_2$ has broadened immensely the scope of their applications as highly enantioselective catalysts in synthesis carried out in $scCO_2$. Interestingly, the activity and selectivity of enzymes can be modulated by changes in the working pressure or temperature, increasing the range of products that a single enzyme can form [42]. The application of $scCO_2$ as a solvent in enzyme-catalyzed reactions has been a matter of considerable research because of its favorable transport properties, which accelerate mass-transfer-limited reactions [43]. There are several reports in the literature on enzymatic hydrolysis of vegetable oil or triacylglycerol in $scCO_2$ solvent, most of them in enzymes immobilized into porous silica supports. The major advantage of immobilized enzymes consisted in an easier separation from the product [44, 45].

10.1.2 Reactions Involving CO_2 as a Reactant

In addition to acting as a solvent with unique physical properties, $scCO_2$ proves to have synthetic utility in a variety of organic and inorganic reactions. CO_2 is an inert molecule in most organic environments; however, it is a Lewis acid and reacts with strong bases, such as amines or phosphines. The use of $scCO_2$ as a labile-protecting group for secondary amines has been reported [46]. Moreover, $scCO_2$ has been used as a C_1 building block, that is, as a carbon source, in the formation of formic acid, methyl formate, and dimethyl formamide [47, 48]. The application of $scCO_2$ is now reasonably established for hydrogenation and hydroformilation. The high degree of solubility of H_2 gas in $scCO_2$ is one factor contributing to the improved results of these reactions in this medium. In addition to carbamate formation in the reaction of CO_2 and amines, the fluid can also form carbonates. An important class of these materials is inorganic carbonates formed by the reaction of CO_2 and alkaline earth metals [49].

10.2 Schiff Base Synthesis in Supercritical CO_2

The richness of amine and organonitrogen functional groups and the potential that they have in natural and synthetic process make them an important class of family in reactive chemistry.

One significant obstacle that chemists encountered in synthetic chemistry when using scCO$_2$ as a solvent is the lack of research carried out in the chemistry of nitrogen. The reason is related with the reaction between amines and CO$_2$ that forms insoluble materials (carbamates), often inadequate for further chemical processing. On other occasions, the drawback of the formation of these insoluble species is at the same time an advantage, for instance, the capacity of mono- and diamines to uptake CO$_2$ that is used in the technology for CO$_2$ capture, as a help to decrease the greenhouse effect. Indeed, amine scrubbing, which has been used to separate CO$_2$ from natural gas and hydrogen since 1930 [50], now is a robust technology and it is used on a large scale for CO$_2$ capture from coal-fired power plants. Recently, the interest in trying new ideas for reactions of amines and imines and scCO$_2$ has increased and turned into a new field of research.

A Schiff base or imine bond (C=N) is the result of the condensation of a carbonyl compound with an amine (Scheme 10.1).

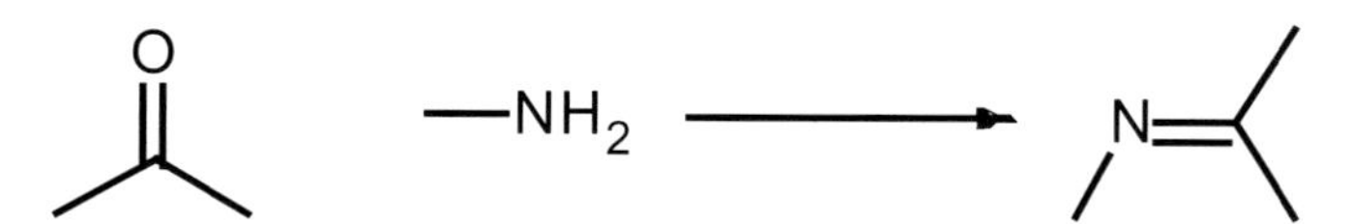

Scheme 10.1 Reaction scheme of imine formation.

The possibilities of using scCO$_2$ as a green solvent for the preparation of Schiff bases and, in particular, macrocyles containing imine groups following the classical reaction has not been extensively explored [51, 52]. However, and despite the disadvantages, scCO$_2$ has a series of characteristics that are expected to facilitate the synthesis of Schiff base molecules and, moreover, the formation of Schiff-based macrocycles [53]. On the one hand, the solvating properties of this fluid can be continuously varied from gas-like to liquid-like values with small changes in the pressure and/or temperature, which can be used to control the thermodynamics and kinetics of the process. Moreover, the fact that scCO$_2$ is an aprotic solvent is also an advantage in the imine synthetic process, because it cannot protonate the amine and there are no labile protons interfering in the reaction. On the other hand, compressed CO$_2$ acts as a Lewis acid, which favors the formation of the acid-catalyzed imine bond.

CO_2 can react directly with primary and secondary amines to produce carbamates through the formation of zwitterion intermediates [54] (Scheme 10.2).

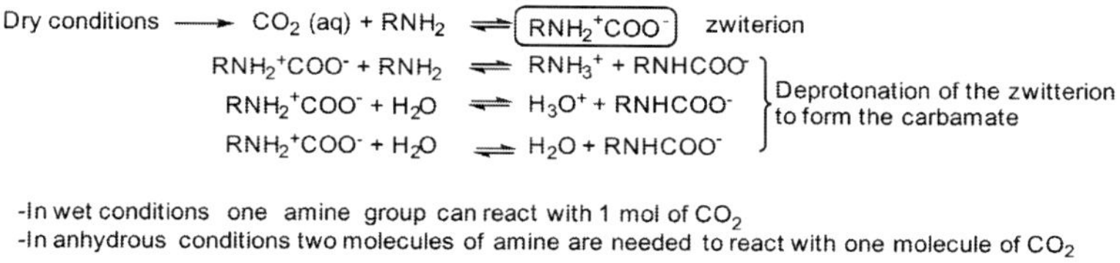

Dry conditions ⟶ $CO_2 (aq) + RNH_2 \rightleftharpoons RNH_2^+COO^-$ zwiterion

$RNH_2^+COO^- + RNH_2 \rightleftharpoons RNH_3^+ + RNHCOO^-$

$RNH_2^+COO^- + H_2O \rightleftharpoons H_3O^+ + RNHCOO^-$

$RNH_2^+COO^- + H_2O \rightleftharpoons H_2O + RNHCOO^-$

Deprotonation of the zwitterion to form the carbamate

-In wet conditions one amine group can react with 1 mol of CO_2
-In anhydrous conditions two molecules of amine are needed to react with one molecule of CO_2

Scheme 10.2 Reaction schemes of behavior of CO_2 and amines in dry and wet conditions.

In an aqueous amine environment, as the one employed for CO_2 capture, the base that deprotonates the zwitterion can be another amine, H_2O or OH^-. On the contrary, in dry conditions, the deprotonation occurs with another amine. Considering the equilibria when working with amines in $scCO_2$, it is advisable to evaluate the extension of the possible reaction between the CO_2 and the amine under used experimental conditions. Another fact to take into account when choosing an amine as a reagent is how hindered is the nitrogen within the bulky structure. It has been shown that introducing steric hindrance by a bulky substituent adjacent to the amino group lowers the stability of the carbamate formed by the CO_2–amine reaction [55]. It is also important to think about the substituents of the organic building blocks. An example of imine formation reaction in $scCO_2$ is shown in Scheme 10.3. In this reaction, a molecule containing two imine bonds was achieved at a pressure of 15 MPa and at 35.5°C of temperature.

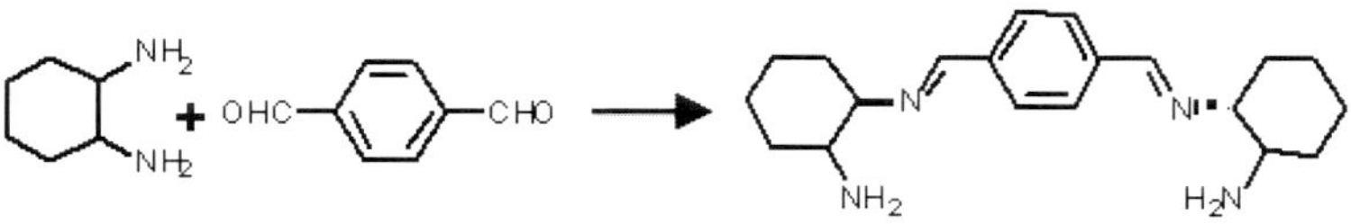

Scheme 10.3 Reaction scheme of a molecule containing two imine groups in $scCO_2$.

In the same way that molecules containing one or two imine bonds can be prepared, polyimine macrocycles can also be synthesized by carefully choosing the processing conditions. Examples of the prepared cyclophanes are shown in Scheme 10.4.

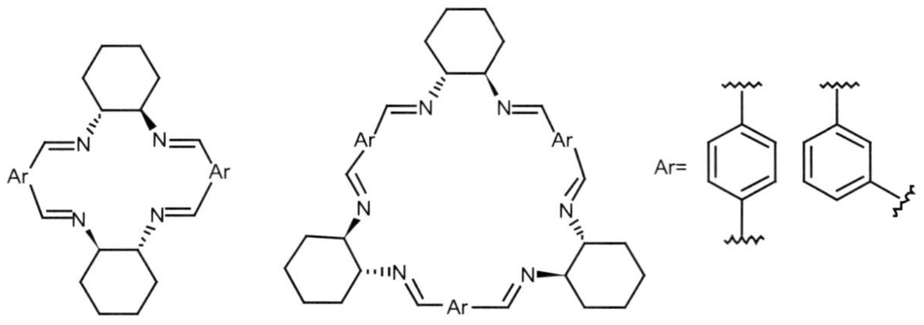

[2+2]

[3+3]

Scheme 10.4 Structures of polyimine macrocycles achieved in $scCO_2$.

These molecules are synthesized using a classic synthetic strategy based on a [2+2] and [3+3] cyclocondensation, where two or three units of (1*R*,2*R*)-diaminocyclohexane react with two or three units of an aromatic dialdehyde, respectively. It should be taken into account that the amine object of study, (1*R*,2*R*)-diaminocyclohexane, is a hygroscopic molecule and it is the adsorbed water that, when in contact with CO_2, produces the acid medium necessary for the catalysis. The preparation of these cycles required a higher pressure and reaction time than in the case of the diimine molecule presented in Scheme 10.3. The pressure and temperature conditions used in Scheme 10.3 were not sufficient to allow the complete macrocyclation; thus a raise of the processing conditions to improve the solvent power of $scCO_2$ was necessary. In the condensation process, as the reaction proceeds to completion, the size of the molecules increases and so does their molecular weight. As the molecule gets heavier, it becomes more insoluble in certain conditions of pressure and temperature. Therefore, the system pressure and the reaction time need to be raised up to 20 MPa and two hours, respectively, to enhance the solvation power of the CO_2, thus helping to dissolve the formed molecules and leading the cyclation process to completion.

The designed supercritical route is not only a greener and safer method than the classical procedure but also a one-stage process that would lead to high yield values, thus allowing a sustainable use of resources. One of the main advantages of this methodology is that the synthesized Schiff bases had an empty core, not filled with solvent molecules, since the $scCO_2$ is eliminated as a gas during depressurization. Hence, the supercritically as-synthesized compounds are ready to participate in host–guest chemistry or to act as selective probe materials. Big efforts are being made to control the reaction and processing conditions in $scCO_2$ amine fine chemistry, but the perspectives of carrying this process from batch work to the bulk hold a promising future.

10.3 The Ship-in-a-Bottle Host–Guest Approach for the Preparation of Hybrid Materials

As the need for new materials for catalytic supports arises, their development have increased and improved. An interesting advance in this area is the creation of catalytic structures composed of reactive clusters uniformly prepared and accommodated in restrictive spaces, such as the cavities of porous materials. In this way, when introducing organic ions with catalytic activity inside restrictive spaces, the migration and/or decomposition under the reaction conditions of these catalytic ions can be avoided. Examples of such catalytic structures are organic cations stabilized inside zeolite X and Y cavities, where the entrapped cations are immobilized inside of the zeolite cages, and strongly stabilized by the negatively charged environment, leading to a very stable structure. Despite the majority of these organic catalysts having the correct size to fit inside the intrazeolite large cavities, they are far too large to be transported throughout the zeolite channels. It is possible to generate bulky organic cations embedded within the large cages of these zeolites by using the ship-in-a-bottle procedure, which consists of constructing the guest molecule by reacting small precursors or building blocks within the zeolite large cavities [56–63].

10.3.1 Encapsulation of Chromophores

A particular case of host–guest synthesis procedures is the encapsulation of chromophores and cationic organic dyes within nanoporous matrices to be used, for instance, in optical applications. Important classes of materials are triphenylpyryl and triphenyltrityl cations used as electron transfer photosensitizers for photochemical reactions involving aromatics, dienes or alkenes, and epoxides and for treating air and water pollutants using solar light. However, an efficient use of these cations is only possible when they are trapped inside of nanoporous supports, since the free cations hydrolyze in water. In this respect, zeolites and aerogels are among the most porous host materials known, with accessible surface areas as large as 1000 m^2g^{-1}. However, for small-pore-opening materials, cations are too large to diffuse through the pore opening, even if they have the proper size to be housed into the internal large cavities. These materials can be incorporated deep in the material following ship-in-a-bottle methods [64–66]. A notorious example is the use of faujasite Y for photochemical applications. Faujasite Y is a zeolite with large cages of 1.2 nm diameter connected by 0.74 nm channels. For pyrylium and tritylium cations, the entrapment inside the cages should be carried out following the ship-in-a-bottle procedure [67, 68], since their sizes are larger than that of the zeolite channels. The encapsulation of these cations have been carried out using conventional solvents using extreme reaction conditions in terms of time, temperature, and organic solvent use [69]. The use of $scCO_2$ for the ship-in-a-bottle impregnation of porous substrates, avoiding at the same time the solid–liquid purification procedure, emerges as a very successful option, mainly due to the high diffusivity and the virtually negligible surface tension of this fluid. A similar procedure is followed to encapsulate the cations into microporous zeolites and mesoporous aerogels [70] (Fig. 10.1).

For both studied matrices, the impregnated cations showed higher hydrolysis stability when encapsulated than in their crystalline form. The cations supported in the mesoporous aerogel must exhibit higher intrinsic photocatalytic activity due to the absence of the light diffusion restrictions occurring in the opaque zeolite. The resulting host–guest assemblies can act as easily recoverable heterogeneous photocatalysts. These examples show another way to carry out chemical reactions using $scCO_2$ as solvent through a C–C bond formation.

Conventionally, the ship-in-a-bottle synthetic approach consists in the adsorption into a thermally dehydrated zeolite of different building blocks dissolved in aprotic organic solvents, such as isooctane, toluene, or cyclohexane, followed by a reflux treatment and finalizing with an exhaustive Soxhlet extraction for product purification [71]. This methodology is commonly limited by the slow diffusion of the building blocks from the liquid phase into the channels, competition between solvent and guest molecules for the adsorbent adsorption sites, very long diffusion times, and the prolonged reacting and processing times. When taking into account the inconveniences that working with organic solvents have, the replacement of these by $scCO_2$ becomes an effective and valuable method to build up and process these hybrid catalytic materials.

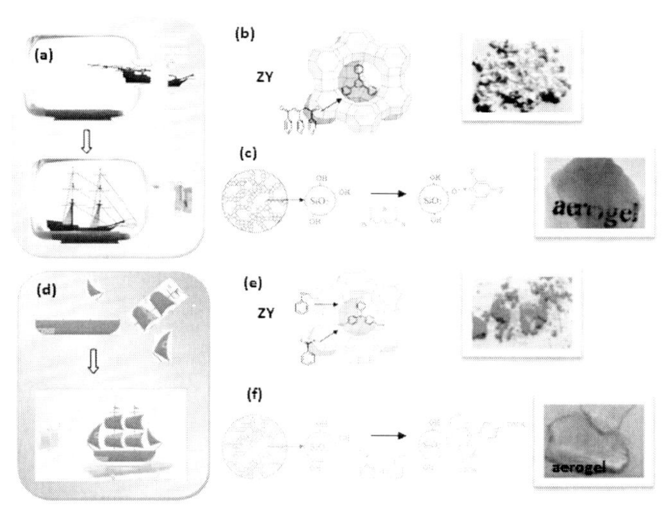

Figure 10.1 Schematic representation of ship-in-a-bottle procedures (a, d) and the $scCO_2$ (T = 313–523 K, P = 15–17.5 MPa) encapsulation reactions of triphenylpyryl cations obtained by the dehydratation and cyclization of the diketone 1,3,5-tripheyl-2-pentene-1,5-dione in zeolite (b) and aerogel (c) and triphenyltrityl cations formed from the reaction between benzaldehyde and anisole in zeolite (e) and aerogel (f).

Impregnation of solid matrixes using $scCO_2$ as a mobile phase has proven to be feasible when the solute compound has an appreciable solubility in $scCO_2$. Moreover, the high diffusivity and the virtually negligible surface tension of this fluid makes $scCO_2$ a successful alternative for the impregnation of microporous substrates, for two reasons—on one side, the competition between the solvent and the solute for the substrate adsorption sites is very much reduced in $scCO_2$ with respect to liquid solvents, and $scCO_2$ gas-like viscosity allows rapid penetration of the dissolved substances into the cavities of microporous materials [72, 73]. To explain how $scCO_2$ is a valuable technology to carry out processes of impregnation of organic molecules in microporous materials and, furthermore, to carry out chemical reactions within this structures, some examples of introduction of organic catalytic cations inside porous rigid structures are presented following. For this, a tridirectional zeolite of the faujasite-Y-type was used as a model study [74]. The compounds chosen as examples for the loading procedure of organic molecules inside the zeolites cages using $scCO_2$ as a solvent are the 2,4,6-triphenylpyrylium (Ph_3Py^+) cation and a collection of different triphenyl trityl cations. The encapsulated cationic samples are described as [(Ph_3Py^+)-Z] for the pyrylium cation and [(RPh_3C^+)-Z] for the different triphenyltrityl cations.

10.3.2 Preparation of 2,4,6-Triphenylpyrylium Encapsulated in Faujasite Y [(Ph_3Py^+)-Z]

Pyrylium salts have proved to be good catalysts for photochemical reactions, but with a poor photostability [75, 76]. Besides, pyrylium cations decompose in aqueous media [77] and must be adsorbed or encapsulated in inorganic supports to increase their chemical stability. Specifically, the encapsulation of the Ph_3Py^+ cation inside faujasite Y is a good example of organic cation stabilized inside an inorganic rigid structure. The Ph_3Py^+ cation fits inside the large cages of the faujasite Y; however, it is too bulky to diffuse through the 0.74 nm channels. Therefore, a good solution for preparing this catalytic structure is by diffusion of a precursor, in this case the diketone 1,3,5-triphenyl-2-pentene-1,5-dione, through the zeolite channels, which after reacting with the acid sites of the zeolite large cavities, formed the Ph_3Py^+ cation (Scheme 10.5).

The supercritical encapsulation experiments of large size molecules into the zeolites cavities needs to be carried out in two steps. These steps consist of first directing the cationic precursor through the connecting channels of the zeolites until they reach the largest cavity, called the diffusion step. The second step implies an intermolecular proton transfer of the acidic sites of the zeolites to the oxygen of one of the carbonyl groups; this protonation leads to an intermolecular reaction leading to the formation of the (Ph_3Py^+) cation. This stage is called the cyclation step.

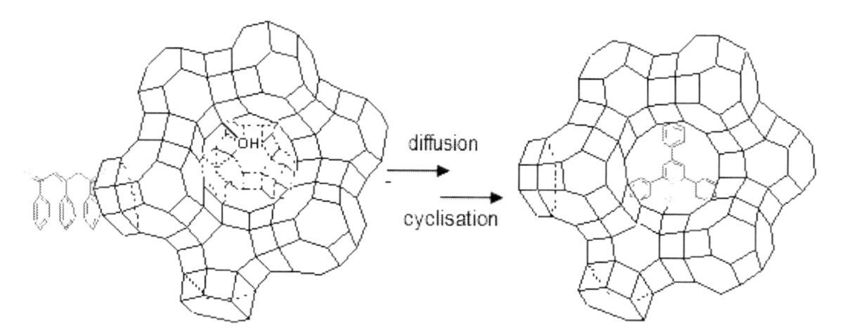

Scheme 10.5 Synthetic ship-in-a-bottle approach of Ph_3Py^+ encapsulated in zeolite [(Ph_3Py^+-Z] through the diffusion of the diketone precursor.

The condition at which the synthetic supercritical process takes place at its maximum of effectiveness, that is, with the highest pyrylium loadings of about 7 wt% occurs, is made in two different temperature steps, low temperature for diffusion and high temperature for cyclisation. The diffusion of the diketone inside of the zeolite channels is carried out at relatively low temperature (60°C) and pressure (150 bar), thus minimizing the Ph_3Py^+ formation, which could lead to the blocking of the zeolite channels. The second step was the cyclization through dehydration at 150°C of the diketone to produce the Ph_3Py^+ cation. The first indication of the successful formation of the organic cation into the zeolite cavities was a change in the substrate color. While the zeolite Y matrix was initially white, the adsorption of Ph_3Py^+ was noticeably visible from the development of yellow color. Diffuse reflectance ultraviolet-visible (UV-Vis) spectra of the solid powders were used to determine the composition of the obtained products, through comparison with a commercial pyrylium cation (Ph_3PyBF_4), not captured in any matrix. The measured diffuse reflectance spectrum of the commercial Ph_3PyBF_4 displayed a broad band between 320 and 530 nm corresponding to an envelope of two chromophores. The same absorption bands are observed for the zeolite-encapsulated Ph_3Py^+, indicating the existence of the cation within the faujasite cages. Also the use of $scCO_2$ proved to be very effective in the purification process. The elimination of the excess of the starting material or formed by products was more energy- and time effective using the $scCO_2$ procedure than by Soxhlet extraction. Moreover, Soxhlet-extracted samples were contaminated with dicloromethane, while $scCO_2$ purified samples were recovered solvent free

10.3.3 Preparation of Triphenyltrityl Cations Encapsulated in Faujasite Y [(RPh_3C^+)-Z]

The ship-in-a-bottle technique in $scCO_2$ has been also used for the preparation of a series of benzyl trityl cations, in particular, the preparation of three zeolite-encapsulated trityl cations: triphenyltrityl, dimethoxytriphenyltrityl, and methoxytriphenyltrityl (Scheme 10.6). Triphenyltrityl cations, with the general formula RPh_3C^+, are widely known organic compounds due to their numerous applications, such as protecting groups, in dye chemistry and in photochemical reactions [78, 79]. The most used method for the preparation of triphenyltrityl compounds by C–C bond formation is the acid-catalyzed Friedel–Crafts reaction of benzaldehydes with benzene or other aromatic compounds [80, 81]. As in the pyrylium case, these cations can be immobilized inside of the zeolite cages and being strongly stabilized by the negatively charged environment, and thus they can act as stable long-life photosensitizers. Since the studied triphenyltrityl cations had a size higher than the diameter of the zeolite channels, the fabrication was based in the diffusion of the selected building blocks, followed by the condensation reaction to form the C–C bond inside of the zeolite large cavieites to form the final organic cation. In this case, benzene, benzaldehyde, anisole, and diphenylcarbinol (Scheme 10.6) were used as organic units to synthesise the different cations. The obtained loadings in each reaction in Scheme 10.6 were estimated from thermogravimetric analysis and are shown in Table 10.1.

Table 10.1 Loadings obtained for the different encapsulated cations [(RPh_3C^+)-Z]

Building blocks	Loading [wt%]
Diphenylcarbinol and benzene	8.1
Benzaldehyde and anisole	7.9
Diphenylcarbinol and anisole	7.9
Benzophenone and anisole	5.9

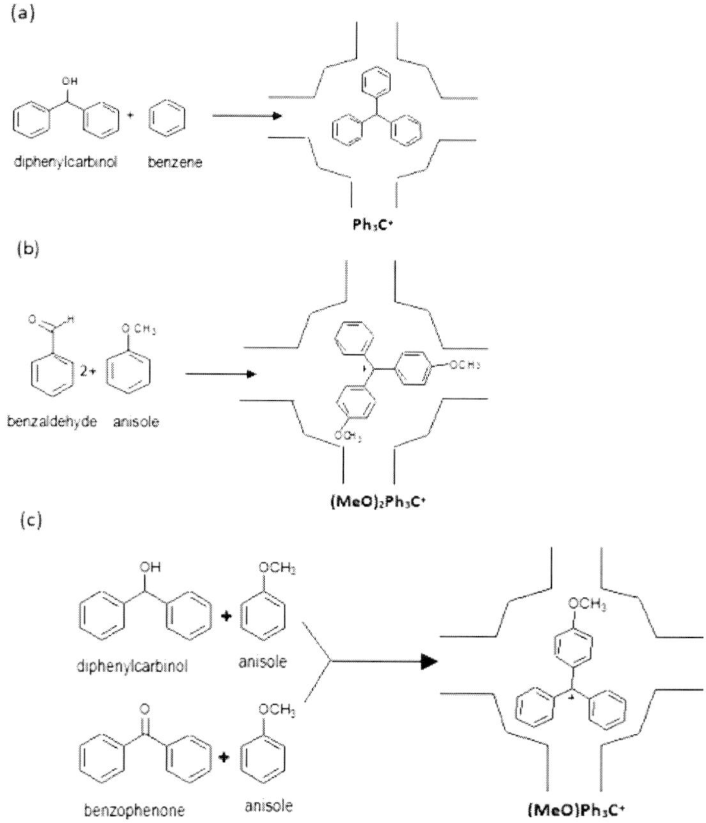

(a)

diphenylcarbinol benzene

Ph_3C^+

(b)

benzaldehyde anisole

$(MeO)_2Ph_3C^+$

(c)

diphenylcarbinol anisole

benzophenone anisole

$(MeO)Ph_3C^+$

Scheme 10.6 Different synthesis protocols of triphenyltrityl cations in zeolite cavities promoted by acid catalytic condensation.

One of the key points that indicate the successful formation of the trityl cation, and that is commonly shown in all of the prepared samples, is the C–Ph stretching band at 1381–1384 cm^{-1} appearing in the Fourier transform infrared spectra. This signal is specifically characteristic of the triphenyltrityl cations, thus indicating their formation inside of the zeolite supercages. The compound $[(Ph_3C^+)\text{-}Z]$ (Scheme 10.6a) is obtained as a yellow powder. In this case, the reluctance of unsusbstituted benzene to undergo electrophilic attack required superacid media for the hydroxyalkylation to occur [82–84]. Hence, zeolites with a high Si/Al ratio must be used. The advantage of using $scCO_2$ as a solvent for the ship-in-a-bottle synthesis is that the acidity of the medium increased due to the interaction of CO_2 with the zeolite adsorbed water, thus facilitating the occurrence of the reaction in zeolites with a Si/Al ratio lower than 2. Sample [$(MeO)_2Ph_3C^+$)-Z] (Scheme 10.6b) is characterized from owing ochre color. The substituents MeO in the formed cation have been described to be in the *para* position and partially penetrating through the windows connecting neighboring cavities [85, 86]. The compound [($MeOPh_3C^+$)-Z], with a dark orange color, was prepared using two different approaches (routes I and II in Scheme 10.6c). A higher yield was obtained following route I with respect to route II (Table 10.1) due to the higher reactivity of diphenylcarbinol alcohol in comparison with benzophenone. Under used experimental conditions the isomerization and disproportionation of anisole and the oxidation of benzaldehyde were also possible processes. However, the kinetics of $R_xPh_3C^+$ formation inside of the zeolite cages was favored and the cation was effectively produced.

10.4 Conclusions

This chapter has revised the latest advances in the synthesis of organic molecules using $scCO_2$ as solvent and processing media. The high flow rates and fast reaction kinetics allow the design of high-throughput processes in small-scale reactors, such as zeolite microreactors. It was shown that imine chemistry is possible in $scCO_2$ and, despite all of the drawbacks, large polyimine macrocycles, with molecular weights ranging from 425 to 637 $gmol^{-1}$, can be formed under $scCO_2$ conditions and in the absence or organic solvents. Moreover, supercritical fluid technology was demonstrated as a successful approach for the impregnation of microporous zeolites with large organic molecules. The formation of the encapsulated final guest (pyrylium and trityl cations) in the interior of the cavities of aerogels and zeolites, using a synthetic method based on a ship-in-a-bottle methodology in $scCO_2$, is a feasible approach.

Acknowledgments

A. Lopez-Periago thanks the MINECO for a Ramón y Cajal Contract (RYC-2012-11588).

References

1. Anastas, P. T., and Warner, J. C. (1998). *Green Chemistry: Theory and Practice* (Oxford University Press, New York, USA).
2. McHungh, M., and Krukonis, U. (1986). *Supercritical Fluids Extractions* (Butterworth, Stockham, MA, USA).
3. Licence, P., Ke, J., Sokolova, M., Rossb, S. K., and Poliakoff, M. (2003). Chemical reactions in supercritical carbon dioxide: from laboratory to commercial plant, *Green Chem.*, **5**, pp. 99–104.
4. Sarbu, T., Styranec, T., and Beckman, E. J. (1999). Non-fluorous polymers with very high solubility in supercritical CO_2 down to low pressures, *Nature*, **405**, pp. 165–168.
5. Jessop, P. G., and Leitner W. (Ed.) (1999). *Chemical Synthesis Using Supercritical Fluids* (Wiley-VCH, Weinheim).
6. Several authors. (1999). Supercritical fluids reactions, *Chem. Rev.*, **99**, pp. 353–634.
7. Niessen H. G., and Woelk K, (2007). Investigations in supercritical fluids, in *In Situ NMR Methods in Catalysis*, eds. Bargon, J., and Kuhn, L. T, *Top. Curr. Chem.*, **276**, pp. 69–110 (Springer-Verlag, Berlin, Heidelberg).
8. Beckman E. J. (2004). Supercritical and near-critical CO_2 in green chemical synthesis and processing, *J. Supercrit. Fluids*, **28**, pp. 121–191.
9. Hyde J. R., Walsh, B., Singh, J., and Poliakoff, M. (2005). Continuous hydrogenation reactions in supercritical CO_2 "without gases," *Green Chem.*, **7**, pp. 357–361.
10. Haumann M., and Riisagar A. (2008). Hydroformylation in room temperature ionic liquids (RTILs): catalyst and process developments, *Chem. Rev.*, **108**, pp. 1474–1497.
11. Du, L., Kelly, J. Y., Roberst, G. W., and DeSimone, J. M. (2009). Fluoropolymer synthesis in supercritical carbon dioxide, *J. Supercrit. Fluids*, **47**, pp. 447–457.
12. Han, X., and Poliakoff, M. (2012). Continuous reactions in supercritical carbon dioxide: problems, solutions and possible ways forward, *Chem. Soc. Rev.*, **41**, pp. 1428–1436.
13. Mayadevi, S. (2012). Reactions in supercritical carbon dioxid, *Ind. J. Chem.*, **51A**, pp. 1298–1305.
14. Randolph, T. W., Clark, D. S., Blanch, H. W., and Prausnitz, J. M. (1998), Enzymatic oxidation of cholesterol aggregates in supercritical carbon dioxide, *Science*, **239**, pp. 387–390.

15. Cooper, A. I., (2000). Synthesis and processing of polymers using supercritical carbon dioxide, *J. Mater. Chem.*, **10**, p. 207.
16. De Vries T. J., Duchateau, R., Vortsman, M. A. G., and Keurentjes, J. T. F. (2000). Polymerisation of olefins catalysed by a palladium complex in supercritical carbon dioxide, *Chem. Commun.*, **10**, pp. 263–264.
17. Beuermann, S., Buback, M., and Jurgens, M. (2003). Free-radical terpolymerization of styrene and two methacrylates in a homogeneous phase containing supercritical CO_2, *Ind. Eng. Chem. Res.*, **42**, pp. 6338–6342.
18. Hadida, S., Super, M. S., Beckman, E. J., and Curran, D. P. (1997). Radical reactions with alkyl and fluoroalkyl (fluorous) tin hydride reagents in supercritical, *J. Am. Chem. Soc*, **119**, pp. 7406–7407.
19. Isaacs, N. S., and Keating, N. (1992). The rates of a Diels-Alder reaction in supercritical carbon dioxide, *J. Chem. Soc., Chem. Commun.*, pp. 876–877.
20. Burk, M. J., Feng, S. G., Gross, M. F., and Tumas,W. (1995). Asymmetric catalytic-hydrogenation reactions in supercritical carbon-dioxide, *J. Am. Chem. Soc.*, **117**, pp. 8277–8278.
21. van Eldik, R., and Klärner, F. G. (2007). Application of high pressure in transition metal-catalyzed reactions, in *High Pressure Chemistry: Synthetic, Mechanistic, and Supercritical Applications*, ed. Reiser, O. (Wiley-VCH Verlag GmbH).
22. Jessop, P. G., Ikariya, T., and Noyori, R. (1999). Homogeneous catalysis in supercritical fluids, *Chem. Rev.*, **99**, pp. 475–493.
23. Leitner, W. (1999). Reactions in supercritical carbon dioxide ($scCO_2$), in *Modern Solvent Systems*, ed. Knochel, P., *Top. Curr. Chem.*, **206**, pp. 107–132 (Springer-Verlag, Berlin, Heidelberg).
24. Jiang, H. F. (2005). Transition metal-catalyzed organic reactions in supercritical carbon dioxide, *Curr. Org. Chem.*, **9**, pp. 289–297.
25. Skouta, R. (2009). Selective chemical reactions in supercritical carbon dioxide, water, and ionic liquids, *Green Chem. Lett. Rev.*, **2**, pp. 121–156.
26. Prajapati, D., and Gohain, M. (2004). Recent advances in the application of supercritical fluids for carbon-carbon bond formation in organic synthesis, *Tetrahedron*, **60**, pp. 815–833.
27. van den Hark, S., Harrod, M., and Moller, P. (1999). Hydrogenation of fatty acid methyl esters to fatty alcohols at supercritical conditions, *J. Am. Oil Chem. Soc.*, **76**, pp. 1363–1370.
28. Leitner, W. (2002). Supercritical carbon dioxide as a green reaction medium for catalysis, *Acc. Chem. Res.*, **35**, pp. 746–756.

29. Oakes, R. S., Clifford, A. A., and Rayner, C. M. (2001). The use of supercritical fluids in synthetic organic chemistry, *J. Chem. Soc., Perkin Trans.*, **1**, pp. 917–941.

30. Poliakoff, M., and King, P. (2001). Phenomenal fluids, *Nature*, **412**, p. 125.

31. DeSimone, J. M., Guan, Z., and Elsbernd, C. S. (1992). Synthesis of fluoropolymers in supercritical carbon dioxide, *Science*, **257**, pp. 945–947.

32. Guan, Z., Combes, J. R., Menceloglu, Y. Z., and DeSimone, J. M. (1993). Homogeneous free radical polymerizations in supercritical carbon dioxide: 1. Thermal decomposition of 2,2′-azobis(isobutyronitrile), *Macromolecules*, **26**, pp. 2663–2669.

33. Kendall, J. L., Canelas, D. A., Young, J. L., and DeSimone, J. M. (1999). Polymerizations in supercritical carbon dioxide, *Chem. Rev.*, **99**, pp. 543–563.

34. Herk, A. M and Manders, B. G. (1997). Propagation rate coefficients of styrene and methyl methacrylate in supercritical carbon dioxide, *Macromolecules*, **30**, pp. 4780–4782.

35. Krukonis, V. (1985). Processing of polymers with supercritical fluids, *Polym. News*, **11**, pp. 7–16.

36. Giessauf, A., Magor, W., Steinberger, D. J., and Marr, R. (1999). A study of hydrolases stability in supercritical carbon dioxide (SC-CO_2), *Enzyme Microb. Technol.*, **24**, pp. 577–583.

37. Zagrobelny, J., and Bright, F. V. (1992). In situ studies of protein conformation in supercritical fluids: trypsin in carbon dioxide, *Biotechnol. Prog.*, **8**, pp. 421–423.

38. Habulin, M., and Knez, Ž. (2001). Activity and stability of lipases from different sources in supercritical carbon dioxide and near-critical propane, *J. Chem. Technol. Biotechnol.*, **76**, pp. 1260–1266.

39. Damar, S., and Balaban, M. O. (2006). Review of dense phase CO_2 technology: microbial and enzyme inactivation, and effects on food quality, *J. Food Sci.*, **71**, pp. R1–R11.

40. Kamihira, M., Taniguchi, M., and Kobayashi, T. (1987). Sterilization of microorganisms with supercritical carbon-dioxide, *Agric. Biol. Chem.*, **51**, pp. 407–412.

41. Balaban, M. O., Arreora, A. G., Marshall, M., Peplow, A., Wei, C. I., and Cornell, J. (1991). Inactivation of pectinesterase in orange juice by supercritical carbon dioxide, *J. Food Sci.*, **56**, pp. 743–746.

42. Mesiano, A. J., Beckman E. J., and Russell, A. J. (1999). Supercritical biocatalysis, *Chem. Rev.*, **99**, pp. 623–633.

43. Oliveira, J. V., and Oliveira, D. (2000). Kinetics of the enzymatic alcoholysis of palm kernel oil in supercritical CO_2, *Ind. Eng. Chem. Res*, **39**, pp. 4450–4454.

44. Hampson, J. W., and Foglia, T. A. (1999). Effect of moisture content on immobilized lipase-catalyzed triacylglycerol hydrolysis under supercritical carbon dioxide flow in a tubular fixed-bed reactor, *J. Am. Oil Chem. Soc.*, **76**, pp. 777–781.

45. Taher, H. al-Zuhair, S. al-Marzouqi, A. H., Haik, Y., and Farid, M. M. (2011). A review of enzymatic transesterification of microalgal oil-based biodiesel using supercritical technology, *Enzyme Res.*, pp. 1–25.

46. Fürstner, A., Ackermann, L., Beck, K., Hori, H., Koch, D., Langemann, K., Liebel, M., Six, C., and Leitmer, W. (2001). Olefin metathesis in supercritical carbon dioxide, *J. Am. Chem. Soc.*, **123**, pp. 9000–9006.

47. Jessop, P. G., Ikariya, T., and Noyori, R. (1994). Homogeneous catalytic-hydrogenation of supercritical carbon-dioxide, *Nature*, **368**, pp. 231–233.

48. Jessop, P. G., Hsiao, Y., Ikariya, T., and Noyori, R. (1996). Homogeneous catalysis in supercritical fluids: hydrogenation of supercritical carbon dioxide to formic acid, alkyl formates and formamides, *J. Am. Chem. Soc.*, **118**, pp. 344–355.

49. Alavi, M. A., and Morsali, A. (2011). Alkaline-earth metal carbonate, hydroxide and oxide nano-crystals: synthesis methods, size and morphologies consideration, in *Nanotechnology and Nanomaterials "Nanocrystal,"* ed. Masuda, Y. (InTech).

50. Kohl, A., and Nielson, R. (1997). *Gas Purification*, 5th Ed. (Gulf).

51. Gawronski, J., Kołbon, H., Kwit, M., and Katrusiak, A. (2000). Designing large triangular chiral macrocycles: efficient [3+3], diamine-dialdehyde condensations based on conformational bias, *J. Org. Chem.*, **65**, pp. 5768–5773.

52. Kuhnert N., Rossignolo, G. M., and Lopez-Periago, A. M. (2003). The synthesis of trianglimines: on the scope and limitations of the [3 + 3] cyclocondensation reaction between (1R,2R)-diaminocyclohexane and aromatic dicarboxaldehydes, *Org. Biomol. Chem.*, **1**, pp. 1157–1170.

53. López-Periago, A. M.,García-González C. A., and Domingo C. (2010). Towards the synthesis of Schiff base macrocycles under supercritical CO_2 conditions, *Chem. Commun.*, **46**, pp. 4315–4317.

54. Choi, S., Drese, J. H., and Jones, C. W. (2009). Adsorbent materials for carbon dioxide capture from large anthropogenic point sources, *ChemSusChem*, **2**, pp. 796–854.

55. Aboudheir, A., Tontiwachwuthikul, P., Chakma, A., and Idem, R. (2003). Kinetics of the reactive absorption of carbon dioxide in high CO_2-loaded concentrated aqueous monoethanolamine solutions, *Chem. Eng. Sci.*, **58**, pp. 5195–5210.

56. Sanjuan, A., Alvaro, M., Aguirre, G., García, H., and Scaiano, J. C. (1998). Intrazeolite photochemistry. 21: 2,4,6-triphenylpyrylium encapsulated inside zeolite Y supercages as heterogeneous photocatalyst for the generation of hydroxyl radical, *J. Am. Chem. Soc.*, **120**, p. 7351.

57. Amat, A. M., Arques, A., Bossmann, S. H., Braun, A. M., Gob, S., and Miranda, M. A. (2003). Camel through the eye-of-a-needle: direct introduction of the TPP^+ Ion inside Y-zeolites by formal ion exchange in aqueous medium, *Angew. Chem., Int. Ed.*, **42**, pp. 1653–1655.

58. Rao, V. J., Prevost, N., Ramamurthy, V., Kojima, M., and Johnston, L. J. (1997). Generation of stable and persistent carbocations from 4-vinylanisole within zeolites, *Chem. Commun.*, **22**, pp. 2209–2210.

59. Cano, M. L., Cozens, F. L., García, H., Martí, V., and Scaiano, J. C. (1996). Intrazeolite photochemistry. 13: photophysical properties of bulky 2,4,6-triphenylpyrylium and tritylium cations within large- and extralarge-pore zeolites, *J. Phys. Chem.*, **100**, pp. 18152–18157.

60. Cano, M. L., Corma, A., Fornés, V., García, H., Miranda, M. A., Baerlocher, C., and Lengauer, C. (1996). Triarylmethylium cations encapsulated within zeolite supercages, *J. Am. Chem. Soc.*, **118**, pp. 11006–11013.

61. Heinrichs, C., and Hölderich, W. F. (1999). Novel zeolitic hosts for "ship-in-a-bottle" catalysts, *Catal. Lett.*, **58**, pp. 75–80.

62. Doménech, A., Ferrer, B., Fornés, V., García, H., and Leyva, A. (2005). Ship-in-a-bottle synthesis of triphenylamine inside faujasite supercages and generation of the triphenylammonium radical ion, *Tetrahedron*, **61**, pp. 791–796.

63. Tao, T., and Maciel, G. E. (1995). ^{13}C NMR observation of the triphenylmethyl cation imprisoned inside the zeolite HY supercage, *J. Am. Chem. Soc.*, **117**, pp. 12889–12890.

64. de Vos, D. E., Feijen, E. J. P., Schoonheydt, R. A., and Jacobs, P. A. (1994). Influences of ligand and of zeolite topology on the structure of CoII Schiff base chelates in faujasite type zeolites, *J. Am. Chem. Soc.*, **116**, pp. 4746–4752.

65. Corma, A., and Garcia, H. (2004). Supramolecular host–guest systems in zeolites prepared by ship-in-a-bottle synthesis, *Eur. J. Inorg. Chem.*, **6**, pp. 1143–1164.

66. Salavati-Niasari, M. ad Sobhani, A. (2008). Ship-in-a-bottle synthesis, characterization and catalytic oxidation of cyclohexane by host

(nanopores of zeolite-Y)/guest (Mn(II), Co(II), Ni(II) and Cu(II) complexes of bis(salicyaldehyde)oxaloyldihydra, *J. Mol. Catal. A: Chem.*, **285**, pp. 58–67.

67. Miranda, M.A., and Garcia, H. (1994). 2,4,6-Triphenylpyrylium tetrafluoroborate as an electron transfer photosensitizer, *Chem. Rev.*, **94**, pp. 1063–1089.
68. Aldag, R. (1990). *Photocromism: Molecules and Systems*, eds. Dürr, H., and Bouas-Laurent, H. (Elsevier, London, UK).
69. Corma, A. (1995). Inorganic solid acids and their use in acid-catalyzed hydrocarbon reactions, *Chem. Rev.*, **95**, pp. 559–614.
70. Murillo-Cremaes, N., López-Periago, A. M., Saurina, J., Roig, A., and Domingo, C. (2010). A clean and effective supercritical carbon dioxide method for the host-guest synthesis and encapsulation of photoactive molecules in nanoporous matrices, *Green Chem.*, **12**, pp. 2196–2204.
71. Sartori, G., and Savage, D. W. (1983). Sterically hindered amines for CO_2 removal from gases, *Ind. Eng. Chem., Fundam.*, **22**, pp. 239–249.
72. Álvaro, M., García, H., Sanjuán, A., and Esplá, M. (1998). Hydroxyalkylation of benzene derivatives by benzaldehyde in the presence of acid zeolites, *Appl. Catal. A: Gen.*, **175**, pp. 105–112.
73. López-Periago, A. M., Fraile, J., García-González, C. A., and Domingo, C. (2009). Impregnation of a triphenylpyrylium cation into zeolite cavities using supercritical CO_2, *J. Supercrit. Fluids*, **50**, pp. 305–312.
74. López-Periago, A. M., García-González, C. A., Saurina, J., and Domingo, C. (2010). Preparation of trityl cations in faujasite micropores through supercritical CO_2 impregnation, *Microp. Mesop. Mater.*, **132**, pp. 357–362.
75. Lercher, J. A., and Jentys, A. (2002). Application of microporous solids as catalysts in *Handbook of Porous Solids*, Vol. 2, eds. Schüth, F., Sing, K. S. W., and Weitkamp, J., pp. 1097–1155 (Wiley-VCH, Weinheim).
76. Fakis, M., Tsigaridas, G., Polyzos, I., Giannetas, V., Persephonis, P., Spiliopoulos, I., and Mikroyannidis, J. (2001). Intensity dependent nonlinear absorption of pyrylium chromophores, *Chem. Phys. Lett.*, **342**, pp. 155–161.
77. Montes-Navajas, P., Teruel, L., Corma, A., and Garcia, H. (2008). Specific binding effects for cucurbituril in 2,4,6-triphenylpyrylium–cucurbituril host–guest complexes: observation of room-temperature phosphorescence and their application in electroluminescence, *Chem. Eur. J.*, **14**, pp. 1762–1768.

78. Williams, A. (1971). The hydrolysis of pyrylium salts. Kinetic evidence for hemiacetal intermediates, *J. Am. Chem. Soc.*, **93**, pp. 2733–2737.

79. Corma, A., and García, H. (2004). Supramolecular host-guest systems in zeolites prepared by ship-in-a-bottle synthesis, *Eur. J. Inorg. Chem.*, **6**, pp. 1143–1164.

80. Kazarian, S.G. (1997). Applications of FTIR spectroscopy to supercritical fluid drying, extraction and impregnation, *Appl. Spectrosc. Rev.*, **32**, pp. 301–348.

81. Kikic, F., and Vecchione, F. (2003). Supercritical impregnation of polymers, *Curr. Opin. Solid State Mater. Sci.*, **7**, pp. 399–440.

82. Uzer, S., Akman, U., and Hortacsu, O. (2006). Polymer swelling and impregnation using supercritical CO_2: a model-component study towards producing controlled-release drugs, *J. Supercrit. Fluids*, **38**, pp. 119–128.

83. Saito, S., Ohwada, T., and Shudo, K. (1995). Friedel-Crafts-type reaction of benzaldehyde with benzene. Diprotonated benzaldehyde as the reactive intermediate, *J. Am. Chem. Soc.*, **117**, pp. 11081–11084.

84. Olah, G. A., Rasul, G., York, C., and Prakash, G. K. S. (1995). Superacid-catalyzed condensation of benzaldehyde with benzene. Study of protonated benzaldehydes and the role of superelectrophilic activation, *J. Am. Chem. Soc.*, **117**, pp. 11211–11214.

85. Corma, A., and García, H. (2000). A unified approach to zeolites as acid catalysts and as supramolecular hosts exemplified, *J. Chem. Soc. Dalton Trans.*, **9**, pp. 1381–1394.

86. Scaiano, J. C., and García, H. (1999). Intrazeolite photochemistry: towards supramolecular control of molecular photochemistry, *Acc. Chem. Res.*, **32**, pp. 783–790.

Chapter 11

Supercritical CO_2 for the Reactive Precipitation of Calcium Carbonate: Uses and Applications to Industrial Processing

Concepción Domingo,[a] Ana M. López,[a] Julio Fraile,[a] and Ana Hidalgo[b]
[a]*Materials Science Institute of Barcelona (CSIC), Campus de la UAB, 08193 Bellaterra, Spain*
[b]*CSIC Delegation in Andalusia, Pabellón del Perú, Avd. Maria Luisa s/n, 41013 Sevilla, Spain*
conchi@icmab.es

The formation of powdered calcite from slurries containing a calcium source and carbon dioxide (industrial carbonation route) is a complex process of considerable importance nowadays. In the absence of additives, the rhombohedral morphology can be obtained in precipitation processes by using solution routes but rarely by the industrial method, where the most common morphology of precipitated calcite is the scalenohedral one. Rhombohedral calcite with a very low degree of agglomeration, high specific surface area, and sizes going from few microns to nanocrystals can be obtained under supercritical conditions. The use of CO_2 in a compressed

Supercritical Fluid Nanotechnology: Advances and Applications in Composites and Hybrid Nanomaterials
Edited by Concepción Domingo and Pascale Subra-Paternault

ISBN 978-981-4613-40-8 (Hardcover), 978-981-4613-41-5 (eBook)
www.panstanford.com

form facilitates the reduction of the reactor size to keep the desired production rate. Moreover, the use of ultrasounds coupled to $scCO_2$ treatment largely improves carbonation process kinetics. The $scCO_2$ carbonation process can be also applied to the in situ precipitation of calcite inside of the pores of cellulose paper or Portland cement, thus increasing the density and reducing water permeability and the pH of the material, which enhances the durability in certain applications. Cement carbonation not only generates a high added value product but also could help in CO_2 capture and storage and, therefore, it is regarded as a sustainable process. Calcium-based CO_2 solid sorbents for CO_2 capture reveal a different efficiency of the carbonation/calcination cycle according to their origin, natural or synthetic, and the synthesis method. Under realistic conditions (decarbonation at 1173 K) and after 25 cycles of CO_2 adsorption/desorption, $scCO_2$-precipitated $CaCO_3$ sorbents have a residual conversion value two times higher than that of natural limestone. All these different aspects of supercritical carbonation will be addressed in this chapter.

11.1 CO_2 Carbonation Reaction

Natural carbonation of portlandite mineral (calcium hydroxide, $Ca(OH)_2$) is a well-known phenomenon associated with the weathering of alkaline rocks, which plays important roles in day-to-day life, such as the control of the fraction of carbon dioxide (CO_2) in the atmosphere or the duration of stainless steel concrete structures placed in humid environments [1, 2]. Besides, the formation of calcium carbonate ($CaCO_3$) by a reaction between $Ca(OH)_2$ and CO_2 is a complex process of considerable importance in the industrial, ecological, geochemical, and biological areas. The main $CaCO_3$ polymorphs are calcite, aragonite, and vaterite [3]. Calcite is the thermodynamically stable phase at room temperature and atmospheric pressure. Aragonite is the high-temperature phase displaying a needle-like morphology and vaterite is the low-temperature phase, often found as spherical aggregates. Calcite has a principal rhombohedral crystal structure, bounded by the (104) face as the most stable surface. Besides a rhombohedral structure, a myriad of morphological variants is possible, underlining the importance of the elongated scalenohedral calcite form, bounded by the (21-1) face.

The carbonation reaction consists in bubbling CO_2 gas through a concentrated aqueous slurry of $Ca(OH)_2$ (Eq. 11.1).

$$Ca(OH)_{2(s,\,H_2O)} + CO_{2(g)} \leftrightarrow CaCO_{3(s)} + H_2O_{(l)} \tag{11.1}$$

The solid–liquid–gas atmospheric carbonation of $Ca(OH)_2$ is a slow process with low carbonation efficiency, mainly due to the low solubility of CO_2 in water. In this respect, accelerated carbonation methods using compressed or supercritical carbon dioxide ($scCO_2$) as a reactant have been described as efficient alternatives to the atmospheric process [4–9]. Formation of $CaCO_3$ involves four main steps: dissolution of $Ca(OH)_2$, formation of carbonate ions, chemical reaction, and crystal growth. The global process is schematized in Fig. 11.1a. Calcite formation has been described as proceeding through two consecutive *fast* and *slow* reaction stages (Fig. 11.1b) [10]. In the first *fast* stage, the heterogeneous precipitation of $CaCO_3$ on the $Ca(OH)_2$ surface occurs. As the reaction proceeds and the conversion to carbonate increases, a layer of $CaCO_3$ develops on the surface of the $Ca(OH)_2$ particles, forming a core–shell structure [11]. The dissolution of $Ca(OH)_2$ occurs in two stages [12]: First, the $Ca(OH)_2$ particles chemically dissolve on the surface (Eq. 11.2, where $K_{ps(2)}$ is the solubility product), and second, Ca^{2+} ions diffuse away from the surface (Eq. 11.3). In the *slow* stage the reaction rate decreases, since the precipitation of $CaCO_3$ inhibited $Ca(OH)_2$ dissolution and $Ca^{2+}{}_s$ diffusion to some extent.

$$Ca(OH)_2 \leftrightarrow Ca^{2+}{}_s + 2OH^- \text{ (solid surface)} \quad K_{ps(2)\_298K} = 10^{-5.3} \tag{11.2}$$

$$Ca^{2+}{}_s + 2OH^- \text{ (solid surface)} \leftrightarrow Ca^{2+}{}_b + 2OH^- \text{ (bulk solution)} \tag{11.3}$$

For the formation of carbonate ions, first the reagent CO_2 needs to enter the water phase and, therefore, solubility and mass transport resistance would be important parameters controlling the reaction rate. Figure 11.1c shows, in the pressure–temperature CO_2 phase diagram, different solubility values of CO_2 in H_2O at several standard working parameters at the vapor, liquid, and supercritical conditions. The resistance applied by the water medium to CO_2

penetration can be stated in terms of density and surface tension. At high pressures, the densities of the water and CO_2 phases are similar and less critical shaking intensity is required to overcome the interfacial tension (Fig. 11.1d).

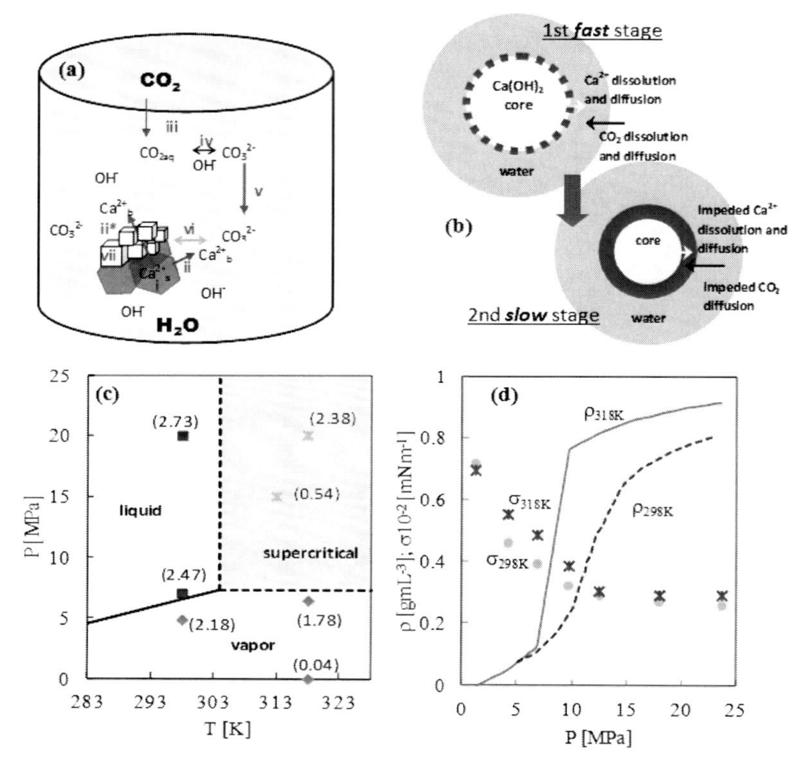

Figure 11.1 (a) Steps occurring in the carbonation reaction: chemical dissolution (i) of $Ca(OH)_2$ hexagonal platelets (⬢) occurs, forming $Ca^{2+}{}_s$ (Eq. 11.2); (ii) and (ii*) indicate the diffusion (→) of $Ca^{2+}{}_s$ to the bulk, thus becoming a reactive $Ca^{2+}{}_b$ cation (Eq. 11.3). CO_2 absorbs in the water phase (iii) to form CO_{2aq} (Eq. 11.4), and the fast formation of carbonate ions (iv) (Eqs. 11.5 and 11.6) is followed by the diffusion (→) of $CO_3{}^{2-}$ (v) and the reaction (•) of $Ca^{2+}{}_b$ with $CO_3{}^{2-}$ (vi) (Eq. 11.8), forming calcite crystals (□) (vii). (b) Schematic representation of the *fast* and *slow* stages of the carbonation reaction in Eq. 11.1. (c) Phase diagram of CO_2, showing the solubility of CO_2 in H_2O (mol%) at different P and T. (d) Density (ρ) and CO_2/water interfacial tension (σ) at 298 and 318 K.

After CO_2 is absorbed in the water phase, it is hydrated and forms CO_{2aq} or carbonic acid (H_2CO_3) (Eq. 11.4). H_2CO_3 subsequently yields bicarbonate (HCO_3^-) (Eq. 11.5) and carbonate (CO_3^{2-}) (Eq. 11.6) ions. These transformations are fast, although only ca. 1% of the absorbed CO_2 turns into carbonate. K_4, K_5, and K_6 are the equilibrium constants, while k_4 and k_5 are the rate constants (s^{-1}), all at 298 K [13].

$$CO_2 + H_2O \leftrightarrow CO_{2aq}\ (\text{or } H_2CO_3) \quad K_4 = 10^{-1.5} \quad k_4 = 10^{-1.8} \quad (11.4)$$

$$H_2CO_3 + OH^- \leftrightarrow HCO_3^- + H_2O \quad K_5 = 10^{-6.3} \quad k_5 = 10^{3.8} \quad (11.5)$$

$$HCO_3^- + OH^- \leftrightarrow CO_3^{2-} + H_2O \quad K_6 = 10^{-10.3} \quad \text{Instantaneous} \quad (11.6)$$

Precipitation proceeds spontaneously (Eq. 11.7), $K_{ps(7)}$ being the solubility product at 298 K for the calcite phase [14].

$$Ca^{2+}{}_{(aq)} + CO_3^{2-}{}_{(aq)} \leftrightarrow CaCO_{3(calcite)} \quad K_{ps(7)\_298K} = 10^{-8.3} \quad (11.7)$$

Using the classical approach, when the carbonation reaction is characterized as *slow,* the reaction rate is calculated from standard kinetics defined by CO_2 diffusion. To characterize the carbonation reaction as *fast,* an enhancement factor contributing to CO_2 absorption should appear, for instance, the use of $scCO_2$ [15].

11.2 Nonconventional $scCO_2$ Coupled to Ultrasonic Stirring for $CaCO_3$ Precipitation

The precipitation of $CaCO_3$ can be carried out from mechanically stirred slurries of $Ca(OH)_2$ via either the diffusion of diluted gaseous CO_2 through an atmospheric reactor (Fig. 11.2a) or via the addition of compressed CO_2 into a high-pressure autoclave [6, 7] (Fig. 11.2b). Further, the use of ultrasound combined with $scCO_2$ has been designed as an alternative stirring method (Fig. 11.2b,c) [16, 17]. The use of power ultrasound in liquids is well known to cause a number of physical effects (turbulence, agglomeration, microstreaming, etc.) as well as chemical effects [18]. These effects arise principally from the phenomenon known as cavitation. Cavitation refers to the formation, growth, and violent collapse of microbubbles in a sonicated liquid due to pressure fluctuations. However, when a liquid is pressurized, the acoustic intensity required to produce cavitation increases and this, generally, places a natural limit on the application of ultrasound to high pressure processes. In principle, the absence of phase boundaries should exclude cavitation phenomena above the critical point. However, the use of power ultrasound coupled with $scCO_2$ has been already successfully applied to accelerate extraction processes in the food industry, for particle size modification in different precipitation methods and for the dissolution of uranium in decontamination of nuclear solid waste procedures [19–23].

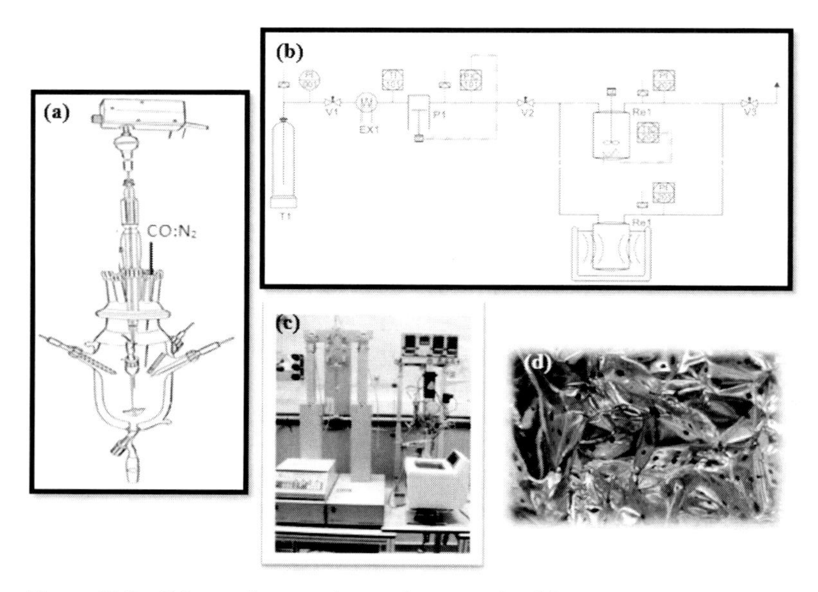

Figure 11.2 Schematic experimental setups for (a) atmospheric carbonation in a Pyrex reactor provided with a vertical stirrer (800 rpm) and a continuous flow of CO_2:N_2 and (b) high-pressure (13–20 MPa) carbonation equipment with vertical stirring (150–400 rpm) and ultrasound agitation (42 Hz, 70 W). Reactors were charged with a $Ca(OH)_2$ slurry of concentration 150–200 g·L^{-1}. Carbonation is carried out at 313–318 K. (c) Picture of the equipment setup used for the supercritical equipment coupled with an ultrasound bath. (d) Strip of aluminum foil after treatment with $scCO_2$ and ultrasounds (20 MPa, 313 K, 70 W, 30 min).

A simple experiment, carried out with a strip of aluminum foil located inside of the reactor, shows the effect of ultrasound power in a system containing exclusively $scCO_2$ [24]. The aluminum foil is stable against the effects of pressurizing and depressurizing the reactor with CO_2. However, after application of ultrasound at supercritical conditions, the aluminum foil suffers severe damage, as shown in Fig. 11.2d. Consequently, it is feasible to consider that a process similar to cavitation is occurring within the supercritical fluid through the formation and subsequent collapse of small bubbles of subcritical gas. Ultrasound could also produce mechanical fluctuations and thermal effects, which create microdisturbances in the supercritical phase.

11.3 Applications of $scCO_2$ Accelerated Carbonation

Powdered $CaCO_3$ is the most produced white pigment worldwide, as it is used as a coating, filler, or extender in the paper, paint, rubber and plastic, textile, detergent, cosmetic, and food industries, among others. In the different industrial applications, strictly defined parameters, related to particle size and particle size distribution, morphology, surface area, polymorphic form, and chemical purity, play a vital role. The main sources of particulate $CaCO_3$ are ground calcium carbonate (GCC) mined from calcite deposits and precipitated calcium carbonate (PCC) produced by means of the chemical reaction of $Ca(OH)_2$ with CO_2.

PCC is more versatile than GCC, since modifying the conditions prevailing during precipitation, the polymorph, the morphology, the size, and some physical properties of the produced powder can be controlled [6, 16, 25–29]. Furthermore, the reaction of portlandite with CO_2 is a valuable process used to tailor Portland calcium-rich cements, determining their hydraulic and mechanical properties and overall permeability [30, 31]. In geological applications, the carbonated cement is used as a borehole cement to enhance the confining properties of wells and plugs for underground disposal of hazardous waste contaminated with heavy metals or even radioactive compounds, as well as for the storage of CO_2 for greenhouse emissions mitigation in oil or gas reservoirs, deep saline cavities, or flood basalts [32, 33]. Moreover, $CaCO_3$ is an important precursor of solid absorbents (CaO and $Ca(OH)_2$) used in processes of in situ removal of CO_2 at high temperature [34–36].

11.3.1 $scCO_2$ in the Production of PCC

In Fig. 11.3, the most common established procedures for the preparation of PCC are schematically depicted.

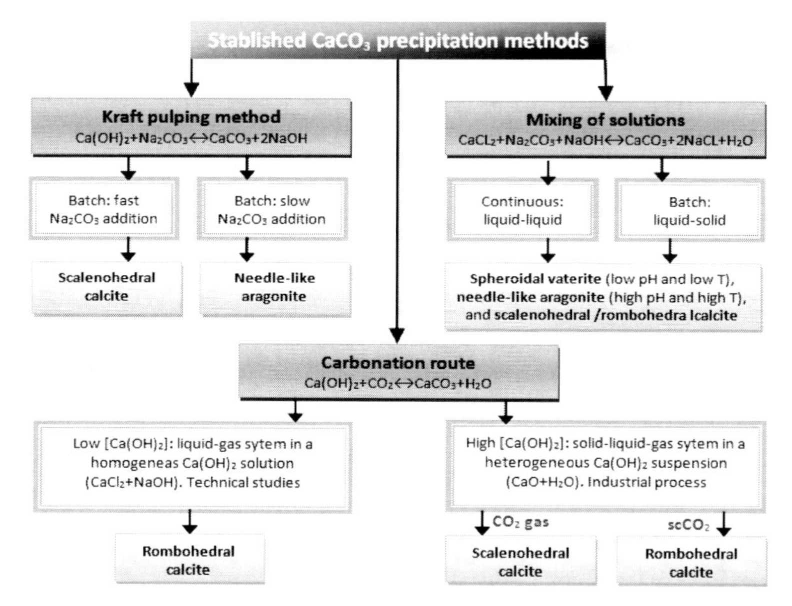

Figure 11.3 Most common established procedures for $CaCO_3$ precipitation. Industrial calcite is usually produced by bubbling CO_2 through a concentrated $Ca(OH)_2$ slurry (carbonation route). $CaCO_3$ is also obtained as a by-product in the causticizing reaction used for the industrial production of sodium hydroxide (Kraft pulping method). Most laboratory studies on the controlled precipitation of $CaCO_3$ are performed by mixing solutions containing low-concentration calcium salts and carbonate ions. Additionally, the majority of technical studies regarding the industrial carbonation route are carried out through the analysis of liquid–gas systems by eluding the precipitation of $Ca(OH)_2$.

The liquid–liquid–solid and liquid–gas–solid reactive processes have been adopted in methodological studies because of the simplicity in the control of process variables. The industrial route involves the calcination of limestone (natural $CaCO_3$) to produce quicklime (CaO) and CO_2, which is followed by the formation of slaked lime (concentrated suspensions of $Ca(OH)_2$) and, finally, of the carbonation reaction (Eq. 11.1). In the absence of additives, the rhombohedral morphology is obtained by using solution routes, but not by the industrial method with CO_2 gas, which gives the scalenohedral form. Isometric particles with rhombohedral morphology are more easily dispersed, which is relevant for their use in composite applications. However, the experimental conditions needed to obtain rhombohedral calcite are low concentration of both CO_2 and Ca^{2+} (liquid–gas system), which is not acceptable from an industrial point of view. Therefore, the morphologic control must be often provided by adding additives [37, 38]. High pressure [39], magnetic fields [40], or microwave irradiation [41] have also been applied for calcite crystals shape control.

The development of new methods for the production of PCC by carbonation of concentrated $Ca(OH)_2$ slurries with different morphologies, avoiding agglomeration and the addition of additives, together with a significant acceleration and intensification of the carbonation process, is considered of great interest. The use of $scCO_2$ as a reagent has demonstrated to produce outstanding improvements for the conversion of $Ca(OH)_2$ to the stable calcite polymorph. Both the reaction rate and the carbonation efficiency are increase by using compressed CO_2 [6, 7]. In the experimental system, the ratio between the height of the X-ray diffraction (XRD) peaks corresponding to calcite polymorph ($2\theta = 29.4°$) and $Ca(OH)_2$ ($2\theta = 18.1°$ or $34.1°$) is often used as an indicator to follow up the carbonation reaction and the composition of the obtained PCC. For the atmospheric procedure, only when the experiments are performed for very long periods of time (>20 h) the majority of the $Ca(OH)_2$ phase is transformed to $CaCO_3$ and its representative peaks almost disappear from the diffractogram (Fig. 11.4a). For shorter CO_2 contact periods, the difficult-to-penetrate $CaCO_3$ layer precipitated onto $Ca(OH)_2$ surface impeded the completion of the reaction (Fig. 11.1b). In the supercritical conditions, the reaction is almost completed in periods of one to two hours (Fig. 11.4a).

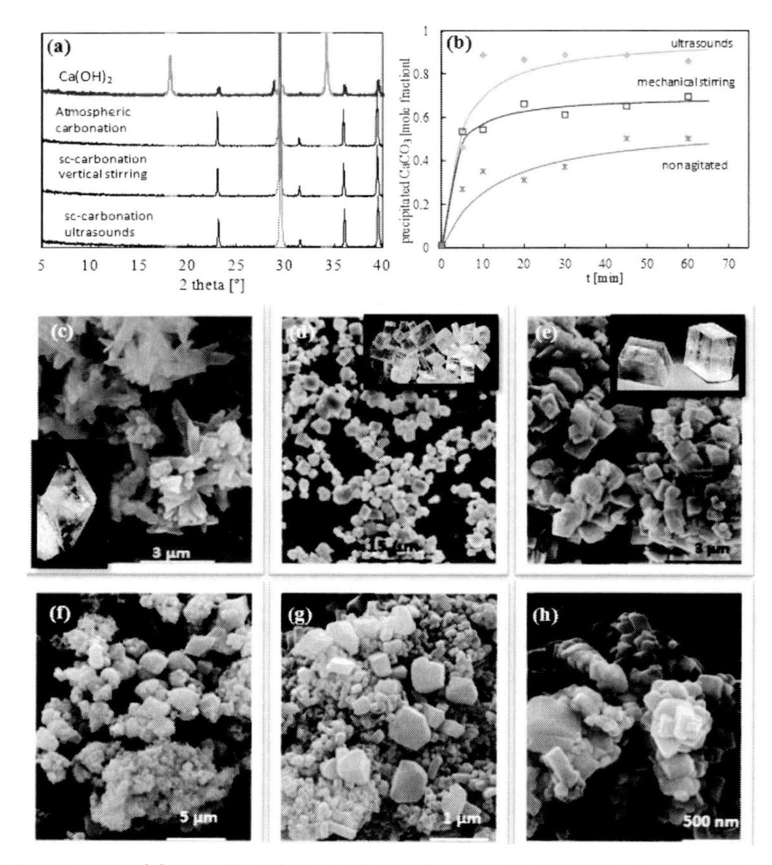

Figure 11.4 (a) XRD (Rigaku Rotaflex RU200 B) patterns of $CaCO_3$ samples obtained under an atmospheric CO_2:N_2 (20:80 vol%) mixture (24 h) and supercritical (2 h) conditions; (b) $scCO_2$ (313 K and 13 MPa) kinetic data (symbols) and data fitting (lines) calculated using the pseudo second-order kinetic model (Eq. 11.1); and SEM (Hitachi S570) pictures, taken at 60 min, of calcite particles precipitated by (c) atmospheric carbonation and $scCO_2$ carbonation of a $Ca(OH)_2$ slurry (200 $g{\cdot}L^{-1}$) with (d) vertical stirring and (e) ultrasounds. The insets show the hexagonal scalenohedral (c) and rhombohedral (d, e) morphologies in calcium carbonate minerals; and (f–h) carbonation of dry $Ca(OH)_2$ giving $CaCO_3$ by addition of water-saturated $scCO_2$.

The rate-limiting step of calcite formation is determined by the initial conditions employed (reagents concentration, stirring method, temperature, etc.). At the high initial reagents concentration used in the supercritical carbonation route, diffusion of either CO_2 or Ca^{2+} in the *fast* and *slow* reaction stages, respectively, would be the determinant process. Hence, stirring efficiency would be one of the main parameters influencing the reaction rate and phase conversion. The effect of the stirring effectiveness on the carbonation kinetics has been analyzed for the scCO$_2$ system using ultrasonic agitation and mechanical stirring and in the absence of agitation [16, 17] (Fig. 11.4b). The degree of $Ca(OH)_2$ conversion to $CaCO_3$ can be fitted using the pseudo-second-order model in Eq. 11.8 [42]:

$$\frac{d[C_{\text{CaCO3,t}}]}{dt} = k_{\text{CaCO3}}(C_{\text{CaCO3,max}} - C_{\text{CaCO3,t}})^2 \tag{11.8}$$

where k_{CaCO3} is the rate constant (mol·min^{-1}) of $CaCO_3$ precipitation and $C_{\text{CaCO3,max}}$ (mol) and $C_{\text{CaCO3,t}}$ (mol) are the number of moles of precipitated $CaCO_3$ after 60 min and at time *t* (min), respectively. Figure 11.4b shows that in the first *fast* stage of the reaction (5–10 min), similar production rates of $CaCO_3$ are obtained using either ultrasonic or mechanical stirring (~0.1 mol·min^{-1}), while the reaction rate is significantly lower for the experiment performed without agitation (~0.03 mol·min^{-1}). In this stage, the carbonation rate is controlled by the mixing of the reagents. Considering together the *fast* and *slow* stages of the carbonation reaction, ultrasound stirring confirms to be clearly superior in regard of the conversion grade, since the overall degrees of conversion after 60 min of carbonation are 85%–90% with ultrasonic agitation, 65%–70% with mechanical stirring and only ~50% for the nonagitated system. In the experiments performed applying scCO$_2$ and ultrasound stirring, the main mechanism responsible of the high attained conversion can be related to the intense agitation caused by acoustic streaming, and from the shearing forces, jets, and shock waves produced by conventional cavitational collapse in the water-rich phase of the $Ca(OH)_2$ slurry [43–45]. Moreover, ultrasonic agitation also causes significant turbulent motion of the suspended solids and collisions between growing crystals that produces abrasive surface effects and delamination of the $CaCO_3$ layer, thus increasing the $Ca(OH)_2$ interfacial area for dissolution. To obtain high carbonation conversions, the ultrasonic agitation needed to be coupled with the presence of a high CO_2 concentration in water; in contrast, by treating a $Ca(OH)_2$ slurry with atmospheric CO_2 and ultrasonic agitation during 60 min only ca. 10 wt% was reacted to $CaCO_3$.

Crystal growth and particle size are primarily influenced by supersaturation, but final crystal habit is influenced by excess species, which can act as additives. The excess concentration of a reagent with respect to the stoichiometric proportion affects the final crystal habit by inducing different growth rates on crystal faces. The stoichiometric (104) face of rhombohedral calcite tends to disappear from the growth morphology when precipitated from solutions with a $[Ca^{2+}]/[CO_3^{2-}] > 1.1–1.2$ [46, 47]. In contrast, the growth of the predominant (21-1) face in scalenohedral calcite is inhibited for solutions under identical conditions of supersaturation but with more stoichiometric $[Ca^{2+}]/[CO_3^{2-}]$ ratios. During the atmospheric carbonation process, the amount of available CO_3^{2-} for carbonation is restricted by the low solubility of CO_2 gas in water (Fig. 11.1c), which leads to the accumulation of an excess of dissolved Ca^{2+} in the growth medium producing scalenohedral particles. The rhombohedral morphology is favored by enhancing the concentration of the CO_3^{2-} anion in the $scCO_2$ medium, which reduces the ratio of Ca^{2+} excess. Scanning electron microscopy (SEM) analysis indicates that during the conventional batch carbonation process carried out in an open vessel the less stable scalenohedral morphology is preferentially precipitated (Fig. 11.4c). Conversely, for the performed high pressure experiments, the favored form is rhombohedral calcite (Fig. 11.4d and Fig. 11.4e for mechanical and ultrasound stirring, respectively). Intergrown stepped rhombohedral aggregates and a population of multinucleated rhombohedral calcite particles of sizes in the range of 1–3 μm are obtained using the $scCO_2$ route, regardless of the stirring method. It is found that micrometric particles are precipitated in systems containing a large quantity of water in the fluid phase, while nanometric calcite is formed when using a reduced water percentage. Previous work performed on the morphological control of calcite precipitated by carbonation of slaked lime has shown that the presence of submicrometric particles points toward calcite precipitation by the $Ca(OH)_2$ surface route (Fig. 11.1b). In this route, nanometric particles of amorphous calcium carbonate are first precipitated on the lime surface. Reorganization of this precursor is a slow process that requires a passage in solution to nucleate dehydrated calcite. In media with a reduced amount of water, the interaction of the new nucleated tiny calcite particles with the substrate surface makes their interfacial free energy to decrease, thus diminishing their tendency to grow and stabilizing the fine particles [17]. Hence, a large population of nanometric particles (ca. 0.1–0.2 μm) is observed in the SEM images for systems precipitated with a reduced amount of water, together with few micrometric crystals (Fig. 11.4f–h).

11.3.2 $scCO_2$ in situ Precipitation of $CaCO_3$ into the Pores of Cellulose Paper

In the specific case of the paper industry, calcium carbonate has been increasingly used as a filler, since ca. 80% of the world paper production has been converted to an alkaline papermaking process [48]. $CaCO_3$, improves several paper properties such as opacity, printing characteristics, luster, and whiteness. In the conventional process, $CaCO_3$ particles are added to the papermaking pulp suspended in water. A common problem with such a system is that a relatively high portion of the suspended filler particles is lost during the water draining of the pulp, and only a small part is retained in the cellulose fibers. Particle shape and particle size distribution are two of the most important filler properties in this process. For instance, paper filled with rhombohedral calcite is more straightforwardly processed than paper filled with scalenohedral calcite, since the former is more easily drained leading to lower filler lost [49]. Additionally, a narrow crystal size distribution is always preferred. In addition, as the cost of the pulp and energy has increased, much effort is now devoted to the development of techniques that facilitate high loading levels. Recently, several improved processes for the production of filler-containing paper have emerged, in which the precipitation of the filler is performed in situ [50]. Among them, the Fiber Loading™ process [51] shows that fiber impregnation with $CaCO_3$ can be accomplished by an in situ reaction between $Ca(OH)_2$ and CO_2 in pressurized reactors (between 0.3 and 1 MPa) previously filled with paper pulp.

Considering that rhombohedral calcite with a low degree of agglomeration is obtained by a $scCO_2$ carbonation process, this procedure has been applied to the formation of calcite inside of the pores of cellulose paper [7]. Using a similar setup and procedure as the one depicted in Fig. 11.2b with mechanical stirring, cellulose cotton paper (Fig. 11.5a) has been impregnated with $CaCO_3$ particles. The diffusion process is slower than in the system without cellulose and, therefore, longer reaction periods are necessary to achieve a considerable conversion of $Ca(OH)_2$ to calcite [52, 53]. Observation of the SEM micrographs indicates that the rhombohedral shape of calcite is precipitated (Fig. 11.5b), and particles bound to the cellulose fibers are evidenced very deep into the pores (Fig. 11.5c). Applications of this process can range from those related to the preparation of high added value papers (e.g., lightweight high-opacity Bible paper) to others related to conservation of cultural paper heritage [54, 55].

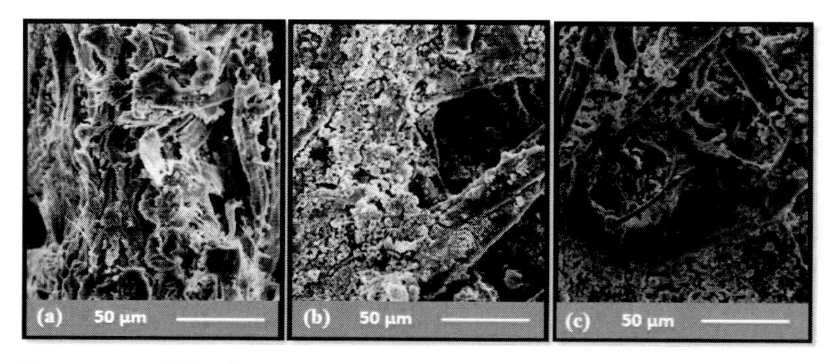

Figure 11.5 SEM photographs of hollow cellulose paper: (a) pristine (Albet 150, pore diameter 30–50 μm) and (b, c) impregnated with rhombohedral calcite using an $scCO_2$ approach. For supercritical processing, the paper is first equilibrated with the $Ca(OH)_2$ slurry for 0.5 h. Then, $scCO_2$ at 20 MPa and 318 K is contacted with the solids for 5 h. $CaCO_3$-impregnated paper is finally washed with deionized water and dried in an oven with circulating air at 373 K.

11.3.3 Enhancement of Portland Cement Properties by $scCO_2$ Carbonation: Application of Cement Carbonation in Waste Disposal

Concrete, based on ordinary Portland cement, is one of the most widely used materials on earth, with applications ranging from construction to waste immobilization [56]. There is a major concern in producing cementitious materials with longer durability than the current products and prepared in a more efficient manner, since the energy consumption in the cement industry is extremely high, being responsible for 6% of the total anthropogenic CO_2 emissions and for 4% of the planet's global warming effect [57]. The main products of Portland cement hydration include $Ca(OH)_2$ (20–25 wt%), calcium silicate hydrates (CSH, 40–60 wt%), and calcium sulfoaluminate hydrates (10–20 wt%) involving ettringite. Some of the main advances in concrete are due to the application of nanotechnology to cementitious materials, with an increase in the knowledge and understanding of basic phenomena in cement at the nanoscale (e.g., structure and mechanical properties of the main hydrate phases, origins of cement cohesion, cement hydration, interfaces in concrete, and mechanisms of degradation) [58]. Concrete is a nanostructured, multiphase, composite material that ages over time. It is composed of an amorphous phase, nanometer to micrometer size crystals, and bound water. The amorphous phase CSH is the glue that holds concrete together and is itself a nanomaterial.

A well-known phenomenon associated with cement-based materials placed in open locations is the gradual carbonation, which consists of the chemical reaction between the CO_2 present in the air and the calcium contained in the water of the pores of the cement paste, giving place to the sparingly soluble $CaCO_3$ mineral [59]. The sources of Ca^{2+} are the previously mentioned hydrated compounds. One of the main consequences of carbonation is the modification of the pH of the pore water from an initial value of ~13 to a value below 9 in the carbonated zone. For common steel reinforced concrete, the reduction in pH caused by carbonation is one of the main causes of concrete structure deterioration over time, since it enables the corrosion of steel [60]. Instead, cement carbonation allows the preparation of structural concrete reinforced with nonexpensive alkali-intolerant products, such as industrial, plant, or animal solid waste [61, 62]. Moreover, this procedure can influence positively the recycling of demolished concrete structures [63] and the production of materials for the immobilization of hazardous products [64–66]. Portland cement can play a major role in providing cost-effective, long-term zonal isolation for CO_2 capture and storage in the form of compressed CO_2, used to mitigate climate change [67–69]. Leakage through wells is one of the major issues when storing CO_2

in depleted oil or gas reservoirs. These wells are mostly cemented with conventional Portland cement, which is carbonated in contact with injected CO_2 [70, 71].

Significant cement carbonation occurs naturally, but over numerous years. The low reaction rate in natural carbonation is due to the low CO_2 concentration in the atmosphere and the slow diffusion of CO_2 into the cement pores [72]. In artificially accelerated carbonation processes, samples are exposed to concentrated and/or compressed CO_2 [73, 74]. The advantages of treating hardened cement pastes with $scCO_2$ are primarily related to the fact that the carbonation process is greatly accelerated [30, 75–80]. This can have a direct effect in industry because, by speeding the carbonation reaction, the process can be commercially feasible for the production of a broad range of desirable cement-based materials. Although most of the materials being studied are aimed at high cost markets, such as encapsulating materials for nuclear disposal, one industrial process that was commercialized by Supramics® involved the combination of $scCO_2$ and fly ash to modify cement for low-cost building materials [81].

In the carbonation process, the reagent CO_2 needs to enter the pore water containing the ion Ca^{2+} and, therefore, mass transport resistance is a very important parameter in the reaction mechanism. In ambient conditions, the amount of CO_2 dissolved in the pore water, and hence, available CO_3^{2-} for carbonation, is restricted by the low solubility of atmospheric CO_2 in water (0.06 mol%) [82]. Natural carbonation is a phenomenon controlled by CO_2 diffusion [83–86]. By using compressed CO_2, the first effect is a huge rise (near 100-fold) of CO_2 solubility in water with the increase in pressure to 20 MPa (2.4 mol% at 318 K). In this case, the reaction acceleration is also due to the ease of penetration and diffusion of the $scCO_2$ into the micropores of the cement paste, providing continuous availability of fresh reactant. Kinetics studies have shown that by raising the pressure in the supercritical conditions the degree of carbonation is not significantly increased [87]. Hence, the supercritical process is not CO_2 diffusion controlled. The overall kinetics for the supercritical cement carbonation reaction begins rapidly but then drastically slows down with time. The reason is related with the progressive filling of the pores with calcite that has a direct impact on mass transfer. The ingress of CO_2 reagent and the migration of Ca^{2+} within the microstructure are both hindered at a significant carbonation extent.

Artificially intensified carbonation of hydrated cement pastes is carried out in high-pressure reactors at temperatures of 313–333 K and pressures of 15–25 MPa and at a low flow rate of $scCO_2$ for a period of time of several hours [75, 76] (Fig. 11.6a). The evolution of the main crystalline phases during carbonation is often determined by means of XRD analysis and thermogravimetric analysis (TGA). In the example shown in Fig. 11.6b, only the cement paste processed under supercritical conditions was almost fully carbonated, since the intensity of the $Ca(OH)_2$ line ($2\theta = 18.1°$) was reduced relative to the $CaCO_3$ peak ($2\theta = 29.4°$). The pattern with regards to the depletion of $Ca(OH)_2$ and formation of $CaCO_3$ observed in the XRD analysis is confirmed by thermogravimetric analysis. The presence of $CaCO_3$ formed during treatment is shown by a weight loss at $T > 900$ K in the TGA profile. TGA curves in Fig. 11.6c indicated that for raw and naturally carbonated samples, the CO_2 started to be expelled from the carbonate at approximately 900 K and ended at 975 K. However, in the $scCO_2$ treated sample, the decarbonation ended at 1050 K. It is suggested in the literature that in the $scCO_2$ treated sample the precipitated calcite is more crystalline than in the atmospherically carbonated product. More crystalline samples would be expected to start undergoing decarbonation at a higher temperature. The weight loss shown in the range 675–800 K is the result of dehydroxylation of $Ca(OH)_2$.

The textural properties of carbonated cement pastes must be evaluated to fully understand the effects of the CO_2 treatment and to assess the suitability of a carbonation technique for specific applications [88]. For example, the engineering properties of concrete are directly influenced by the number, type, and size of pores. The total pore volume affects the strength and elasticity of concrete, whereas concrete permeability is influenced by pore size and connectivity. The pore structure of Portland composite cement pastes is strongly influenced by exposure to carbonation. This aspect is studied using the adsorption isotherms of N_2 and Hg. Low-temperature N_2 adsorption–desorption analysis is treated by Brunauer–Emmett–Teller (BET) theory and it is considered suitable to accurately estimate the volume of medium capillary, gel pores, and micropores (<50 nm) [89]. Mercury intrusion porosimetry (MIP) is used for macroporosity (>50 nm) analysis [90].

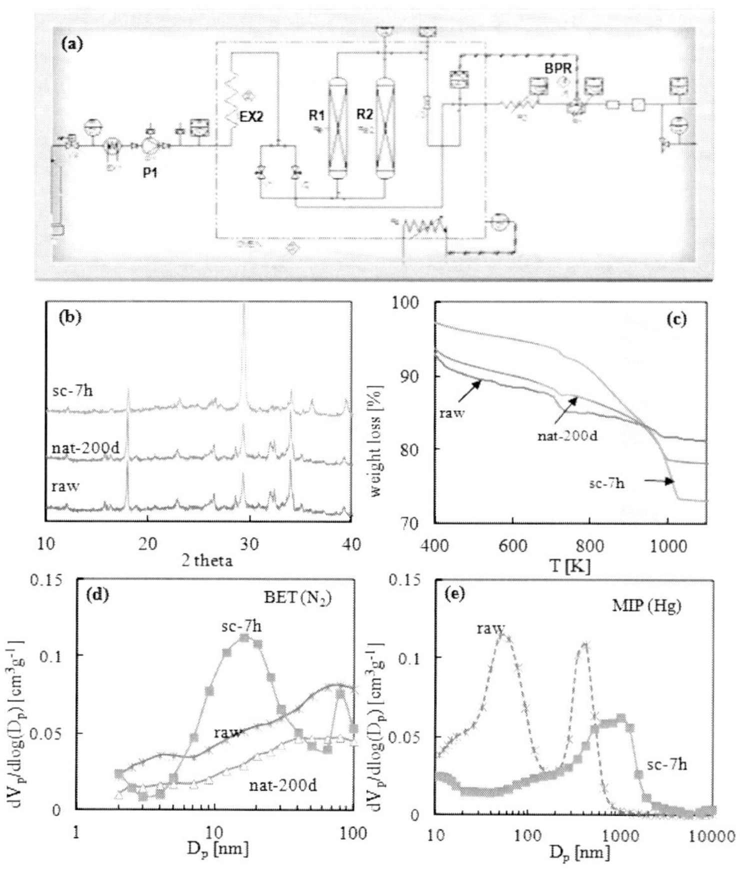

Figure 11.6 Processing and characterization of raw and treated cement pastes (standard Portland cement): (a) process flow diagram of the supercritical equipment. CO_2 is compressed by means of membrane pump P1 and heated in the heat exchanger EX2. Carbonation is carried out in the tubular reactors R1 and R2. System pressure is controlled with the back pressure regulator BPR. Samples were carbonated either naturally during 200 d at 293 K (nat-200d) or supercritically at 20 MPa and 318 K for 7 h (sc-7h); (b) XRD patterns; (c) TGA profiles (SEIKO 320U); (d) BET (ASAP 2000 Micromeritics INC); and (e) MIP (Micromeritics Autopore IV 9500) pore diameter distribution on a pore volume basis.

The overall reduction in the total N_2 pore volume in cement pastes with carbonation is associated with the deposition of $CaCO_3$ inside the pores [91]. Calcite precipitation help to fill pore space and densify the cement paste, since the molar volume of this mineral is 17 vol% larger than that of portlandite. The reduction in pore volume is more appreciable for natural carbonation (50 vol%) and less significant for supercritical carbonation (25 vol%), even when $scCO_2$ treated samples reached a large degree of carbonation. This fact has been related to the facility of $scCO_2$ for fast diffusion to the interior of the cement paste grains through the pore channels. In the initial stages of the reaction, $scCO_2$ invades the entire sample homogeneously, causing an initial degree of pervasive carbonation on the pore walls. Conversely, the carbonation front with atmospheric CO_2 penetrates slowly into the interior of the concrete and it is mostly concentrated on the exterior pores hindering the access of N_2 to the interior of the sample during BET measurements, which goes to show that the lowest pore volume measured for natural carbonated samples could be only an apparent result. Only $scCO_2$ carbonated samples developed measurable microporosity, indicating that during supercritical carbonation the precipitated carbonate is adhered to pore walls, thus reducing pore diameter. Results obtained using BET analysis showed that the fine pore size distribution profile of a hardened cement paste carbonated under atmospheric conditions has a similar shape than that of the raw material (Fig. 11.6d). Supercritically carbonated samples often have a significant portion of the porosity attributed to mesopores. In the area of macroporosity, the global effect of supercritical carbonation is an important decrease of total macropore volume (Fig. 11.6e). Only the percentage of very large pores increases significantly after supercritical carbonation, but this is likely due to microcracking occurring during depressurization and dehydration.

Concrete carbonation increases cement paste density and reduces the overall permeability, conferring a great resistance to swelling and shrinkage, which are all desirable properties for matrices intended to immobilize solid waste. Hazardous waste, contaminated with heavy metals and other inorganic pollutants or even radioactive compounds, can be stabilized within carbonated concrete to reduce the toxicity and to minimize the release into the environment. The use of carbonated concrete has been proposed in primary and secondary containment structures in nuclear power plants (Fig. 11.7a), which require concrete with uniform density, low shrinkage, low permeability, and high durability. Moreover, carbonated cement composites have been selected to develop radioactive waste containment structures [92–94]. In the case of low-level radioactive waste, attention is paid to the use of cementitious materials for immobilization and stabilization [95]. The possibilities of enhancement of concrete mechanical properties and durability by treatment with $scCO_2$ have been already described for applications in cement waste stabilization [96–99].

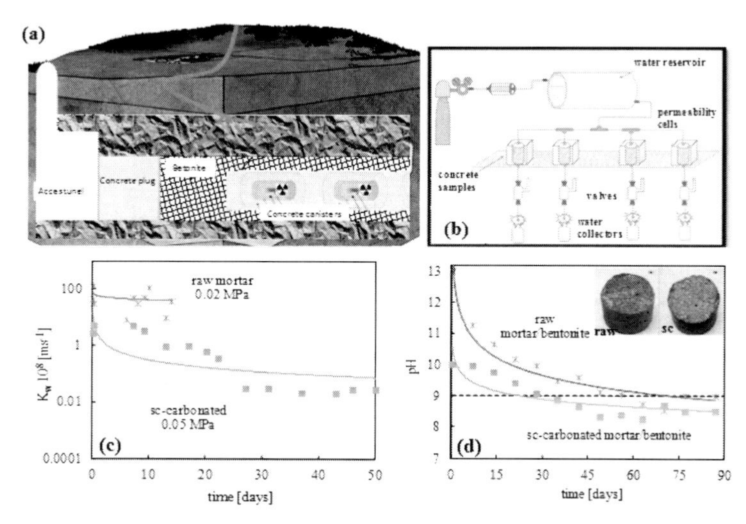

Figure 11.7 Evolution of the permeability and leaching processes of cementitious materials in contact with bentonite and groundwater: (a) sketch of a deep geological repository placed in a granitic rock; (b) experimental setup used for the permeability measurements; (c) permeability behavior; and (d) variation of the solution pH (black discontinuous line indicates the pH below which the bentonite barrier is stable). The inset shows the phenolphthalein test results: purple color for the raw material (highly basic pH) and gray cement color for the supercritically carbonated sample (near neutral pH).

The risks derived from handling nuclear fuel, including storage and treatment of nuclear waste, together with the development of a strong environmental social sensitivity, have driven numerous researches activities in the field of radioactive energy safety. Deep geological storage of high-level nuclear waste is, at present, an important element in such a scenario. Most designs for high-level nuclear radioactive waste confinement, particularly in long-term deep and humid geological underground disposal, envisage the presence of carbonated cementitious materials for their use as a barrier, liner, and encasement vessels in conjunction with clay (bentonite) barriers (Fig. 11.7a) [32, 100, 101]. For this application, permeability, defined as the movement of a fluid through a porous medium under an applied pressure, is considered as the most important property of cement composites governing its long-term durability [102–104]. Moreover, the interest is focused on the preparation of cement pastes with a pH similar to that of the bentonite pore water (pH = 7–9) that will not cause instability when placed in contact with the clay [105].

The modifications in the microstructure of a hardened cement paste after it has reacted with CO_2 (Fig. 11.6d,e) affect its permeability properties [32, 75]. The permeability is studied in mortar composites, constituted of cement paste usually with the addition of pozzolands and sand in a proportion of ca. 50 wt% [106, 107]. The porosity and interconnectivity of pores in the cement paste and the microcracks in the composite, especially at the paste-aggregate interface, influence the permeability. In the setup shown in Fig. 11.7b, the water permeability test is applied to mortar cylinders of 30 mm diameter and 20 mm height. Samples are first saturated at 0.02 MPa water pressure during 24 h. The gradient of hydraulic pressure used to measure the permeability of noncarbonated composites is of 0.02 MPa applied during 10 days, while this value must be increased to 0.05 MPa in supercritically processed products to have a significant water flow that is then maintained for 50 days. The volume of water collected after passing through the inner core of the cement cylinders is related to the coefficient of water permeability, K_w ($m \cdot s^{-1}$), by applying Darcy's law [108], expressed in Eq. 11.9:

$$K_w = \frac{QL}{A\Delta h} \tag{11.9}$$

where Q (m^3s^{-1}) is the rate of water flow, L (m) is the specimen thickness, A (m^2) is the cross-sectional area of the specimen, and Δh (m) is the hydraulic head drop through the specimen.

The water permeability test shows that the supercritically carbonated composite had a significantly increased water permeability resistant behavior when compared to the pristine sample (Fig. 11.7c) [109].

Standard Portland cements produce leachates of high or very high pH, frequently above 13 that are very rich in alkalis and calcium. The hyperalkaline solution would react with the bentonite in the proximity of the concrete, inducing bentonite dissolution and the release of structural cations. Carbonation produces a decrease of pore water pH to values of 7–9, which contributes to limit the propagation of an alkaline plume toward other components of the barrier. To analyze the effect of supercritical carbonation in the stability of the interface bentonite/cement, mixtures of both components were altered using commercial mineral water taken from a granitic formation (pH = 7.9). Experiments are designed to allow an intense interaction between solids and granitic solution: water was renewed weekly, samples were used in powder, and temperature was raised to 353 K to accelerate the reactions. This test, performed during three months, is considered representative of long-term interaction of solids and pore fluids at the concrete/bentonite interface. The interaction text shows that the pH of the leachates decreases progressively from an initial alkaline value of either 13 or 10 for the raw or the supercritically carbonated sample, respectively, to a final pH equilibrium value of 8.5–9, similar for both studied samples (Fig. 11.7d). However, there are important differences between supercritically carbonated and noncarbonated samples. First, before reaching the steady state pH, on average, the pH of the leachates from the carbonated sample is one unit lower than the values measured for the raw material. Furthermore, carbonated products reached the equilibrium pH faster than noncarbonated samples.

11.3.4 Supercritical CO_2-Precipitated Calcite in the Capture and Storage of CO_2

The removal of anthropogenic CO_2 from combustion processes with solid sorbents has been highlighted in recent years as a potential way to reduce greenhouse gas emissions in industry. For postcombustion capture processes, where CO_2 is a minor component (<15 vol%) of the off-gases, the most representative of the used solid absorbents are calcium-based materials, including CaO and $Ca(OH)_2$, which form $CaCO_3$ (Eq. 11.1) [110]. The $CaCO_3$ is, then, regenerated to high-surface area and high-porosity CaO, this being the starting material in CO_2 absorption processes (Eq. 11.10) [111].

$$CaO + CO_2 \rightarrow CaCO_3 \tag{11.10}$$

Factors describing the quality of CO_2 sorbents include fast absorption and desorption kinetics, large regenerability maintaining a high absorption capacity, and multicycle stability. The use of natural limestone for CO_2 absorption is limited by the fast decrease of the reversibility of the carbonation reaction during the carbonation/calcination cycles, mainly attributed to pore blocking and absorbent sintering [112, 113]. The reversibility decreases when the calcination is carried out at a high temperature [114] (>900°C) or in the presence of CO_2 [115]. In general, both the absorption capacity and the kinetics increase to a considerable extent by decreasing mineral particle size. However, energy consumption associated with the grinding of reactant minerals could make up 75% of the total energy costs of the process [116]. In light of this, two different research pathways have been developed to increase sorbent efficiency. In the first strategy, the natural limestone is modified to prevent sintering [117], either by adding additives and doping with compounds that increase mass diffusion or by supporting CaO in an inert matrix (Al_2O_3 or TiO_2). In the second approach, the objective is to produce PCC micro- or nanoparticles with improved capture characteristics [118, 119]. It is worth mentioning that in all cases, it is at the expense of increasing the cost of the sorbent when comparing with the use of natural limestone.

Previous investigations show that calcite with rhombohedral morphology could be precipitated using compressed CO_2 as a reagent (Fig. 11.2b), either in the form of micro- or nanoparticles (Fig. 11.4d–h). It is found that a system of micrometric particles (1–3 μm) of $CaCO_3$ surrounding a $Ca(OH)_2$ core ($PSC_{Ca(OH)2}$ 83:27 wt% $CaCO_3$:$Ca(OH)_2$) was precipitated in setups containing a large quantity of water in the fluid phase (Fig. 11.4d,e), while nanometric calcite with a CaO core (PSC_{CaO} 82:28 wt% $CaCO_3$:CaO) was formed by carbonating dry CaO with wet $scCO_2$ (Fig. 11.4f–h). To demonstrate the possible advantages in CO_2 capture of the $scCO_2$-precipitated rhombohedral particles, the absorption behavior is compared with results obtained for the atmospherically prepared scalenohedral PCC (Fig. 11.4c) and grinded natural carbonate (GCC) [34]. The cyclic carbonation/calcination reactions are studied by using a magnetic suspension balance (Fig. 11.8a). Carbonation is carried out at 1023 K under a CO_2:N_2 flow. Experiments are first performed at mild experimental conditions using the isothermal CO_2 pressure swing absorption mode, in which the calcination is performed at 1023 K with N_2. $CaCO_3$ can be used as a sorbent in these mild conditions only for some specific applications, for instance, in the absorption of CO_2 in advanced hydrogen production methods [120]. Cyclic behavior is further analyzed under more realistic carbonation industrial conditions corresponding to 1173 K, short reaction times, and a continuous flow of CO_2, in which drawbacks related to loss of sorbent activity are more significant [121, 122]. The decomposition of the carbonate to the oxide is always 100%, but the reactivity of the oxide so formed and the extent of recarbonation depend on the chemical, textural, and morphological characteristics of the solid source of Ca^{2+}, as well as on the carbonation experimental conditions (temperature, concentration of CO_2, gas flow, reaction time, etc.). Microbalance measured conversion degrees (X_w in weight) are represented in Fig. 11.8b,c, in which absorption data for synthetic carbonates in the literature [115, 119, 121–127] and the curve derived for micronized natural limestone [113] are also displayed. Different decay tendencies of CO_2 absorption capacity in the long carbonation–decomposition series are observed for the three studied samples. At the calcination temperature of 1023 K (Fig. 11.8b), an enhanced reversibility of the reaction was observed for sample $PSC_{Ca(OH)2}$ constituted by rhombohedral micrometric particles compared to the nanometric powder (PSC_{CaO}) or the scalenohedral needles ($PA_{Ca(OH)2}$). Indeed, after 25 cycles and under mild experimental conditions of carbonation/calcination, sample $PSC_{Ca(OH)2}$ has one of the highest values of residual carbonation conversion described in the literature (X_w = 0.65). Conversely, at 1173 K calcination temperature the behavior of $PSC_{Ca(OH)2}$ and PSC_{CaO} samples was similar (Fig. 11.8c), with a residual carbonation conversion value at cycle 25 of ca. 0.3.

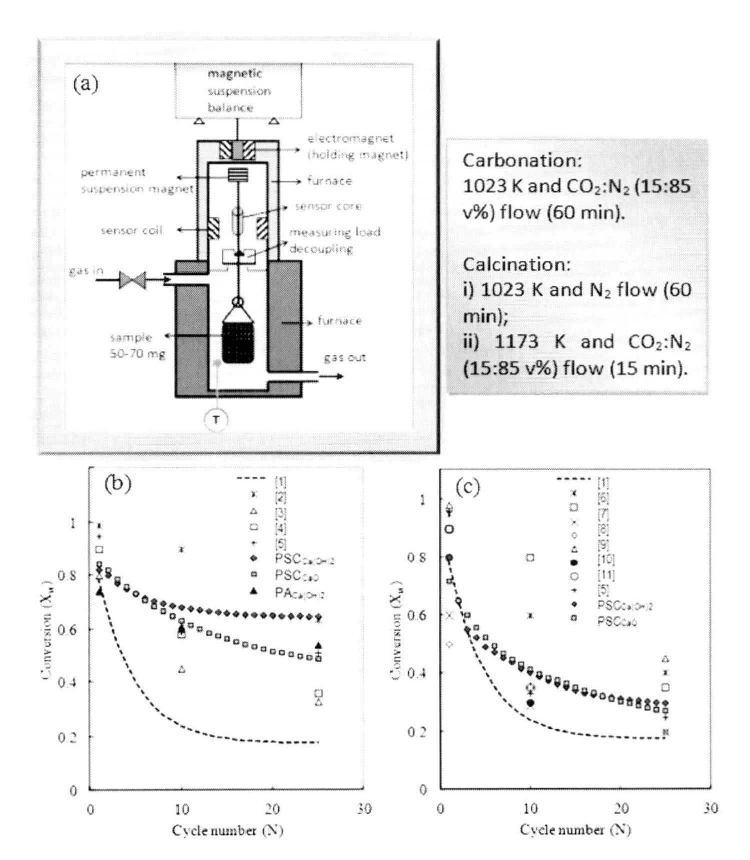

Figure 11.8 (a) Scheme of an integrated microbalance system (Rubotherm) and measurement experimental conditions and (b, c) conversion degrees (X_w in weight) for supercritically synthesized samples, as described in the text: 1, limestone curve derived from Abanades et al. [113], and literature data at cycles 1, 10, and 25; 2, for CaO derived from calcium acetate hydrate [123]; 3, for monodispersed PCC (5 μm) [121]; 4, for nanosized $CaCO_3$ (40 nm) [115]; 5, for nanosized $CaCO_3$ [122]; 6, for CaO derived from calcium acetate hydrate [127]; 7, for nanometric $CaCO_3$ [124]; 8 and 9, for micro- and nanometric PCC [125]; 10, for micrometric PCC (3 μm) [126]; and 11, for nanosized $CaCO_3$ (40 nm) [119].

Literature results show that natural limestone sorbent activity decays from ca. 0.7–0.8 at cycle 1 to ca. 0.2 at cycle 10 and to ca.

0.16 at cycle 25. Hence, even at 1173 K the residual carbonation conversion value for the synthetic material is two times higher than for the natural product. Replacing limestone with synthetic sorbents will be only economically feasible provided CO_2 uptake is higher than that of limestone for a larger number of cycles. Starting from nascent CaO, the carbonation reaction (Eq. 11.8) is generally initiated with a *fast* chemical solid–gas reaction occurring on the surface, which is followed by a *slow* diffusion process. The end of the *fast* reaction period is marked by the formation of a product layer around a CaO core. The *slow* reaction is controlled by the diffusion of the carbon dioxide through the newly formed carbonate layer to the interior of the particles. Cycle conversion profiles show three different reaction steps, assigned to the *fast* (I) and *slow* (II) carbonation reaction, and to the fast (III) decarbonation process (Fig. 11.9a,b). The diffusion of reactants is often the rate-limiting step in slow gas–solid-state reactions and, hence, the reaction rate in the *slow* step (II) can be modeled by the general form of Jander's equation [128] with a reaction order of $n = 2$ (Eq. 11.11):

$$k_2 t = \{1 - (1 - X_m)^{1/3}\}^n \quad (11.11)$$

where t (s) is the time, X_m is the molar conversion ratio (C_{CaCO3}/C^0_{CaO}), C_{CaCO3} being the concentration of $CaCO_3$ (mol·m^{-3}) at the reaction time and C^0_{CaO} the initial concentration of CaO (mol·m^{-3}), and k_2 is the apparent reaction rate constant (s^{-1}). For $n = 1$, a contracting volume equation is obtained, $k_1 t = 1 - (1 - X_m)^{1/3}$, in which the progress of the product layer from the surface to the inner crystal is considered to be rate-limiting and, thus, can be applied to estimate the reaction rate in the first *fast* reaction step (I). The decomposition of $CaCO_3$ in step (III) can be described by a phase boundary controlled reaction where the rate controlling step is the transfer across the contracting interface, described by the contracting volume equation with $X = (1 - X_m)$ and k_3 as the apparent decarbonation rate constant. For cycle 1, the highest k_1 reaction rate constant is obtained for sample PSC_{CaO}, reflecting the nanometric character of the initially precipitated particles (Fig. 11.9c). However, at cycle 10 the behavior of the rapid reaction is reversed, showing higher values of k_1 for sample $PSC_{Ca(OH)2}$ than for nanometric, but sinterized, PSC_{CaO}. In average, the k_1 values for samples supercritically prepared are ca. 2.5 times higher than those of the atmospherically prepared sample $PA_{Ca(OH)2}$. The values of k_1 are relatively stable with cycles for the micrometric samples, and only decrease somewhat for the nanometric one. Similar reaction rates (k_2 = 0.6–1 10^{-6} s^{-1}) are observed for the three studied samples during the diffusion-controlled step (Fig. 11.9d).

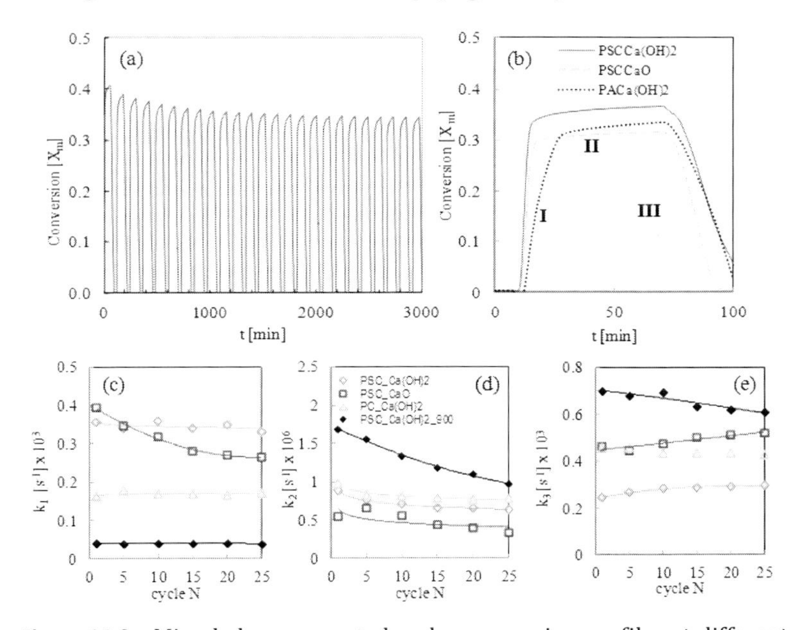

Figure 11.9 Microbalance reported molar conversion profiles at different calcination conditions (750°C): (a) 25 absorption/desorption cycles carried out for the micrometric sample $PSC_{Ca(OH)2}$; (b) micro-, nano-, and elongated $CaCO_3$ particles in cycle 25 of operation; and evolution of the reaction rate with cycles: (c) step (I) k_1, (d) step (II) k_2, and (e) step (III) k_3.

Surprisingly, the slowest decarbonation rate k_3 values are found for sample $PSC_{Ca(OH)2}$ (Fig. 11.9e). This rate determines the minimum cycle time required for complete regeneration of the solid sorbent. The conversion profile recorded for the $PSC_{Ca(OH)2}$ sample calcined at 1173 K with 15% CO_2 in N_2 indicates that the carbonation reaction rate k_1 is reduced in 1 order of magnitude when compared to data at 1023 K under pure N_2 (Fig. 11.9c). In contrast, at 1173 K the values of k_2 (Fig. 11.9d) and k_3 (Fig. 11.9e) significantly increase. From a practical point of view, the supercritically prepared samples have the advantage of showing an accelerated conversion rate in the *fast* reaction period. For these samples, the transition was reached in less than 5 min at 1023 K, while more than 15 min were necessary for the conventionally prepared samples. This fact could facilitate the design of continuous processes for the supercritically prepared samples. By increasing the calcination temperature to 1173 K, the time for carbonation also increases to ca. 10 min.

Tubular furnace treated samples under the mild carbonation conditions (Fig. 11.10a) can be used to perform an extensive investigation on the morphological and textural characteristics, since calcined and carbonated samples are recovered in a suitable weight amount [34]. In the performed series of experiments, calcined samples are labeled as $F(N)_a$ and after recarbonation as $F(N)_b$, N being the cycle number. Samples are taken from the holder for analysis after the temperature dropped to ambient conditions, while either before or after recarbonation, at a selected number of cycles: samples $F(1)_a$, $F(1)_b$, $F(2)_b$, $F(3)_b$, $F(5)_b$, $F(8)_b$, and $F(13)_b$ taken at cycles 1, 1, 2, 3, 5, 8, and 13, respectively, that is, following the Fibonacci sequence. Figures 11.10b and 11.10c show the SEM analysis of the raw materials and those recovered from the tubular furnace experiments performed for samples $PSC_{Ca(OH)2}$ and PSC_{CaO}, respectively. Precipitated $PSC_{Ca(OH)2}$ is constituted by intergrown stepped rhombohedral aggregates and a population of multinucleated rhombohedral calcite particles of sizes in the range 1–3 μm. Sample PSC_{CaO} shows a large population of agglomerated nanometric particles (100–200 nm) together with a few micrometric crystals. A first observation of the SEM images indicated that the morphological characteristics of the parent $CaCO_3$ particles remained after calcination, and the existence of the former shape can be still readily recognized after 13 cycles. Hence, for sample $PSC_{Ca(OH)2}$ shown in Fig. 11.10b it is clearly observed that the thermal decomposition of calcite single crystals resulted in pseudomorphs that fully preserve the external shape of the rhombohedral faces. In contrast, for samples PSC_{CaO} the nanometric nature of the initial $CaCO_3$ particles disappear, giving place to visible micrometric granules formed by the sintering of the nanometric units (Fig. 11.10c).

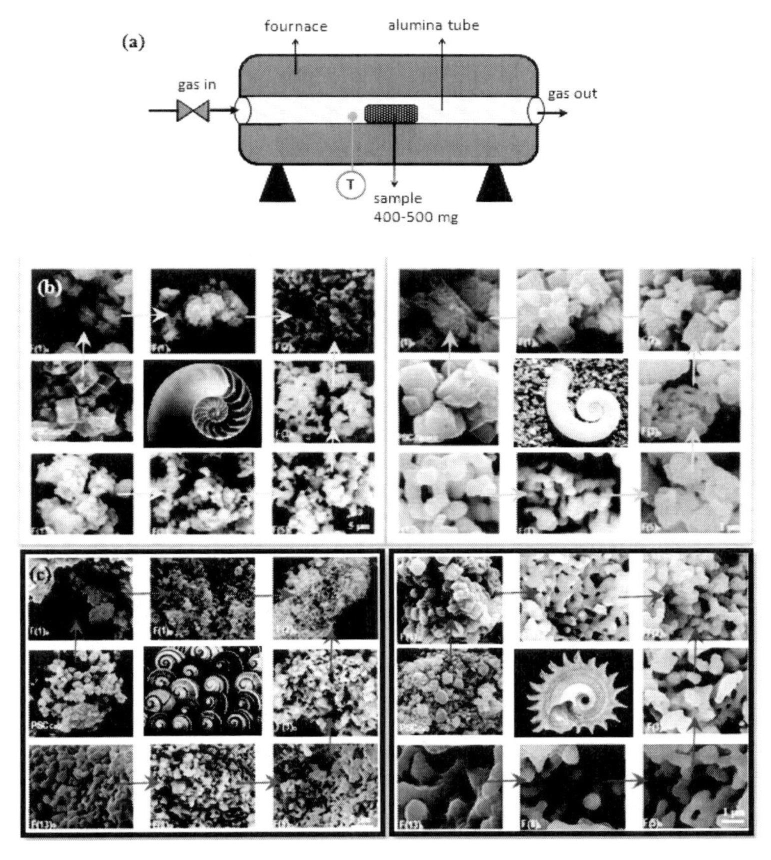

Figure 11.10 (a) Tubular furnace (Carbolite 3216) surrounding an alumina tube and SEM images at two magnifications (x20000 and x80000) of (b) micrometric $PSC_{Ca(OH)2}$ and (c) nanometric PSC_{CaO} recovered from the tubular furnace in experiments at 700°C and N_2 flow during calcination. Samples were recovered at cycle 1 after both calcination ($F(1)_a$) and carbonation ($F(1)_b$) and at cycles 2 ($F(2)_b$), 3 ($F(3)_b$), 5 ($F(5)_b$), 8 ($F(8)_b$), and 13 ($F(13)_b$) after carbonation.

After the first calcination (sample $F(1)_a$), the crystals of nascent CaO are arranged in small nanoparticles with very small pores in between [129]. For sample $PSC_{Ca(OH)2}$ the characteristics rods with parallel alignment can be clearly observed. When the CaO in $F(1)_a$ is submitted to recarbonation, meso- and macroporosity appear, but the network of small pores could be observed only in the first two or three cycles (samples $F(2)_b$). SEM analysis reveals that when a high cycle number is reached, the grains increase in size, as a consequence of sintering, and macropores are formed at the expenses of micropores. After cycles 5–8, the different systems maintain a network of large voids (0.5–1 μm).

Acknowledgments

The financial support of the Spanish government under project BIOREG MAT2012-35161 is gratefully acknowledged.

References

1. Simons, S. J. R., Hills, C. D., Carey, P. J., and Fernández Bertos, M. (2004). A review of accelerated carbonation technology in the treatment of cement-based materials and sequestration of CO_2, *J. Hazard. Mater.*, **112**, pp. 193–205.
2. Wendt, C. H., Butt, D. P., Joyce Jr., E. L., Sharp, D. H., and Lackner, K. S. (1995). Carbon dioxide disposal in carbonate minerals, *Energy*, **20**, pp. 1153–1170.
3. Tai, C. Y., and Chen, F. B. (1998). Polymorphism of $CaCO_3$ precipitated in a constant-composition environment, *AIChE J.*, **44**, pp. 1790–1798.
4. Gu, W., Bousfield, D. W., and Tripp, C. P. (2006). Formation of calcium carbonate particles by direct contact of Ca $(OH)_2$ powders with supercritical CO_2, *J. Mater. Chem.*, **16**, pp. 3312–3317.
5. Montes-Hernández, G., and Renard, F. (2011). Co-utilisation of alkaline solid waste and compressed-or-supercritical CO_2 to produce calcite and calcite/Se^0 red nanocomposite, *J. Supercrit. Fluids*, **56**, pp. 48–55.
6. Domingo, C., García-Carmona, J., Loste, E., Fanovich, A., Fraile, J., and Gómez-Morales J. (2004). Control of calcium carbonate morphology by precipitation in compressed and supercritical carbon dioxide media, *J. Cryst. Growth*, **271**, pp. 268–273.
7. Domingo, C., Loste, E., Gómez-Morales, J., García-Carmona, J., and Fraile, J. (2006). Calcite precipitation by a high-pressure CO_2 carbonation route, *J. Supercrit. Fluids*, **36**, pp. 202–215.
8. Regnault, O., Lagneau, V., and Schneider, H. (2009). Experimental measurement of portlandite carbonation kinetics with supercritical CO_2, *Chem. Geol.*, **265**, pp. 113–121.

9. Van Ginneken, L., Dutré, V., Adriansens, W., and Weyten, H. (2004). Effect of liquid and supercritical carbon dioxide treatments on the leaching performance of a cement- stabilized waste form, *J. Supercrit. Fluids*, **30**, pp. 175–178.
10. Shih, S. M., Ho, C.-S., Song, Y.-S., and Lin, J.-P. (1999). Kinetics of the reaction of $Ca(OH)_2$ with CO_2 at low temperature, *Ind. Eng. Chem. Res.*, **38**, pp. 1316–1322.
11. Kawano, J., Shimobayashi, N., Kitamura, M., Shinoda, K., and Aikawa, N. (2002). Formation process of calcium carbonate from highly supersaturated solution, *J. Cryst. Growth,* **237–239**, pp. 419–423.
12. Giles, D. E., Ritchie, I. M., and Xu, B. A. (1993). The kinetics of dissolution of slaked lime, *Hydrometallurgy*, **32**, pp. 119–122.
13. Babak, V. N., Babak, T. B., and Kholpanov, L. P. (2001). Absorption of carbon dioxide by alkali and aminoalcohol solutions in two-phase systems, *Theor. Found. Chem. Eng.*, **35**, pp. 557–560.
14. Elfil, H., and Roques, H. (2001). Role of hydrate phases of calcium carbonate on the scaling phenomenon, *Desalination*, **137**, pp. 177–186.
15. Jung, W. M., Kang, S. H., Kim, W.-S., and Choi, C. K. (2000). Particle morphology of calcium carbonate precipitated by gas-liquid reaction in a Couette-Taylor reactor, *Chem. Eng. Sci.*, **55**, pp. 733–747.
16. Lopez-Periago, A. M., Pacciani, R., Garcia-Gonzalez, C., Vega, L. F., and Domingo, C. (2010). A breakthrough technique for the preparation of high-yield precipitated calcium carbonate, *J. Supercrit. Fluids*, **52**, pp. 298–305.
17. Lopez-Periago, A. M., Pacciani, R., Vega, L. F., and Domingo, C. (2011). Monitoring the effect of mineral precursor, fluid phase CO_2-H_2O composition and stirring on $CaCO_3$ crystallization in a supercritical-ultrasound carbonation process, *Cryst. Growth Des.*, **11**, pp. 5324–5332.
18. Santos, R. M., François, D., Mertens, G. Elsen, J., and van Gerven, T. (2013). Ultrasound-intensified mineral carbonation, *Appl. Therm. Eng.*, **57**, pp. 154–163.
19. Enokida, Y., Abd El-Fatah, S., and Wai, C. M. (2002). Ultrasound-enhanced dissolution of UO in supercritical CO_2 containing a co-philic complexant of tri-n-butylphosphate and nitric acid, *Ind. Eng. Chem. Res.*, **41**, pp. 2282–2286.
20. Riera, E., Golás, Y., Blanco, A., Gallego, J. A., Blasco, M., and Mulet, A. (2004). Mass transfer enhancement in supercritical fluids extraction by means of power ultrasound, *Ultrason. Sonochem.*, **11**, pp. 241–244.

21. Hu, A., Zhao, S., Liang, H., Qiu, T., and Chen, G. (2007). Ultrasound assisted supercritical fluid extraction of oil and coixenolide from adlay seed, *Ultrason. Sonochem.*, **14**, pp. 219–224.

22. Kim, S. H., Yuvaraj, H., Jeong, Y.-T., Park, C., Kim, S. W., and Lim, K. T. (2009). The effect of ultrasonic agitation on the stripping of photoresist using supercritical CO_2 and co-solvent formulation, *Microelectr. Eng.*, **86**, pp. 171–175.

23. Chattopadhyay, P., and Gupta, R. B. (2001). Production of antibiotic nanoparticles using supercritical CO_2 as antisolvent with enhanced mass transfer, *Ind. Eng. Chem. Res.*, **40**, pp. 3530–3539.

24. Balachandran, S., Kentish, S. E., Mawson, R., and Ashokkumar, M. (2006). Ultrasonic enhancement of the supercritical extraction from ginger, *Ultrason. Sonochem.*, **13**, pp. 471–479.

25. Uebo, K., Yamazaki, R., and Yoshida, K. (1992). Precipitation mechanism of calcium carbonate fine particles in a three-phase reactor, *Adv. Powder Technol.*, **3**, pp. 71–76.

26. Kitamura, M., Konno, H., Yasui, A., and Masuoka, H. (2002). Controlling factors and mechanism of reactive crystallization of calcium carbonate polymorphs from calcium hydroxide suspensions, *J. Cryst. Growth*, **236**, pp. 323–332.

27. Kim, W.-S., Hirasawa, I., and Kim, W-S. (2004). Polymorphic change of calcium carbonate during reaction crystallization in a batch reactor, *Ind. Eng. Chem. Res.*, **43**, pp. 2650–2657.

28. Xanthos, M. (2005). *Functional Fillers for Plastics*, 2nd Ed. (Wiley-VCH, Germany).

29. Mathur, K. K., and Vanderheiden, D. B. (2001). Fillers: types, properties and performance, in *Polymers, Modifiers and Additives*, eds. Lutz, Jr., J. T., and Grossman, R. F., pp. 125–172 (Marcel Dekker, New York, USA).

30. García-González, C. A., Hidalgo, A., Andrade, C., Alonso, M. C., Fraile, J., López-Periago, A. M., and Domingo, C. (2006). Modification of composition and microstructure of Portland cement pastes as a result of natural and supercritical carbonation procedures, *Ind. Eng. Chem. Res.*, **45**, pp. 4985–4992.

31. Huet, B., Tasoti, V., and Khalfallah, I. (2011). A review of Portland cement carbonation mechanisms in CO_2 rich environment, *Energy Proc.*, **4**, pp. 5275–5282.

32. Huertas, F. J., Hidalgo, A., Rozalén, M. L., Pellicione, S., Domingo, C., García-González, C. A., Andrade, C., and Alonso, C. (2009). Interaction of bentonite with supercritically carbonated concrete, *Appl. Clay Sci.*, **42**, pp. 488–496.

33. Schaef, H. T., Ilton, E. S., Qafoku, O., Martin, P. F., Felmy, A. R., and Rosso, K. M. (2012). In situ XRD study of Ca^{2+} saturated montmorillonite (STX-1) exposed to anhydrous and wet supercritical carbon dioxide, *Int. J. Greenhouse Gas Control*, **6**, pp. 220–229.

34. López-Periago, A. M., Fraile, J., López-Aranguren, P., Vega, L. F., and Domingo, C. (2013). CO_2 capture efficiency and carbonation/calcination kinetics of micro and nanosized particles of supercritically precipitated calcium carbonate, *Chem. Eng. J.*, **226**, pp. 357–366.

35. Choi, S., Drese, J. H., and Jones, C. W. (2009). Adsorbent materials for carbon dioxide capture from large anthropogenic point sources, *ChemSusChem*, **2**, pp. 796–854.

36. Olajire, A. A. (2013). A review of mineral carbonation technology in sequestration of CO_2, *J. Pet. Sci. Eng.*, **109**, pp. 364–392.

37. Cölfen, H. (2003). Precipitation of carbonates: recent progress in controlled production of complex shapes, *Curr. Opin. Colloid Interface Sci.*, **8**, pp. 23–31.

38. El-Sheikh, S. M., El-Sherbiny, S., Barhoum, A., and Deng, Y. (2013). Effects of cationic surfactant during the precipitation of calcium carbonate nano-particles on their size, morphology, and other characteristics, *Colloids Surf., A*, **422**, pp. 44–49.

39. Mathur, V. K. (2001). High speed manufacturing process for precipitated calcium carbonate employing sequential pressure carbonation, *US Patent* 6,251,356

40. Chibowski, E., Holysz, L., Szczes, A., and Chibowski, M. (2003). Precipitation of calcium carbonate from magnetically treated sodium carbonate solution, *Colloids Surf., A*, **225**, pp. 63–73.

41. Rodríguez-Clemente, R., and Gómez-Morales, J. (1996). Microwave precipitation of $CaCO_3$ from homogeneous solutions, *J. Cryst. Growth*, **169**, pp. 339–346.

42. Montes, G., Renard, F., Geoffroy, N., Charlet, L., and Pironon, J. (2007). Calcite precipitation from CO_2-H_2O-$Ca(OH)_2$ slurry under high pressure of CO_2, *J. Cryst. Growth*, **308**, pp. 228–236.

43. Goldfarb, D. L., Corti, H. R., Marken, F., and Compton, R. G. J. (1998). High-pressure sonoelectrochemistry in aqueous solution: soft cavitation under CO_2, *Phys. Chem. A*, **102**, pp. 8888–8893.

44. Kuijpers, M. W. A., van Eck, D., Kemmere, M. F., and Keurentjes, J. T. F. (2002). Cavitation-induced reactions in high-pressure carbon dioxide, *Science*, **298**, pp. 1969–1970.

45. Kojima, Y., Yamaguchi, K., and Nishimiya, N. (2010). Effect of amplitude and frequency of ultrasonic irradiation on morphological

characteristics control of calcium carbonate, *Ultrason. Sonochem.*, **17**, pp. 617–620.

46. Jung, T., Kim, W-S., and Choi, C. K. (2004). Effect of nonstoichiometry on reaction crystallization of calcium carbonate in a Couette-Taylor reactor, *Cryst. Growth Des.*, **4**, pp. 491–495.
47. García-Carmona, J., Gómez-Morales, J., and Rodríguez-Clemente, R. (2003). Rhombohedral-scalenohedral calcite transition produced by adjusting the solution electrical conductivity in the system $Ca(OH)_2$-CO_2-H_2O, *J. Colloid Interface Sci.*, **261**, pp. 434–440.
48. Asuncion, J. (2003). *The Complete Book of Papermaking* (Lark Books Editor).
49. Singh, R., Lavrykov, S., and Ramarao, B. V. (2009). Permeability of pulp fiber mats with filler particles, *Colloids Surf., A*, **333**, pp. 96–107.
50. Sykes, M. S., Klungness, J. H., Tan, F., and AbuBakr, S. M. (1998). Value added mechanical pulps for lightweight, high-opacity paper, in *Tappi Pulping Conference Proceedings*, pp. 539–544.
51. Klungness, J. H., Caulfield, D. F., Sachs, I. B., Sykes, M. S., Tan, F., and Shilts, R. W. (1993). *Method for Fiber Loading a Chemical Compound*, US Patent No. 5,223,090.
52. Wang, H., Huang, W. H., and Hana, Y. (2013). Diffusion-reaction compromises the polymorphs of precipitated calcium carbonate, *Particuology*, **11**, pp. 11301–11308.
53. Montes-Hernandez, G., Fernandez-Martinez, A., Charlet, L., Tisserand, D., and Renard, F. (2008). Textural properties of synthetic nano-calcite produced by hydrothermal carbonation of calcium hydroxide, *J. Cryst. Growth*, **310**, pp. 2946–2953.
54. Cai, H. H., Li, S-D., Tian, G-R., Wang, H-B., and Wang, J-H. (2003). Reinforcement of natural rubber latex film by ultrafine calcium carbonate, *J. Appl. Polym. Sci.*, **87**, pp. 982–985.
55. Giorgi, R., Dei, L., Ceccato, M., Schettino, C., and Baglioni, P. (2002). Nanotechnologies for conservation of cultural heritage: paper and canvas deacidification, *Langmuir*, **18**, pp. 8198–8203.
56. Taylor, H. F. W. (1990). *Cement Chemistry* (Academic Press, New York, USA).
57. Phair, J. W. (2006). Green chemistry for sustainable cement production and use, *Green Chem.*, **8**, pp. 763–780.
58. Sanchez, F., and Sobolev, K. (2010). Nanotechnology in concrete: a review, *Const. Build. Mater.*, **24**, pp. 2060–2071.

59. Hidalgo, A., Petit, S., Domingo, C., Alonso, C., and Andrade, C. (2007). Microstructural characterization of leaching effects in cement pastes due to neutralisation of their alkaline nature part I: Portland cement pastes, *Cem. Concr. Res.*, **37**, pp. 63–70.

60. Keysar, S., Hasson, D., Semiat, R., and Bramson, D. (1997). Corrosion protection of mild steel by a calcite layer, *Ind. Eng. Chem. Res.*, **36**, pp. 2903–2908.

61. Purnell, P., and Beddows, J. (2005). Durability and simulated ageing of new matrix glass fibre reinforced concrete, *Cem. Concr. Compos.*, **27**, pp. 875–879.

62. Hermawan, D., Hata, T., Umemura, K., Kawai, S., Kaneko, S., and Kuroki, Y. (2000). New technology for manufacturing high-strength cement-bonded particleboard using supercritical carbon dioxide, *J. Wood Sci.*, **88**, pp. 46–50.

63. Liu, L., Ha, J., Hashida, T., and Teramura, S. (2001). Development of a CO_2 solidification method for recycling autoclaved lightweight concrete waste, *J. Mater. Sci. Lett.*, **20**, pp. 1791–1794.

64. Lange, L. C., Hills, C. D., and Poole, A. B. (1997). Effect of carbonation on properties of blended and non-blended cement solidified waste forms, *J. Hazard. Mater.*, **52**, pp. 193–198.

65. Macias, A. Kindness, A., and Glasser, F. P. (1997). Impact of carbon dioxide on the immobilization potential of cemented wastes: chromium, *Cem. Concr. Res.*, **27**, pp. 215–221.

66. Alba, N., Vázquez, E., Gassó, S., and Baldasano, J. M. (2001). Stabilization/solidification of MSW incineration residues from facilities with different air pollution control systems. Durability of matrices versus carbonation, *Waste Manage.*, **21**, pp. 313–319.

67. Santra, A., and Sweatman, R. (2011). Understanding the long-term chemical and mechanical integrity of cement in a CCS environment, *Energy Proc.*, **4**, pp. 5243–5250.

68. Zhang, M., and Bachu, S. (2011). Review of integrity of existing wells in relation to CO_2 geological storage: what do we know?, *Int. J. Greenhouse Gas Control*, **5**, pp. 826–840.

69. Fernández Bertos, M., Simons, S. J. R., Hills, C. D., and Carey, P. J. (2004). A review of accelerated carbonation technology in the treatment of cement-based materials and sequestration of CO_2, *J. Hazard. Mater. B.*, **112**, pp. 193–205.

70. Rimmele, G., Barlet-Gouedard, V., Porcherie, O., Goffe, B., and Brunet, F. (2008). Heterogeneous porosity distribution in Portland cement exposed to CO_2-rich fluids, *Cem. Concr. Res.*, **38**, pp. 1038–1048.

71. Barlet-Gouedard, V., Rimmele, G., Porcherie, O., Quisel, N., and Desroches, J. (2009). A solution against well cement degradation under CO_2 geological storage environment, *Int. Greenhouse Gas Control*, **3**, pp. 206–216.

72. Bukowski, J. M., and Berger, R. L. (1979). Reactivity and strength development of CO_2 activated nonhydraulic calcium silicates, *Cem. Concr. Res.*, **9**, pp. 57–68.

73. Al-Kadhimi, T. K. H., Banfill, P. F. G., Millard, S. G., and Bungey, J. H. (1996). An accelerated carbonation procedure for studies on concrete, *Adv. Cem. Res.*, **8**, pp. 47–59.

74. Soroushian, P., Aouadi, F., Chowdhury, H., Nossoni, A., and Sarwar, G. (2004). Cement-bonded straw board subjected to accelerated processing. *Cem. Concr. Compos.*, **26**, pp. 797–802.

75. García-González, C. A., Hidalgo, A., Fraile, J., López-Periago, A. M., Andrade, C., and Domingo, C. (2007). Porosity and water permeability study of supercritically carbonated cement pastes involving mineral additions, *Ind. Eng. Chem. Res.*, **46**, pp. 2488–2496.

76. García-González, C. A., el Grouh, N., Hidalgo, A., Fraile, J., López-Periago, A.M., Andrade, C., and Domingo, C. (2008). New insights on the use of supercritical carbon dioxide for the accelerated carbonation of cement pastes, *J. Supercrit. Fluids*, **43**, pp. 500–509.

77. Hidalgo, A., Domingo, C., Garcia, C., Petit, S., Alonso, C., and Andrade, C. (2008). Microstructural changes induced in Portland cement-based materials due to natural and supercritical carbonation, *J. Mater. Sci.*, **43**, pp. 3101–3111.

78. Hartmann, T., Paviet-Hartmann, P., Rubin, J. B., Fitzsimmons, M. R., and Sickafus, K. E. (1999). The effect of supercritical carbon dioxide treatment on the leachability and structure of cemented radioactive waste-forms, *Waste Manage.*, **19**, pp. 355–361.

79. Fernández-Carrasco, L., Rius, J., and Miravitlles, C. (2008). Supercritical carbonation of calcium aluminate cement, *Cem. Concr. Res.*, **38**, pp. 1033–1037.

80. Farahi, E., Purnell, P., and Short, N. R. (2013). Supercritical carbonation of calcareous composites: influence of curing, *Cem. Concr. Comp.*, **43**, pp. 48–53.

81. Jones, R. H. (1996). *Cement Treated with High-Pressure CO_2*, US005518540A Patent.

82. Diamond, L. W., and Akinfiev, N. N. (2003). Solubility of CO_2 in water from -1.5 to 100 °C and from 0.1 to 100 MPa: evaluation of literature

data and thermodynamic modeling, *Fluid Phase Equilib.*, **208**, pp. 265–290.

83. Bakharev, T., Sanjayan, J. G., and Cheng, Y. (2001). Resistance of alkali-activated slag concrete to carbonation, *Cem. Concr. Res.*, **31**, pp. 1277–1283.
84. Goñi, S., Gaztañaga, M. T., and Guerrero, A. (2002). Role of cement type on carbonation attack, *J. Mater. Res.*, **17**, pp. 1834–1842.
85. Groves, G. W., Rodway, D. I., and Richardson, I. G. (1990). The carbonation of hardened cement pastes, *Adv. Cem. Res.*, **3**, pp. 117–125.
86. Ngala, V. T., and Page, C. L. (1997). Effects of carbonation on pore structure and diffusional properties of hydrated cement pastes, *Cem. Concr. Res.*, **27**, pp. 995–1007.
87. van Gerven, T., van Baelen, D., Dutré, V., and Vandecasteele, C. (2004). Influence of carbonation and carbonation methods on leaching of metals from mortars, *Cem. Concr. Res.*, **34**, pp. 149–156.
88. Basheer, L., Basheer, P. A. M., and Long, A. E. (2005). Influence of coarse aggregate on the permeation, durability and the microestructure characteristics of ordinary Portland cement concrete, *Constr. Build. Mater.*, **19**, pp. 682–690.
89. Juenger, M. C., and Jennings, H. M. (2001). The use of nitrogen adsorption to assess the microstructure of cement paste, *Cem. Concr. Res.*, **31**, pp. 883–892.
90. Vocka, R., Gallé, C., Dubois, M., and Lovera, P. (2000). Mercury intrusion porosimetry and hierarchical structure of cement pastes, theory and experiment, *Cem. Concr. Res.*, **30**, pp. 521–527.
91. Johannesson, B., and Utgenannt, P. (2001). Microestructural changes caused by carbonation of cement mortar, *Cem. Concr. Res.*, **31**, pp. 925–931.
92. Maki, Y., and Ohnuma, H. (1992). Application of concrete to the treatment and disposal of radioactive waste in Japan, *Nucl. Eng. Des.*, **138**, pp. 179–188.
93. Trotignon, L., Devallois, V., Peycelon, H., Tiffreau, C., and Bourbon, X. (2007). Predicting the long term durability of concrete engineered barriers in a geological repository for radioactive waste, *Phys. Chem. Earth*, **32**, pp. 259–274.
94. Ulm, F. J., Heukamp, F. H., and Germaine, J. T. (2002). Residual design strength of cement-based materials for nuclear waste storage systems, *Nucl. Eng. Des.*, **211**, pp. 51–60.

95. Rice, G., Miles, N., and Farris, S. (2007). Approaches to control the quality of cementitious PFA grouts for nuclear waste encapsulation, *Powder Technol.*, **174**, pp. 56–59.

96. Garrabrants, A. C., Sanchez, F., and Kosson, C. D. (2004). Changes in constituent equilibrium leaching and pore water characteristics of a Portland cement mortar as a result of carbonation, *Waste Manage.*, **24**, pp. 19–34.

97. Purnell, P., Short, N. R., and Page, C. L. (2001). Super-critical carbonation of glass-fiber reinforced cement. Part 1. Mechanical testing and chemical analysis, *Composites A*, **32**, pp. 1777–1787.

98. Short, N. R., Purnell, P., and Page, C. L. (2001). Preliminary investigations into the supercritical carbonation of cement pastes, *J. Mater. Sci.*, **36**, pp. 35–41.

99. Short, N. R., Bough, A. R., Seneviratne, A. M. G., Purnell, P., and Page, C. L. (2004). Preliminary investigations of the phase composition and fine pore structure of super-critically carbonated cement pastes, *J. Mater. Sci.*, **39**, pp. 5683–5687.

100. Clarke, W. L. (1996). Safe disposal of nuclear waste, *Sci. Technol. Rev.*, 3, pp. 6–16.

101. Buzzi, O., Boulon, H. M., Deleruyelle, F., and Besnus, F. (2007). Hydromechanical study of rock–mortar interfaces, *Phys. Chem. Earth*, **32**, pp. 820–831.

102. Banthia, N., Biparva, A., and Mindess, S. (2005). Permeability of concrete under stress, *Cem. Concr. Res.*, **35**, pp. 1651–1655.

103. Khan, M. I., and Lynsdale, C. J. (2002). Strength, permeability, and carbonation of high-performance concrete, *Cem. Concr. Res.*, **32**, pp. 123–131.

104. Gaucher, E. C., and Blanc, P. (2006). Cement/clay interactions: a review; experiments, natural analogues, and modeling, *Waste Manage.*, **26**, pp. 776–788.

105. Faucon, P., Adenot, F., Jacquinot, J. F., Petit, J. C., Cabrillac, R., and Jorda, M. (1998). Long-term behavior of cement pastes used for nuclear waste disposal: review of physicochemical mechanisms of water degradation, *Cem. Concr. Res.*, **28**, pp. 847–857.

106. Fernández, R., Cuevas, J., Sánchez, L., Vigil de la Villa, R., and Leguey, S. (2006). Reactivity of the cement–bentonite interface with alkaline solutions using transport cells, *Appl. Geochem.*, **21**, pp. 977–992.

107. Huertas, F. J., Caballero, E., Jiménez de Cisneros, C., Huertas, F., and Linares, J. (2001). Kinetics of montmorillonite dissolution in granitic solutions, *Appl. Geochem.*, **16**, pp. 397–407.

108. Diamond, S. (1999). Aspects of concrete porosity revisited, *Cem. Concr. Res.*, **29**, pp. 1181–1188.

109. Bary, B., and Sellier, A. (2004). Coupled moisture-carbon dioxide-calcium transfer model for carbonation of concrete, *Cem. Concr. Res.*, **34**, pp. 1859–1872.

110. Abanades, J. C. (2002). The maximum capture efficiency of CO_2 using a carbonation/calcination cycle of $CaO/CaCO_3$, *Chem. Eng. J.*, **90**, pp. 303–306.

111. Barker, R. (1973). The reversibility of the reaction $CaCO_3$ $CaO+CO_2$, *J. Appl. Chem. Biotech.*, **23**, pp. 733–742.

112. Bouquet, E., Leyssens, G., Schonnenbeck, C., and Gilot, P. (2009). The decrease of carbonation efficiency of CaO along calcination–carbonation cycles: experiments and modeling, *Chem. Eng. Sci.*, **64**, pp. 2136–2146.

113. Abanades, J. C., and Alvarez, D. (2003). Conversion limits in the reaction of CO_2 with lime, *Energy Fuels*, **17**, pp. 308–315.

114. Mess, D., Sarofim, A. F., and Longwell, J. P. (1999). Product layer diffusion during the reaction of calcium oxide with carbon dioxide, *Energy Fuels*, **13**, pp. 999–1005.

115. Florin, N. H., and Harris, A. T. (2008). Screening CaO-based sorbents for CO_2 capture in biomass gasifiers, *Energy Fuels*, **22**, pp. 2734–2742.

116. Abanades, J. C., Rubin, E. S., and Anthony, E. J. (2004). Sorbent cost and performance in CO_2 capture systems, *Ind. Eng. Chem. Res.*, **43**, pp. 3462–3466.

117. Manovic, V., and Anthony, E. J. (2010). Lime-based sorbents for high-temperature CO_2 capture. A review of sorbent modification methods, *Int. J. Environ. Res. Public Health*, **7**, pp. 3129–3140.

118. Luo, C., Shen, Q., Ding, N., Feng, Z., Zheng, Y., and Zheng, C. (2012). Morphological changes of pure micro- and nano-sized $CaCO_3$ during a calcium looping cycle for CO_2 capture, *Chem. Eng. Tech.*, **35**, pp. 1–9.

119. Florin, N. H., and Harris, A. T. (2009). Reactivity of CaO derived from nano-sized $CaCO_3$ particles through multiple CO_2 capture-and-release cycles, *Chem. Eng. Sci.*, **64**, pp. 187–191.

120. Yu, F. C., and Fan, L. S. (2011). Kinetic study of high-pressure carbonation reaction of calcium-based sorbents in the calcium looping process (CLP), *Ind. Eng. Chem. Res.*, **50**, pp. 11528–11536.

121. Grasa, G. S., Abanades, J. C., Alonso, M., and Gonzalez, B. (2008). Reactivity of highly cycled particles of CaO in a carbonation/calcination loop, *Chem. Eng. J.*, **137**, pp. 561–567.

122. Grasa, G., González, B., Alonso, M., and Abanades, J. C. (2007). Comparison of CaO-based synthetic CO_2 sorbents under realistic calcination conditions, *Energy Fuels*, **21**, pp. 3560–3562.

123. Lu, H., Reddy, E. P., and Smirniotis, P. G. (2006). Calcium oxide based sorbents for capture of carbon dioxide at high temperatures, *Ind. Eng. Chem. Res.*, **45**, pp. 3944–3949.

124. Lysikov, A. I., Salanov, A. N., and Okunev, A. G. (2007). Change of CO_2 carrying capacity of CaO in isothermal recarbonation-decomposition cycles, *Ind. Eng. Chem. Res.*, **46**, pp. 4633–4638.

125. Liu, W., Low, N. W., Feng, B., and Wang, G. Diniz da Costa, J. C. (2010). Calcium precursors for the production of CaO sorbents for multicycle CO_2 Capture, *Environ. Sci. Tech.*, **44**, pp. 841–847.

126. Materic, V., Sheppard, C., and Smedley, S. I. (2010). Effect of repeated steam hydration reactivation on CaO-based sorbents for CO_2 capture, *Environ. Sci. Tech.*, **44**, pp. 9496–9501.

127. Luo, C., Zheng, Y., Zheng, C., Yin, J., Qin, C., and Feng, B. (2013). Manufacture of calcium-based sorbents for high temperature cyclic CO_2 capture via sol-gel process, *Int. J. Greenhouse Gas Control*, **12**, pp. 193–199.

128. Aihara, M., Nagai, T., Matsushita, J., Negishi, Y., and Ohya, H. (2001). Development of porous solid reactant for thermal-energy storage and temperature upgrade using carbonation/decarbonation reaction, *Appl. Energy*, **69**, pp. 225–238.

129. Alvarez, D., and Abanades, J. C. (2005). Pore-size and shape effects on the recarbonation performance of calcium oxide submitted to repeated calcination/recarbonation cycles, *Energy Fuels*, **19**, pp. 270–278.

Chapter 12

Polymer Processing Using Supercritical Fluid–Based Technologies for Drug Delivery and Tissue Engineering Applications

Ana Rita C. Duarte, João F. Mano, and Rui L. Reis
3B's Research Group—Biomaterials, Biodegradables and Biomimetics, University of Minho, Headquarters of the European Institute of Excellence on Tissue Engineering and Regenerative Medicine, AvePark, 4806-909 Taipas, Guimarães, Portugal
ICVS/3B's—PT Government Associate Laboratory, Braga/Guimarães, Portugal
aduarte@dep.uminho.pt

From the use of botanical plants in early human civilizations through synthetic chemistry and biotechnology, drug research has always passionate scientists creating exciting challenges to a large number of researchers from different fields, thus promoting a collaborative effort between polymer scientists, pharmacologists, engineers, chemists, and medical researchers. Worldwide, there is an increasing concern on health care that creates a major opportunity for development of new pharmaceutical formulations. Ageing

Supercritical Fluid Nanotechnology: Advances and Applications in Composites and Hybrid Nanomaterials
Edited by Concepción Domingo and Pascale Subra-Paternault

ISBN 978-981-4613-40-8 (Hardcover), 978-981-4613-41-5 (eBook)
www.panstanford.com

populations worried about the quality of life in the older years are actively seeking for new, more effective and patient compliant drug delivery devices. This has been the driving force for the continuous growth of the research made on delivery devices, which has become a powerful technique in health care. It has been recognized for long that simple pills or injections may not be the suitable methods of administration of a certain active compound. These medications present several problems and/or limitations, like poor drug bioavailability and systemic toxicity, derived essentially from pharmacokinetic and other carrier limitations and low solubility of the drugs in water. Therefore and to overcome these drawbacks, clinicians recommend frequent drug dosing, at high concentrations in order to overcome poor drug bioavailability but causing a potential risk of systemic toxicity. Polymer science has open new strategies for drug delivery systems. This chapter overviews of possible strategies involving polymer modification and processing for controlled drug delivery and drug delivery in tissue engineering.

12.1 Controlled Drug Delivery Systems

A controlled drug release system consists in a drug carrier capable of releasing the bioactive agent in a specific location at a specific rate [1]. The main purpose of these controlled release systems is to achieve a more effective therapy, that is, a system with a delivery profile that would yield a high blood level of the drug over a long period of time, avoiding large fluctuations in drug concentration and reducing the need of several administrations. Furthermore, these systems often improve the drug performance, provide patient compliance, and prolong drug stability. It is of particular interest the key role that materials have in the development of these new drug delivery systems, from polymers to ceramics or even metals [2–4]. When a pharmaceutical agent is encapsulated within, or attached to, a polymer or lipid, drug safety and efficacy can be greatly improved and new therapies are possible. This has been the driving force for active study of the design of these materials, intelligent delivery systems and approaches for delivery through different administration routes [5]. Drug delivery systems are usually classified according to the mechanism that controls the release of the active compound.

There are three primary mechanisms by which active agents can be released from a delivery system: diffusion, degradation (erosion, chemical reaction), and swelling (solvent activation) [6, 7]. Any or all of these mechanisms may occur in a given release system.

Diffusion-controlled systems are the most common ones. Two types of diffusion-controlled systems have been developed, presenting the same basic principle: diffusion occurs when a drug or other active agent passes through the polymer that forms the controlled-release device. One type of diffusion-controlled release system corresponds to a reservoir device in which the bioactive compound (drug) forms a core surrounded by an inert diffusion barrier. These systems include membranes, capsules, microcapsules, liposomes, and hollow fibers. In this case, drug diffusion through the polymer matrix is the rate limiting step, and release rates are determined by the choice of polymer and its consequent effect on the diffusion and partition coefficient of the drug to be released [7]. The second type is a monolithic device in which the active substance is dispersed or dissolved in an inert polymer. These are possibly the most common devices for controlled drug delivery since they are relatively easy to fabricate and there is no danger of an accidental high dosage that could result from the rupture of the membrane on the reservoir device. The dosage release properties may be dependent upon the solubility of the drug in the polymer matrix, or in the case of porous matrixes also the tortuosity of the network, dependent on whether the drug is dispersed or dissolved in the polymer [8].

Biodegradable materials degrade within the body as a result of natural biological processes, eliminating the need to remove a drug delivery system after release of the active agent has been completed. Chemically controlled systems can be achieved using bioerodible or pendant chain systems, that is, either by polymer degradation or cleavage of the drug from the polymer [9]. Polymer degradation can be defined as the conversion of a material that is insoluble in water into metabolites that are water soluble. In ideal bioerodible systems, the drug is homogeneously distributed in the polymer, just like in the matrix devices. As the polymer surrounding the drug is eroded, the drug is released. In the case of pendant chain system, the drug is covalently bound to the polymer and it is released by bond scission due to water or enzymes. Furthermore, degradation may take place through bulk hydrolysis, in which the polymer degrades in a fairly uniform manner throughout the matrix or it may occur only at the surface of the polymer, resulting in a release rate that is proportional to the surface area of the drug delivery system [10].

12.1.1 Particle Formation/Encapsulation

Many technologies have been proposed to prepare polymeric particles for controlled drug release [11, 12]. In particular, supercritical fluid (SCF) technology presents many possibilities for particle formation and/or encapsulation. The differences between the processing techniques that have been reported are a result of the interactions and phase behavior of the active compounds with the SCF [13–15]. A brief description of the main characteristics of each process is listed in Table 12.1.

Table 12.1 Summary of the supercritical fluid techniques for particle formation and/or encapsulation

Technique	RESS (Rapid Expansion from Saturated Solutions)	SAS (Supercritical Anti-Solvent)	PGSS (Particles from Gas Saturated Solutions)
Role of SCF	Solvent	Anti-solvent	Solute
Basic principle	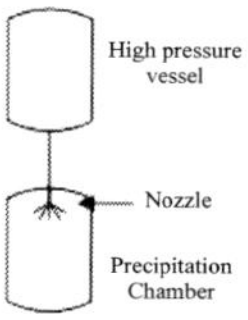 High pressure vessel; Nozzle; Precipitation Chamber The compound(s) are dissolved in the SFC phase and the solution is expanded into a low pressure vessel.	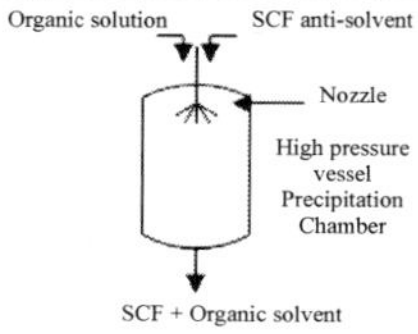 Organic solution; SCF anti-solvent; Nozzle; High pressure vessel Precipitation Chamber; SCF + Organic solvent The compound(s) are dissolved in an organic solution. The SCF acts as an anti-solvent promoting the precipitation of the solute.	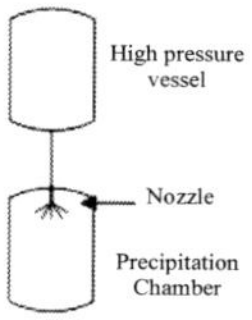 High pressure vessel; Nozzle; Precipitation Chamber The SCF is dissolved in the melted solution and the solution is expanded into a low pressure vessel.

In a rapid expansion from saturated solutions (RESS) process, the SCF acts as a solvent. In this technique, the active compound is dissolved in the SCF phase and the solution is expanded into a low-pressure vessel. Although this process is highly advantageous as no organic solvents are involved it requires high gas/solute ratios and high operating pressure and temperature, as the solubility of the compounds is usually low. The preparation of delivery systems using this technique is limited by the poor solubility of polymers in the fluid phase, and therefore it has not been widely used in the preparation of controlled delivery systems [15].

The supercritical antisolvent (SAS) technique uses the SCF as an antisolvent, taking advantage of the poor solubility of high-molecular-weight or polar compounds in SCFs. In this process the compound or mixture of compounds are dissolved in an organic solution and the SCF acts as an antisolvent promoting the precipitation of the solute. The principle of the process is to decrease the solvent power of the liquid by the addition of an antisolvent in which the solute is insoluble. This process broadens the applicability of the technique for the development of controlled drug delivery systems as numerous active compounds and polymers may be processed in a single-step operation after determining the appropriate solvent for the mixture. Lopez-Periago et al. report the preparation of polymethylmethacrylate (PMMA) particles loaded with triflusal and the preparation of poly(lactic acid) fibers by supercritical antisolvent [16, 17]. Furthermore SAS extends the applicability of SCF technology to the development of encapsulated systems. Encapsulation processes by supercritical precipitation techniques have been reviewed by Cocero et al. in a manuscript where different techniques and the mechanisms behind them are discussed in detail [18]. Hybrid materials can be prepared by coprecipitation of the active compound and the coating material. In this case a physical mixture of the active compound and drug carrier, such as polymer, is obtained and the interactions between them lead to a controlled release of the pharmaceutical agent. On the other hand, the possibility to coprecipitate different solutions through a coaxial nozzle has been explored and it is reported as supercritical enhanced dispersion of solutions (SEDS) in different papers [19–21]. In this case, the active compound is injected through the inner nozzle capillary, the polymer is injected through the middle capillary and the antisolvent flows through the outer part. This nozzle design offers the possibility to precipitate particles of the core material in the matrix of the coating material, which nucleates around the first particles to be precipitated forming a thin shell or a capsule around the active ingredient. The initial concentrations of both active ingredient and carrier, as well as the flow ratio of the two solutions, will determine the final morphology of the particles as well as the encapsulation efficiency. SAS processing can be carried out under mild operating conditions, that is, near ambient temperatures. This technique allows the production of particles in the nanometer size and the particle size and morphology can be easily controlled. The main disadvantages are related with the difficulty in scaling up the process, as different thermodynamic and hydrodynamic effects need to be taken into account. The use of organic solvent represents also a disadvantage of the process, since it has to be rigorously controlled in the case of pharmaceutical applications and may be present in residual amounts in the product.

In particles from gas-saturated solutions (PASS) the SCF acts as a solute [22]. In this technique, the SCF is dissolved in the melted solution and the solution is expanded into a low-pressure vessel. This method does not require the use of organic solvents, and it can be easily scaled up and has a high production capacity. Nonetheless, the high temperatures required to process some polymers may compromise its application for the processing of thermosensitive active compounds. For example, protein-loaded lipid microparticles were produced and reported by Salas et al. [23]. Casettari et al. report the use of PASS for the development of mucoadhesive particles form chitosan and poly(lactic acid) for gastrointestinal drug delivery [24]. Falconer et al. present a multivariate study on the effect of different operating conditions for the preparation of progesterone-loaded gelucire particles and the results demonstrate that not only the independent variables of pressure and temperature influence the result but also the interaction between them influence the yield of the process [25]. Other examples, such as the encapsulation of caffeine or *trans*-chalcone in lipid carriers have been presented by de Sousa et al. [26, 27]. Garcia-Gonzalez et al. report the encapsulation of agents with different polarity by PASS and the encapsulation of inorganic particles for topical administration [28, 29].

12.1.2 Impregnation

The preparation of drug release products often necessitates the use of a mobile phase to dissolve and carry the drug component, which also swells and stretches the polymer matrix, facilitating the diffusion of the drug, and increasing the rate of impregnation. Conventionally, the preparation of these systems involves three steps: solubilization of the pharmaceutical in an appropriate solvent, diffusion of the pharmaceutical through the polymer, and elimination of the residual solvent. The dispersion of active compounds within a finished or semifinished matrix that will serve as a carrier takes advantage of the solubility of pharmaceutical compounds in supercritical conditions. The drug is dissolved in carbon dioxide and diffuses into the bulk of the matrix, and when the system is depressurized the gas rapidly diffuses out of the polymer, leaving the drug absorbed or entrapped within the polymeric matrix and warranting the complete removal of solvent, without exposing polymers and drugs to high temperatures, which may degrade them. In Fig. 12.1, a schematic diagram of the process is presented.

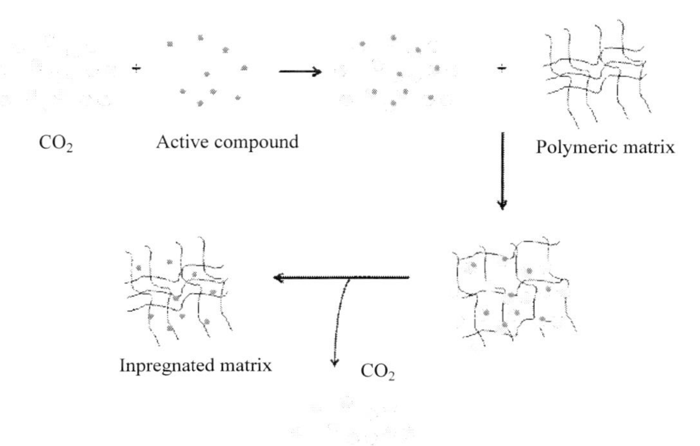

Figure 12.1 Schematic representation of the supercritical fluid impregnation process.

SCF impregnation is a process that requires the knowledge of the interactions between a polymer-active compound and an SCF. Kazarian et al. [30–33] distinguish two mechanisms of impregnation assisted by SCFs. The first mechanism corresponds to a simple deposition of the compound when the fluid leaves the swollen matrix. In this case, the solute is solubilized in carbon dioxide and the polymer is exposed to this solution for a predetermined period followed by depressurization of the system. When the system is depressurized, carbon dioxide molecules quickly leave the polymer matrix, leaving the solute trapped inside. As reported by Kazarian et al., this mechanism concerns mostly solutes with a relatively high solubility in the fluid and it is specific to impregnations carried out on a matrix subjected to swelling upon exposure to SCFs. The second effect, not specific to SCFs, corresponds to chemical interactions (like van der Walls interactions) between the solute and the matrix, which would favor the preferential partitioning of the solute with the polymer phase [32].

Kikic and Vecchione reviewed in 2003 the potential applications of polymer impregnation [34]. Since then, this technique has been employed in the preparation of a large number of drug delivery devices, which can be administered through different routes. Ophthalmic drug delivery systems have been developed for the treatment of diseases as glaucoma. Braga et al. report the development of hydrogels as ophthalmic drug delivery systems in which chitosan derivatives were impregnated with flurbiprofen and timolol maleate [35]. The same research group describes the preparation of therapeutic contact lenses after the impregnation of commercially available contact lenses with different active compounds [36–39]. The authors report that SCF impregnation does not compromise the integrity of the contact lenses, which provides major advantages over conventional impregnation techniques. The development of intraocular drug delivery systems has also gained attention, especially in the postoperative treatment of cataract surgery [40–42]. The preparation of transdermal drug delivery systems has also been reported in the literature. In this application the possibility to homogeneously disperse the active compound in the matrix, the ease of diffusion of the drug into the bulk of the material, and the fact that no organic solvents are used are particularly relevant. In the work of Argemi et al. [43], transdermal patches were impregnated with naproxen. The membranes prepared have shown a sustainable drug release up to 24 h. The preparation of wound dressings impregnated with two natural bioactive compounds is another example of a drug delivery device prepared using supercritical solvent impregnation reported in the literature [44].

Supercritical impregnation of polymeric matrices in the form of powders has also been widely explored. Examples of the systems studied and reported in the past few years are the case of poly(vinylpyrrolidone) impregnated with ketoprofen [45], hydroxypropylmethyl cellulose impregnated with indomethacin [46], impregnation of ibuprofen and timolol maleate in poly(ε-caprolactone) [47, 48], impregnation of paclitaxel, roxithromycin, and 5-fluorouracil in poly(lactic acid) particles [49–51], among others. Duarte et al. have compared the release of naproxen from ethylcellulose/methylcellulose particles prepared by different techniques [52]. Although in some cases the yield of SCF impregnation is not as high as the yield achieved by conventional techniques, such as solvent evaporation or soaking, the technique still presents major advantages.

12.1.3 Molecular Imprinting

Molecular imprinting is a technique that allows the design of a precise macromolecular structure able to recognize specific molecules [53, 54]. The mechanism underlying molecular imprinting is similar to the enzyme substrate concept. The principle of preparation of molecularly imprinted polymers (MIPs) is schematically represented in Fig. 12.2.

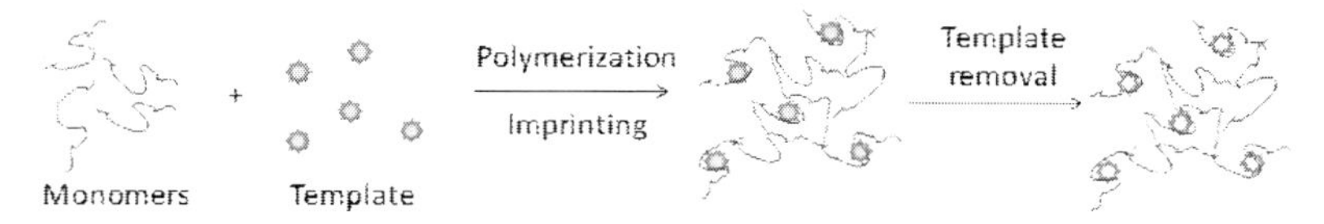

Figure 12.2 Schematic representation of the preparation of moleculary imprinted polymers (MIPs).

A network with specific conformational and structural sites is formed by polymerization and crosslinking of the monomer around the template. After polymerization, the template is leached out, providing macromolecular cavities for the entrapment of the particular molecule used. The envisaged applications for this technology greatly surpass the pharmaceutical field and encompass analytical applications, such as biosensors, immunoassays, separation media, and affinity supports, among others. Although the mechanism is relatively simple the optimization of is more complicated due to the contribution of several variables involved in the process, such as the functional monomer(s), the type of crosslinker, the ratio between monomer and crosslinker, and the ratio between monomer and template.

The polymerization can be carried out under supercritical conditions, using carbon dioxide as a reaction media or following conventional polymerization routes. Duarte et al. have reported the preparation of molecularly imprinted poly(diethylene glycol dimethacrylate) with salicylic and acetylsalicylic acids [55]. Results indicate that the amount of drug impregnated is significantly higher when a template molecule is present during the polymerization step. Other examples of MIPs have been reported following the same approach and using propanolol, ibuprofen, or flufenamic acid as model drugs [56–58]. Flufenamic acid was impregnated in a thermoresponsive drug delivery system based on polyisopropylacrylamide that was prepared using the molecular imprinting approach. This work represents a step forward in the development of complex delivery systems using clean technologies. In another work, Kobayashi et al. report the preparation of an uracil imprinted membrane of poly(styrene-co-maleic acid), demonstrating the flexibility of the technology for the preparation of different types of substrates [59].

12.1.4 Externally Triggered Delivery Devices

Smart drug delivery systems have been object of intense research [60–64]. The ability to release a bioactive compound according to a physiological need in a spatiotemporal-controlled manner may be the answer to avoid fluctuations and high concentrations of drugs that induce undesired side effects. The possibility to trigger the release of a drug by external stimuli would be highly beneficial [62, 65]. Several mechanisms have been described in the literature. Table 12.2 presents the mechanisms of action behind several external stimuli. Temperature-sensitive drug delivery systems are by far the most widely studied. Particularly interesting are hydrogels containing poly(*N*-isopropylacrylamide) (PNIPAAm). Hydrogels based on PNIPAAm present various applications from drug delivery, cell encapsulation, and cell culture surfaces. This polymer presents, in aqueous solutions, a low critical solution temperature (LCST) around 32°C, which makes it extremely interesting for applications in the biomedical field. Below the LCST, in aqueous solutions, it presents a flexible extended coil conformation, which makes it hydrophilic. Close to the LCST it becomes hydrophobic as polymer chains collapse and aggregate into a globular structure. Furthermore, copolymerization of NIPAAm with other monomers may change the overall hydrophilicity of the polymer. Conventional methods of polymerization involve the use of organic solvents and often require the use of crosslinking agents, which might be toxic. Temtem et al. have reported the successful polymerization of PNIPAAm in supercritical carbon dioxide [66]. The process proposed allows the in situ polymerization of PNIPAAm, leading to the development of smart-drug delivery devices in a single-step process [67].

Table 12.2 Mechanism of action of different stimuli used as triggers for controlled drug release

External stimuli	Mechanism of action
Thermal	Change in temperature > change in polymer–polymer and water–polymer interactions > change in polymer conformation and solubility > change in swelling > drug release
pH	Change in pH > swelling > drug release
Ionic strength	Change in ionic strength > change concentration of ions inside drug delivery device > change in swelling > drug release
Chemical species	Electron-donating compounds > formation of charge/transfer complex > change in swelling > drug release
Enzyme mediated	Change in swelling of the matrix > enzyme activity promoted over substrate > degradation of substrate by enzymatic cleavage > drug release
Magnetic	Magnetic field applied > change in pores in matrix > change in swelling > drug release
Electrical	Electrical field applied > membrane charging > electrophoresis of charged drug > change in swelling > drug release
Ultrasound irradiation	Ultrasound irradiation > temperature increase > drug release

Certain polysaccharides and polymers respond to pH changes as is the case with chitosan, alginate, and hyaluronic acid as examples of natural polymers and polyacrylic acid as a synthetic pH-responsive polymer [68]. Chitosan owes its pH-sensitive behavior to the large amount of amino groups present in its chains and swells in acidic pH, while polyacrylic acid, on the other hand, due to the presence of acidic groups swells in basic media. Temtem et al. refer to the preparation of a dual-stimuli-responsive matrix based on chitosan and PNIPAAm [69]. Chitosan is a pH-responsive polymer, while PNIPAAm is thermosensitive. The results demonstrate the possibility to control polymer swelling by either pH or temperature, and drug release can be modulated according to the different stimuli applied. Other examples of externally triggered pulse-wise drug delivery devices have been developed for specific applications, nonetheless SCF technology has, to our knowledge, not yet been reported in the development of such systems.

12.2 Drug Delivery in Tissue Engineering Applications

The concept of tissue engineering has long surpassed the idea of a merely inert support for cell attachment and growth. Tissue engineering is a promising therapeutic approach that involves the edges of a triangle in which materials, active principles, and cells all play an important role (Fig. 12.3) [12, 70–72].

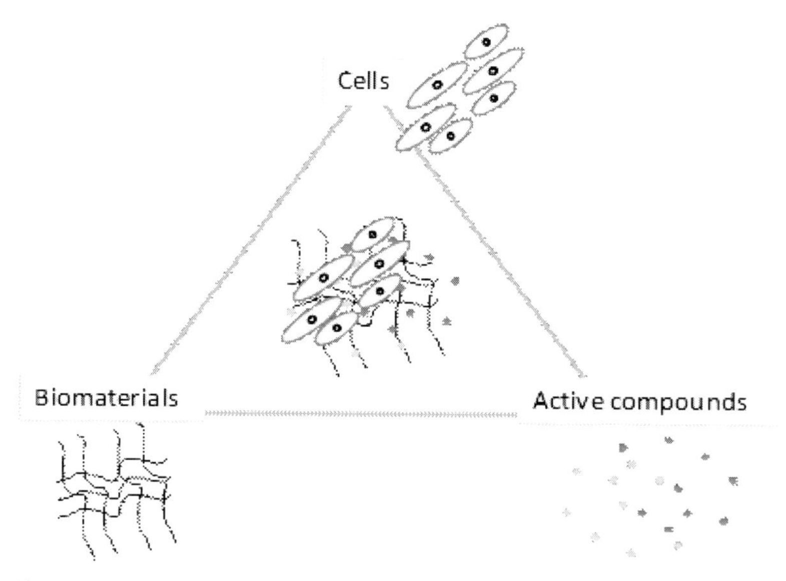

Figure 12.3 Tissue engineering: a combinatory approach of biomaterials, cells, and active compounds.

In this sense, the preparation of delivery systems able to sustain the release of biologically active molecules is a major challenge. Within the tissue engineering field, not only does a controlled release need to be achieved, but the polymer architecture also requires very particular features for each given tissue [73]. 3D architectures or scaffolds should present adequate surface properties, both chemically and topographically, as these characteristics will ultimately dictate cell adhesion to the surface. Furthermore, they should present adequate porosity, mean pore size, and interconnectivity between the pores to promote cell penetration and ensure oxygen and nutrient diffusion into the bulk of the matrix, as well as waste retrieval. Matrices must also have appropriate mechanical properties to withstand mechanical forces and maintain physical integrity, the materials must be biocompatible, and the degradation products must be noncytotoxic.

The major challenge in tissue engineering arises in the optimization of polymer-processing techniques [74]. A variety of different processing techniques have been developed and include fiber bonding, freeze drying, solvent casting and particle leaching, wet spinning, particle aggregation, electrospinning, 3D potting, and SCF technology, among others (Fig. 12.4) [75]. The choice of the most suitable polymer-processing technique depends greatly on the characteristics of the polymer itself, particularly its solubility in aqueous or organic solutions and its thermal properties as these will ultimately determine the feasibility to successfully produce matrices with the desired features. Processing thermosensitive bioactive compounds requires, however, the use of mild processing conditions and the reduction of the amount of organic solvents used [76, 77]. This presents an increase challenge in materials processing for tissue engineering and regenerative medicine.

The application of green technologies for the preparation of structures for tissue engineering has gained much attention in the past 10 years and a number of publications have been reported describing a variety of different techniques aiming to pursue the development of a single-step technology able to produce a material with all the desired properties. From the use of carbon dioxide as a drying agent in SCF drying to the use of carbon dioxide as a plasticizing agent in gas foaming and sintering or the use of carbon dioxide as an antisolvent in the supercritical assisted phase inversion method, almost all encounter a way to satisfy most of the characteristics requested. A summary of the features of the different technologies and some examples of drug delivery systems for tissue engineering and regenerative medicine applications are listed in Table 12.3 [78–88]. Hydrogel foaming is a more recent technique that has been explored for the preparation of porous scaffolds and in this process hydrogels foaming involves the dissolution of carbon dioxide in the water phase present on the hydrogel that will promote foaming of the structure upon depressurization [89, 90]. Sintering is a technique that occurs at near critical conditions and relies on the slight plasticization of the polymer particles that are fused together, creating a 3D environment [91, 92]. Although these techniques have not yet been reported for the preparation of drug delivery systems, their mild processing conditions foresee interesting developments in this field, especially in the impregnation of proteins, growth factors, and cells.

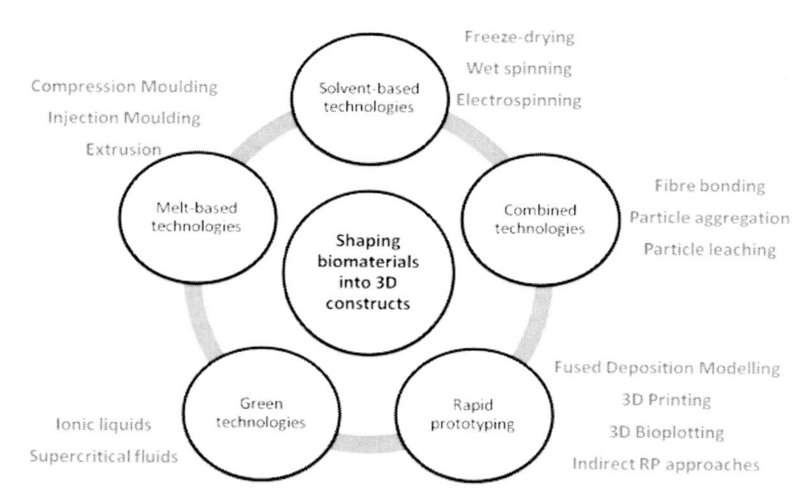

Figure 12.4 Summary of different polymer-processing methodologies employed in scaffold fabrication for tissue engineering and regenerative medicine.

Table 12.3 Summary of different supercritical fluid-based techniques used in the preparation of drug delivery systems in tissue engineering applications

Technique	Principle	Examples of drug delivery systems prepared
Foaming	The polymer is exposed to carbon dioxide at the saturation pressure and temperature, which plasticizes the polymer and reduces the glass transition temperature. Upon depressurization, thermodynamic instability causes supersaturation of the carbon dioxide dissolved in the polymeric matrix, and hence, nucleation of the cells occurs.	Poly(D,L-lactide) and poly(D,L-lactide-co-glycolide) impregnated with indomethacin [78] Poly(methyl methacrylate)-poly(L-lactic acid) foam loaded with ibuprofen [79] Poly(D,L-lactide-co-glycolide) foam with encapsulated growth factors [80] 3D architectures of poly(D,L-lactide-co-glycolide) as vehicles for DNA delivery [81] Vascular endothelial growth factor encapsulated in poly(lactic-co-glycolic acid) (PLGA) scaffolds [82, 83]
Phase inversion	The polymer is dissolved in an organic solvent and placed in contact with a nonsolvent (CO_2), which causes the solution to be phase-separated, creating a 3D porous structure.	Starch–poly(lactic acid) blend impregnated with dexamethasone [84] Poly(vinylidene fluoride-co-hexafluoropropylene) loaded with amoxicillin [85] Polymethylmethacrylate loaded with amoxicilin [86]
Drying	Supercritical drying is a drying technique that does not compromise the integrity of the structure as there are no phase boundaries, that is, phase transitions involved in the process.	Chitosan matrices impregnated with camptothecin and griseofulvin [87] Chitin scaffolds loaded with dexamethasone [88]

12.3 Conclusions

The preparation of drug delivery systems is intimately related with polymer modification and the design of new processes able to produce systems that meet most of the requirements of an ideal delivery system. Different techniques, from particle formation to impregnation to polymerization, have been explored and are reported in the literature. In the future the combination and integration of different techniques may see exciting perspectives as a single technique may not be enough for the development of a drug delivery system that meets all the required features. The integration of different technologies could provide interesting developments in shaping biomaterials into various constructs, opening a wide range of opportunities for the preparation of enhanced polymeric materials as structural supports for drug delivery.

Acknowledgments

The authors acknowledge the funding from the European Union Seventh Framework Programme (FP7/2007-2013) under grant agreement number REGPOT-CT2012-316331-POLARIS and from the project "Novel Smart and Biomimetic Materials for Innovative Regenerative Medicine Approaches" RL1 - ABMR - NORTE-01-0124-FEDER-000016 cofinanced by North Portugal Regional Operational Programme (ON.2 – O Novo Norte), under the National Strategic Reference Framework (NSRF), through the European Regional Development Fund (ERDF).

References

1. Peppas, N. A. (2013). Historical perspective on advanced drug delivery: how engineering design and mathematical modeling helped the field mature, *Adv. Drug Delivery Rev.*, **65**, pp. 5–9.
2. Mehnert, W., and Mader, K. (2012). Solid lipid nanoparticles production, characterization and applications, *Adv. Drug Delivery Rev.*, **64**, pp. 83–101.
3. Park, J. S., Lee, J. H., Shin, H. S., Lee, T. W., Kim, M. S., Khang, G., Rhee, J. M., Lee, H. K., and Lee, H. B. (2007). Biodegradable polymer microspheres for controlled drug release, *Tissue Eng. Regen. Med.*, **4**, pp. 347–359.

4. Freiberg, S., and Zhu, X. (2004).Polymer microspheres for controlled drug release, *Int. J. Pharm.*, **282**, pp. 1–18.
5. Langer, R. (1990). New methods of drug delivery, *Science*, **249**, pp. 1527–1533.
6. Siepmann, J., and Peppas, N. A. (2012). Modeling of drug release from delivery systems based on hydroxypropyl methylcellulose (HPMC), *Adv. Drug Delivery Rev.*, **64**, pp. 163–174.
7. Siepmann, J., and Gopferich, A. (2001). Mathematical modeling of bioerodible, polymeric drug delivery systems, *Adv. Drug Delivery Rev.*, **48**, pp. 229–247.
8. Pillai, O., and Panchagnula, R. (2001). Polymers in drug delivery, *Curr. Opin. Chem. Biol.*, **5**, pp. 447–451.
9. Nair, L. S., and Laurencin, C. T. (2007). Biodegradable polymers as biomaterials, *Prog. Polym. Sci.*, **32**, pp. 762–798.
10. Gopferich, A. (1996). Mechanisms of polymer degradation and erosion, *Biomaterials*, **17**, pp. 103–114.
11. Lima, A. C., Sher, P., and Mano, J. F. (2012). Production methodologies of polymeric and hydrogel particles for drug delivery applications, *Expert Opin. Drug Delivery*, **9**, pp. 231–248.
12. Santo, V. E., Gomes, M. E., Mano, J. F., and Reis, R. L. (2013). Controlled release strategies for bone, cartilage, and osteochondral engineering: part I; recapitulation of native tissue healing and variables for the design of delivery systems, *Tissue Eng. Part B*, **19**, pp. 308–326.
13. Mishima, K. (2008). Biodegradable particle formation for drug and gene delivery using supercritical fluid and dense gas, *Adv. Drug Delivery Rev.*, **60**, pp. 411–432.
14. Qiao, J. C., Ha, X. L., Guan, P., Zhao, Y., and Tian, W. (2008). Preparation and development of pharmaceutical microcapsules, *Prog. Chem.*, **20**, pp. 171–181.
15. Yeo, S. D., and Kiran, E. (2005). Formation of polymer particles with supercritical fluids: a review, *J. Supercrit. Fluids*, **34**, pp. 287–308.
16. Lopez-Periago, A. M., Vega, A., Subra, P., Argemi, A., Saurina, J., Garcia-Gonzalez, C. A., and Domingo, C. (2008). Supercritical CO_2 processing of polymers for the production of materials with applications in tissue engineering and drug delivery, *J. Mater. Sci.*, **43**, pp. 1939–1947.
17. Andanson, J. M., Lopez-Periago, A. M., Garcia-Gonzalez, C. A., Domingo, C., and Kazarian, S. G. (2009). Spectroscopic analysis of triflusal impregnated into PMMA from supercritical CO_2 solution, *Vib. Spectrosc.*, **49**, pp. 183–189.

18. Cocero, M. J., Martin, A., Mattea, F., and Varona, S. (2009). Encapsulation and co-precipitation processes with supercritical fluids: fundamentals and applications, *J. Supercrit. Fluids*, **47**, pp. 546–555.

19. Franceschi, E., de Cesaro, A. M., Feiten, M., Ferreira, S. R. S., Dariva, C., Kunita, M. H., Rubira, A. F., Muniz, E. C., Corazza, M. L., and Oliveira, J. V. (2008). Precipitation of beta-carotene and PHBV and co-precipitation from SEDS technique using supercritical CO_2, *J. Supercrit. Fluids*, **47**, pp. 259–269.

20. He, W. Z., Jiang, Z. H., Suo, Q. L., and Li, G. M. (2010). Mechanism of dispersing an active component into a polymeric carrier by the SEDS-PA process, *J. Mater. Sci.*, **45**, pp. 467–474.

21. Wang, Y. L., Dave, R.N., and Pfeffer, R. (2004). Polymer coating/encapsulation of nanoparticles using a supercritical anti-solvent process, *J. Supercrit. Fluids*, **28**, pp. 85–99.

22. Elvassore, N. Flaibani, M., Bertucco, A., and Caliceti, P. (2003). Thermodynamic analysis of micronization processes from gas-saturated solution, *Ind. Eng. Chem. Res.*, **42**, pp. 5924–5930.

23. Salmaso, S., Elvassore, N., Bertucco, A., and Caliceti, P. (2009). Production of solid lipid submicron particles for protein delivery using a novel supercritical gas-assisted melting atomization process, *J. Pharm. Sci.*, **98,** pp. 640–650.

24. Casettari, L., Castagnino, E., Stolnik, S., Lewis, A., Howdle, S. M., and Illum, L. (2011). Surface characterisation of bioadhesive PLGA/chitosan microparticles produced by supercritical fluid technology, *Pharm. Res. Dordr.*, **28**, pp. 1668–1682.

25. Falconer, J. R., Wen, J. Y., Zargar-Shoshtari, S., Chen, J. J., Mohammed, F., Chan, J. D., and Alany, R. G. (2012). The effects of supercritical carbon dioxide processing on progesterone dispersion systems: a multivariate study, *AAPS PharmSciTech*, **13**, pp. 1255–1265.

26. de Sousa, A. R. S., Silva, R., Tay, F. H., Simplicio, A. L., Kazarian, S. G., and Duarte, C. M. M. (2009). Solubility enhancement of trans-chalcone using lipid carriers and supercritical CO_2 processing, *J. Supercrit. Fluids*, **48**, pp. 120–125.

27. de Sousa, A. R. S., Simplicio, A. L., de Sousa, H. C., and Duarte, C. M. M. (2007). Preparation of glyceryl mono stearate-based particles by PGSS((R)): application to caffeine, *J. Supercrit. Fluids*, **43**, pp. 120–125.

28. Garcia-Gonzalez, C. A., Argemi, A., de Sousa, A. R. S., Duarte, C. M. M., Saurina, J., and Domingo, C. (2010). Encapsulation efficiency of solid lipid hybrid particles prepared using the PGSS (R) technique and

loaded with different polarity active agents, *J. Supercrit. Fluids*, **54**, pp. 342–347.

29. Garcia-Gonzalez, C. A., de Sousa, A. R. S., Argemi, A., Periago, A. L., Saurina, J., Duarte, C. M. M., and Domingo, C. (2009). Production of hybrid lipid-based particles loaded with inorganic nanoparticles and active compounds for prolonged topical release, *Int. J. Pharm.*, **382**, pp. 296–304.

30. Kazarian, S. G., and Martirosyan, G. G. (2002). Spectroscopy of polymer/drug formulations processed with supercritical fluids: in situ ATR-IR and Raman study of impregnation of ibuprofen into PVP, *Int. J. Pharm.*, **232**, pp. 81–90.

31. Kazarian, S. G., Briscoe, B. J., and Lawrence, C. J. (1999). Supercritical enhanced processing, *Polym. Process Eng.*, **99**, pp. 28–36.

32. Kazarian, S. G., Vincent, M. F., West, B. L., and Eckert, C. A. (1998). Partitioning of solutes and cosolvents between supercritical CO_2 and polymer phases, *J. Supercrit. Fluids*, **13**, pp. 107–112.

33. Kazarian, S. G., West, B. L., Vincent, M. F., and Eckert, C. A. (1997). Spectroscopic method for in situ analysis of supercritical fluid extraction and impregnation of polymeric matrices, *Am. Lab.*, **29**, p. B18.

34. Kikic, I., and Vecchione, F. (2003). Supercritical impregnation of polymers, *Curr. Opin. Solid State Matter*, **7**, pp. 399–405.

35. Braga, M. E. M., Pato, M. T. V., Silva, H. S. R. C., Ferreira, E. I., Gil, M. H., Duarte, C. M. M., and de Sousa, H. C. (2008). Supercritical solvent impregnation of ophthalmic drugs on chitosan derivatives, *J. Supercrit. Fluids*, **44**, pp. 245–257.

36. Braga, M. E. M., Yanez, F., Alvarez-Lorenzo, C., Concheiro, A. Duarte, C. M. M., Gil, M. H., and de Sousa, H. C. (2010). Improved drug loading/release capacities of commercial contact lenses obtained by supercritical fluid assisted molecular imprinting methods, *J. Controlled Release*, **148**, pp. E102–E104.

37. Costa, V. P., Braga, M. E. M., Duarte, C. M. M., Alvarez-Lorenzo, C., Concheiro, A., Gil, M. H., and de Sousa, H. C. (2010). Anti-glaucoma drug-loaded contact lenses prepared using supercritical solvent impregnation, *J. Supercrit. Fluids*, **53**, pp. 165–173.

38. Costa, V. P., Braga, M. E. M., Guerra, J. P., Duarte, A. R. C., Duarte, C. M. M., Leite, E. O. B., Gil, M. H., and de Sousa, H. C. (2010). Development of therapeutic contact lenses using a supercritical solvent impregnation method, *J. Supercrit. Fluids*, **52**, pp. 306–316.

39. Braga, M. E. M., Costa, V. P., Pereira, M. J. T., Fiadeiro, P. T., Gomes, A. P. A. R., Duarte, C. M. M., and de Sousa, H. C. (2011). Effects of operational conditions on the supercritical solvent impregnation of acetazolamide in balafilcon A commercial contact lenses, *Int. J. Pharm.*, **420**, pp. 231–243.

40. Masmoudi, Y., Ben Azzouk, L., Forzano, O., Andre, J. M., and Badens, E. (2011). Supercritical impregnation of intraocular lenses, *J. Supercrit. Fluids*, **60**, pp. 98–105.

41. Gonzalez-Chomon, C., Braga, M. E. M., de Sousa, H. C., Concheiro, A., and Alvarez-Lorenzo, C. (2012). Antifouling foldable acrylic IOLs loaded with norfloxacin by aqueous soaking and by supercritical carbon dioxide technology, *Eur. J. Pharm. Biopharm.*, **82**, pp. 383–391.

42. Duarte, A. R. C., Simplicio, A. L., Vega-Gonzalez, A., Subra-Paternault, P., Coimbra, P., Gil, M. H., de Sousa, H. C., and Duarte, C. M. M. (2007). Supercritical fluid impregnation of a biocompatible polymer for ophthalmic drug delivery, *J. Supercrit. Fluids*, **42**, pp. 373–377.

43. Argemi, A., Ellis, J. L., Saurina, J., and Tomasko, D. L. (2011). Development of a polymeric patch impregnated with naproxen as a model of transdermal sustained release system, *J. Pharm. Sci.*, **100**, pp. 992–1000.

44. Dias, A. M. A., Braga, M. E. M., Seabra, I. J., Ferreira, P., Gil, M. H., and de Sousa, H. C. (2011). Development of natural-based wound dressings impregnated with bioactive compounds and using supercritical carbon dioxide, *Int. J. Pharm.*, **408**, pp. 9–19.

45. Manna, L., Banchero, M., Sola, D., Ferri, A., Ronchetti, S., and Sicardi, S. (2007). Impregnation of PVP microparticles with ketoprofen in the presence of supercritical CO_2, *J. Supercrit. Fluids*, **42**, pp. 378–384.

46. Gong, K., Rehman, I. U., and Darr, J. A. (2008). Characterization and drug release investigation of amorphous drug-hydroxypropyl methylcellulose composites made via supercritical carbon dioxide assisted impregnation, *J. Pharm. Biomed.*, **48**, pp. 1112–1119.

47. Natu, M. V., Gil, M. H., and de Sousa, H. C. (2008). Supercritical solvent impregnation of poly(epsilon-caprolactone)/poly(oxyethylene-b-oxypropylene-b-oxyethylene) and poly(epsilon-caprolactone)/poly(ethylene-vinyl acetate) blends controlled release applications, *J. Supercrit. Fluids*, **47**, pp. 93–102.

48. Yoganathan, R., Mammucari, R., and Foster, N. R. (2010). Impregnation of ibuprofen into polycaprolactone using supercritical carbon dioxide, *J. Phys.: Conf. Ser.*, **215**, pp. 1–5 (012087).

49. Yoda, S., Sato, K., and Oyama, H. T. (2011). Impregnation of paclitaxel into poly(DL-lactic acid) using high pressure mixture of ethanol and carbon dioxide, *RSC Adv.*, **1**, pp. 156–162.

50. Yu, J. P., Guan, Y. X., Yao, S. J., and Zhu, Z. Q. (2011). Preparation of roxithromycin-loaded poly(L-lactic acid) films with supercritical solution impregnation, *Ind. Eng. Chem. Res.*, **50**, pp. 13813–13818.

51. Zhan, S. P., Chen, C., Zhao, Q. C., Wang, W. J., and Liu, Z. J. (2013). Preparation of 5-Fu-loaded PLLA microparticles by supercritical fluid technology, *Ind. Eng. Chem. Res.*, **52**, pp. 2852–2857.

52. Duarte, A. R. C., Costa, M. S., Simplicio, A. L., Cardoso, M. M., and Duarte, C. M. M. (2006). Preparation of controlled release microspheres using supercritical fluid technology for delivery of anti-inflammatory drugs, *Int. J. Pharm.*, **308**, pp. 168–174.

53. Alvarez-Lorenzo, C., and Concheiro, A. (2004). Molecularly imprinted polymers for drug delivery, *J. Chromatogr. B*, **804**, pp. 231–245.

54. Mayes, A. G., and Whitcombe, M. J. (2005). Synthetic strategies for the generation of molecularly imprinted organic polymers, *Adv. Drug Delivery Rev.*, **57**, pp. 1742–1778.

55. Duarte, A. R. C., Casimiro, T., Aguiar-Ricardo, A., Simplicio, A. L., and Duarte, C. M. M. (2006). Supercritical fluid polymerisation and impregnation of molecularly imprinted polymers for drug delivery, *J. Supercrit. Fluids*, **39**, pp. 102–106.

56. da Silva, M. S., Nobrega, F. L., Aguiar-Ricardo, A., Cabrita, E. J., and Casimiro, T. (2011). Development of molecularly imprinted co-polymeric devices for controlled delivery of flufenamic acid using supercritical fluid technology, *J. Supercrit. Fluids*, **58**, pp. 150–157.

57. da Silva, M. S., Viveiros, R., Morgado, P. I., Aguiar-Ricardo, A., Correia, I. J., and Casimiro, T. (2011). Development of 2-(dimethylamino)ethyl methacrylate-based molecular recognition devices for controlled drug delivery using supercritical fluid technology, *Int. J. Pharm.*, **416**, pp. 61–68.

58. Ye, L., Yoshimatsu, K., Kolodziej, D., Francisco, J. D., and Dey, E. S. (2006). Preparation of molecularly imprinted polymers in supercritical carbon dioxide, *J. Appl. Polym. Sci.*, **102**, pp. 2863–2867.

59. Kobayashi, T., Leong, S. S., and Zhang, Q. Q. (2008). Using polystyrene-co-maleic acid for molecularly imprinted membranes prepared in supercritical carbon dioxide, *J. Appl. Polym. Sci.*, **108**, pp. 757–768.

60. Mano, J. F. (2008). Stimuli-responsive polymeric systems for biomedical applications, *Adv. Eng. Mater.*, **10**, pp. 515–527.

61. Gil, E. S., and Hudson, S. M. (2004). Stimuli-responsive polymers and their bioconjugates, *Prog. Polym. Sci.*, **29**, pp. 1173–1222.

62. Qiu, Y., and Park, K. (2012). Environment-sensitive hydrogels for drug delivery, *Adv. Drug Delivery Rev.*, **64**, pp. 49–60.

63. Hoffman, A. S. (2013). Stimuli-responsive polymers: biomedical applications and challenges for clinical translation, *Adv. Drug Delivery Rev.*, **65**, pp. 10–16.

64. Alarcon, C. D. H., Pennadam, S., and Alexander, C. (2005). Stimuli responsive polymers for biomedical applications, *Chem. Soc. Rev.*, **34**, pp. 276–285.

65. Lehner, R., Wang, X. Y., Wolf, M., and Hunziker, P. (2012). Designing switchable nanosystems for medical application, *J. Controlled Release*, **161**, pp. 307–316.

66. Temtem, M., Casimiro, T., Mano, J. F., and Aguiar-Ricardo, A. (2007). Green synthesis of a temperature sensitive hydrogel, *Green Chem.*, **9**, pp. 75–79.

67. Duarte, A. R. C., Mano, J. F., and Reis, R. L. (2011). Thermosensitive polymeric matrices for three-dimensional cell culture strategies, *Acta Biomater.*, **7**, pp. 526–529.

68. Prabaharan, M., and Mano, J. F. (2006). Stimuli-responsive hydrogels based on polysaccharides incorporated with thermo-responsive polymers as novel biomaterials, *Macromol. Biosci.*, **6**, pp. 991–1008.

69. Temtem, M., Barroso, T., Casimiro, T., Mano, J. F., and Aguiar-Ricardo, A. (2012). Dual stimuli responsive poly(N-isopropylacrylamide) coated chitosan scaffolds for controlled release prepared from a non residue technology, *J. Supercrit. Fluids*, **66**, pp. 398–404.

70. Malafaya, P. B., Silva, G. A., Baran, E. T., and Reis, R. L. (2002). Drug delivery therapies II. Strategies for delivering bone regenerating factors, *Curr. Opin. Solid State Matter*, **6**, pp. 297–312.

71. Malafaya, P. B., Silva, G. A., Baran, E. T., and Reis, R. L. (2002). Drug delivery therapies I - General trends and its importance on bone tissue engineering applications, *Curr. Opin. Solid State Matter*, **6**, pp. 283–295.

72. Santo, V. E., Gomes, M. E., Mano, J. F., and Reis, R. L. (2012). From nano- to macro-scale: nanotechnology approaches for spatially controlled delivery of bioactive factors for bone and cartilage engineering, *Nanomedicine*, **7**, pp. 1045–1066.

73. Malafaya, P. B., Silva, G. A., and Reis, R. L. (2007). Natural-origin polymers as carriers and scaffolds for biomolecules and cell delivery in tissue engineering applications, *Adv. Drug Delivery Rev.*, **59**, pp. 207–233.

74. Mano, J. F., Silva, G. A., Azevedo, H. S., Malafaya, P. B., Sousa, R. A., Silva, S. S., Boesel, L. F., Oliveira, J. M., Santos, T. C., Marques, A. P., Neves, N. M., and Reis, R. L. (2007). Natural origin biodegradable systems in tissue engineering and regenerative medicine: present status and some moving trends, *J. R. Soc. Interface*, **4**, pp. 999–1030.

75. Liu, C., Xia, Z., and Czernuszka, J. T. (2007). Design and development of three-dimensional scaffolds for tissue engineering, *Chem. Eng. Res. Des.*, **85**, pp. 1051–1064.

76. Duarte, A. R. C., Mano, J. F., and Reis, R. L. (2009). Perspectives on: supercritical fluid technology for 3D tissue engineering scaffold applications, *J. Bioact. Compat. Polym.*, **24**, pp. 385–400.

77. Duarte, A. R. C., Mano, J. F., and Reis, R. L. (2008). Supercritical fluids: an emerging technology for the preparation of scaffolds for tissue engineering, *Tissue Eng. Part A*, **14**, pp. 782–782.

78. Cabezas, L. I., Fernandez, V., Mazarro, R., Gracia, I., de Lucas, A., and Rodriguez, J. F. (2012). Production of biodegradable porous scaffolds impregnated with indomethacin in supercritical CO_2, *J. Supercrit. Fluids*, **63**, pp. 155–160.

79. Velasco, D., Benito, L., Fernandez-Gutierrez, M., san Roman, J., and Elvira, C. (2010). Preparation in supercritical CO_2 of porous poly(methyl methacrylate)-poly(L-lactic acid) (PMMA-PLA) scaffolds incorporating ibuprofen, *J. Supercrit. Fluids*, **54**, pp. 335–341.

80. Hile, D. D., Amirpour, M. L., Akgerman, A., and Pishko, M. V. (2000). Active growth factor delivery from poly(D,L-lactide-co-glycolide) foams prepared in supercritical CO_2, *J. Controlled Release*, **66**, pp. 177–185.

81. Nof, M., and Shea, L. D. (2002). Drug-releasing scaffolds fabricated from drug-loaded microspheres, *J. Biomed. Mater. Res.*, **59**, pp. 349–356.

82. Kanczler, J. M., Ginty, P. J., White, L., Clarke, N. M. P., Howdle, S. M., Shakesheff, K. M., and Oreffo, R. O. C. (2010). The effect of the delivery of vascular endothelial growth factor and bone morphogenic protein-2 to osteoprogenitor cell populations on bone formation, *Biomaterials*, **31**, pp. 1242–1250.

83. Kanczler, J. M., Barry, J., Ginty, P., Howdle, S. M., Shakesheff, K. M., and Oreffo, R. O. C. (2007). Supercritical carbon dioxide generated vascular endothelial growth factor encapsulated poly(DL-lactic acid) scaffolds induce angiogenesis in vitro, *Biochem. Biophys. Res. Commun.*, **352**, pp. 135–141.

84. Duarte, A. R. C., Mano, J. F., and Reis, R. L. (2009). Preparation of chitosan scaffolds loaded with dexamethasone for tissue engineering applications using supercritical fluid technology, *Eur. Polym. J.*, **45**, pp. 141–148.

85. Cardea, S., Sessa, M., and Reverchon, E. (2010). Supercritical phase inversion to form drug-loaded poly(vinylidene fluoride-co-hexafluoropropylene) membranes, *Ind. Eng. Chem. Res.*, **49**, pp. 2783–2789.

86. Reverchon, E., Cardea, S., and Rappo, E. S. (2006). Production of loaded PMMA structures using the supercritical CO_2 phase inversion process, *J. Membr. Sci.*, **273**, pp. 97–105.

87. Ji, C. D., Barrett, A., Poole-Warren, L. A., Foster, N. R., and Dehghani, F. (2010). The development of a dense gas solvent exchange process for the impregnation of pharmaceuticals into porous chitosan, *Int. J. Pharm.*, **391**, pp. 187–196.

88. Silva, S. S., Duarte, A. R. C., Mano, J. F., and Reis, R. L. (2013). Design and functionalization of chitin-based microspheres scaffolds, *Green Chem.*, **15**, pp. 3252–3258.

89. Tsioptsias, C., Paraskevopoulos, M. K., Christofilos, D., Andrieux, R., and Panayiotou, C. (2011). Polymeric hydrogels and supercritical fluids: the mechanism of hydrogel foaming, *Polymer*, **52**, pp. 2819–2826.

90. Ji, C. D., Annabi, N., Khademhosseini, A., and Dehghani, F. (2011). Fabrication of porous chitosan scaffolds for soft tissue engineering using dense gas CO_2, *Acta Biomater.*, **7**, pp. 1653–1664.

91. Singh, M., Sandhu, B., Scurto, A., Berkland, C., and Detamore, M. S. (2010). Microsphere-based scaffolds for cartilage tissue engineering: using subcritical CO_2 as a sintering agent, *Acta Biomater.*, **6**, pp. 137–143.

92. Alves, A., Duarte, A. R. C., Mano, J. F., Sousa, R. A., and Reis, R. L. (2012). PDLLA enriched with ulvan particles as a novel 3D porous scaffold targeted for bone engineering, *J. Supercrit. Fluids*, **65**, pp. 32–38.

Chapter 13

An Integrated Supercritical Extraction and Impregnation Process for Production of Antibacterial Scaffolds

María A. Fanovich,[a] Jasna Ivanovic,[b] and Philip T. Jaeger[c]

[a]*Institute of Materials Science and Technology (INTEMA), CONICET, J. B. Justo 4302, B7608FDQ, Mar del Plata, Argentina*

[b]*University of Belgrade, Faculty of Technology and Metallurgy, Karnegijeva 4, 11000 Belgrade, Serbia*

[c]*Institute of Thermal Process Engineering, Hamburg University of Technology, Eißendorfer Str. 38, 21073 Hamburg, Germany*

mafanovi@fi.mdp.edu.ar

This chapter deals with integrated processes using carbon dioxide in supercritical conditions for the development of functionalized materials. The basics of the involved process units are presented to allow a general understanding, and for enabling to extend the design to a variety of applications. A conceptual design is proposed by integrating three steps for producing scaffolds with antibacterial properties. A setup for an integrated $scCO_2$ extraction unit of natural compounds with posterior impregnation on solid matrices is described in detail including a final step of foaming, as a way of

Supercritical Fluid Nanotechnology: Advances and Applications in Composites and Hybrid Nanomaterials
Edited by Concepción Domingo and Pascale Subra-Paternault

ISBN 978-981-4613-40-8 (Hardcover), 978-981-4613-41-5 (eBook)
www.panstanford.com

product formulation, by making use of the same supercritical fluid as being used for the prior extraction and impregnation steps. The strategy followed is based on minimizing the loss of extract matter in the tubes, vessels, and heat exchangers of the equipment by directly using the $scCO_2$–extract solution for impregnation, avoiding the low efficiency of extract recovery and energy consumption occurring in a separation step forced by pressure reduction. Results of production of functionalized poly(caprolactone) scaffolds with natural compounds extracted from Patagonian Usnea lichen are presented as an example. To establish appropriate operating conditions for each of the processing steps, supercritical extraction of Usnea as well as sorption kinetics and resulting material properties have been studied in detail before arriving at an optimized overall process.

13.1 Introduction

Since the beginning of the eighties of the past century, supercritical extraction from solid natural resources is finding a growing number of industrial applications. Starting from food industry and stimulants (coffee, hops), a large variety of almost any lipid-containing seeds, spices or herbs has been extracted to obtain products of high purity and quality. Most of these high-price products are applied in cosmetics or pharmaceutical industry [1–3]. A distinction is made depending on whether the main aim is to extract a valuable component or to retrieve some undesired substance in order to increase the value of the source material, like in the case of extraction of *aflotoxines* from cork. Apart from extraction, high pressure can also be applied for modifying the original properties of the (natural) source material or to formulate an intermediate product. Treatment of rice with compressed or supercritical carbon dioxide ($scCO_2$) generates certain properties that have advantages for posterior processing [4]. It has been proved that carbon dioxide induces swelling of plant cell structures, which in its turn improves fluid accessibility to extractable substances and, therefore, enhances the yield during the extraction process [5]. One important drawback is the energetic balance of high pressure extraction [6]. Principally, a high electrical input is required for condensing the CO_2 after the product separation step. To overcome this important disadvantage, closed solvent loops should be considered and big pressure steps avoided. Consecutive processing steps have been designed in a way for improving the energetic balance in industrial processes, like decaffeination of green coffee beans, especially when the pressure level can be maintained at isobaric conditions throughout the process. To separate the extractable from the compressed solvent, an adsorption process can be included. However, this usually implies a further separation step for regenerating the adsorbent. Hence, the separation of the extract from the supercritical solvent needs to be envisaged in a smart way. One possibility is bonding the extracted substance to a substrate, guided in a certain way that the formulation of an end product is included in the process. A number of challenges need to be overcome:

(i) Different kinetics of the materials in the different process steps
(ii) Alteration of substrate properties
(iii) Technical challenges, such as isobaric pumps (hermetic pumps)
(iv) Product retrieval

This chapter shows how different processing mechanisms may fit together to an integrated process, concentrating it on three steps: extraction, sorption, and formulation (Fig. 13.1). Different examples are given for research, pilot, and industrial-scale processes, focusing on a novel process for manufacturing antibacterial scaffolds or devices based on phytoextracts and biopolymers.

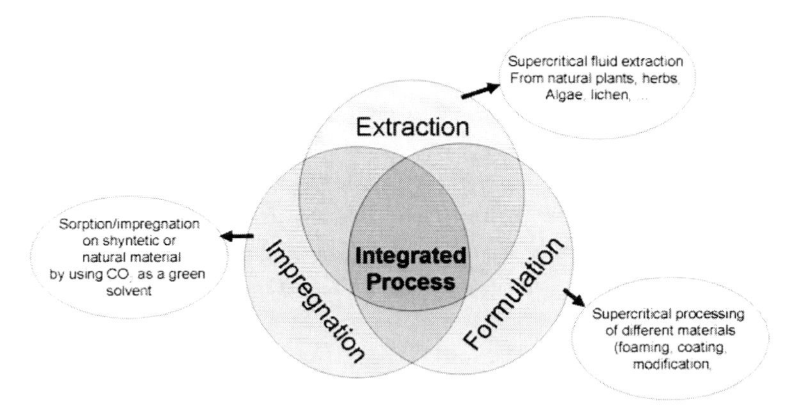

Figure 13.1 Schematic representation of an integrated process by using supercritical carbon dioxide as a solvent.

13.2 Supercritical Extraction Processes from Natural Products

Supercritical fluid extraction (SFE) has been proposed for a wide variety of substances. Most common vegetal materials have been investigated with focus on one of two possible goals: either the production of natural extracts without the use of traditional organic solvents or the removal of undesired components. In both cases, a tendency to the use of techniques that avoid or minimize damages to the environment is recognized. The products obtained by SFE are free from toxic residues and, generally, possess higher quality than the products obtained by conventional techniques [2]. Even though being an expensive process, in terms of the high pressure resistant equipment needed and the energetic balance, SFE is hardly to be omitted from any laboratory or production plant dealing with natural extracts of high quality. The process itself is recognized as clean and product friendly and, therefore, increases the company´s reputation. As mentioned before, a number of drawbacks have to be overcome in the future in order to develop this process as a sustainable alternative to conventional liquid solvent-based processes. In the first place, efficiency of the process in terms of product yield must be enhanced. One important point in this respect is pretreatment of the solid source material. Different ways of pretreatment are applied, which can be subdivided into mechanical, thermal, and chemical methods [7]. Also innovative pretreatment methods, such as enzymatic action, rapid decompression for micronizing, and swelling with compressed CO_2, have been developed but not yet been applied at a large scale [8]. All in all, SFE remains as a process that mostly finds its application in high-priced products. Recent tendencies strongly direct toward natural extracts for being used in health care, especially in antibacterial applications. The literature offers data from a wide variety of substances being extracted under supercritical conditions [9–34]. Table 13.1 summarizes the extraction of bioactive compounds from different sources with antibacterial properties.

Table 13.1 Summary of the published works on the extraction by SFE of bioactive compounds with antibacterial properties from natural sources

Material/Extract of interest	Extraction conditions	Yield (%)	Ref.
Clove bud–oregano/ Eugenol, carvacrol, thymol	CO_2: 10 MPa, 40°C, 0,62kg/h, 5 h	2–5	[9]
Ramulus cinnamoni/ Volatile oil	CO_2: 23–41 MPa, 40°C–50°C	-	[10]
Black cumin *(Nigella sativa)*/Essential oil	CO_2: 40 MPa, 40°C, 25 g/min, 30 min	-	[11]
Hyssop (*Hyssopus officinalis L.*)/ Sabinene (S), iso-pinocamphene (i-P), pinocamphene (P)	CO_2 + methanol (1.5 % v/v) 10 MPa, 55°C, 30 min (dynamic)	S: 4.2–17.1, i-P: 0.9–16.5, P: 0.7–13.6	[12]
Basil (*Ocimum gratissimum*)/ Eugenol	CO_2: 10–30 MPa, 40°C	1–1.8	[13]
Cajanus cajan (L.) Huth/Cajaninstilbene acid	Not reported	-	[14]
Myrtle (*Myrtus communis* L.) leaves/ Phenolic compounds	CO_2: 23 MPa, 45°C, 0.3 kg·h^{-1}	2–2.5	[15]
"Ban-Zhi-Lian" (*Scutellaria barbata* D. Don)/Oleanolic acid (OA), ursolic acid (UA)	CO_2: 27.6 MPa, 55°C, 2.1 mL/min + 14.1% aqueous ethanol (80/20 v/v), ultrasound, 50 min	OA: 1.4 × 10^{-3} UA: 5.9 × 10^{-3}	[16]
Tulsi (*Ocimum sanctum Linn.*)/ Eugenol	CO_2: 40 MPa, 70°C, 1 h static	0.46	[17]
Ginger (*Zingiber officinale Roscoe*) roots/Gingerols and shogaols	CO_2: 25 MPa, 60°C	2.62	[18]

(*Continued*)

Table 13.1 (*Continued*)

Material/Extract of interest	Extraction conditions	Yield (%)	Ref.
Lichen *Usnea barbata*/Usnic acid	CO_2: 30 MPa, 25°C	2.08	[8]
Patagonian lichen *Usnea lethariiformis*/Usnic acid, difractaic acid	CO_2: 30 MPa, 40°C	1.0	[19]
Grape pomace (*Vitis vinifera*)/Gallic acid, p-OH-benzoic acid, vanillic acid, epicathechin	CO_2: 30 MPa, 50°C–60°C	–	[20]
Rosmarinus officinalis/Rosmarinic acid	Not reported	–	[21]
Nigella sativa seeds/ Essential oils, thymoquinone	CO_2: 30 MPa, 40°C (pretreated at 9 MPa)	21–26	[22]
Cordia verbenacea DC (Borraginaceae)/ Artemetin, β-sitosterol, α-humulene, and β-caryophyllene	CO_2: 30 MPa, 30°C–50°C with cosolvent	5–8.6	[23]
Piper regnellii var. pallescens/ Neolignans (conocarpan, eupomatenoid-3, eupomatenoid-5, and eupomatenoid-6)	CO_2: 10–25 MPa; 40°C–60°C	0.6–2.4	[24]
Agaricus brasiliensis mushrooms/ Linoleic acid palmitic acid	CO_2: 30 MPa, 50°C	1.2	[25]

Material/Extract of interest	Extraction conditions	Yield (%)	Ref.
Achyrocline satureioides/ Aromatic compounds	CO_2: 12 MPa, 30°C	1.26	[26]
Vetiveria zizanioides/ Essential oils	CO_2: 19 MPa, 50°C and 15% ethanol	5.9	[27]
Shiitake (*Lentinula edodes*) mushroom/ Palmitic acid, linoleic acid, and ergosterol	CO_2: 20 MPa, 40°C, 15% ethanol	3.81	[28]
Marjoram (*Origanum majorana* L.)/ Terpinen-4-ol c-terpinene, linalool, a-terpineol, a-terpinolene, a-terpinene, b-caryophyllene, and spathulenol, cis-Sabinene hydrat	CO_2: 45 MPa, 50°C, 7 kg/h, 245 min	3.8	[29]
Spearmint (*Mentha spi-cata* L.) leaves/ Catechins	CO_2: 20 MPa, 60°C and 60 min	6.0	[30]
Pulverized crude propolis/Phenolic acids, flavonoids, terpenes, and sesquiterpenes	CO_2: 15 MPa, 40°C, 5% ethanol	24.8	[31]
Alpinia oxyphylla Miq./Extract with multibiofunctions	CO_2: 40 MPa, 2 h	–	[32]

(*Continued*)

Table 13.1 (*Continued*)

Material/Extract of interest	Extraction conditions	Yield (%)	Ref.
Dictyoperis membranacea/ Volatiles	CO_2: 9.1 MPa, 40°C, 30 min	15.4	[33]
Spirulina platensis/ γ-Linolenic acid	CO_2: 40 MPa, 40°C, ethanol, 1 h	>90	[34]

SFE from diverse herbs, algae, lichen, and other natural materials usually results in complex mixtures of various components, which have a strong variety in the composition and content of the bioactive substances of interest. In many cases, the complete composition is unknown. However, this aspect could be also of high interest, since the potential synergy of different known and unknown bioactive agents in the extracts is found to enhance the antibacterial properties. Most extractions are carried out at a laboratory level. For high-value products, such as pharmaceutical compound materials, small-scale equipment may already work profitably, whereas for mass products, such as the one needed for hop extracts, the production scale must be large. While it appears that the extraction method is simple, it is essential to study each plant material individually, because the pretreatment of feed material and optimal extraction conditions depend on the structure and composition of each specific starting product (Table 13.1). It can be summarized that extensive research has been carried out with a large variety of plant materials that shows SFE to be effective in recovering bioactive compounds.

The antibacterial activities of plant extracts and oils can be useful for the preservation of raw and processed food, in the pharmaceutical industry, and as alternative medicines and natural therapies. Today, the resistance to multiple antibacterial substances that has been observed worldwide after long term use of antibiotics is a public health problem. Many strains of *Staphylococcus aureus* have developed resistance to antibiotics (i.e., methicillin-resistant *Staphylococcus aureus*, or MRSA), creating a serious problem in medical microbiology. Hence, great interest is focused on extracts that have bactericidal action on *S. aureus*. As an example of isolation of natural antibacterial agents against MRSA, the extraction kinetics of usnic acid from a Patagonian lichen is described in this section. The effect of pretreatment on the extraction yield of this bioactive substance is shown in Fig. 13.2. The yield of lichen extracts isolated at 30 MPa and 40°C is clearly improved by mechanical pretreatment, although not much difference becomes evident among both used milling methods, blade and planetary mill. After a consumption of $scCO_2$ of 12–14 kg_{CO2}/kg_{solid} a similar yield of around 1% was obtained, based on the original mass of the source material. The pressure and temperature of an optimized extraction process depend on both the properties of the raw material and the solubility of the extract. When the separation of the extract from the solvent is performed by lowering the solubility within the solvent, the separator must be operated at lower pressure than the extraction itself. For highly volatile components, the pressure needs to be released to very low values in order to precipitate these substances within the separator. Successive recompression of the solvent requires expensive compressors and decreases dramatically the energetic efficiency of the process.

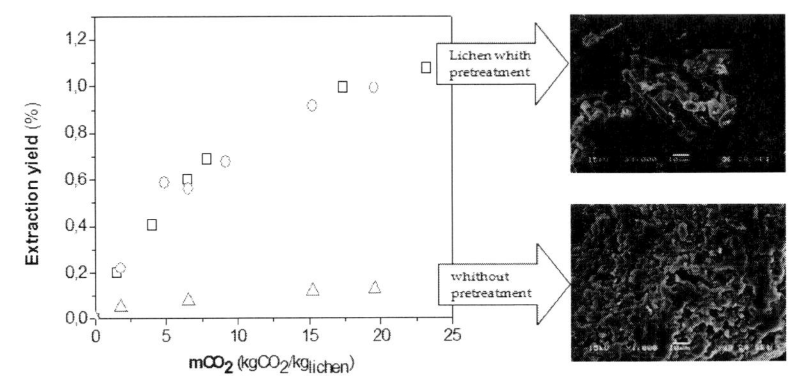

Figure 13.2 Extraction yields (wt%) by using $scCO_2$ at P = 30 MPa and T = 40°C from nonpretreated (Δ) lichen and milled with blade mill (BM) (○) or planetary mill (PM) (□).

13.3 Supercritical Sorption/Impregnation Processes

$scCO_2$ is not only known as a possible solvent for extraction of valuable compounds but also known for its high diffusion ability in organic matter. The latter property is used for impregnating solid matrices with bioactive agents [35–37]. Figure 13.3 shows a scheme of the process, mainly consisting of two steps, (i) solubilization of the solute and (ii) impregnation of the matrix.

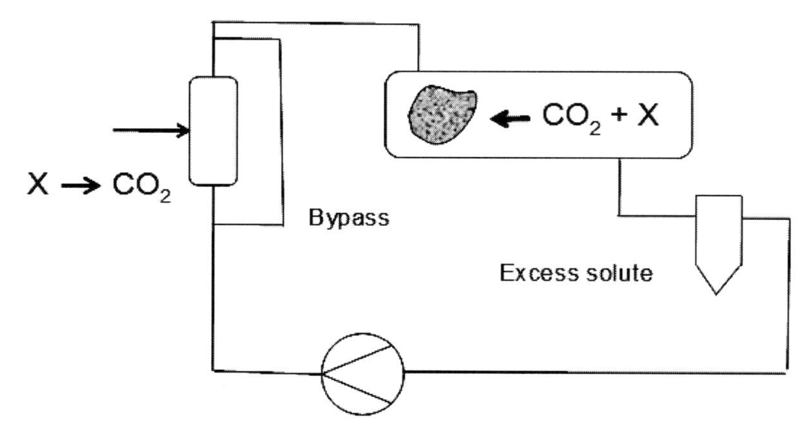

Figure 13.3 Scheme of the supercritical impregnation process.

The impregnation process requires the design to be performed adjusting the solvent capacity and the required amount for the impregnation, while taking into account the different mass transfer resistances. In amorphous polymers, for example, poly(ethylene terephthalate) (PET) materials, the glass transition temperature (T_g) is reduced due to the absorbing CO_2, which facilities operation at moderate temperatures [38]. Biopolymers behave in an analogous way [39]. In the process, the solute is placed in a so-called saturation vessel. For adjusting the amount of solute, various possibilities of continuously feeding into the process have been investigated [40, 41]. In the continuous approaches, the feed must be provided as a solution or slurry (suspension) but at the cost of additional mass transfer resistance through the used solvent, for example, poly(ethylene glycol).

Figure 13.4 shows the effect of different process parameters on the change in solute volume during CO_2 sorption under pressure in poly(caprolactone) (PCL) samples. While the change in volume during exposure to carbon dioxide does not show a clear tendency as a function of pressure within the investigated range, it is clearly shown that at low temperatures the swelling is more pronounced (Fig. 13.4). Swelling of the polymer by $scCO_2$ effectively increases the diffusion coefficient of the solute by several orders of magnitude. When the system is depressurized, the dissolved fluid readily diffuses out of the polymer, while the solute is expected to be trapped in the matrix. Different molecules can be impregnated into a polymer matrix at temperatures low enough to avoid thermal degradation and loss of bioactive properties [42]. For other applications, such as the dyeing of synthetic fibers, this process has reached industrial scale [43]. Currently, studies are conducted on the application of diverse plant extracts and other herbal oils for impregnation of textiles [44].

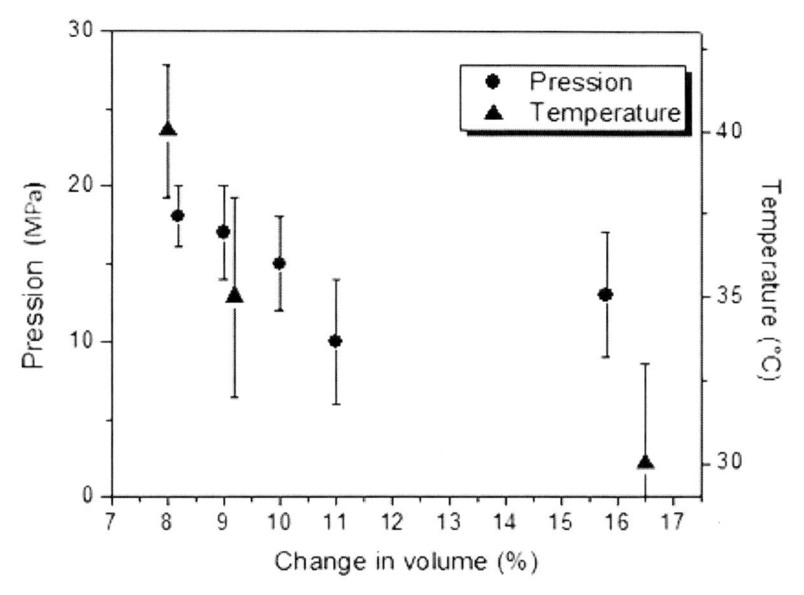

Figure 13.4 Effect of pressure (P = 15 MPa, •) and temperature (T = 35°C, ▲) on change in volume (%) of poly(caprolactone) samples.

In Table 13.2, some examples of the incorporation of bioactive agents into a carrier matrix by using $scCO_2$ are listed. Most of

Table 13.2 Examples of works of impregnation or incorporation of a bioactive agent into different matrixes by using $scCO_2$

Solute/Matrix	Conditions	Impregnation (wt%)	Ref.
Thymol/Cotton gauze	15.5 MPa, 35°C	19,6	[45]
Natural origin quercetin and thymol/*N*-carboxybuthylchitisan, agarosa	10–20 MPa, 30°C–50°C Ethanol (10% v/v) for quercetin	–	[46]
Thymol/Linear low-density polyethylene films	7–12 MPa 40°C, 4 h	1.48–3.81	[47]
Dexamethasone/ Poly(caprolactone) and silica nanoparticles	Supercritical foaming mixing 14–25 MPa, 35°C, 2–14 h	tunable yields	[48]
Ciannamaldehyde/ Cassava starch	15–25 MPa, 35°C, 3 and 15 h impregnation times, 1 and 10 MPa·min^{-1} depressurization rates	1–2.5	[49]
Gentamicin/Alginate–pectin blends	Supercritical assisted atomization	20–30	[50]
Gentamicin/Chitosan	20 MPa, 60°C with ethanol, flow 4.9 g·min^{-1}, 6 h	–	[51]
Juca (*Libidibia ferrea*) extract/*N*-carboxybutyl chitosan dressing	Two step Extraction 25 MPa, 50°C Impregnation 27 MPa, 50°C	0.4–0.5	[52]

the solutes shown in Table 13.2 are commercial or synthetic substances. The process related with Juca material [52] refers to a natural extract combined with an impregnation process, although considered as two separate steps. It is noteworthy in this work that a slightly higher pressure is proposed for the impregnation, rather deposition, although this is not clearly explained. The processes described in Refs. [48, 50] do not apply the impregnation step. In these applications, the solute is physically mixed with the substrate or carrier. Hence, the concentration of the solute could be established independently from the solubility and diffusivity within the matrix of the carrier material.

13.4 Formulation of a Scaffold

An ideal scaffold for bone tissue engineering should be biocompatible and osteoconductive and it should provide structural support to the newly formed bone [53, 54]. These materials, obtained from natural or synthetic compounds, must have a proper degradation rate and some required characteristics related with morphological features for good performance. A scaffold should have an interconnected porous structure, sufficient mechanical strength, and proper cell–scaffold interaction. The pore size and morphology and the degree of porosity are very important parameters in tissue engineering. For instance, an interpenetrating network of pores in the range of 100–500 μm is required to allow vascularization and tissue in-growth [55]. A variety of techniques has been developed to fabricate porous scaffolds of biopolymers; among these techniques, foaming with $scCO_2$ is superior to other methods for producing solvent-free porous structures. Hence, the use of $scCO_2$ is of high interest as an alternative green solvent for processing biodegradable and biocompatible polymers in pharmaceutical and medical applications [56–59].

The foaming process by using $scCO_2$ can be divided into three steps: (i) sorption of $scCO_2$ until saturation, (ii) nucleation of foam bubbles, and (iii) growth of foam bubbles. Karimi et al. [60] reported a detailed study of the influence of the $scCO_2$ pressure, temperature, rate of quenching, etc., on the microstructure of PCL foams. In the analysis, foaming is described as a process that only occurs when the sample is melted. When the foaming is produced at low depressurization rates (slow quenching), the pressure decay rate can be idealized as an isothermal procedure. The foaming temperature determines the width of the pore size distribution. The foaming process follows a complex mechanism, mainly three factors being involved in the process: (i) release of CO_2 from the polymeric matrix during nucleation and growth of bubbles, (ii) expansion of CO_2 inside the bubbles with decreasing pressure, and (iii) increase of the matrix viscosity. The balance between these factors determines the resulting pore morphology.

Depressurization starts from a melted state of the polymer and continues toward the melting line, which can be determined from differential scanning calorimetry (DSC) experiments at elevated pressure (Fig. 13.5) [19]. The extent of supersaturation when arriving at the melting line, that is, the pressure difference down to this point, determines the foam structure significantly. Table 13.3 shows the effect of different process parameters on the mean pore diameter of several PCL samples measured after depressurization. A slightly increasing pore size can be observed with pressure. The reason is related with the fact that a higher amount of gas is dissolved in the polymer at higher pressures, which results in a larger void space during foaming. Bubbles growth is enhanced by including an increased amount of CO_2 present in their vicinity. The pore size is mainly influenced by temperature and depressurization rate. A fast depressurization rate results in an enhanced nucleation rate, because supersaturation is high at the onset of nucleation. Hence, a high number of bubbles is formed with less time for growing and coalescing. To interpret the effect of temperature on the pore size, the plasticizing behavior needs to be accounted for. During the step where the bubbles grow, the polymer needs to remain in a fluid state, which is probably the case at high temperatures where the viscosity is reduced. Even at temperatures below the melting temperature, at atmospheric pressure, bubbles are formed giving evidence of a change in the melting behavior during exposure to compressed carbon dioxide.

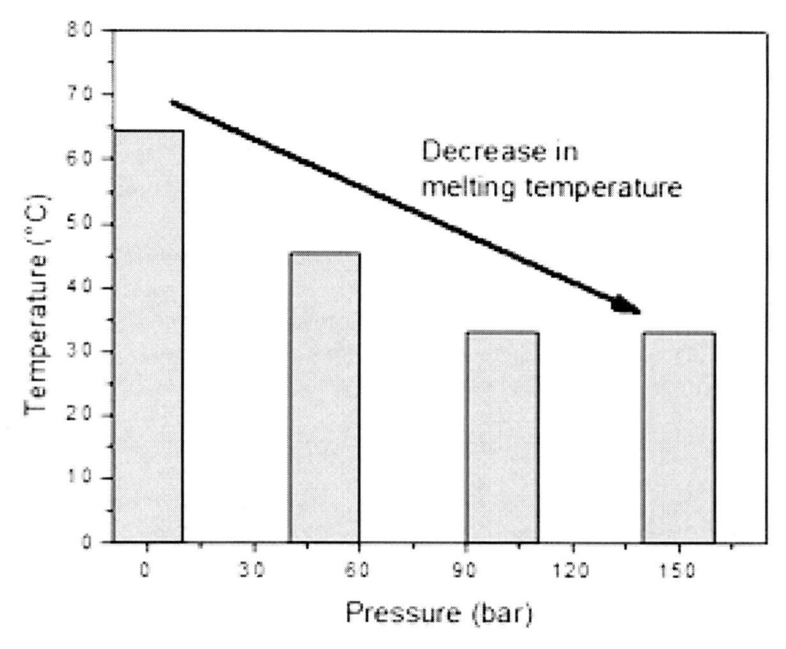

Figure 13.5 Variation in the melting temperature of poly(caprolactone) with CO_2 pressure determined from high-pressure DSC analysis data [19].

Table 13.3 Effect of pressure, temperature, and depressurization rate on the mean pore diameter of supercritical processed poly(caprolactone) samples

$T = 35°C$, $dP/dt = 1\ MPa \cdot min^{-1}$		$P = 15\ MPa$, $dP/dt = 1\ MPa \cdot min^{-1}$		$P = 15\ MPa$, $T = 35°C$	
P (MPa)	Mean pore diameter (μm) ± d.s	T (°C)	Mean pore diameter (μm) ± d.s	dP/dt ($MPa \cdot min^{-1}$)	Mean pore diameter (μm) ± d.s
10	150 ± 50	30	50 ± 20	0.1	460 ± 30
15	240 ± 60	35	230 ± 20	0.5	340 ± 60
17	390 ± 50	40	560 ± 100	1	230 ± 50
18	200 ± 40			2	150 ± 40

13.5 Integrated Process for Production of Functionalized Materials

The integrated extraction and impregnation process, including posterior formulation, is here illustrated by the example of functionalization of PCL scaffolds with a natural antibacterial agent. The strategy is based on minimizing the loss of extract mass in the tubes, vessels and exchangers of the equipment by directly using the $scCO_2$–extract solution for impregnation, avoiding the low efficiency of extract recovery in a (conventional) separation step carried out by pressure reduction. The basic idea of the proposed process is presented in Fig. 13.6 for a natural extract obtained from a Patagonian lichen and a commercial PCL matrix.

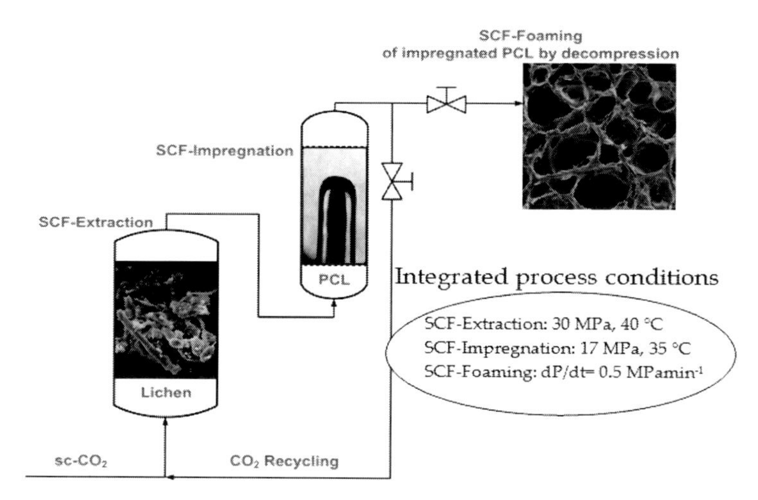

Figure 13.6 Concept of an integrated process, including extraction, impregnation, and formulation of the final product.

An integrated process seems promising to improve the useful amount of extract incorporated into the PCL matrix. It is important to note that the solubility of the extract in $scCO_2$ changes according to the conditions in each vessel. The extraction variables are temperature, pressure, particle size, moisture content of the raw material, extraction time, CO_2 flow rate, and solvent-to-source material ratio, which all need to be optimized for an efficient process. In general, extraction yield increases with pressure due to the increase of the solubility in $scCO_2$ and decreases with temperature due to the decrease of CO_2 density. However, the antibacterial activity of extracts on specific bacteria may be adversely affected. This fact leads to a compromised situation to select the extraction conditions. In the results shown below, pressure and temperature used for extraction were fixed at 40°C and 30 MPa, according to the published antibacterial activities of usnea extracts against MRSA [61]. The proposed process design for in-line impregnation of solids with antibacterial lichen extracts after SFE is shown in Fig. 13.7. This laboratory-scale unit has been designed for an integrated extraction-impregnation process and was extended by closing the solvent $scCO_2$ cycle in a way that the extractor (E, 500 mL) and the adsorption column (A, 100 mL) can be operated at different temperatures and pressures independently from each other.

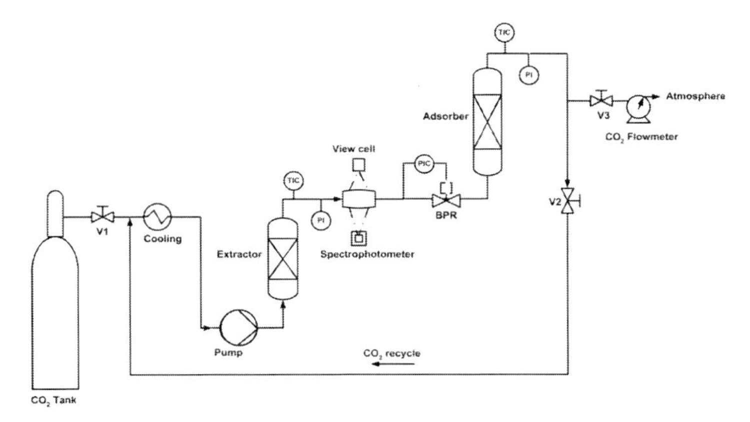

Figure 13.7 Setup of a proposed process that integrates extraction, impregnation, and foaming steps.

The extractor vessel is filled with the raw material from which a target substance is to be extracted. The adsorption column is filled with solid to be impregnated by the extract from the prior extraction step. The CO_2 is pumped into the extractor until the required pressure is reached. For the supercritical extraction of the *Usnea* lichen, working conditions of 30 MPa and 40°C were found as an optimum [19]. Temperature and pressure of the adsorption step need to be adjusted to the desired transport properties of the impregnation, taking into account the changing properties of the carrier matrix, as described above. For PCL, DSC analysis results indicated an optimum temperature of 35°C for not completely melting the polymer, but still enabling posterior foaming (Fig. 13.5). The integrated procedure is characterized by two variables of time, t_1 and t_2, representing the kinetics of the extraction and the impregnation respectively. In the experimental design study, t_1 is the time of continuous extraction-impregnation in a single passing mode at given conditions (extraction: 30 MPa/40°C; impregnation: 15–17 MPa/35°C), and t_2 represents the time of recycling of the solution at 15–17 MPa/35°C through both steps. Product obtained from the recycling procedure is compared to product obtained in "single-passing mode." A valve between the extractor and the view cell is used for adjusting the pressure in both steps. The impregnated products are assigned according to the process parameters: PCLt_1/t_2. Thus, the obtained samples were named PCL3/0, PCL1/1, PCL2/1, and PCL2/4. For example, PCL2/1 means that CO_2 was passed through the equipment without recycling during t_1 = 2 h, and afterward, recycling was carried out over t_2 = 1 h at homogeneous pressure throughout the equipment.

In the proposed process, the variation in pressure and temperature within the impregnation vessel (A), taking into account the conditions established within the extractor vessel (E), is the main factor that ensures an environment supersaturated with solute in the neighborhood of the polymeric matrix. Table 13.4 shows that impregnation of *Usnea* extract achieves values in the range of 0.7–2.8 wt%. Recycling of the scCO_2–extract solution was beneficial to increase the amount of extract within the matrix. A procedure with two hours for extraction-impregnation (t_1) ensures the maximum amount of extract in solution with scCO_2. A recycle time (t_2) of only one hour was sufficient to achieve an acceptable impregnation value. A recycling time (t_2) greater than two hours was not beneficial, probably due to the deposition of the extract on the walls and piping of the equipment when reducing the temperature and pressure and also to dilution during the recycling process. Recycling with a scCO_2–extract solution with less concentration of extract can lead to "washing" of the impregnated polymeric matrix. Finally, the product is obtained by depressurization in a controlled manner in order to obtain a defined matrix structure (solid polymer foam). Table 13.4 summarizes the characteristics of the scCO_2 impregnated PCL. For comparison, scCO_2 treated PCL samples, without impregnation, are also shown.

Table 13.4 Parameters of processing (t_1, t_2); porosity (ε), mean pore size, and impregnation

Sample	Procedure	Porosity (ε) (%)	Mean pore size (μm)	Impregnation (wt%)
PCL	[62]	70	340	–
PCL3/0	E-VC t_1: 3 h t_2: 0	64	342	0.7
PCL1/1	E-A t_1: 1 h t_2: 1 h	72	290	1.5
PCL2/1	E-A t_1: 2 h t_2: 1 h	40	270	2.8
PCL2/4	E-A t_1: 2 h t_2: 4 h	72	365	2.2

The results of the antibacterial screening of the tested functionalized PCL scaffolds are presented in Fig. 13.8. The impregnated PCL samples (PCL3/0, PCL1/1, PCL2/1, PCL2/4) show a bactericidal effect on *Listeria innocua* (Fig. 13.8a), reducing the viability of inoculated cells by more than 99%. The most effective sample is PCL2/1, the one with the highest percentage of impregnation (2.8 wt%). However, the bactericidal effect is not directly proportional to the percentage of impregnation, as observed for samples PCL3/0 and PCL1/1. PCL3/0 is found to be only superficially impregnated, since the processing time was not sufficient for complete diffusion of the solute into the matrix. This superficial impregnation can be the cause of the greater bactericidal effect of sample PCL3/0 on *L. innocua* with respect to the PCL1/1 sample. Figure 13.8b shows the effect of the percentage of impregnation and the inoculums size on the antibacterial activity against the MRSA strain. The tested samples showed a higher antimicrobial activity against MRSA ATCC 43300 when a lower bacterial inoculum level (2×10^4 CFU/mL) was assayed. With a low inoculum size, growth reductions were more pronounced for most samples. In this case, a minimum amount of 1.5 wt% of impregnation was needed for producing a bacteriostatic effect (PCL1/1, PCL2/1, PCL2/4).

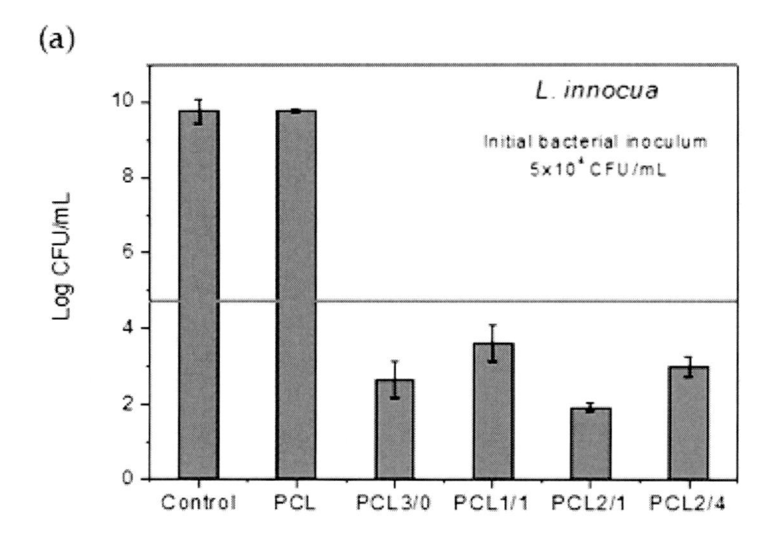

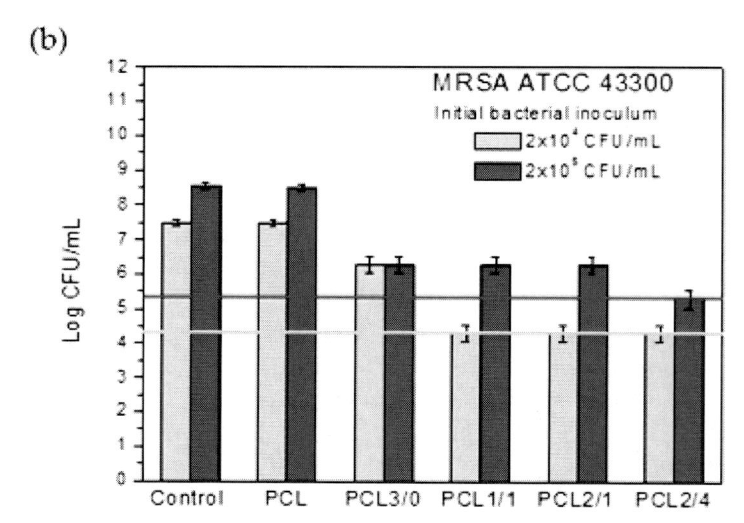

Figure 13.8 Growth inhibition by functionalized PCL scaffolds: (a) *L. innocua*, PCL concentration tested: 200 mg·mL^{-1}; and (b) MRSA, PCL concentration tested: 100 mg·mL^{-1}. Vertical bars represent means of three replicates ± standard deviation. Control: culture medium without sample; PCL: culture medium + PCL without impregnation.

13.6 Conclusions and Remarks

The extraction of antibacterial substances and its adsorption on polymeric materials by an appropriate combination of extraction, impregnation, and foaming conditions is feasible. A "three step in one" $scCO_2$ impregnation of biopolymers process provides an efficient method for tailoring the chemistry and morphology of the aforementioned type of scaffolds by simultaneously obtaining the desired composition and microstructure. Polymeric materials exhibit a wide range of interactions with $scCO_2$; while some do not interact in any way, others exhibit pronounced interaction that can reach the solution or material degradation. This opens up a range of possibilities for the design of the final product in the proposed process involving extraction, impregnation, and formulation.

References

1. Herrero, M., Cifuentes, A., and Ibañez, E. (2006). Sub- and supercritical fluid extraction of functional ingredients from different natural sources: plants, food-by-products, algae and microalgae; a review, *Food Chem.*, **98**, pp. 136–148.
2. Martínez, J. L. (2008) *Supercritical Fluid Extraction of Nutraceuticals and Bioactive Compounds* (CRC Press, Taylor and Francis, USA).
3. Sovová, H. (2012). Steps of supercritical fluid extraction of natural products and their characteristic times, *J. Supercrit. Fluids*, **66,** pp. 73–79.
4. Sharif, M., Kamran, R., Syed, S. H., and Paraman, I. (2014). Characterization of supercritical fluid extrusion processed rice–soy crisps fortified with micronutrients and soy protein, *LWT: Food Sci. Technol.*, **56**, pp. 414–420.
5. Stamenic, M., Zizovic, I., Eggers, R., Jaeger, P., Heinrich, H., Rój, E., Ivanovic, J., and Skala, D. (2010). Swelling of plant material in supercritical carbon dioxide, *J. Supercrit. Fluids*, **52**, pp. 125–133.
6. Farías-Campomanes, A. M., Rostagno, M. A., and Meireles, M. A. A. (2013). Production of polyphenol extracts from grape bagasse using supercritical fluids: yield, extract composition and economic evaluation, *J. Supercrit. Fluids*, **77**, pp. 70–78.
7. Zacchi, P., and Eggers, R. (2008). High–temperature pre-conditioning of rapeseed: a polyphenol-enriched oil and the effect of refining, *Eur. J. Lipid Sci. Technol.*, **110**, pp. 111–119.

8. Ivanovic, J., Meyer, F., Misicc, D., Asanin, J., Jaeger, P., Zizovic, I., and Eggers, R. (2013). Influence of different pre-treatment methods on isolation of extracts with strong antibacterial activity from lichen Usnea barbata using carbon dioxide as a solvent, *J. Supercrit. Fluids*, **76**, pp. 1– 9.

9. Ivanovic, J., Dimitrijevic-Brankovic, S., Misic, D., Ristic, M., and Zizovic, I. (2013). Evaluation and improvement of antioxidant and antibacterial activities of supercritical extracts from clove buds, *J. Funct. Foods*, **5**, pp. 416–423.

10. Liang, M. T., Yang, C. H., Li, S. T., Yang, C. S., Chang, H. W., Liu, C. S., Cham, T. M., and Chuang, L. Y. (2008). Antibacterial and antioxidant properties of Ramulus cinnamomi using supercritical CO_2 extraction, *Eur. Food Res. Technol.*, **227**, pp. 1387–1392.

11. AlHaj, N. A., Shamsudin, M. N., Zamri, H. F., and Abdullah, R. (2008). Extraction of essential oil from using supercritical carbon dioxide: study of antibacterial activity, *Am. J. Pharmacol. Toxicol.*, **3**, pp. 225–228.

12. Kazazi, H., Rezaei, K., Ghotb-Sharif, S. J., Emam-Djomeh, Z., and Yamini, Y. (2007). Supercritical fluid extraction of flavors and fragrances from Hyssopus officinalis L. cultivated in Iran, *Food Chem.*, **105**, pp. 805–810.

13. Leal, P. F., Chaves, F. C. M., Ming, L. C., Petenate, A. J., and Meireles, M. A. A. (2006). Global yields, chemical compositions and antioxidant activities of clove basil (Ocimum gratissimum L.) extracts obtained by supercritical fluid extraction, *J. Food Proc. Eng.*, **29**, pp. 547–559.

14. Zu, Y.-G., Liu, X.-L., Fu, Y.-J., Wu, N.-K., and Wink, M.-Y. (2010). Chemical composition of the SFE-CO_2 extracts from Cajanus cajan (L.) Huth and their antimicrobial activity in vitro and in vivo, *Phytomedicine*, **17**, pp. 1095–1101.

15. Pereira, P., Bernardo-Gil, M. G., Cebola, M. J., Mauricio, E., and Romano, A. (2013). Supercritical fluid extracts with antioxidant and antimicrobial activities from myrtle (Myrtus communis L.) leaves. Response surface optimization, *J. Supercrit. Fluids*, **83**, pp. 57–64.

16. Yang, Y.-C., Wei, M.-C., Hong, S.-J., Huang, T.-C., and Lee, S.-Z. (2013). Development/optimization of a green procedure with ultrasound-assisted improved supercritical carbon dioxide to produce extracts enriched in oleanolic acid and ursolic acid from Scutellaria barbata D. Don, *Ind. Crops Prod.*, **49**, pp. 542–553.

17. Ghosh, S., Chatterjee, D., Das, S., and Bhattacharjee, P. (2013). Supercritical carbon dioxide extraction of eugenol-rich fraction from

Ocimum sanctum Linn and a comparative evaluation with other extraction techniques: process optimization and phytochemical characterization, *Ind. Crops Prod.*, **47**, pp. 78–85.

18. Mesomo, M. C., Corazza, M. L., Ndiaye, P. M., Dalla Santa, O. R., Cardozo, L., and Scheer, A. P. (2013). Supercritical CO_2 extracts and essential oil of ginger (Zingiber officinale R.): chemical composition and antibacterial activity, *J. Supercrit. Fluids*, **80**, pp. 44–49.
19. Fanovich, M. A., Ivanovic, J., Misic, D., Alvarez, M. V., Jaeger, P., Zizovic, I., and Eggers, R. (2013). Development of polycaprolactone scaffold with antibacterial activity by an integrated supercritical extraction and impregnation process, *J. Supercrit. Fluids*, **78**, pp. 42–53.
20. Oliveira, D. A., Salvador, A. A., Smânia Jr., A., Smânia, E. F. A., Maraschin, M., and Ferreira, S. R. S. (2013). Antimicrobial activity and composition profile of grape (Vitis vinifera) pomace extracts obtained by supercritical fluids, *J. Biotechnol.*, **164**, pp. 423–432.
21. Zibetti, A., WüstAydi, A., Livia, M., Arauco Bolzan, A., and Barth, D. (2013). Solvent extraction and purification of rosmarinic acid from supercritical fluid extraction fractionation waste: economic evaluation and scale-up, *J. Supercrit. Fluids*, **83**, pp. 133–145.
22. Piras, A., Rosa, A., Marongiu, B., Porcedda, S., Falconieri, D., Dessì, M. A., Ozcelik, B., and Koca, U. (2013). Chemical composition and in vitro bioactivity of the volatile and fixed oils of Nigella sativa L. extracted by supercritical carbon dioxide, *Ind. Crops Prod.*, **46**, pp. 317–323.
23. Michielin, E. M. Z., Salvador, A. A., Riehl, C. A. S., Smânia Jr., A., Smânia, E. F. A., and Ferreira, S. R. S. (2009). Chemical composition and antibacterial activity of Cordia verbenacea extracts obtained by different methods, *Bioresource Technol.*, **100**, pp. 6615–6623.
24. Ortega Terra Lemos, C., Augusto dos Santos Garcia, V., Menoci Gonçalves, R., Correa Ramos Leal, I., Dias Siqueira, V. L., Cardozo Filho, L., and Ferreira Cabral, V. (2012). Supercritical extraction of neolignans from Piper regnelli var. pallescens, *J. Supercrit. Fluids*, **71**, pp. 64–70.
25. Mazzutti, S., Ferreira, S. R. S., Riehl, C. A. S., Smania Jr., A., Smania, F. A., and Martínez, J. (2012). Supercritical fluid extraction of Agaricus brasiliensis: antioxidant and antimicrobial activities, *J. Supercrit. Fluids*, **70**, pp. 48–56.
26. Figueiró Vargas, R. M., Salli Tavares Barroso, M., Góes Neto, R., Scopel, R., Alves Falcão, M., Finkler da Silva, C., and Cassel, E. (2013). Natural products obtained by subcritical and supercritical fluid extraction from Achyrocline satureioides (Lam) D.C. using CO_2, *Ind. Crops Prod.*, **50**, pp. 430–435.

27. Thai Danh, L., Truong, P., Mammucari, R., and Foster, N. (2010). Extraction of vetiver essential oil by ethanol-modified supercritical carbon dioxide, *Chem. Eng. J.*, **165**, pp. 26–34.
28. Good Kitzberger, C. S., Smânia Jr., A., Curi Pedrosa, R., and Salvador Ferreira, S. R. (2007). Antioxidant and antimicrobial activities of shiitake (Lentinula edodes) extracts obtained by organic solvents and supercritical fluids, *J. Food Eng.*, **80**, pp. 631–638.
29. Vági, E., Simándi, B., Suhajda, Á., and Héthelyi, É. (2005). Essential oil composition and antimicrobial activity of Origanum majorana L. extracts obtained with ethyl alcohol and supercritical carbon dioxide, *Food Res. Int.*, **38**, pp. 51–57.
30. Gadkari, P. V., and Balaraman, M. (2014). Catechins: sources, extraction and encapsulation; a review, *Food Bioprod. Process*, http://dx.doi.org/10.1016/j.fbp.2013.12.004.
31. Biscaia, D., and Ferreira, S. R. S. (2009). Propolis extracts obtained by low pressure methods and supercritical fluid extraction, *J. Supercrit Fluids*, **51**, pp. 17–23.
32. Lee, C.-C., Chiu, C.-C., Liao, W.-T., Wu, P.-F., Chen, Y.-T., Huang, K.-C., Chou, Y.-T., Wen, Z.-H., and Wang, H.-M. (2013). Alpinia oxyphylla Miq. bioactive extracts from supercritical fluid carbon dioxide extraction, *Biochem. Eng. J.*, **78**, pp. 101–107.
33. el Hattab, M., Culioli, G., Piovetti, L., Eddine Chitour, S., and Valls, R. (2007). Comparison of various extraction methods for identification and determination of volatile metabolites from the brown alga Dictyopteris membranacea, *J. Chromat. A*, **1143**, pp. 1–7.
34. Sajilata, M. G., Singhal, R. S., and Kamat, M. Y. (2008). Supercritical CO_2 extraction of γ-linolenic acid (GLA) from Spirulina platensis ARM 740 using response surface methodology, *J. Food Eng.*, **84**, pp. 321–326.
35. Braga, M. E. M., Pato, M. T. V., Silva, H. S. R. C., Ferreira, E. I., Gil, M. H., Duarte, C. M. M., and de Sousa, H. C. (2008). Supercritical solvent impregnation of ophthalmic drugs on chitosan derivatives, *J. Supercrit. Fluids*, **44**, pp. 245–257.
36. Natu, M. V., Gil, M. H., and de Sousa, H. C. (2008). Supercritical solvent impregnation of poly(ε-caprolactone)/poly(oxyethylene-b-oxypropylene-b-oxyethylene) and poly(ε-caprolactone)/ poly (ethylene-vinyl acetate) blends for controlled release applications, *J. Supercrit. Fluids*, **47**, pp. 93–102.
37. Costa, V. P., Braga, M. E. M., Guerra, J. P., Duarte, A. R. C., Leite, E. O. B., Duarte, C. M. M. Gil, M. H., and de Sousa, H. C. (2010). Development of

therapeutic contact lenses using a supercritical solvent impregnation method, *J. Supercrit. Fluids*, **52**, pp. 306–316.

38. van Schnitzler, J., and Eggers, R. (1999). Mass transfer in polymers in a supercritical CO_2 atmosphere, *J. Supercrit. Fluids*, **16**, pp. 81–92.
39. Takahashi, S., Hassler, J. C., and Kiran, E. (2012). Melting behavior of biodegradable polyesters in carbon dioxide at high pressures, *J. Supercrit. Fluids*, **72**, pp. 278– 287.
40. Eggers, R., van Schnitzler, J., and Kempe, T. (2005). Continuous dosing of soilds in high pressure systems, *Proc. ISASF 2005,* Trieste, E22.
41. Voges, S., Jaeger, P., and Eggers, R. (2007). Solid bed properties under high gas pressure, *Chem. Eng. Technol.*, **30**, pp. 709–714.
42. López-Periago, A., Argemí, A., Andanson, J. M., Fernández, V., García-González, C. A., Kazarian, S. G., Saurina, J., and Domingo, C. (2009). Impregnation of a biocompatible polymer aided by supercritical CO_2: evaluation of drug stability and drug–matrix interactions, *J. Supercrit. Fluids*, **48**, pp. 56–63.
43. Eggers, R., van Schnitzler, J., Huber, R., and Worner, G. (1999). *Process for Dyeing a Textile Substrate in At Least One Supercritical Fluid*, US Patent No. 5958085.
44. Shahid-ul-Islam, M. S., and Mohammad, F. (2013). Perspectives for natural product based agents derived from industrial plants in textile applications: a review, *J. Cleaner Prod.*, **57**, pp. 2–18.
45. Milovanovic, S., Stamenic, M., Markovic, D., Radetic, M., and Zizovic, I. (2013). Solubility of thymol in supercritical carbon dioxide and its impregnation on cotton gauze, *J. Supercrit. Fluids*, **84**, pp. 173–181.
46. Dias, A. M. A., Braga, M. E. M., Seabra, I. J., Ferreira, P., Gil, M. H., and de Sousa, H. C. (2011). Development of natural-based wound dressings impregnated with bioactive compounds and using supercritical carbon dioxide, *Int. J. Pharm.*, **408**, pp. 9–19.
47. Torres, A., Romero, J., Macan, A., Guarda, A., and Galotto, M. J. (2014). Near critical and supercritical impregnation and kinetic release of thymol in LLDPE films used for food packaging, *J. Supercrit. Fluids*, **85**, pp. 41–48.
48. de Matos, M. B. C., Piedade, A. P., Alvarez-Lorenzo, C., Concheiro, A., Braga, M. E. M., and de Sousa, H. C. (2013). Dexamethasone-loaded poly(β-caprolactone)/silica nanoparticles composites prepared by supercritical CO_2 foaming/mixing and deposition, *Int. J. Pharm.*, **456**, pp. 269–281.

49. de Souza, A. C., Dias, A. M. A., de Sousa, H. C., and Tadini, C. C. (2014). Impregnation of cinnamaldehyde into cassava starch biocomposite films using supercritical fluid technology for the development of food active packaging, *Carbohydr. Polym.*, **102**, pp. 830–837.
50. Aquino, R. P., Auriemma, G., Mencherini, T., Russo, P., Porta, A., Adami, R., Liparoti, S., Della Porta, G., Reverchon, E., and del Gaudio, P. (2013). Design and production of gentamicin/dextrans microparticles by supercritical assisted atomisation for the treatment of wound bacterial infections, *Int. J. Pharm.*, **440**, pp. 188–194.
51. Temtem M., Silva, L. M. C., Andrade, P. Z., dos Santos, F., Lobato da Silva, C., Cabral, J. M. S., Abecasis, M. M., and Aguiar-Ricardo, A. (2009). Supercritical CO_2 generating chitosan devices with controlled morphology. Potential application for drug delivery and mesenchymal stem cell culture, *J. Supercrit. Fluids*, **48**, pp. 269–277.
52. Dias, A. M. A., Rey-Rico, A., Oliveira, R. A., Marceneiro, S., Alvarez-Lorenzo, C., Concheiro, A., Júnior, R. N. C., Braga, M. E. M., and de Sousa, H. C. (2013). Wound dressings loaded with an anti-inflammatory jucá (Libidibia ferrea) extract using supercritical carbon dioxide technology, *J. Supercrit. Fluids*, **74**, pp. 34–45.
53. Bose, S., Roy, M., and Bandyopadhyay, A. (2012). Recent advances in bone tissue engineering scaffolds, *Trends Biotechnol.*, **30**, pp. 546–554.
54. Correlo, V. M., Oliveira, J. M., Mano, J. F., Neves, N. M., and Reis R. L. (2010). Natural origin materials for bone tissue engineering: properties, processing and performance, in *Textbook on Principles of Regenerative Medicine*, 2nd Ed., pp. 557–586.
55. Jones, J. R. (2009). New trends in bioactive scaffolds: the importance of nanostructure, *J. Eur. Ceram. Soc.*, **29**, pp. 1275–1281.
56. Cooper, A. I. (2003). Porous materials and supercritical fluids, *Adv. Mater.*, **15**, pp. 1049–1059.
57. Davies, O. R., Lewis, A. L., Whitaker, M. J., Tai, H., Shakesheff, K. M., and Howdle, S. M. (2008). Applications of supercritical CO_2 in the fabrication of polymer systems for drug delivery and tissue engineering, *Adv. Drug Delivery Rev.*, **60**, pp. 373–387.
58. Reverchon, E., Cardea, S., and Rapuano, C. (2008). A new supercritical fluid-based process to produce scaffolds for tissue replacement, *J. Supercrit. Fluids*, **45**, pp. 365–373.
59. Sauceau, M., Fages, J., Common, A., Nikitine, C., and Rodier, E. (2011). New challenges in polymer foaming: a review of extrusion processes assisted by supercritical carbon dioxide, *Prog. Polym. Sci.*, **36**, pp. 749–766.

60. Karimi, M., Heuchel, M., Weiggel, T., Schossig, M., Hofmann, D., and Lendlein, A. (2012). Formation and size distribution of pores in poly(ε-caprolactone) foams prepared by pressure quenching using supercritical CO_2, *J. Supercrit. Fluids*, **61**, pp. 175–190.

61. Zizovic, I., Ivanovic, J., Misic, D., Stamenic, M., Djordjevic, S., Kukic-Markovic, J., and Petrovic, S. D. (2012). SFE as a superior technique for isolation of extracts with strong antibacterial activities from lichen Usnea barbata L., *J. Supercrit. Fluids*, **72**, pp. 7–14.

62. Fanovich, M. A., and Jaeger, P. (2012). Sorption and diffusion of compressed carbon dioxide in polycaprolactone for the development of porous scaffolds, *Mater. Sci. Eng. C*, **32**, pp. 961–968.

Chapter 14

Compressed Fluids, Porous Polymers and Tissue Engineering

Aurelio Salerno and Concepción Domingo
Materials Science Institute of Barcelona (CSIC),
Campus de la UAB s/n, 08193 Bellaterra, Spain
asalerno@icmab.es

One of the major goals of tissue engineering is the development of 3D porous biodegradable scaffolds able to stimulate cells growth and to induce tissue healing and self-repair. Compressed fluids, such as carbon dioxide, are excellent candidates for biomaterials processing and porous scaffold fabrication as they allow the development of clean and toxic-free processes for scaffold fabrication that are beneficial to cells and biological tissues. The aim of this chapter is to provide to the reader an overview about current strategies to design and fabricate porous polymeric scaffolds for tissue engineering by means of compressed fluids. Furthermore, the materials/processing/structure relationship of scaffolds prepared by supercritical fluids plus combined approaches is described to suggest possible strategies for advanced scaffold manufacturing.

Supercritical Fluid Nanotechnology: Advances and Applications in Composites and Hybrid Nanomaterials
Edited by Concepción Domingo and Pascale Subra-Paternault

ISBN 978-981-4613-40-8 (Hardcover), 978-981-4613-41-5 (eBook)
www.panstanford.com

14.1 Introduction to Biomaterials and Tissue Engineering Scaffolds

Tissue engineering (TE) is an increasingly growing and multidisciplinary research field holding the promise to develop novel therapeutic treatments for tissue and organ loss and failure, which are two of the major human health problems. To achieve this challenging goal, TE proposes a variety of strategies that are based on the appropriate combination of stem cells, biomaterials, and molecular cues [1–3]. In particular, in these approaches, cells isolated from a donor tissue are transplanted onto a biocompatible and biodegradable substrate, named *scaffold*. The scaffold acts as a temporary synthetic analogue of the extracellular matrix (ECM) of native tissue, stimulating stem cell growth, proliferation, and differentiation and progressively degrading until the new tissue is formed [1–3]. The new tissue formation is induced by maintaining the cell-seeded scaffold in appropriate bioreactors before implantation into the patient (in vitro strategy) or grafted back directly into the patient to function as the introduced replacement tissue (in vivo strategy).

Stem cells are present in many adult tissues and are an attractive cell type for the regeneration of damaged tissues in clinical applications. Indeed, these cells are undifferentiated cells and are able to self-renew with a high proliferative capacity. Furthermore, stem cells possess the potential of differentiation in any other cells type, such as osteoblasts, chondrocytes and fibroblasts [4, 5]. Although stem cells can be obtained from different sources, mesenchymal stem cells are the most used in TE as they can be extracted from several tissues, including bone marrow, umbilical cord blood, and adipose tissue [5].

A biomaterial is defined as "a nonviable material used in a medical device, intended to interact with biological systems" [6]. Biomaterials are very important components of TE strategies, as they are the main constituents of porous scaffolds. Moreover, biomaterial scaffolds are necessary to enhance the ability of stem cells to induce appropriate tissue regeneration in vitro, as well as the restoration of a tissue diseased function when directly implanted in vivo [7–9]. This is because stem cells are anchorage dependent and will die after transplantation if an appropriate adhesion substrate is not provided. Simultaneously, as observed in the natural tissues, cells act in synergy with the ECM that provide three-dimensionality and direct cell-to-cell communication and transmit multiple biophysical and biochemical stimuli for cell adhesion, migration, proliferation, and differentiation, among others [9].

It is also important to note that the regenerative potential of stem cells and scaffolds has been dramatically improved in recent years by using appropriate molecular cues that are able to stimulate cell responses appropriately [7]. Such examples are growth factors able to stimulate cell growth as well as differentiation agents able to induce proper stem cell differentiation and biosynthesis. Therefore, the next generation of biomaterials and scaffolds will have to not only provide adequate mechanical and structural support for stem cells but also deliver molecules able to actively guide and control stem cell attachment, migration, proliferation, and differentiation.

Biophysical and biochemical properties and requirements for scaffolds in TE have been extensively reviewed and recent examples include aspects of degradation, mechanical properties, and drug delivery [1–3, 8]. These can be summarized as follows:

(i) Provide a bioactive surface to promote cell–material interactions and the biological recognition by the host.
(ii) Support the development of a 3D tissue by a pore structure suitable for cell adhesion, proliferation, migration, and differentiation.
(iii) Induce functional construct vascularization and development via the correct spatial and temporal presentation of topological and biochemical cues.
(iv) Ensure a mechanical function stimulating stem cell differentiation and biosynthesis as well as supporting structurally cells and external loading.

Apart from the properties of the materials, mainly in terms of biocompatibility, biodegradation, and bioactivity, the 3D architecture of the scaffold is very important when attempting to meet these requirements and to mimic the structure and functions of the native ECM. Controlling cell behavior and tissue regeneration by tailoring the scaffold's pore structure is a critical step in the development of the next generation of bioactive TE scaffolds. In particular, morphological and structural properties, such as porosity, pore size distribution and interconnectivity, pore wall morphology, and surface area, were reported to have a profound impact on the biological response of a scaffold material [2, 3, 8]. Pores are necessary for new tissue formation aided by scaffolds because they allow migration and proliferation of stem cells in three dimensions, as well as scaffold vascularization [10]. In addition, a porous surface improves mechanical integration and stability between the implant and the surrounding tissue. Regarding pore size, there are several literature investigations demonstrating that pores of 5 μm size are necessary for tissue vascularization, while pores of 5–15 μm are optimal for fibroblast growth, between 20–125 μm for regeneration of skin, and in the range of 100–350 μm for regeneration of bone tissue [2, 8].

All of these specifications indicate that scaffold design and manufacture must find the optimal balance between materials and processing techniques in order to achieve required morphological, structural, and functional properties for the specific application.

The goal of this chapter is to provide to the reader an overview about current materials and technologies for the design and fabrication of porous scaffolds for TE. Furthermore, special emphasis will be devoted to compressed fluid-based techniques, which enable the fabrication of scaffolds with pore structure features resembling those observed in the ECM of native tissues.

14.2 Overview of Porous Scaffold Materials and Fabrication Techniques

14.2.1 Materials

Biomaterials are key component of TE strategies, as they are the main components of scaffolds and biomedical devices. Ideally, biomaterials for TE applications must fulfill a series of properties. These include (1) biocompatibility, intended as the capability to perform with an appropriate host response in a specific application; (2) biodegradability, intended as the property to degrade without producing toxic bi-products; (3) processablility, intended as the possibility to manufacture biomedical devices and porous scaffolds of desired internal structure and external shape; (4) sterilizability, by using process technologies appropriate for biological uses; and, ultimately, (5) mechanical properties tailored for the required application.

Biomaterial scaffolds used to restore the structure and function of damaged tissue and organ have evolved greatly in the past decades due to the widespread knowledge accumulated on materials design, processing, and characterization of scaffold–cell interactions. Part of this evolution has been the development of novel scaffold materials, compatible with cells and tissues and resulting from contemporary advances in the fields of materials science and molecular biology. Different materials such as metals, ceramics, chemically synthesized polymers, natural polymers, and combinations of these materials to form composites were used depending on the specific application [2, 3, 6, 8]. Metals are the most used biomaterials to replace structural components of the human body. This is because they are very reliable from the viewpoint of mechanical performance. In particular, metals possess tensile strength, fatigue strength, and fracture toughness properties that make them excellent candidates for the fabrication of medical devices for the replacement of hard tissue such as artificial hip joints, bone plates, coronary stents, and dental implants [11]. Polymers are mostly made of organic components and are characterized by macromolecular properties comparable to lipids, proteins, and polysaccharides, which are key functional organic components of the ECM [12, 13]. Further advantages of polymeric biomaterials are their relatively simple processability and their broad range of application, spanning from nondegradable implants to controlled degradable biomedical devices [14]. Both synthetic polymers and biologically derived (or natural) polymers were extensively investigated as biodegradable polymeric biomaterials. Materials from natural resources, such as collagen and gelatin, possess the great advantage of biological recognition because of the presence of receptor-binding ligands inside their chemical structure [6, 9]. Conversely, synthetic biomaterials, such as poly(lactic acid) (PLA) and poly(caprolactone) (PCL) may overcome the problems related to purification, immunogenicity, and pathogen transmission of natural materials and may also provide a greater control over the final scaffolds' properties [3, 14].

Due to their chemical similarity to the inorganic phase of bone, ceramic biomaterials such as calcium phosphates (e.g., hydroxyapatite and α- and β-tricalciumphosphate) have been more intensively investigated in respect to their possible application as bone scaffolds [2, 10]. These materials are bioactives and osteoconductives and are able to bond directly to bone [10]. Ceramic biomaterials in the form of micro- and nanoparticles, as well as metal oxides, are also used as dispersed fillers inside natural and/or synthetic polymeric matrixes to design and fabricate multiphase composites, taking advantage of the single components. In particular, in such cases, composite scaffolds evidenced improved mechanical properties, degradation kinetics, and biological response if compared to polymeric scaffolds [13, 15–17].

14.2.2 Fabrication Techniques

Several fabrication processes have been developed to obtain 3D porous scaffolds from biocompatible and biodegradable materials. Among them, thermodynamic-based processing of polymeric solutions, such as gas foaming and phase separation, is the most used method to create porous scaffolds with controlled 3D pore structures [2, 3, 8]. The gas-foaming process uses blowing agents, mainly carbon dioxide (CO_2), nitrogen (N_2), and mixtures of these two fluids, which are usually solubilized at constant temperature and high pressure inside a polymeric matrix. When the solubilization of the blowing agent into the polymer is completed, the polymer/blowing agent solution is brought to the supersaturated state either by increasing temperature (temperature-induced phase separation) or by reducing pressure (pressure-induced phase separation). This step brings the nucleation and growth of gas bubbles into the polymeric matrix. The decrease of the blowing agent inside of the polymeric matrix and the concomitant increase of matrix stiffness allow for the final pore structure stabilization [3, 8, 15, 18]. One of the great advantages of this technique is related to the possibility of fine control of the porous network of the scaffold, avoiding the use of organic solvents. Indeed, the presence of residues of these chemicals inside the scaffold may be harmful for cells and nearby tissue and may inactivate the biological signals incorporated into the polymeric matrix. The modulation of the operating parameters, namely, pressure, temperature, and the depressurization profile, allows for the fine control of the scaffold pore structure. Gas foaming is usually applied to synthetic polymers, such as PLA and PCL and their composites. Conversely, a limited number of works have been reported on gas-foamed scaffolds from natural polymers, as they do not allow for the easy diffusion, solubilization, and foaming of the blowing agent inside the polymeric matrix.

Polymeric scaffolds can be obtained by phase separation caused by temperature change or antisolvent addition, using either conventional liquids or supercritical fluids [8, 19, 20]. Phase separation from a polymer–solvent solution is based on the thermodynamic demixing of the system into a polymer-rich and a polymer-poor phases, for instance, by cooling the solution down to a decomposition curve. The removal of the solvent within the polymer-lean phase by solvent evaporation, sublimation, or solvent/nonsolvent exchange, allows reaching an open-pore network, whereas the polymer that composes the polymer-rich phase solidifies in the final structure of the scaffold. Porous polymeric scaffolds with a wide range of porosity and pore size distribution, down to nanometric-scale resolution, can be fabricated by optimizing the polymer/solvent choice and cooling temperature profile during phase separation [8, 19, 20]. Although this technique can be applied to both synthetic and natural biomaterials, there is still the need of replacing organic solvents [20].

Reverse-templating techniques utilize percolating porogen agents, characterized by proper concentration and desired size and shape, to imprint ordered arrays of pores within a continuous matrix. The solvent casting/particulate leaching method is the most commonly mentioned one in fabrication of scaffold for TE [8]. The method involves mixing water soluble or easy removable particles into a biodegradable polymer/solvent solution or a polymer melt. The system is then solidified by extracting the solvent or by cooling and, finally, the porogen is leached out from the matrix by soaking in an appropriate solvent. Several materials have been used as porogen agents to achieve porosity architecture within polymeric scaffolds. The selection of the templating agent has been constrained to the compatibility of the particulate removal method (solvent, heating, chemical treatment) with polymer characteristics. Sodium chloride, sugar, gelatin, or hydrocarbon particles are widely used in particulate-leaching process, salt crystals being the most widespread particles used for this method [8, 15, 21]. Furthermore, pore size and shape are easily controlled by selecting the size and shape of the porogen [8]. The materials choice is critical in this technique, as the polymer must be soluble in an organic solvent that, at the same time, is a poor solvent for the porogen. Furthermore, as in the case of phase separation, the solvent must be biocompatible or easy to remove from the scaffold. In some cases, residual solvents in scaffolds may be removed with supercritical carbon dioxide (scCO_2), allowing for the retention of the pore architecture.

In the past decade, techniques such as lithography and injection have been developed to increase the control over scaffold geometry and to achieve high reproducibility and large-scale manufacturing. This is achieved by using externally controlled setups, which allow creating ordered arrays of pores with desired shapes and characteristics within biomaterial scaffolds [22]. The main disadvantages of these approaches are the high cost of equipment, the sometimes limited capacity to achieve three-dimensionality, and, from a biomedical point of view, low compatibility with the processing of biomolecules. As an alternative, electrospinning and molecular self-assembly are based on the physical and/or chemical assembly of individual atoms, molecules, or supramolecular building blocks to form useful constructs [23, 24].

14.3 Supercritical Fluids, Biomaterials Processing, and Porous Scaffold Manufacturing

As defined by Darr and Poliakoff in 1999, a supercritical fluid is "any substance, the temperature and pressure of which are higher than their critical values, and which has a density close to or higher than its critical density" [25]. Supercritical fluids have unique properties that may enhance many types of process operations, such as low surface tension, low viscosity, high diffusivity, and density-dependent solvent power [26]. In particular, physicochemical properties of supercritical fluids, such as density, viscosity, and diffusivity, are intermediate between those of liquids and gases and are continuously adjustable from gas to liquid with small pressure and temperature variations. This is because, from a microscopic point of view, the fluid is characterized by local fluctuations of density and by simultaneous regions of both gas and liquid local densities [27]. An additional advantage of using supercritical fluids arises from the fact that they may replace many environmentally harmful solvents currently used in polymer processing. For example, $scCO_2$, which is by far the most widely used supercritical fluid, is relatively cheap, nontoxic, and nonflammable and can be used to process biomaterials, cells, and heat-labile molecules for both TE and health care applications [28, 29].

The motivation for using $scCO_2$ in polymer and biomaterial processing rely first on the need of using benign solvents. As previously discussed, $scCO_2$ has unique properties that could be very benefit in polymer processing. For instance, sorption of $scCO_2$ into polymers can be used to "plasticize" the polymer by decreasing its glass transition and/or melting temperatures. The plasticization of polymers is the result of the ability of CO_2 molecules to interact, for example, with the basic sites in polymer molecules, such in the case of the interaction between CO_2 and the carbonyl group in polyesters [28]. As a direct consequence, the segmental and chain mobility of the polymer increase, while the chain-to-chain interactions decrease. These effects finally lead to the decrease of polymer viscosity, which can be beneficial for enhancing polymer extrusion and blending, as well as the increase of the diffusion of molecules and compounds through polymer matrices, such in the case of impregnation and extraction processes [25–28, 30, 31]. Biomaterials processes using $scCO_2$ can be classified into three main groups according to the role of $scCO_2$ in the process: solvent, antisolvent, and solute.

The first group of processes utilizes $scCO_2$ as a solvent. Such examples are supercritical fluid extraction (SFE) and rapid expansion from supercritical solution (RESS) [20, 32–34]. Extraction of target compounds from solid and liquid matrices by means of $scCO_2$ is attracting increasing attention in the biomedical field, especially for the fabrication of porous scaffolds for TE and for the purification and sterilization of biomedical implants [20, 35–37]. In particular, as it will be discussed in detail in the next section of this book chapter, SFE processes can be used for the fabrication of porous nanostructured scaffolds by means of $scCO_2$ drying of organic and inorganic hydrogels [20, 36]. At the same time, sterilization of heat-sensitive devices, such as endoscopes, and purification of porous implants are crucial to prevent the patient's infection [35, 37]. Indeed, the ability of $scCO_2$ to destroy various species of bacteria as well as to be used for the purification of natural implants can make this fluid an important alternative to traditional sterilization and purification agents in the biomedical field. The RESS technique is commonly used for the synthesis of polymeric particles and fibers. This process is based on the initial dissolution of the biomaterial in $scCO_2$, followed by the expansion of the solution to a lower pressure through an appropriate nozzle. As a direct consequence, the temperature of the system is lowered and the solvent power of $scCO_2$ decreases, leading to biomaterial precipitation [32]. The as-obtained particles and fibers can be useful for the fabrication of porous scaffolds and drug delivery carriers.

The second group of processes belongs to the supercritical antisolvent (SAS) approach. In this approach, a liquid solution containing the biomaterial is put in contact with $scCO_2$. By using processing conditions suitable to ensure the complete miscibility of $scCO_2$ with the solvent and negligible solubility of the solute in $scCO_2$, the solution goes to a supersaturated state and the solute precipitate [32, 34]. Different biomedical devices, in the form of particles and scaffolds, can be produced by SAS processes depending on the biomaterial/solvent system and the operating parameters, such as temperature, pressure, and depressurization rate [32, 34, 38].

The third application of $scCO_2$ is as a solute agent for the gas-foaming process. Gas foaming is based on the solubilization of $scCO_2$ at high pressures inside the polymer, followed by pressure quench and pore nucleation and growth [18, 29]. In particular the next part of the chapter is focused on two $scCO_2$-based approaches for scaffold fabrication, (i) gas foaming and (ii) phase separation combined with $scCO_2$ drying.

14.4 Porous Scaffold Fabrication by Means of Gas Foaming–Based Approaches

$scCO_2$ foaming is one of the most used processing techniques for the fabrication of biodegradable polymeric foams with controlled pore structures suitable as TE scaffolds [2, 3, 8]. Indeed, this technique offers the great advantage of achieving a fine control over the biophysical and biochemical features of the scaffolds by the appropriate selection of the operating conditions. Furthermore, scaffold fabrication can be achieved at low temperature and avoiding the use of organic solvents that may be harmful for cells and nearby tissue and may inactivate the biological signals eventually incorporated into the polymeric matrix [2, 3, 8].

Figure 14.1 reports a scheme of the different steps and operating conditions of the $scCO_2$ foaming process. As shown in Fig. 14.1, when processing a biomaterial with the gas-foaming technique, the first step is the selection of the most appropriate material and the optimization of its composition and physical and chemical properties. Once the material has been appropriately designed, it can be subjected to the foaming process. Polymer plasticization is carried out by saturation with $scCO_2$ at constant temperature and pressure, typically in the ranges of 35°C–45°C and 10–30 MPa, respectively. When the solubilization of the $scCO_2$ into the polymer is completed, the system is brought to the supersaturated state by reducing the pressure to atmospheric, with the effect of inducing the nucleation and growth of the pores into the polymeric matrix. To modulate the final pore structure of the scaffolds, it is essential to select appropriately the pressure, temperature and time of saturation as well as the profile of blowing agent venting [3, 8, 18, 39–41].

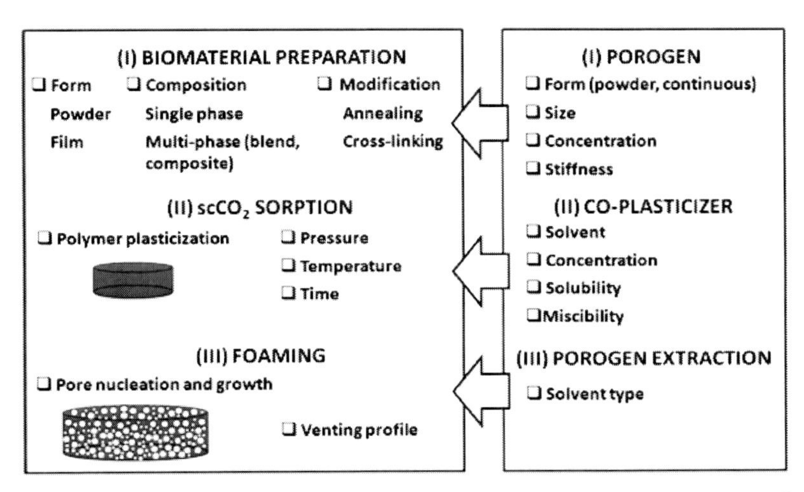

Figure 14.1 Scheme and operating conditions of $scCO_2$ foaming-based processes.

A biomaterial can be processed by gas foaming, starting from a particulate of bulky form, while its composition can be characterized by a single phase or by a homogeneous or heterogeneous mixture of two or more components. Blending different materials has been reported as an efficient way to control the foaming process, as well as to improve the final microstructural and biological response of porous scaffolds [3, 13, 15, 21, 29, 39]. For instance, it is reported that the use of heterogeneous blends, composed of a synthetic polymer, such as PCL, and a natural polymer, such as gelatin or zein, facilitates the foaming behavior of the system in a wide range of operating temperatures [21, 39]. Furthermore, multiphase synthetic/natural polymeric blends can improve the control over scaffold hydrophilicity, degradation kinetics, and bioactivity [13, 39].

Several works reported about the fabrication of porous scaffolds by means of $scCO_2$ foaming [29, 39–41]. Completely amorphous polymers are excellent candidates for this process, as they allow the blowing agent to easily diffuse within the polymeric matrix and "plasticize" the material to render it rubbery [41]. Conversely, the $scCO_2$ foaming of semicrystalline polymers is more difficult, owing to the different $scCO_2$ solubility and diffusivity within the amorphous and crystalline domains of the polymer [29]. Furthermore, the presence of nonplasticized crystal regions, maintaining a high viscosity/stiffness, hinders, when possible, the formation of porous scaffolds with uniform morphology and pore structure [29]. Controlling the starting properties of the biomaterial by applying appropriate thermal treatments to control polymer crystallization can be an effective way to enhance semicrystalline polymer foaming and scaffold fabrication [29].

Although gas foaming allowed the preparation of polymeric scaffolds with definite porosities and pore sizes, the achievement of a high degree of interconnectivity is often impaired by a combination of rheological and processing limitations, which do not allow complete pore opening during foaming and lead to the formation of a nonporous external skin [3, 8, 18, 40, 41]. Reduced pore interconnectivity represents a great limitation for gas-foamed scaffolds, as open pores are necessary to allow the 3D cell colonization and the diffusion of nutrients and metabolic waste in the entire pore structure. The scheme of Fig. 14.1 shows different ways to improve the foaming properties of a polymeric biomaterial and to enhance the interconnectivity of the final scaffold. In particular, this can be achieved by blending the biomaterial with a solid porogen before foaming and by further extracting the porogen from the foamed scaffold by soaking the sample in an appropriate solvent [3, 15, 42]. Alternatively, it is also demonstrated that the pore structure and interconnectivity of the scaffold can be enhanced by adding an appropriate plasticizer to $scCO_2$ during the foaming step [43, 44]. The plasticizer is typically a solvent of the same biomaterial, such as acetone or ethyl lactate, which creates a binary mixture with $scCO_2$ at appropriate temperature and pressure. By using binary mixtures as blowing agents, polymer plasticization can be enhanced and foaming may result in porous scaffolds with enhanced pore interconnectivity [43, 44].

Figure 14.2 reports some representative scanning electron microscopy (SEM) images showing the morphology of porous scaffolds that can be obtained with the two aforementioned approaches. The morphology of PCL scaffolds prepared by using NaCl particles with a mean size of 5 μm and with concentrations equal to 30 or 60 wt% and prepared at a saturation temperature higher than the PCL melting point (70°C) is reported in Fig. 14.2a and Fig. 14.2b, respectively. The scaffolds are characterized by a uniform morphology and highly interconnected porosity. Furthermore, by increasing the particle concentration from 30 to 60 wt% the porosity and pore size decrease from 87% and 70 μm to 79% and 25 μm. This effect depends on the increase in the stiffness of the PCL-NaCl composites that restrict the pore growth, as well as to the enhanced nucleation of the pores [42].

As shown in Fig. 14.2c,d, totally different morphologies may be achieved by using a continuous porogen for the preparation of the scaffold. In particular, in this case, the scaffolds are prepared by melt-blending PCL and gelatin, followed by gas foaming and selective gelatine extraction from the foamed sample [21, 45]. To modulate the final properties of PCL scaffolds, two different foaming temperatures are selected. A foaming temperature of 44°C was used to fabricate a PCL scaffold with a double-scale pore size distribution (Fig. 14.2c). This scaffold is provided of larger and elongated pores of the order of several hundreds of microns, created by the removal of the continuous porogen (gelatin), interconnected to the smaller and rounded pores obtained by the foaming step. Conversely, a foaming temperature of 70°C was used to fabricate a PCL scaffold, characterized by a pore size distribution on a single scale [21, 45].

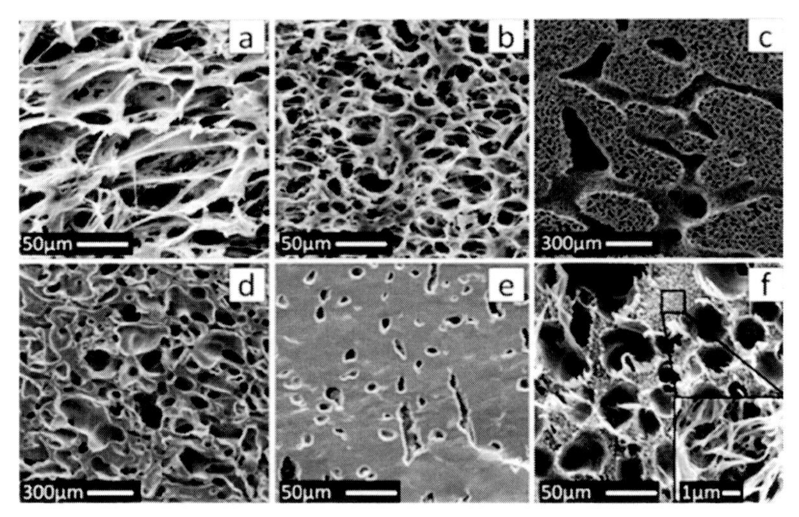

Figure 14.2 SEM images of the cross section of porous PCL scaffolds prepared by $scCO_2$ foaming-based approaches (a, b) by combining foaming and NaCl leaching [42], (c, d) by combining foaming and selective polymer extraction from a co-continuous blend [21, 45], and (e, f) by using binary mixtures of $scCO_2$ and ethyl lactate as foaming agents [44].

Improving polymer plasticization by adding proper plasticizers, such as acetone or ethyl lactate, to $scCO_2$ are alternative solutions to enhance the foaming of semicrystalline polymers, especially when the process is carried out at temperatures lower than the polymer melting point. Indeed, a small quantity of plasticizer can be dissolved in $scCO_2$ at an appropriate temperature and pressure to improve its solubilization inside the polymeric matrix and, consequently, its capability to decrease polymer melting down to the selected operating temperature [43, 44]. To corroborate these considerations, in Fig. 14.2e and Fig. 14.2f the morphologies of PCL scaffolds prepared by using pure CO_2 and a mixture of CO_2 and ethyl lactate (0.2 mol%) as a blowing agent, respectively, are compared. As shown, the PCL scaffold prepared by using $scCO_2$/ethyl lactate as a blowing agent is characterized by enhanced foaming and by the presence of a nanofibrous structure inside the pore walls (high-magnification SEM image reported in the inset of Fig. 14.2f) [44]. This peculiar effect is ascribable to the ability of the blowing agent mixture to diffuse inside the amorphous regions of the crystallites of the polymer and the consequent stretching of the crystalline domains during pores nucleation and growth [44].

Although blending a biomaterial with a porogen agent can allow improving the interconnectivity of the scaffold's pores, this second phase can also have a strong impact on the foaming behavior of the system. For instance, as shown in Fig. 14.3, the final porosity and mean pore size of foamed scaffolds are directly correlated to the viscoelastic properties of the material and, in turn, to the NaCl concentration. The addition of inorganic fillers to a polymer melt influences its flow properties. Several factors, such as particle size and shape, filler content, and interactions between the phases, have complex influences on the viscoelastic properties of the composites [15, 42]. Figure 14.3a shows the frequency-dependent elastic (G') and viscous (G'') moduli for PCL/NaCl composites containing 30 and 60 wt% NaCl. As shown, at low frequencies the moduli increase by several orders of magnitude with the increase of the NaCl amount and a solid-like feature emerges for the highly filled samples, with G' higher than G'' and both the moduli weakly dependent on ω. This effect indicates that, for a 60 wt% concentration the NaCl, particles are completely percolating inside the polymeric matrix. As a direct consequence, the highly filled composite results in a porous scaffold with lower values of porosity (Fig. 14.3b) and mean pore size (Fig. 14.3c) in all the 25°C–40°C range of foaming temperature investigated. Furthermore, it is observed that foaming temperatures ranging from 28°C to 32°C are the most appropriate to ensure sufficient sample foaming, while avoiding pore collapse [46].

An emerging trend in the TE field is the design and fabrication of continuous graded osteochondral scaffolds for the simultaneous treatment of compromised cartilage and subchondral bone in articular joint defects [46]. Indeed, these scaffolds may provide simultaneously the appropriate biomimetic 3D environments for the growth of the osseous and cartilaginous compartments and of the gradual, continuous interface between bone and cartilage. As previously showed, the concentration of NaCl strongly affects the porosity and pore size of the resulting foamed scaffolds. Therefore, it could be expected that, by creating a spatial gradient of NaCl particles concentration inside the polymeric matrix, scaffolds with spatial gradients of porosity and pore size can be achieved. Figure 14.4 shows the microstructure of a PCL scaffolds prepared using a PCL/NaCl composite, with 30/60 wt% NaCl spatial gradients.

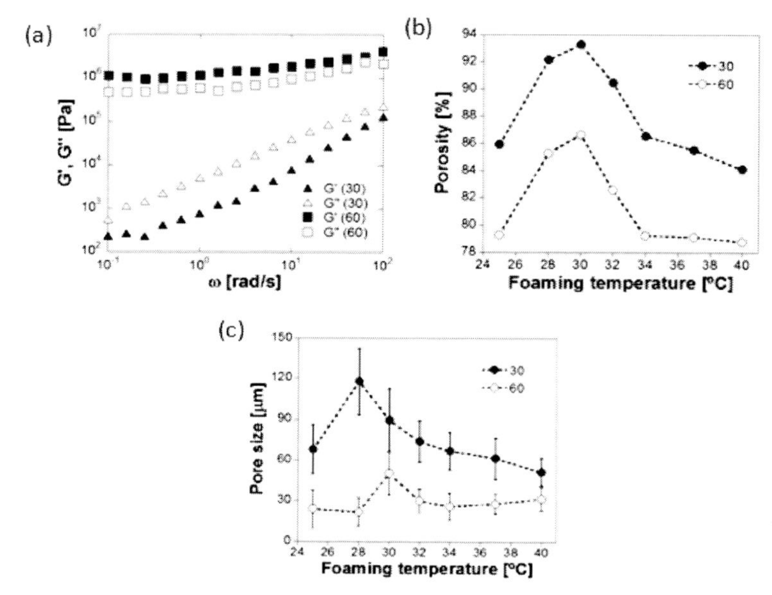

Figure 14.3 Effect of the concentration of NaCl particles on (a) the rheological properties of the PCL/NaCl composites, (b) porosity, and (c) mean pore size of resulting PCL scaffolds [42, 46].

As shown, the foamed scaffold is characterized by a spatial gradient of porosity and pore size. In particular, the region of the sample with NaCl concentration of 30 wt% (upper zone of SEM micrograph reported in Fig. 14.4) showed higher porosity and mean pore size if compared to the region of the sample with NaCl concentration of 60 wt% (lower zone of SEM micrograph reported in Fig. 14.4) [42, 46]. All of these results demonstrated the great potential of the gas-foaming technique in designing porous biodegradable polymeric scaffolds with controlled morphology and pore structure features and avoiding the use of operating conditions that can negatively affect the final scaffold biocompatibility.

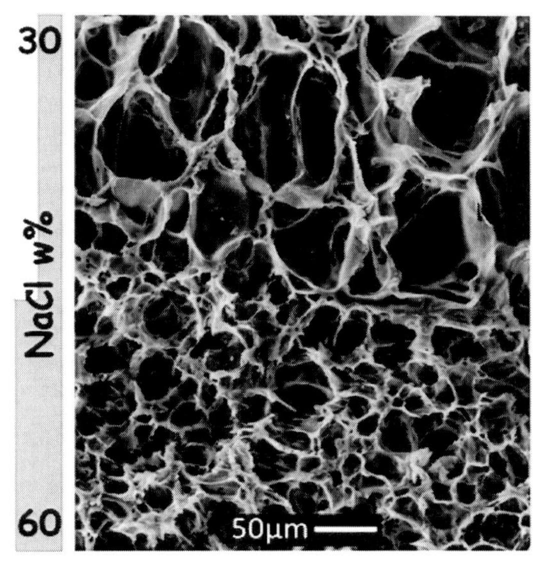

Figure 14.4 SEM micrograph of anisotropic PCL scaffolds prepared by using a PCL/NaCl composite with 30/60 wt% gradient of NaCl particles [42, 46].

14.5 Porous Scaffold Fabrication by Means of Phase Separation and $scCO_2$ Drying Approaches

The pursuit of nanometer-scale porous scaffolds characterized by a 3D fibrous structure represents a new realm of matter of current research in TE. Indeed, nanometer-scale fibrous scaffolds of various biocompatible materials are able to mimic the collagen structure of the ECM, enhancing cell/material cross talking at the interface and promoting cell adhesion, proliferation, and differentiation [20, 47–49]. Phase separation is one of the most versatile approaches that can be used to fabricate nanometer scale fibrous scaffolds with controlled morphology and pore structure [20, 48, 49]. As discussed in the first part of this chapter, phase separation is based on the thermodynamic demixing of a homogeneous polymeric solution into a polymer-rich and a polymer-poor phase, for instance, by cooling the solution down to a demixing curve [20, 48, 49]. Figure

14.5 reports the scheme of the thermal induced phase separation combined with $scCO_2$ drying, which was recently proposed for the fabrication of nanofibrous PLA materials [20]. As shown, the process consisted of four steps. In the first step the polymer is dissolved in ethyl lactate at 70°C under magnetic stirring for eight hours to prepare a homogeneous solution with PLA concentrations in the 3–5.5 wt% range. Subsequently, the solution is cooled to the gelation temperature and the gels further soaked in ethanol to extract the ethyl lactate and to induce polymer precipitation. Finally, the as-obtained alcogels are dried by using $scCO_2$ at 19 MPa, 39°C for 1.5 h. The evolution of the structure and nanometer-scale properties of PLA scaffolds prepared by the thermal induced phase separation strongly depends on polymer crystallization during solution gelation as well as gel drying [20, 50]. For instance, it was demonstrated that fibres formation depends on the initial condensation of amorphous nanoparticles of the polymer, followed by the progressive crystallization from the gel [50]. Consequently, the final morphological, structural and thermal properties of the scaffolds are dependent on material crystallization, which, in turn, is correlated to the temperature profile during quenching.

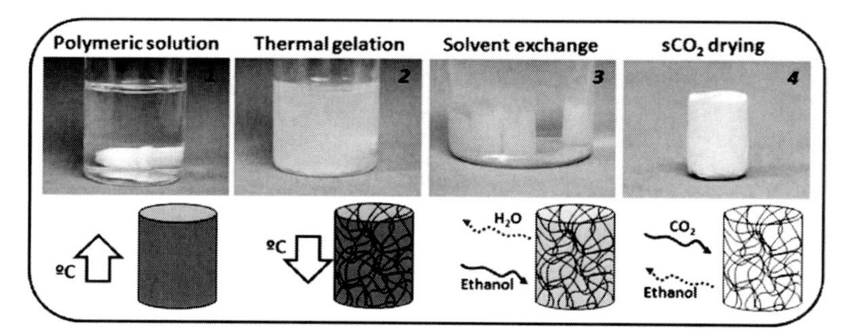

Figure 14.5 Scheme of the thermal-induced phase separation and $scCO_2$-drying combined process used for the fabrication of nanometer-scale fibrous PLA scaffolds.

Once polymer crystallization and solution gelation are achieved, the solvent is extracted by a nonsolvent, such as ethanol, and the gel dried. This is a critical step, as polymeric gels formed upon drying in air often provide white and dense collapsed samples as a consequence of the capillary force induced by ethanol removing. By using $scCO_2$ it is possible to dry the gel through the formation of a supercritical mixture of the CO_2 and the ethanol, taking advantage of the absence of surface tension in the mixture, which can be easily eliminated by venting the vessel [20].

Figure 14.6 reports the morphological and structural properties of nanometer-scale fibrous PLA scaffolds prepared by combining

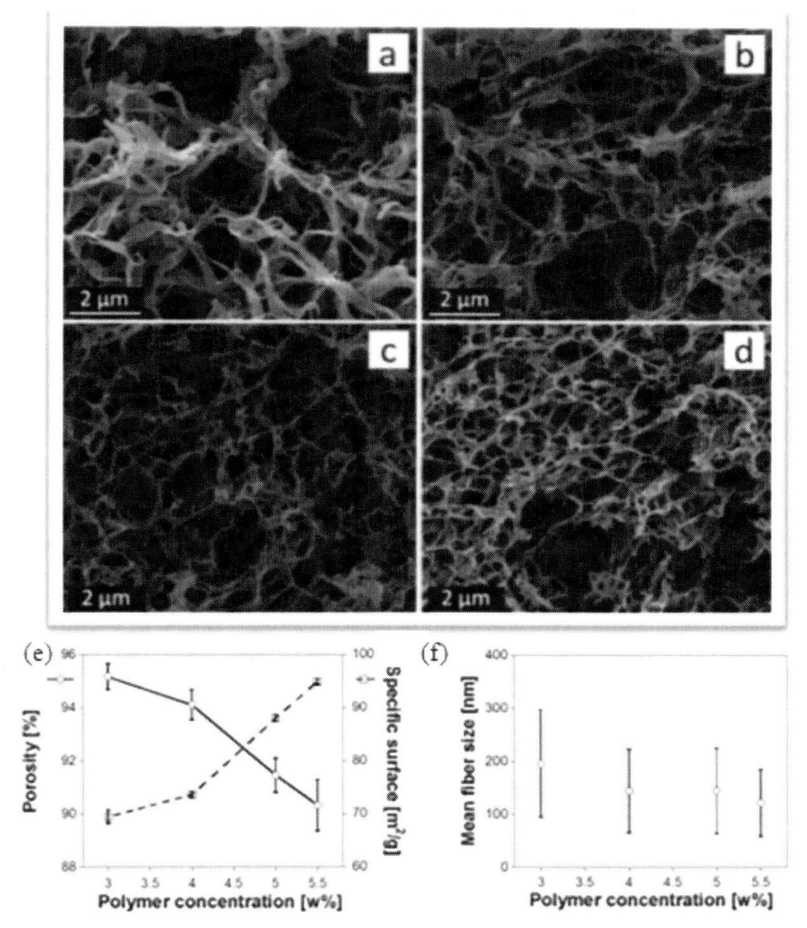

Figure 14.6 SEM micrographs showing the morphology of nanometer-scale fibrous PLA scaffolds as a function of polymer concentration in the initial solution: (a) 3 wt%, (b) 4 wt%, (c) 5 wt%, and (d) 5.5 wt%. (e) Porosity and specific surface of PLA scaffolds as a function of polymer concentration in the initial solution. (f) Mean fiber size of PLA scaffolds as a function of polymer concentration in the initial solution [20].

phase separation and $scCO_2$ drying processes. The results show that the scaffolds are characterized by a nanometer-scale fibrous structure, with fibre density increasing with the increase of the concentration of polymer in the starting solution (compare Figs. 14.6a–14.6d).

The possibility to fabricate porous scaffolds characterized by a nanofibrous structure suitable for cell attachment and a porous network of a pore size of hundreds of microns designed for 3D cell colonization and scaffold vascularization is very important in TE [1, 2, 8, 15]. To prepare scaffolds with this kind of pore structure, phase separation is combined with porogen leaching and $scCO_2$ drying. In particular, micrometric gelatin particles can be used as a particulate porogen. The morphology of PLA scaffolds obtained by this approach is shown in Fig. 14.7.

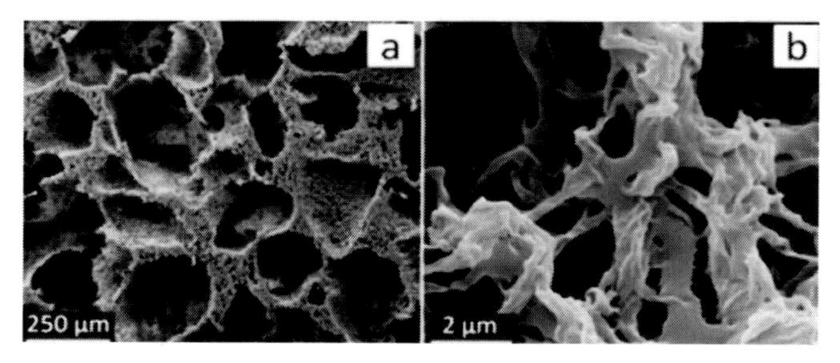

Figure 14.7 SEM micrographs showing the morphology of nanometer-scale fibrous PLA scaffolds prepared by combining the phase separation method with the porogen-leaching process.

It is noticeable that the addition of the particulate porogen allowed for the fabrication of multiscaled porous PLA scaffolds with large pores, in the order of 300 μm, replicating the size of the starting gelatin particles (Fig. 14.7a). Furthermore, by optimizing the phase separation steps, it was possible to induce high pore interconnectivity as well as to recreate a nanometer-scale fibrous architecture on the pore walls (Fig. 14.7b).

14.6 Conclusions

This book chapter reported an overview about the use of compressed fluids (mainly $scCO_2$) for the fabrication of polymeric scaffolds for

TE applications. In the first part, it described the basic principles of scaffold design and fabrication and provided a summary of materials and processing techniques that have been developed and are currently used for the fabrication of 3D porous scaffolds for TE. The central part of the work focused on the use of compressed fluids in TE approaches and biomaterials processing for porous scaffold fabrication. In particular, two main techniques, namely, gas foaming and phase separation combined with $scCO_2$ drying, are described. Their use for the fabrication of porous polymeric scaffolds with controlled morphological and pore structure features are also discussed. This book chapter aims at providing the reader with a series of important knowledge in the field of biomaterials design and scaffold manufacture and to exploit possible strategies toward the development of next generation of multifunctional scaffolds for TE.

Acknowledgments

Aurelio Salerno gratefully acknowledges the CSIC for financial support through a JAE-DOC contract cofinancied by the FSE. The authors also gratefully acknowledge the financial support of Ministerio de Economía y Competitividad through the research project "BIOREG" (MAT2012-35161).

References

1. Tsang, V. L., and Bhatia, S. N. (2004). Three-dimensional tissue fabrication, *Adv. Drug Delivery Rev.*, **56**, pp. 1635–1647.
2. Yang, S., Leong, K., Du, Z., and Chua, C. (2001). The design of scaffolds for use in tissue engineering. Part I. Traditional factors, *Tissue Eng.*, **7**, pp. 679–689.
3. Salerno, A., Netti, P. A., Di Maio, E., and Iannace, S. (2009). Engineering of foamed structures for biomedical application, *J. Cell. Plast.*, **45**, pp. 103–117.
4. Caplan, A. I. (2007). Adult mesenchymal stem cells for tissue engineering versus regenerative medicine, *J. Cell Phys.*, **213**, pp. 341–347.
5. Kern, S., Eichler, H., Stoeve, J., Klüter, H., and Bieback, K. (2006). Comparative analysis of mesenchymal stem cells from bone marrow, umbilical cord blood, or adipose tissue, *Stem Cells*, **24**, pp. 1294–1301.

6. Ratner, B. D., and Bryant, S. J. (2004). Biomaterials: where we have been and where we are going, *Annu. Rev. Biomed. Eng.*, **6**, pp. 41–75.
7. Biondi, M., Ungaro, F., Quaglia, F., and Netti, P. A. (2008). Controlled drug delivery in tissue engineering, *Adv. Drug Delivery Rev.*, **60**, pp. 229–242.
8. Guarino,V., Causa, F., Salerno, A., Ambrosio, L., and Netti, P. A. (2008). Design and manufacture of microporous polymeric materials with hierarchal complex structure for biomedical application, *Mater. Sci. Tech.*, **24**, pp. 1111–1117.
9. Lutolf, M. P., and Hubbell, J. A. (2005). Synthetic biomaterials as instructive extracellular microenvironments for morphogenesis in tissue engineering, *Nat. Biotech.*, **23**, pp. 47–55.
10. Karangeorgiou, V., and Kaplan, D. (2005). Porosity of 3D biomaterial scaffolds and osteogenesis, *Biomaterials*, **26**, pp. 5474–5491.
11. Niinomi, M. (2008). Metallic biomaterials, *J. Art. Org.*, **11**, pp. 105–110.
12. Freed, L. E., Vunjack-Novakovic, G., Biron, R. J., Eagles, D. B., Lesnoy, D. C., Barlow, S. K., and Langer, R. (1994). Biodegradable polymer scaffolds for tissue engineering, *Nat. Biotech.*, **12**, pp. 689–693.
13. Salerno, A., Oliviero, M., Di Maio, E., Netti, P. A., Rofani, C., Colosimo, A., Guida, V., Dallapiccola, B., Palma, P., Procaccini, E. Berardi, A. C., Velardi, F., Teti, A., and Iannace, S. (2010). Design of novel three-phase PCL/TZ–HA biomaterials for use in bone regeneration applications, *J. Mater. Sci. Mater. Med.*, **21**, pp. 2569–2581.
14. Woodruff, M. A., and Hutmacher, D. W. (2010). The return of a forgotten polymer-Polycaprolactone in the 21st century, *Prog. Polym. Sci.*, **35**, pp. 1217–1256.
15. Salerno, A., Zeppetelli, S., Di Maio, E., Iannace, S., and Netti, P. A. (2011). Processing/structure/property relationship of multi-scaled PCL and PCL–HA composite scaffolds prepared via gas foaming and NaCl reverse templating, *Biotech. Boeng.*, **108**, pp. 963–976.
16. Murugan, R., and Ramakrishna, S. (2005). Development of nanocomposites for bone grafting, *Comp. Sci. Tech.*, **65**, pp. 2385–2406.
17. Salerno, A., Zeppetelli, S., Oliviero, M., Battista, E., Di Maio, E., Iannace, S., and Netti, P. A. (2012). Microstructure, degradation and in vitro MG63 cells interactions of a new poly(ε-caprolactone), zein, and hydroxyapatite composite for bone tissue engineering, *J. Bioact. Comp. Polym.*, **27**, pp. 210–226.
18. Jacobs, L. J. M., Kemmere, M. F., and Keurentjes, J. T. F. (2008). Sustainable polymer foaming using high pressure carbon dioxide: a

review on fundamentals, processes and applications, *Green Chem.*, **10**, pp. 731–738.

19. Schugens, C., Maquet, V., Grandfils, C., Jerome, R., and Teyssie, P. (1996). Polylactide macroporous biodegradable implants for cell transplantation. Preparation of polylactide foams by liquid-liquid phase separation, *J. Biomed. Mater. Res.*, **30**, pp. 449–461.
20. Salerno, A., and Domingo, C. (2013). Making microporous nanometre-scale fibrous PLA aerogels with clean and reliable supercritical CO_2 based approaches, *Microp. Mesop. Mater.*, **184**, pp. 162–168.
21. Salerno, A., Oliviero, M., Di Maio, E., Iannace, S., and Netti, P. A. (2009). Design of porous polymeric scaffolds by gas foaming of heterogeneous blends, *J. Mater. Sci. Mater. Med.*, **20**, pp. 2043–2051.
22. Lee, K., Kim, R. H., Yang, D., and Park, S. H. (2008). Advances in 3D nano/microfabrication using two-photon initiated polymerization, *Prog. Polym. Sci.*, **33**, pp. 631–681.
23. Xie, J., Li, X., and Xia, Y. (2008). Putting electrospun nanofibers to work for biomedical research, *Macromol. Rapid Commun.*, **29**, pp. 1775–1792.
24. Hartgerink, J. D, Beniash, E., and Stupp, S. I. (2002). Peptide-amphiphile nanofibers: a versatile scaffold for the preparation of self assembling materials, *Proc. Natl. Acad. Sci. U S A*, **99**, pp. 5133–5138.
25. Darr, J. A., and Poliakoff, M. (1999). New directions in inorganic and metal-organic coordination chemistry in supercritical fluids, *Chem. Rev.*, **99**, pp. 495–541.
26. Yoganathan, R. B., Mammucari, M., and Foster, N. R. (2010). Dense gas processing of polymers, *Polym. Rev.*, **50**, pp. 144–177.
27. Cansell, F., Aymonier, C., and Loppinet-Serani, A. (2003). Review on materials science and supercritical fluids, *Curr. Opin. Solid State Mater. Sci.*, **7**, pp. 331–340.
28. Kazarian, S. G. (2000). Polymer processing with supercritical fluids, *Polym. Sci.*, **42**, pp. 78–101.
29. Salerno, A., Di Maio, E., Iannace, S., and Netti, P. A. (2011). Solid-state supercritical CO_2 foaming of PCL and PCL-HA nano-composite: effect of composition, thermal history and foaming process on foam pore structure, *J. Supercrit. Fluids*, **58**, pp. 158–167.
30. Lian, Z., Epstein, S. A., Blenk, C. W., and Shine, A. D. (2006). Carbon dioxide-induced melting point depression of biodegradable semicrystalline polymers, *J. Supercrit. Fluids*, **39**, pp. 107–117.

31. Hile, D. D., Amirpour, M. L., Akgerman, A., and Pishko, M. V. (2000). Active growth factor delivery from poly(D,L-lactide-coglycolide) foams prepared in supercritical CO_2, *J. Controlled Release*, **66**, pp. 177–185.

32. Sanli, D., Bozbag, S. E., and Erkey, C. (2012). Synthesis of nanostructured materials using supercritical CO_2: part I. Physical transformations, *J. Mater. Sci.*, **47**, pp. 2995–3025.

33. Reverchon, E., Adami, R., Cardea, S., and Della Porta, G. (2009). Supercritical fluids processing of polymers for pharmaceutical and medical applications, *J. Supercrit. Fluids*, **47**, pp. 484–492.

34. Reverchon, E., Adami, R. (2006). Nanomaterials and supercritical fluids, *J. Supercrit. Fluids*, **37**, pp. 1–22.

35. Zhang, J., Davis, T. A., Matthews, M. A., Drews, M. J., LaBerge, M., and An, Y. H. (2006). Sterilization using high-pressure carbon dioxide, *J. Supercrit. Fluids*, **38**, pp. 354–372.

36. Giray, S., Bal, T., Kartal, A. M., Kızılel, S., and Erkey, C. (2012). Controlled drug delivery through a novel PEG hydrogel encapsulated silica aerogel system, *J. Biomed. Mater. Res. Part A*, **100A**, pp. 1307–1315.

37. Bi, L., Li, D., Liu, M., Jin, J., Iv, R., Huang, Z., and Wang, J. (2010). The influence of approaches for the purification of natural cancellous bone grafts: morphology, microstructure, composition, strength and biocompatibility study, *Mater. Lett.*, **64**, pp. 2056–2059.

38. Tsivintzelis, I., Pavlidou, E., and Panayiotou, C. (2007). Porous scaffolds prepared by phase inversion using supercritical CO_2 as antisolvent: I. Poly(L-lactic acid), *J. Supercrit. Fluids*, **40**, pp. 317–322.

39. Salerno, A., Zeppetelli, S., Di Maio, E., Iannace, S., and Netti, P. A. (2010). Novel 3D porous multi-phase composite scaffolds based on PCL, thermoplastic zein and ha prepared via supercritical CO_2 foaming for bone regeneration, *Comp. Sci. Tech.*, **70**, pp. 1838–1846.

40. Woods, H. M., Silva, M. M. C. G., Nouvel, C., Shakesheff, K. M., and Howdle, S.M. (2004). Materials processing in supercritical carbon dioxide: surfactants, polymers and biomaterials, *J. Mater. Chem.*, **14**, pp. 1663–1678.

41. Tai, H., Mather, M. L., Howard, D., Wang, W., White, L. J., Crowe, J. A., Morgan, S. P., Chandra, A., Williams, D. J., Howdle, S. M., and Shakesheff, K. M. (2007). Control of pore size and structure of tissue engineering scaffolds produced by the supercritical fluid processing, *Eur. Cells Mater.*, **14**, pp. 64–77.

42. Salerno, A., Iannace, S., and Netti, P. A. (2008). Open-pore biodegradable foams prepared via gas foaming and microparticulate templating, *Macromol. Biosci.*, **8**, pp. 655–664.

43. Kiran, E. (2010). Foaming strategies for bioabsorbable polymers in supercritical fluid mixtures. Part I. Miscibility and foaming of poly(L-lactic acid) in carbon dioxide + acetone binary fluid mixtures, *J. Supercrit. Fluids*, **54**, pp. 296–307.

44. Salerno, A., and Domingo, C. (2013). A clean and sustainable route towards the design and fabrication of biodegradable foams by means of supercritical CO_2/ethyl lactate solid-state foaming, *RSC Adv.*, **3**, pp. 17355–17363.

45. Salerno, A., Guarnieri, D., Iannone, M., Zeppetelli, S., and Netti, P. A. (2010). Effect of micro- and macroporosity of bone tissue three-dimensional-poly(e-caprolactone) scaffold on human mesenchymal stem cells invasion, proliferation, and differentiation in vitro, *Tissue Eng. Part A*, **16**, pp. 2661–2673.

46. Salerno, A., Iannace, S., and Netti, P. A. (2012). Graded biomimetic osteochondral scaffold prepared via CO_2 foaming and micronized NaCl leaching, *Mater. Lett.*, **82**, pp. 137–140.

47. Dvir, T., Timko, B. P., Kohane, D. S., and Langer, R. (2011). Nanotechnological strategies for engineering complex tissues, *Nat. Nanotech.*, **6**, pp. 13–22.

48. Liu, X., and Ma, P. X. (2009). Phase separation, pore structure, and properties of nanofibrous gelatin scaffolds, *Biomaterials*, **30**, pp. 4094–4103.

49. Yang, F., Murugan, R., Ramakrishna, S., Wang, X., Ma, Y., and Wang, S. (2004). Fabrication of nano-structured porous PLLA scaffold intended for nerve tissue engineering, *Biomaterials*, **25**, pp. 1891–1900.

50. Shao, J., Chen, C., Wang, Y., Chen, X., and Du, C. (2012). Early stage evolution of structure and nanoscale property of nanofibers in thermally induced phase separation process, *React. Funct. Polym.*, **72**, pp. 765–772.

Chapter 15

Polymer Nanocomposites and Nanocomposite Foams in Compressed CO_2

David L. Tomasko and Hrishikesh R. Munj
Department of Chemical and Biomolecular Engineering, Ohio State University, Columbus, OH 43210, USA
tomasko.1@osu.edu

Polymer-based nanocomposites have extraordinary physical, chemical, and mechanical properties, which makes them potential candidates to replace most of the conventional materials in several fields. the nanofiller–polymer interface in the composite plays an essential role in altering properties. Fine dispersion of nanoparticles in the polymer matrix results in a large interfacial region and improved properties. High-pressure CO_2 has specific molecular interactions with many polymers and acts as a green polymer-processing agent. CO_2 can assist in polymer nanocomposite processing in two major ways, (1) improved dispersion of nanoparticles in the polymer matrix and (2) foaming of polymer nanocomposites. CO_2-aided plasticization of polymer matrices helps in better distribution of

Supercritical Fluid Nanotechnology: Advances and Applications in Composites and Hybrid Nanomaterials
Edited by Concepción Domingo and Pascale Subra-Paternault

ISBN 978-981-4613-40-8 (Hardcover), 978-981-4613-41-5 (eBook)
www.panstanford.com

nanofillers. Also, the rapid depressurization method causes breaking of agglomerated nanofillers for complete dispersion. In the case of nanocomposite foaming processes, nanofillers act as heterogeneous nucleating agents to yield small and uniform cell size distribution. Both processes dispersion of nanoparticles and foaming involve several complex factors that ultimately affects the characteristics of the product. This chapter gives a brief idea about nanofiller–polymer–CO_2 interactions, thermodynamics, and processing aspects in the CO_2-based processing of polymer nanocomposites.

15.1 Introduction to Polymer Nanocomposites

Polymers play an essential role in chemical, biomedical, materials, and pharmaceutical industries. Although several kinds of polymers have been used for various applications, certain modifications are necessary to improve properties of polymers for specific applications. Crosslinking, copolymerization, polymer blending, and addition of functional groups by reactions are some of the common ways to modify polymer behavior [1]. Addition of ceramics, metals or minerals as fillers can generate numerous materials, called polymer composites, with unique physical properties [2–4]. On the basis of the size of filler elements polymer composites can be categorized into microcomposites and nanocomposites. General definitions for these composites can be made as "a multicomponent system in which the major constituent is a polymer and the minor constituent has at least one dimension below 100 um and 100 nm for microcomposite and nanocomposites respectively." A number of studies have proved that when size of a material is <100 nm, they exhibit unique optical, electrical, and mechanical properties [5, 6]. Surface area/volume is a critical factor in deciding effectiveness and properties of the filler element. As shown in Fig. 15.1, this ratio is dominated by diameter for spheres and fibers and by thickness for the platelets. Changing these critical dimensions from micron to nanorange affects the surface area/volume ratio by 3 orders of magnitude [7]. Incorporation of such nanoparticles in the polymer matrix imparts their properties to the polymer nanocomposites (PNCs). Due to extraordinary properties of PNCs as compared to conventional alternatives, they are future of polymer-based industry.

In this chapter, we will discuss different aspects related to the high-pressure CO_2-assisted processing of PNCs.

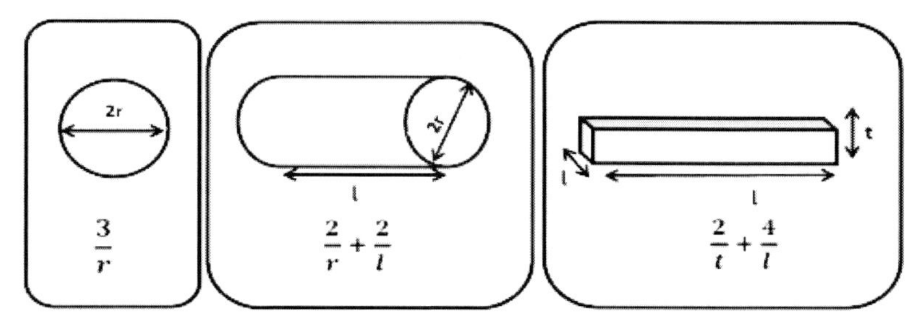

Figure 15.1 Surface-area-to-volume ratio for nanoparticles with a particulate (or spherical) shape (left), a cylindrical shape (center), and a lamellar shape (right) [7].

15.1.1 Polymer–Nanoparticle Interface

Various articles have been published that proved either experimentally or theoretically that the nanofiller–polymer interface affects the viscoelastic properties of the composite. Relaxation characteristics of PNCs depend on volume and network of interfaces in the composite [8]. As polymer chains approach nanoparticles in a composite, the radius of gyration increases for low loadings. The behavior of polymer chains near nanofiller depends on the distance from the particles and is highly dependent on entropic interactions [9–11]. Figure 15.2 shows polymer chain arrangements in the presence of nanoparticles. The interface region plays a key role in stress carrying potentials of carbon nanotubes (CNTs) nanocomposites and redistribution of internal stresses during CNT debonding in the composite [12]. It was shown that the incorporation of CNTs in PNCs increases modulus but the effective loss factor of the composite decreases with increase filler volume fraction due to anomalous behavior of interface in the PNCs [13]. Latest investigations from research about the interfacial region reveal strong energetic interactions between polymer chains and the nanoparticle surface, which leads to alteration of thermomechanical properties [14, 15]. Several new techniques are under development to measure the interface thickness and properties of PNCs [16].

However, a few studies have concluded no effect of the interface region on the properties of PNCs [17]. Recently, a 3D finite-element model has been developed that predicts nanocomposite interface properties based on filler geometry and volume fraction [18]. In the case of filler-polymer properties, spherical nanoparticles have insignificant effects on composite properties as compared to cylindrical and disc fillers. However, spherical nanoparticles have the greatest effect on the interface properties as compared to cylindrical and disc-shaped fillers. In the case of cylindrical and disc fillers, a critical aspect ratio exists beyond which composite properties remain unchanged. This critical aspect ratio is dependent on polymer–filler interactions and volume fraction [19].

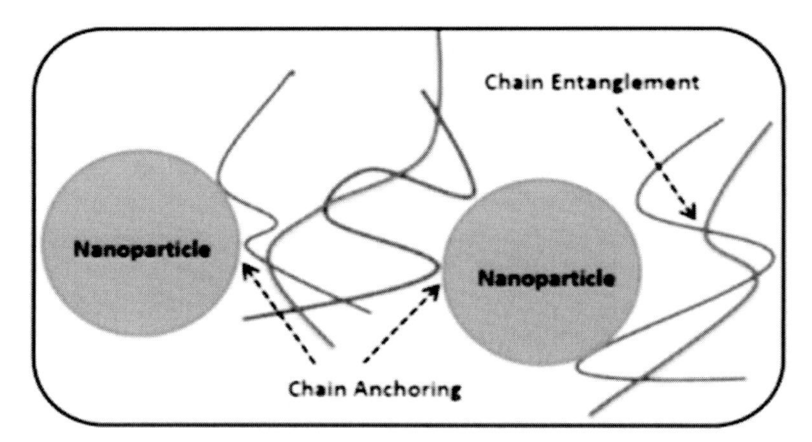

Figure 15.2 Schematic of the interface region in polymer nanocomposites [20].

15.1.2 Dispersion of Nanoparticles in a Polymer Matrix

Considering all research on PNCs, the distribution of nanoparticles in the composite is of the greatest importance. In the case of nanoparticles, uniform dispersion is a challenge due to strong interactions among them leading to agglomeration. Dispersion in the PNCs allows categorizing them into three groups [21]: microcomposites, intercalated nanocomposites, and exfoliated nanocomposites (Fig. 15.3). In the case of microcomposites, polymer chains are unable to penetrate within layers of nanoparticles. Hence, more loading is required to achieve a large interface. When polymer chains penetrate the layered structure of nanoparticles to a limited extent, intercalated nanocomposites are obtained. These composites have larger interface and better properties than microcomposites for the same loading. Ideal nanocomposites are exfoliated nanocomposites where nanoparticles are well mixed in the polymer matrix and layers between the nanoparticles are expanded due to complete penetration by polymer chains. Various studies have investigated thermodynamics behind the dispersion of nanocomposites in the polymer matrix [22]. A simple lattice-based model was used to understand dispersion that concluded enthalpy of the mixing controls the dispersion in spite of loss of conformational entropy.

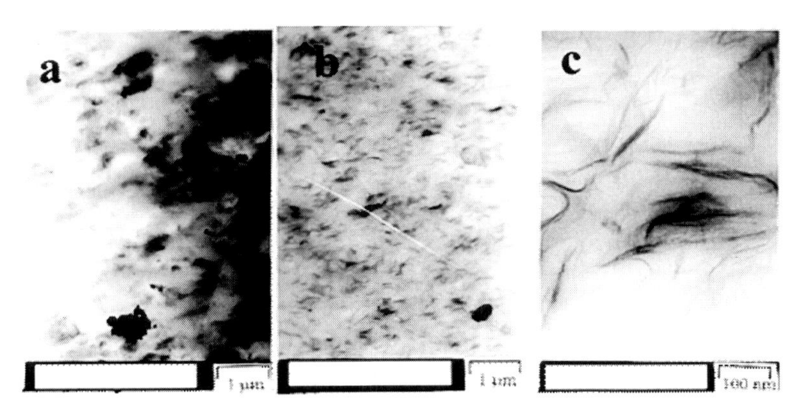

Figure 15.3 Three states of nanoparticle dispersion in PNCs: (a) aggregation, (b) intercalation, and (c) exfoliation [23].

15.1.3 Nanofiller Surface Chemistry

Manipulation of the surface chemistry of nanoparticles aids in reducing particle–particle interactions (agglomeration) and increases particle–polymer compatibility to cause better dispersion in the polymer matrix. The easiest way to understand the surface modification of inorganic nanofillers is addition of organic layer around the particles to enhance chemical interactions between polymer chains and nanoparticles. Several physical and chemical methods have been proposed to alter the surface chemistry of nanoparticles [24]. Although numerous studies have been carried out to change the surface properties to achieve specific interactions between polymer and nanoparticles, a substantial review of this literature is out of the scope of this chapter.

15.2 Fundamentals of Polymer Nanocomposite Foams

The foaming process involves two primary steps, nucleation and growth. Nucleation is the beginning of the phase separation from a homogeneous polymer/blowing agent mixture. Then during the growth phase, nuclei grow to reach thermodynamic and mechanical equilibrium. Various physical characteristics of the polymer such as viscosity, gas solubility, surface tension, and glass transition temperature have prominent influence on the nucleation and growth. In addition, these properties are interrelated, which makes foaming a complex phenomenon [25]. In the later part of the chapter we will discuss processing aspects of foaming processes in depth.

Various techniques have been investigated to foam nanocomposites that have improved properties along with low density. Heterogeneous nucleation is involved in nanocomposite foaming where small quantities of nanoparticles in the polymer act as nucleation sites. Foams are classified according to cell size or average diameter of cells, surface area (m^2/g), and cell density (number of cells per unit volume, cells/cm^3). Microcellular foams (cell size < 10 mm, surface area 10–20 m^2/g, and cell density 10^7–10^9 cells/cm^3) and ultracellular foams (cell size < 0.1 mm, surface area 100–400 m^2/g, and cell density between 10^9–10^{12} cells/cm^3) are two major categories in foams based on these physical characteristics [26]. Nanocomposites aid in producing ultracellular foams under moderate conditions due to the presence of nanoparticles.

15.2.1 Supercritical CO_2 in Nanocomposite Foaming

Supercritical carbon dioxide ($scCO_2$) has been used as a solvent for polymer processing due to its unique interactions with polymers at the molecular level. CO_2 acts as a Lewis acid (electron acceptor) in the presence of polymers and as a result has considerable solubility in many polymers. However, the low polarity of CO_2 results in low solubility in highly polar polymers. Furthermore, free volume and stiffness of the polymer backbone are essential factors in determining CO_2 solubility. $scCO_2$ has been used as a green blowing agent for foaming to replace hazardous solvents and gases. Various methods like antisolvent phase separation, crystallization of swollen/crosslinked polymers, and direct foaming have been used to produce foams using CO_2 [27]. Furthermore, CO_2 has been used in batch as well as continuous foaming processes. However, low solubility and high diffusivity of CO_2 in pure polymers reduce the ability to control the foaming process. In the case of nanocomposites, fine distribution of nanoparticles not only serves as heterogeneous nucleation sites but also decreases gas diffusivity in the composite [28].

15.2.2 Effect of Nanoparticles on Foaming

Nanoparticles act as heterogeneous nucleation sites and lower the energy barrier for nucleation as compared to homogeneous nucleation. Several factors like nanoparticle shape/size, loading, and surface treatment play a major role in determining the foam morphology [29, 30]. Although the tiny size of nanoparticles results in a higher density of potential nucleation sites, it reduces the nucleation efficiency, defined by the ratio of the measured cell density to the potential nucleant density. In most nanocomposite foams, it is observed that cell density is lower than the potential nucleant density, indicating that either nanoparticles are poorly dispersed or nanoparticles are not favored for nucleation energetically [31]. The nanoparticles affect rheological behavior of the composite by increasing viscosity. This assists in preventing cell coalescence and therefore reduced cell size during foaming [32].

15.2.2.1 Effect of shape/size

As the size of the filler is reduced it results in more nucleation sites for same loading. However, beyond a certain critical size the curvature effect becomes significant and there is an increase in the energy barrier for nucleation. This combined effect results in increasing the pore density but reducing the nucleation efficacy [33]. The aspect ratio of nanoparticles is an essential factor that controls cell density. Polymethylmethacrylate (PMMA)/multiwalled carbon nanotube (MWCNT) nanocomposites show saturation pressure can influence the cell density for different aspect ratios. For lower pressures, a higher aspect ratio results in lower cell density since nucleation occurs only at the ends of such nanoparticles [34], whereas in the case of higher pressures, sidewalls act as nucleation sites and a higher cell density is obtained (Fig. 15.4).

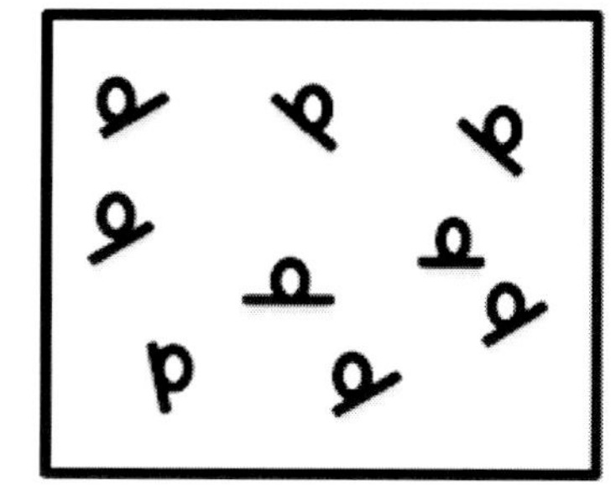

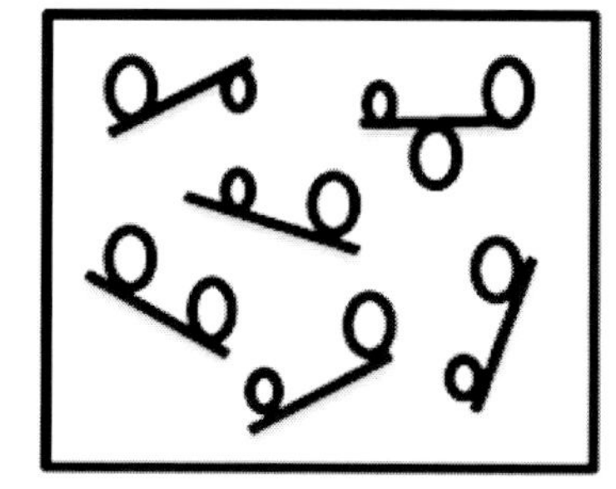

Figure 15.4 Expected nucleation mechanism for CNT-based PNCs. (a) A small CNT acts as a single nucleation site, and (b) a long CNT acts as a site for multiple nucleations [34].

15.2.2.2 Effect of distribution

A fine dispersion of nanoparticles can assist in the formation of nucleation centers. Nonuniform distribution of nanoparticles causes polymer-rich and particle-rich regions, which results in broad or multimodal cell size distribution [35]. Intercalated and exfoliated polystyrene (PS)/clay nanocomposite foams have shown that exfoliated nanocomposite foams have higher nucleation rates. Similar observation was made for low-density polyethylene (LDPE)/ clay nanocomposite where unintercalated and intercalated foams were compared (Fig. 15.5). Although agglomerates show a lower energy barrier, nucleation efficiency was low in these regions. A lower energy barrier results in quicker gas escape during foaming and results in a lower expansion ratio. In the case of better dispersion, gas escape is inhibited and a high expansion ratio is achieved [36]. This suggests nucleation efficiency of nanoparticles is higher in the separated form than the agglomerated form [37].

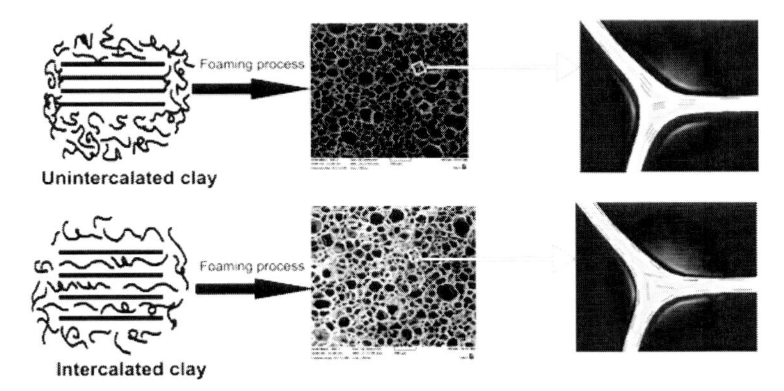

Figure 15.5 Scematic presentation of the mutual effect of clay dispersion and the foaming process [36].

15.2.2.3 Effect of loading

The effect of clay loading has been investigated for poly(D,L-lactic acid) (PLA)/clay nanocomposites. The study concludes that the mean pore diameter decreases, whereas bulk foam density and cell density increase with increase in loading until 5 wt%. With a further increment to 15 wt% there was no change in above characteristics (Fig. 15.6) [38]. Similar results found in the case of poly(caprolactone) (PCL)/clay nanocomposites where 5% loading of clay has improvement in the foam morphology. As clay concentration is increased above 5%, foam morphology approaches pure PCL foams [39]. At a lower concentration of nanofillers, individual particles act as the nucleation site. Hence, cell density increases and cell size reduces linearly with the nanofillers' concentration until a critical limit. Beyond a critical concentration, cell coalescence becomes a dominant factor and there is no further improvement in the foam morphology. Diffusivity of CO_2 in the PNCs is greatly affected without changing the solubility of CO_2. Thus, during the bubble growth phase, mass transfer of CO_2 from polymer to bubble is controlled by clay loading. The early stage of bubble growth is controlled by viscosity, whereas the later stage is dominated by diffusion of CO_2 [40].

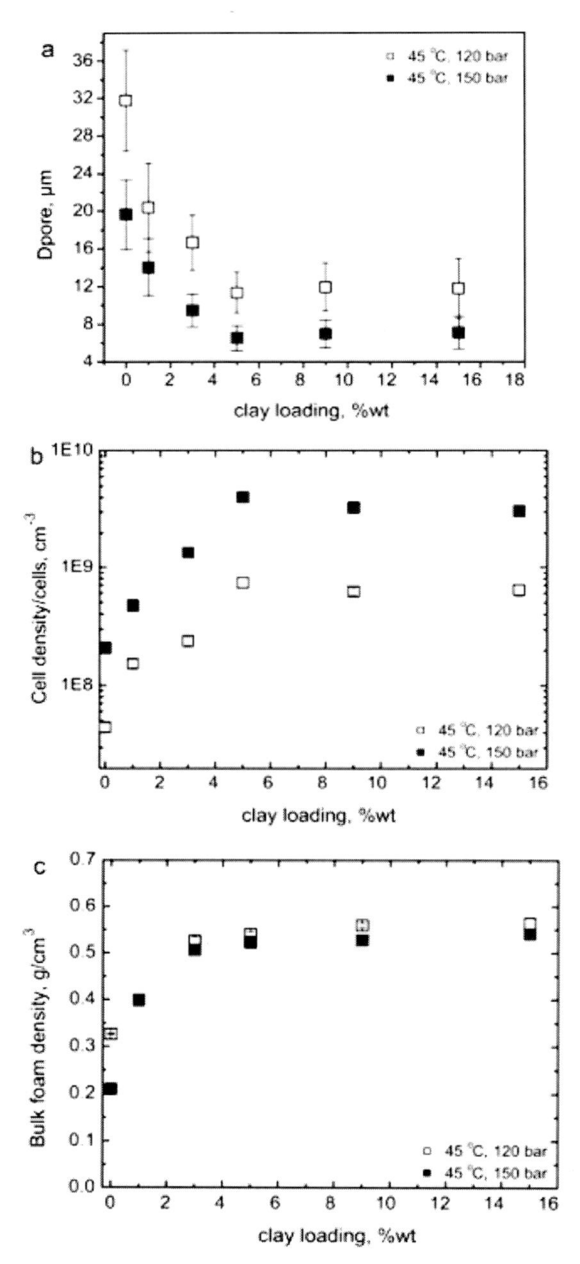

Figure 15.6 Effect of clay loading on the (a) average pore diameter, (b) the population cell density, and (c) the bulk foam density [38].

15.2.2.4 Effect of surface chemistry

The surface chemistry of nanoparticles has most prominent effect on the foaming process. Tailoring the surface chemistry of the nanoparticles provide a great opportunity to finely tune the energy barrier for nucleation and hence control foam morphology [33]. Strength of the interface determines the free-energy barrier for nucleation. A weak interface enhances the nucleation process by forcing the interface apart easily [41]. Influence of the MWCNT aspect ratio cell density can be reduced by surface modification to tune the critical free energy for nucleation. This results in similar nucleation at the ends and the walls of MWCNTs [42]. Reduction in surface free energy and critical nucleus radius is achieved through surface modification of silica nanoparticles with fluoroalkanes [33]. Since CO_2 has high affinity toward fluorinated groups, it further reduced nucleation free energy. In another study of PLA/clay nanocomposites, the effect of the length of the surfactant chain was investigated. As organic surfactant chain length is increased, interaction between the inorganic clay and organic polymer matrix increases. An increased chain length shows significant reduction in the cell size [38]. Grafting CO_2-philic groups on the nanoparticles increases CO_2–nanoparticle interactions and hence the nucleation rate [27].

15.3 Thermodynamic Aspects in Nanocomposite Foams

In this section we will discuss the thermodynamic aspects of the nanocomposite foaming process. As explained earlier, the free-energy barrier for heterogeneous nucleation is lower than homogeneous nucleation. The free-energy barrier is an essential factor in deciding cell size distribution. In the case of homogeneous nucleation, there is limited control over cell size distribution since only operation parameters can be tuned. Solubility limitations of homogeneous nucleations are specific to the polymer/blowing agent system. Heterogeneous nucleation avoids these limitations by using micro-/nanoscaled nucleating agents. Equations 15.1 and 15.2 show nucleation rates for homogeneous and heterogeneous nucleation processes.

$$N_{\text{Homo}} = f_0 C_0 e^{\frac{-nG_{\text{Homo}}}{kT}} \tag{15.1}$$

$$N_{\text{Hetero}} = f_1 C_1 e^{\frac{-nG_{\text{Hetero}}}{kT}} \tag{15.2}$$

N_{Homo} and N_{Hetero} : Nucleation rates for homogeneous and heterogeneous nucleation processes, respectively

f_0 and f_1: Frequency factors representing the frequency that gas molecules joining the embryo nucleus in homogeneous and heterogeneous nucleation, respectively

C_0: Concentration of the gas molecules

C_1: Concentration of heterogeneous nucleation sites

nG_{Homo}: Gibbs free energy associated with the formation of a nucleus

nG_{Hetero}: Gibbs free energy for heterogeneous nucleation

Particles can reduce the free-energy barrier for cell nucleation process, increase the nucleation rate, and aid in reducing spread of cell size distribution [43]. In the case of nanocomposites, the interface acts as a high-energy region as compared to pure polymer. Gibbs free energy for nucleation in the interface is lower causing preferential nucleation in this region. Nucleation efficiency can be enhanced using nucleating agents that energetically favor nucleation and are dispersed uniformly in the polymer matrix. Dispersion of nanoparticles reflects the area of contact between nanoparticles and polymer/CO_2 [31]. In the case of exfoliated nanocomposites a higher interfacial area exists for CO_2 adsorption and cell nucleation. The polymer plays a key role in deciding the nucleation rate. A polymer with higher affinity for CO_2 can reduce interfacial tension between gas and particle and subsequently decrease the contact angle. This results in reduction of the nucleation free energy and boosts the nucleation rate [27]. Equations 15.3 and 15.4 indicate the relation of Gibbs free energy of nucleation with various factors in homogeneous and heterogeneous nucleation processes, respectively.

$$nG_{\text{Homo}} = \frac{16\pi\sigma^3}{3nP^2} \tag{15.3}$$

$$nG_{\text{Hetero}} = \frac{16\pi\sigma^3}{3nP^2}\frac{f(m,w)}{2} \tag{15.4}$$

σ: Interfacial tension at the gas bubble–polymer interface
ΔP: Pressure exerted by the gas on the cell walls
$\frac{f(m,w)}{2}$: Gibbs free-energy reduction factor. If $\frac{f}{2}$ is less than 1, nanoparticles favor kinetics and thermodynamics of gas bubble formation [41]. $f(m, w)$ is calculated using Eq. 15.5.

$$f(m,w) = 1 + \left(\frac{1-mw}{g}\right)^3 + w^3\left[2 - 3\left(\frac{w-m}{g}\right) + \left(\frac{w-m}{g}\right)^3\right] + 3mw^2\left[\frac{(w-m)}{g} - 1\right] \quad (15.5)$$

$m = \cos\theta$
θ: Contact angle (Fig. 15.7)
$w = R/r^*$: Relative curvature of the nucleant surface
R: Filler radius
r^*: Critical radius of the nucleus = $2\sigma/\Delta P$
$g = (1 + w^2 - 2mw)^{0.5}$

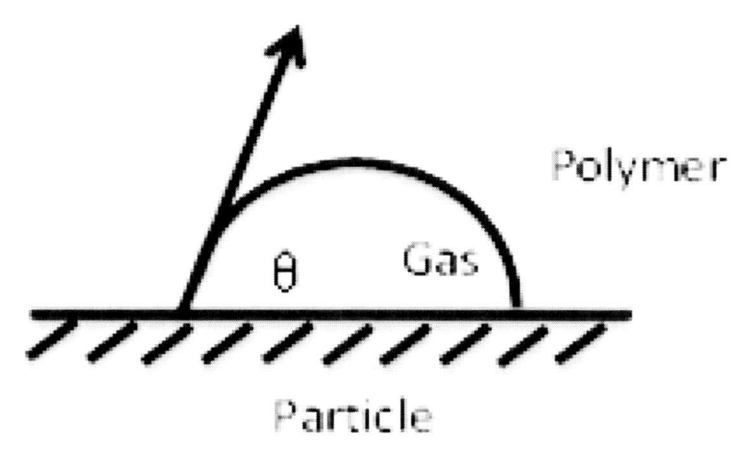

Figure 15.7 Schematics of the polymer–gas–particle interface [35].

f varies with the surface curvature and the contact angle between the polymer, gas, and nanoparticle. Figure 15.8 represents the effect of both curvature ($1/w$) and contact angle (m) on the reduction factor. A small surface curvature and a small contact angle result in higher energy reduction and increased nucleation rates. m is unity in the case of homogeneous nucleation. Due to poor wetting of nanoparticles m is lower than unity in the case of nanocomposites

[44]. According to the above equations increasing the surface curvature leads to a higher nucleation free-energy barrier. Hence flat surfaces are more efficient that other geometries in the case of a constant contact angle [33].

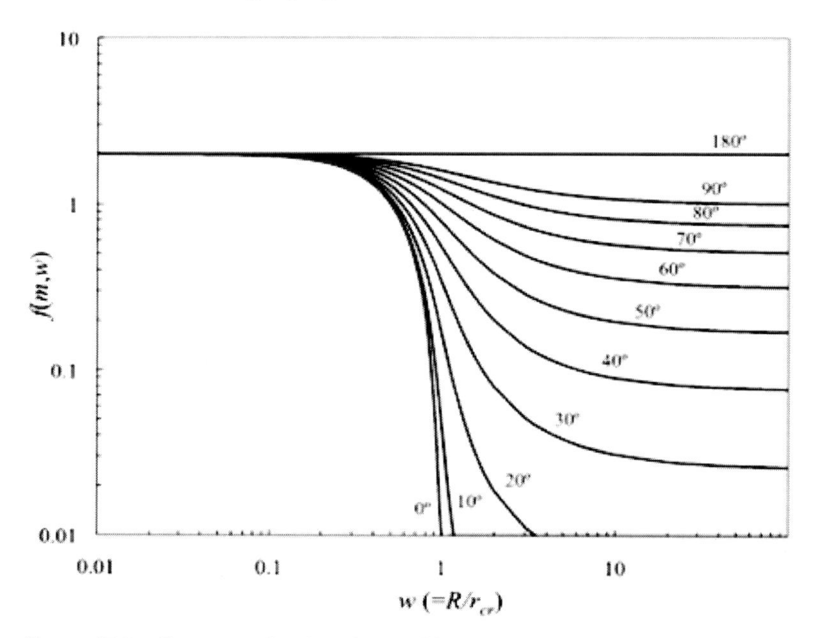

Figure 15.8 Energy reduction factor (f) as a function of relative surface curvature ($1/w$) and contact angle (θ) [33].

Classical nucleation theory is still a common empirical approach for modeling and calculating nucleation rates. However, the use of the bulk interfacial tension in classical theory instills a thermodynamic inconsistency that creates large errors in predicted nucleation rates. Recently, a scaling theory approach has been explored for homogeneous nucleation rates in polymer–CO_2 systems [45]. Scaling theory exploits the fundamental basis of classical theory but relies on a scaled value of surface tension rather than a constant. Through a relatively straightforward correlation to the phase diagram of the system of interest, a more self-consistent calculation of nucleation rates is possible. Unfortunately, this scaling approach has not yet been extended to heterogeneous nucleation of the type discussed above.

15.4 CO_2–Nanoparticle Interactions

Adsorption of CO_2 on nanoparticles is considerably lower as compared to solubility in the polymer [46]. However, surface modification of nanoparticles can enhance CO_2 adsorption and wettability. The surface chemistry of MWCNTs was modified using polymeric dendrimers, which results in not only better dispersion but also an increase in nucleation efficiency [47]. PS brushes were grafted on nanoparticles to enhance the nucleation rate in PS nanocomposites [48]. In another study, functionalized graphene oxide was used to increase CO_2 interactions and reduce the free-energy barrier for nucleation [49]. Silica nanoparticles modified with fluoroalkanes reduced the surface free energy and increased CO_2 interactions due to presence of fluorine groups [33].

15.5 CO_2-Assisted Dispersion of Nanoparticles in Polymer Matrices

In this section, we will discuss the effect of high-pressure CO_2 on dispersion of nanoparticles in the PNCs to achieve a more uniform distribution. Several studies have investigated the intercalation/exfoliation capability of CO_2 in different polymer matrices. There are three major approaches to achieve nanoparticle distribution in the composites: intercalation of prepolymer or polymer from solution, melt intercalation, and in situ intercalative polymerization [50]. Among these methods, melt intercalation is considered most viable from an industrial perspective. Basic procedures of CO_2-aided nanoparticle dispersion can be seen in Fig. 15.9 [50]. CO_2 is contacted with nanoparticle via direct exposure or with the polymer composite and the following mechanism is proposed for the intercalation. During the second step of soaking, CO_2 and polymer chains diffuse between nanoparticle layers. CO_2 causes a reduction of viscosity and increase in mobility of polymer chains, which enhances polymer diffusion. Molecular interactions between nanoparticles and CO_2/polymer restrict the extent of diffusion. Depressurization of CO_2 results in separation of layers due to gas expansion. During expansion, polymer chains are trapped in between nanoparticle platelets, preventing reformation of layers. As interactions between nanoparticles and CO_2 play an important role, several attempts have been made to exploit their impact. Natural clays have been modified with CO_2-philic groups to obtain a high level of CO_2 interaction and sorption inside the layers. If nanofillers are inorganic in nature, it causes hurdles to delamination inside organic polymer matrices. CO_2 is advantageous in such cases where high mobility of polymer chains in the presence of CO_2 assists dispersion. The viscoelastic response of poly(vinylmethylether) (PVME)/clay nanocomposites was studied to explore the effect of CO_2 on dispersion of clay. CO_2 shows a significant effect on the intercalation and thus crossover frequency and storage moduli of PVME/clay nanocomposites [51]. High-pressure CO_2 has been used to delaminate graphite (GR) to form graphene. It is considered to be one of the most viable processes for large quantity graphene production with high purity [52, 53]. Another study shows that when CO_2 is used as a polymerization medium. It allows higher clay loading and extracts residual monomer resulting in a ready-to-use product [54].

Polymerization of D,L-lactide was carried out with clays in the presence of CO_2 to achieve a higher degree of exfoliation [55]. Most of the organoclay galleries are composed of intercalated surfactants (tail-to-tail bound molecules). It has been shown that scCO_2 alters the surface chemistry of organoclays. This results in plasticization and possible cation exchange of tail-to-tail bound molecules. CO_2 shows gallery expansion where surfactant is exposed (paraffin complex arrangement). Both pressure and temperature are effective in surface modification. Also after CO_2 treatment, Lewis acid/base sites are occupied due to CO_2 sorption [56]. An optimum concentration level of CO_2 has been observed for dispersion of clay in polypropylene (PP), which has a positive influence on rheological properties and microstructure. A linear trend is observed in degree of intercalation and the basal spacing between silicate layers with increasing CO_2 concentration in the composite [57, 58]. Recently it was shown that addition of scCO_2 during melt-compounding enhances dispersion of clay in a ternary nanocomposite and improves the mechanical properties [59]. Molecular dynamics simulation also confirms scCO_2 induces gallery swelling of organoclays [60].

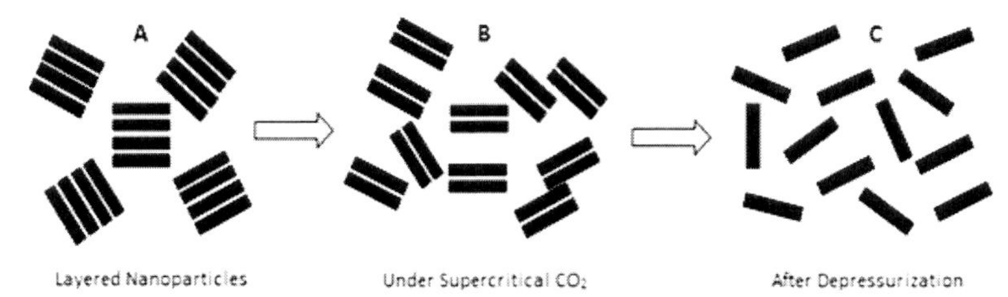

Figure 15.9 Supercritical CO_2-assisted dispersion of nanoparticles. (A) Nanoparticles and the polymer are mixed where nanoparticles are in agglomerated form. (B) Supercritical CO_2 aids in increasing spacing between nanoparticles. (C) Rapid depressurization results in complete dispersion of nanoparticles in the polymer matrix [50].

Along with dispersion of clays, CO_2 has been used to disperse metal and metal oxide nanoparticles in polymers [61–66]. scCO_2 has been used to coat polymers on CNTs, which shows CO_2 is a potential green route to change polymer morphology in the presence of CNT [67]. Recently, scCO_2) was used to achieve CNT expansion. Scanning electron microscopy shows a prominent increase in volume and delamination of CNT due to scCO_2 treatment. Furthermore, the degree of CNT expansion linearly increases with CO_2 pressure and exposure time [68]. In a different study, ultrasound was used in a scCO_2 medium to enhance nanoparticle dispersion [69]. Finally, three different methods for preparing scCO_2-based nanocomposites—blending, sol–gel, and in situ polymerization—have been recently reviewed, showing the viability of these techniques for achieving a high degree of dispersion [70].

15.6 Representative Examples of Nanocomposite Foams

15.6.1 Thermoplastic Nanocomposite Foams

15.6.1.1 Polyethylene

CO_2 has limited solubility in polyethylene (PE) relative to other polymers due to a lack of interactive functional groups (e.g., carbonyl groups). Hence, it is difficult to foam PE efficiently using high-pressure CO_2. Nonetheless, the foaming potential of scCO_2

has been explored for high-density polyethylene (HDPE)/clay nanocomposites. Results show that an increase in clay content affects the rheological properties of the nanocomposite. For a lower clay concentration viscosity is low, which causes cell coalescence. As clay content is increased nanocomposites become more rigid, which prevents cell coalescence and a smaller cell size is obtained [71]. $scCO_2$ has been used to foam LDPE/clay nanocomposites in a continuous extrusion process. Beyond a critical clay concentration, clay agglomeration was observed resulting in increase of cell density [72]. HDPE/palladium (Pd) nanocomposites were foamed using $scCO_2$ to prepare porous polymer-stabilized metals for advanced heterogeneous catalytic reactions [73]. A PE–octene elastomer/clay nanocomposite has been foamed using $scCO_2$. Microcellular foams with small cell size and high cell density were achieved with $scCO_2$ processing [74].

15.6.1.2 Polypropylene

Similar to PE, the solubility of the CO_2 in PP is low due to lack of favorable interactions. Still, CO_2 foaming of PP copolymer/clay nanocomposites has been investigated. PP copolymers can be foamed easily under milder conditions and allow process flexibility as compared to pure PP foaming using CO_2 [75]. The foaming process seems to enhance the nanoparticle dispersion in the matrix by exfoliation of clay agglomerates. In the case of PP-based nanocomposites, nanoparticles play a dual role of heterogeneous nucleating agents during foaming as well as crystallization sites during cooling of foamed structures [76]. In another batch foaming study, high-melt-strength PP and PP/clay nanocomposites foams were explored. The degree of cell coalescence and closed cell formation was lowest when foaming temperatures were comparable to melting temperatures. When the foaming temperature was higher than the melting temperature, the degree of cell coalescence and collapse was high. It was speculated that optimum clay loading for PP nanocomposites is equivalent to the percolation threshold for the nanocomposites. Furthermore, the strain-hardening effect was found to be quick and pronounced for PP nanocomposites, which helps in controlling foam cell size [77]. PP nanocomposite foams have been investigated with visual observation of cell nucleation and growth in situ. This study reveals that cell density increases and bubble growth decreases with increase in clay content. Also, the crystal structure of PP has a significant impact on the foaming process [78]. As discussed earlier, clay particles enhance crystallization and reduce cell coalescence and collapse [79]. In summary, addition of nanoparticles enhances the foamability of PP as compared to pure PP. However, the degree of enhancement is not as high as that of amorphous polymers [80].

15.6.1.3 Polymethylmethacrylate

PMMA has high solubility of CO_2 due to interactions of CO_2 with carbonyl groups and a flexible polymer backbone. Thus, foaming of PMMA with scCO_2 is an easy process as compared to PE or PP. PMMA/MWCNT nanocomposites were foamed using CO_2 and the effect of subsequent hot water immersion was investigated. Bubble density of pure PMMA was smaller than composite foams due to a heterogeneous nucleation driven mechanism. The effect of two different aspect ratios of MWCNTs was explored, which shows shorter tubes results in higher bubble density than longer tubes. Ends of tubes are considered nanoscale cylindrical pores that trap CO_2. If MWCNTs are treated chemically to alter surface chemistry, adsorption of CO_2 at the ends can be tuned. Also ends of MWCNTs are considered to have a lower energy barrier for heterogeneous nucleation as compared to sidewalls. The ratio of bubble density ($D_{\text{low aspect ratio}}/D_{\text{high aspect ratio}}$) for lower-to-higher-aspect-ratio MWCNTs was found to be related to saturation pressure. Although bubble density increases with increasing saturation pressure, CO_2 sorption is same in pure PMMA and composite. It is speculated that due to CO_2 sorption at the ends of tubes, the free-energy barrier for foaming is altered. This reflects in reducing $D_{\text{low aspect ratio}}/D_{\text{high aspect ratio}}$ with increasing CO_2 pressure. Homogeneous nucleation has an almost negligible role in nanocomposites foaming. If these foams were treated with hot water (65°C) after depressurization, the polymer softens and CO_2 solubility reduces. This results in supersaturation of the composite and phase separation occurs, causing CO_2 to move into existing bubbles. Hence, hot water treatment increases bubble size but reduces bubble density due to coalescence [34]. In another PMMA/MWCNT foaming study, effect of dispersion of MWCNTs on foam morphology was explored. If nanofillers are not dispersed properly in the polymer matrix, it results in filler-rich and filler-lean zones and causes bimodal foam morphology. Appropriate surface treatment can be adopted to achieve more uniform dispersion and unimodal bubble size [35]. Another batch foaming study shows, dispersion of CNTs in PMMA depends on CNT concentration and surface modification [29]. PMMA/Silica nanoparticles foams were explored for effect of particle size and surface chemistry of nanoparticles. It was found that heterogeneous nucleation was effected by both factors. Reducing filler size increases number of nucleation sites for the same loading but increases the energy barrier due to curvature effects. This causes an increase in bubble density but reduction in nucleation efficiency. Surface modification can reduce the surface free energy and critical nucleus radius. This results in increased cell density and smaller cell size compared to bare silica [33, 81].

15.6.1.4 Polystyrene

PS–CO_2 is one of the most well-studied systems in order to understand and control the foaming process. However, in the case of PS nanocomposite foams these studies cannot be directly extended since heterogeneous nucleation is dominant in nanocomposite systems. In a PS block copolymer, only exfoliated (as opposed to aggregated) layered silicates can contribute in tuning the cell size and cell density [82]. PS/graphene and PS/graphene oxide nanocomposites foamed using scCO_2 show improvements in the foam characteristics as compared to PS foams [83]. Surface modified graphene has higher efficiency to increase nucleation rate as compared to graphene oxide [49, 83]. To improve thermal properties of the PS foams, three different carbon-based nanofillers (activated carbon [AC], carbon nanofibers [CNFs], and GR) were investigated in an extrusion foaming process. In these foams the cell morphology is different in the center as compared to near the surface when CNFs and GR were used. This difference can be explained by understanding the temperature distribution in the extrudate. The central core of the extrudate takes longer time to cool down as compared with the surface. Thus, center of extrudate has more elongated and broken cells due to more gas diffusion and cell growth. CNF- and GR-based PS nanocomposites have distorted cell morphology. This problem can be avoided using water as cosolvent and AC nanofillers [84]. In another study, effects of clay loading on gas escape rate and cell morphology of PS/clay nanocomposites showed that beyond an optimum loading of clay, aggregates are formed in the matrix, leading to less resistance for gas escape and resulting in large cells [85]. In addition, solubility and mixing issues of nanoparticles with PS matrix is discussed in this study. Remixing the nanoparticles in PS by extruding multiple times in the extruder improves nanoparticles mixing in the PS nanocomposites. Various methods (direct compounding, sonication, and in situ polymerization) of graphite dispersion in the PS matrix have been explored. On the basis of the nanoparticle dispersion and foam characteristics, in situ polymerization method was found to be the most effective [86]. Furthermore, surface modification of nanofillers with polymeric additives allows better dispersion in PS and higher nucleation efficiency [87].

15.6.1.5 Polyurathane

Thermoplastic polyurethanes (TPUs) have elastic properties and reprocessing capability due to the absence of covalent crosslinking. Extrusion foaming of TPUs using CO_2 has been explored to optimize processing variables investigating several challenges involved to achieve stable operation and design of the screw extruder [88]. In another study, TPU was foamed using CO_2 in a batch process to investigate effect of hardness of the polymer on the foam characteristics [89]. CO_2-assisted foaming of TPU/clay nanocomposites shows that a two-step foaming (with high foaming temperature ~150°C and followed by water bath immersion) and one-step foaming process (low foaming temperature ~70°C) can be used to obtain small cell size and high cell density [90]. Open pore TPU scaffolds have been prepared using CO_2-based foam injection molding process to examine effect of various parameters on the foam morphology. Two parameters have a prominent effect on the foam morphology, gas content and degree of weight reduction in the injection molding process [91]. In another study, TPU/clay porous fibers were fabricated using $scCO_2$-assisted extrusion process. Optimization of several operating factors and clay content has yielded foamed fibers with microcellular morphology [92]. Although various studies have explored CO_2-assisted foaming of TPU, this procedure is not applicable to thermosetting polyurethane (PU). In most of the hard PU nanocomposites a reactive foaming process is used [93–96].

15.6.1.6 Poly(caprolactone)

Two different routes were explored to foam PCL/clay nanocomposites with $scCO_2$. In the first case, a PCL/clay mixture is added in a high-pressure reactor along with $scCO_2$ to achieve good dispersion and foaming in a one-step process. Another case involves two distinct steps of clay dispersion without CO_2 followed by CO_2 foaming. The study concluded that one-step dispersion and the foaming process result in poor clay dispersion and nonuniform structures [97]. In the case of semicrystalline polymers like PCL, good nanoparticle distribution cannot be achieved with CO_2 alone. Instead, mixtures of CO_2/organic solvents (e.g., ethanol) can enhance nanoparticle distribution and foaming in a one-step process. PCL/hydroxyapatite (HA) nanocomposites have been foamed using $scCO_2$ and results show that controlling thermal history of nanocomposite and depressurization rate gives control over homogeneous nucleation and bimodal cell size distribution [98]. In another study, PCL/HA nanocomposites were foamed using a combination of gas foaming and particle leaching for tissue engineering. A CO_2/N_2 ratio and salt concentration give flexibility to achieve the desired porous structure for cell adhesion and proliferation [99]. However, inclusion of HA has not shown significant change in the CO_2-assisted foaming of PCL [100].

15.6.2 Nanocomposite Foaming Processes

15.6.2.1 Batch foaming

There are two major categories to carry out foaming process, continuous foaming and batch foaming. In the case of batch foaming, the polymer is saturated with an appropriate foaming agent under required pressure and temperature. Rapid depressurization results in the foaming of the polymer. There are two ways to carry out batch foaming. If the saturation temperature is above the glass transition temperature (T_g) then depressurization results in supersaturation and foaming. The foam characteristics depend on saturation pressure, saturation temperature, and pressure drop rate. However, if the saturation temperature is not above T_g, nucleation cannot be achieved just with high pressure drop rates. Instead, after depressurization, the temperature is elevated by placing the sample in the temperature bath to produce foams. In this case, process has higher flexibility to tune the saturation and operating conditions. Nevertheless, it results in thick skin, since elevating the temperature of the gas-saturated polymer affects the diffusion process. The magnitude and rate of temperature change also become important. Usually, the batch foaming process is carried out well below the melting temperatures of polymers. Although batch foaming is flexible in terms of tuning foam quality, saturation times can vary from hours to days, which limits productivity. A semibatch foaming approach has been investigated to increase the productivity [101]. Foam injection molding is a special case of batch foaming to produce foams with specific shapes. The process has commercialized by Trexel (MuCell Molding Technology) is comprised of a plasticizing unit, injection unit, hydraulic unit, clamping unit, and gas delivery unit [102]. Foam characteristics can be controlled with proper mixing inside the unit and operating conditions. Multiple studies have explored various aspects of injection molding of nanocomposite foams using CO_2 [103, 104]. Injection speed is a critical parameter in the operation, which controls mechanical properties of the foams and avoids melt fracture [104]. Although the injection molding process for nanocomposites is not very different from a conventional process, the change in rheological behavior due to presence of nanofillers becomes an issue in the operation [105].

15.6.2.2 Continuous foaming

Extrusion foaming is the most widely used continuous foaming technique in the industry. A typical extrusion foaming process consists of extruder with a blowing agent injection opening and shaping die. A polymer added in the extruder is mixed with the blowing agent using mechanical mixing achieved by a single or twin screw. This polymer/blowing agent homogeneous mixture is then passed through a shaping die with a pressure drop to induce nucleation and create foamed structures with specific shapes. The extrusion foaming process involves several interdependent operating variables like temperature of extruder, pressure, screw speed, etc. (Fig. 15.10). Stable operation is achieved by optimizing these factors to yield required foam characteristics. Extruded foams exiting the die continue to expand until the temperature drops below T_g. In some cases, the extrudate is cooled rapidly to restrict growth and maintain the structure of the foams [106]. Several studies have addressed advantages and challenges in nanocomposite extrusion foaming. All studies conclude that tuning the operating conditions is the key to achieve stable operation and required morphology for nanocomposite foams. In the case of PNCs, the presence of nanofillers greatly affects the rheology of the composite, which ultimately impacts nucleation and growth. Although most of the polymers mentioned in this chapter have been investigated for continuous foaming process, due to significant effect of nanofillers more work is required to better understand nanocomposites of these polymers.

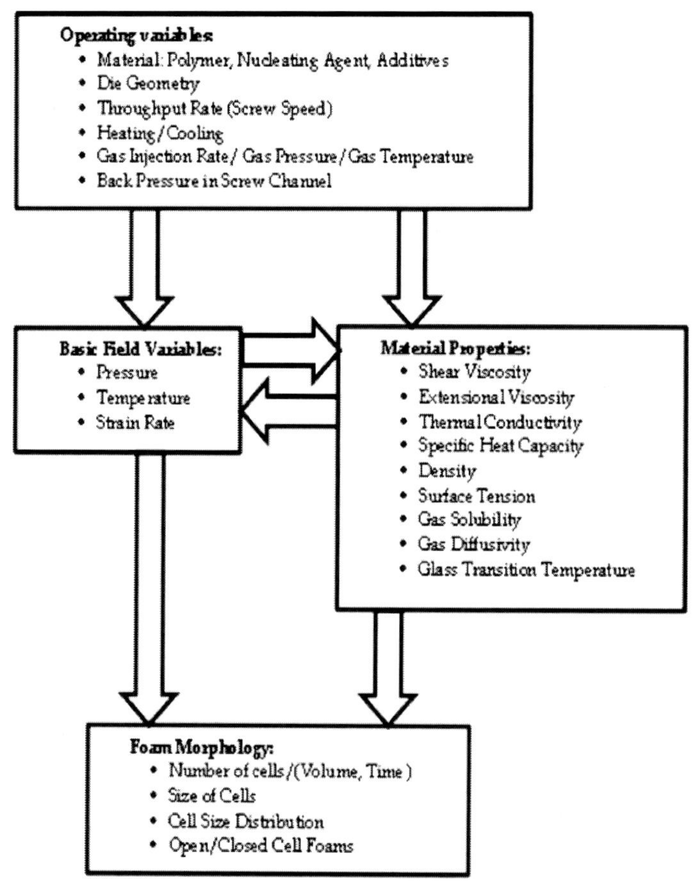

Figure 15.10 Summary of variables affecting the final foam morphology in the foam extrusion process [107].

15.7 Summary

PNCs are receiving significant attention in the field of engineering due to their extraordinary properties. Incorporation of nanofillers shows substantial increase in polymer–filler interface region in the composite as compared to macro-/microfillers. The polymer–filler interface plays a critical role in improving composite properties. Thus, substitution of microfillers with nanofillers gives noticeable improvements in the PNC properties. The polymer–filler interface region has been explored experimentally as well as theoretically. However, more work is needed in this area to conclusively predict polymer interface thickness and its effect on properties. Distribution pattern of nanofillers in the polymer matrix is a key parameter to control interface and composite properties. Intercalated nanocomposites (partial dispersion of nanofillers) show inferior characteristics of composites when compared to exfoliated nanocomposites (complete dispersion of nanofillers). scCO_2 has been used as a green solvent for polymer processing and various studies have used this benign approach to process PNCs as well. In the presence of scCO_2, polymer matrix shows plasticization and a lower glass transition temperature. This facilitates dispersion of nanofillers in the nanofillers in the composite. Furthermore, studies have predicted that rapid depressurization of CO_2 can cause rupture of aggregated nanofillers and aid in exfoliation. The field of PNC foams is developing rapidly due to the ability to replace several conventional materials with lightweight foams with superior properties. CO_2-assisted foaming of PNCs appears to be one of the best ways to fabricate nanocomposite foams with controlled morphology. This book chapter has summarized thermodynamic and processing aspects of nanocomposite foaming using scCO_2. The presence of nanofillers in a polymer matrix alters several viscoelastic properties of the composite, which ultimately affects the foaming process. Although considerable studies have been performed in this field to understand and predict the nanocomposite foaming process, there is a need to investigate this process further and apply existing knowledge to the pilot/commercial scale.

Acknowledgements

This work is supported by a research grant from the National Science Foundation under Grant EEC-0914790. Any opinions, findings, and conclusions or recommendations expressed in this material are those of the authors and do not necessarily reflect the views of the National Science Foundation.

References

1. Abbasi, F., Hamid, M., and Ali-Asgar, K. (2001). Modification of polysiloxane polymers for biomedical applications: a review, *Polym. Int.,* **50**, pp. 1279–1287.
2. Taboas, J. M., Maddox, R. D., Krebsbach, P. H., and Hollister, S. J. (2003). Indirect solid free form fabrication of local and global porous, biomimetic and composite 3D polymer-ceramic scaffolds, *Biomaterials,* **24**, pp. 181–194.
3. Shahinpoor, M., and Kim, K. J. (2001). Ionic polymer-metal composites: I. Fundamentals, *Smart Mater. Struct.,* **10**, pp. 819–833.
4. Vlakh, E. G., Panarin, E. F., Tennikova, T. B., Suck, K., and Kasper, C. (2005). Development of multifunctional polymer–mineral composite materials for bone tissue engineering, *J. Biomed. Mater. Res., Part A,* **75**, pp. 333–341.
5. Wang, C. C., Wallace, C. H., Chunhui, D., Dixon, D. F, Wei, E. I., Feng-Xian, X., Fei, H., and Yong, C. (2012). Optical and electrical effects of gold nanoparticles in the active layer of polymer solar cells, *J. Mater. Chem.,* **22**, pp. 1206–1211.
6. Gajewicz, A., Puzyn, T., Rasulev, B., Leszczynska, D., and Leszczynski, J. (2011). Metal oxide nanoparticles: size-dependence of quantum-mechanical properties, *Nanosci. Nanotechnol. Asia,* **1**, pp. 53–58.
7. Thostenson, E. T., Li, C., and Chou, T. W. (2005). Nanocomposites in context, *Comp. Sci. Technol.,* **65**, pp. 491–516.
8. Qiao, R., and Brinson, L. C. (2009). Simulation of interphase percolation and gradients in polymer nanocomposites, *Comp. Sci. Technol.,* **69**, pp. 491–499.
9. Meth, J. S., and Lustig, S. R. (2010). Polymer interphase structure near nanoscale inclusions: comparison between random walk theory and experiment, *Polymer,* **51**, pp. 4259–4266.

10. Ghanbari, A., Ndoro, T. V., Leroy, F., Rahimi, M., Böhm, M. C., and Müller-Plathe, F. (2011). Interphase structure in silica–polystyrene nanocomposites: a coarse-grained molecular dynamics study, *Macromolecules,* **45**, pp. 572–584.

11. Ndoro, T. V., Böhm, M. C., and Müller-Plathe, F. (2011). Interface and interphase dynamics of polystyrene chains near grafted and ungrafted silica nanoparticles, *Macromolecules,* **45**, pp. 171–179.

12. Needleman, A., Borders, T. L., Brinson, L. C., Flores, V. M., and Schadler, L. S. (2010). Effect of an interphase region on debonding of a CNT reinforced polymer composite, *Comp. Sci. Technol.,* **70**, pp. 2207–2215.

13. Bonakdar, M., Seidel, G. D., and Inman, D. J. (2012). Effect of nanoscale fillers on the viscoelasticity of polymer nanocomposites, in *Proceedings of the 53rd AIAA/ASME/ASCE/AHS/ASC Structures, Structural Dynamics and Materials Conference Honolulu*, Hawaii, USA.

14. Chen, F., Clough, A., Reinhard, B. M., Grinstaff, M. W., Jiang, N., Koga, T., and Tsui, O. K. C. (2013). Glass transition temperature of polymer–nanoparticle composites: effect of polymer–particle interfacial energy, *Macromolecules,* **46**, pp. 4663–4669.

15. Natarajan, B., Li, Y., Deng, H., Brinson, L. C., and Schadler, L. S. (2013). Effect of interfacial energetics on dispersion and glass transition temperature in polymer nanocomposites, *Macromolecules,* **46**, pp. 2833–2841.

16. Zammarano, M., Maupin, P. H., Sung, L. P., Gilman, J. W., McCarthy, E. D., Kim, Y. S., and Fox, D. M. (2011). Revealing the interface in polymer nanocomposites, *ACS Nano,* **5**, pp. 3391–3399.

17. Harms, S., Rätzke, K., Faupel, F., Schneider, G. J., Willner, L., and Richter, D. (2010). Free volume of interphases in model nanocomposites studied by positron annihilation lifetime spectroscopy, *Macromolecules,* **43**, pp. 10505–10511.

18. Wang, H. W., Zhou, H. W., Peng, R. D., and Mishnaevsky Jr, L. (2011). Nanoreinforced polymer composites: 3D FEM modeling with effective interface concept, *Comp. Sci. Technol.,* **71**, pp. 980–988.

19. Mortazavi, B., Bardon, J., and Ahzi, S. (2011). Interphase effect on the elastic and thermal conductivity response of polymer nanocomposite materials: 3D finite element study, *Comput. Mater. Sci.,* **69**, pp. 100–106.

20. Ciprari, D., Jacob, K., and Tannenbaum, R. (2006). Characterization of polymer nanocomposite interphase and its impact on mechanical properties, *Macromolecules,* **39**, pp. 6565–6573.

21. Liu, J., Boo, W. J., Clearfield, A., and Sue, H. J. (2006). Intercalation and exfoliation: a review on morphology of polymer nanocomposites reinforced by inorganic layer structures, *Mater. Manufact. Processes,* **21**, pp. 143–151.

22. Krishnamoorti, R. (2007). Strategies for dispersing nanoparticles in polymers, *MRS Bull.,* **32**, pp. 341–347.

23. Bousmina, M. (2006). Study of intercalation and exfoliation processes in polymer nanocomposites, *Macromolecules,* **39**, pp. 4259–4263.

24. Rong, M. Z., Zhang, M. Q., and Ruan, W. H. (2006). Surface modification of nanoscale fillers for improving properties of polymer nanocomposites: a review, *Mater. Sci. Technol.,* **22**, pp. 787–796.

25. Sauceau, M., Fages, J., Common, A., Nikitine, C., and Rodier, E. (2011). New challenges in polymer foaming: a review of extrusion processes assisted by supercritical carbon dioxide, *Prog. Polym. Sci.,* **36**, pp. 749–766.

26. Thomas, S. (2009). *Natural Fibre Reinforced Polymer Composites: From Macro to Nanoscale* (Old City, Philadelphia, USA).

27. Zeng, C., Han, X., Lee, L. J., Koelling, K. W., and Tomasko, D. L. (2003). Polymer–clay nanocomposite foams prepared using carbon dioxide, *Adv. Mater.,* **15**, pp. 1743–1747.

28. Lee, L. J., Zeng, C., Cao, X., Han, X., Shen, J., and Xu, G. (2005). Polymer nanocomposite foams, *Comp. Sci. Technol.,* **65**, pp. 2344–2363.

29. Zeng, C., Hossieny, N., Zhang, C., Wang, B., and Walsh, S. M. (2013). Morphology and tensile properties of PMMA carbon nanotubes nanocomposites and nanocomposites foams, *Comp. Sci. Technol.,* **82**, pp. 29–37.

30. Sorrentino, L., Aurilia, M., Cafiero, L., and Iannace, S. (2011). Nanocomposite foams from high-performance thermoplastics, *J. Appl. Polym. Sci.,* **122**, pp. 3701–3710.

31. Shen, J., Zeng, C., and Lee, L. J. (2005). Synthesis of polystyrene–carbon nanofibers nanocomposite foams, *Polymer,* **46**, pp. 5218–5224.

32. Chaudhary, A. K., and Jayaraman, K. (2011). Extrusion of linear polypropylene–clay nanocomposite foams, *Polym. Eng. Sci.,* **51**, pp. 1749–1756.

33. Goren, K., Chen, L., Schadler, L. S., and Ozisik, R. (2010). Influence of nanoparticle surface chemistry and size on supercritical carbon dioxide processed nanocomposite foam morphology, *J. Supercrit. Fluids,* **51**, pp. 420–427.

34. Chen, L., Ozisik, R., and Schadler, L. S. (2010). The influence of carbon nanotube aspect ratio on the foam morphology of MWNT/PMMA nanocomposite foams, *Polymer*, **51**, pp. 2368–2375.

35. Zeng, C., Hossieny, N., Zhang, C., and Wang, B. (2010). Synthesis and processing of PMMA carbon nanotube nanocomposite foams, *Polymer*, **51**, pp. 655–664.

36. Seraji, S. M., Razavi Aghjeh, M. K., Davari, M., Salami Hosseini, M., and Khelgati, S. (2011). Effect of clay dispersion on the cell structure of LDPE/clay nanocomposite foams, *Polym. Comp.*, **32**, pp. 1095–1105.

37. Han, X., Zeng, C., Lee, L. J., Koelling, K. W., and Tomasko, D. L. (2003). Extrusion of polystyrene nanocomposite foams with supercritical CO_2, *Polym. Eng. Sci.*, **43**, pp. 1261–1275.

38. Tsimpliaraki, A., Tsivintzelis, I., Marras, S. I., Zuburtikudis, I., and Panayiotou, C. (2011). The effect of surface chemistry and nanoclay loading on the microcellular structure of porous poly (D, L lactic acid) nanocomposites, *J. Supercrit. Fluids*, **57**, pp. 278–287.

39. Liu, H., Han, C., and Dong, L. (2010). Study of the biodegradable poly (ε-caprolactone)/clay nanocomposite foams, *J. Appl. Polym. Sci.*, **115**, pp. 3120–3129.

40. Taki, K., Yanagimoto, T., Funami, E., Okamoto, M., and Ohshima, M. (2004). Visual observation of CO_2 foaming of polypropylene-clay nanocomposites, *Polym. Eng. Sci.*, **44**, pp. 1004–1011.

41. Famili, M. H. N., Janani, H., and Enayati, M. S. (2011). Foaming of a polymer–nanoparticle system: effect of the particle properties, *J. Appl. Polym. Sci.*, **119**, pp. 2847–2856.

42. Chen, L., Goren, B. K., Ozisik, R., and Schadler, L. S. (2012). Controlling bubble density in MWNT/polymer nanocomposite foams by MWNT surface modification, *Comp. Sci. Technol.*, **72**, pp. 190–196.

43. Zhai, W., Yu, J., Wu, L., Ma, W., and He, J. (2006). Heterogeneous nucleation uniformizing cell size distribution in microcellular nanocomposites foams, *Polymer*, **47**, pp. 7580–7589.

44. Ema, Y., Ikeya, M., and Okamoto, M. (2006). Foam processing and cellular structure of polylactide-based nanocomposites, *Polymer*, **47**, pp. 5350–5359.

45. Guo, Z., Burley, A. C., Koelling, K. W., Kusaka, I., Lee, L. J., and Tomasko, D. L. (2012). CO_2 bubble nucleation in polystyrene: experimental and modeling studies, *J. Appl. Polym. Sci.*, **125**, pp. 2170–2186.

46. Guo, Z., Burley, A. C., Koelling, K. W., Kusaka, I., Lee, L. J., and Tomasko, D. L. (2010). Comparison of nanoclay and carbon nanofiber particles on

rheology of molten polystyrene nanocomposites under supercritical carbon dioxide. *J. Appl. Polym. Sci.*, **116**, pp. 1068–1076.

47. Yang, J., Huang, L., Zhou, Y., Chen, F., and Zhong, M. (2013). Multiwalled Carbon nanotubes grafted with polyamidoamine (PAMAM) dendrimers and their influence on polystyrene supercritical carbon dioxide foaming, *J. Supercrit. Fluids*, **82**, pp. 13–21.
48. Yang, J., Huang, L., Zhang, Y., Chen, F., and Zhong, M. (2013). Mesoporous silica particles grafted with polystyrene brushes as a nucleation agent for polystyrene supercritical carbon dioxide foaming, *J. Appl. Polym. Sci.*, **130**, pp. 4308–4317.
49. Li, C., Yang, G., Deng, H., Wang, K., Zhang, Q., Chen, F., and Fu, Q. (2013). The preparation and properties of polystyrene/functionalized graphene nanocomposite foams using supercritical carbon dioxide, *Polym. Int.*, **62**, pp. 1077–1084.
50. Horsch, S., Serhatkulu, G., Gulari, E., and Kannan, R. M. (2006). Supercritical CO_2 dispersion of nano-clays and clay/polymer nanocomposites, *Polymer*, **47**, pp. 7485–7496.
51. Manitiu, M., Horsch, S., Gulari, E., and Kannan, R. M. (2009). Role of polymer–clay interactions and nano-clay dispersion on the viscoelastic response of supercritical CO_2 dispersed polyvinylmethylether (PVME)–clay nanocomposites, *Polymer*, **50**, pp. 3786–3796.
52. Pu, N. W., Wang, C. A., Sung, Y., Liu, Y. M., and Ger, M. D. (2009). Production of few-layer graphene by supercritical CO_2 exfoliation of graphite, *Mater. Lett.*, **63**, pp. 1987–1989.
53. Zheng, X., Xu, Q., Li, J., Li, L., and Wei, J. (2012). High-throughput, direct exfoliation of graphite to graphene via a cooperation of supercritical CO_2 and pyrene-polymers, *RSC Adv.*, **2**, pp. 10632–10638.
54. Urbanczyk, L., Calberg, C., Stassin, F., Alexandre, M., Jérôme, R., Jérôme, C., and Detrembleur, C. (2008). Synthesis of PCL/clay masterbatches in supercritical carbon dioxide, *Polymer*, **49**, pp. 3979–3986.
55. Urbanczyk, L., Ngoundjo, F., Alexandre, M., Jérôme, C., Detrembleur, C., and Calberg, C. (2009). Synthesis of polylactide/clay nanocomposites by *in situ* intercalative polymerization in supercritical carbon dioxide, *Eur. Polym. J.*, **45**, pp. 643–648.
56. Thompson, M. R., Liu, J., Krump, H., Kostanski, L. K., Fasulo, P. D., and Rodgers, W. R. (2008). Interaction of supercritical CO_2 with alkyl-ammonium organoclays: changes in morphology, *J. Colloid Interface Sci.*, **324**, pp. 177–184.

57. Zhao, Y., and Huang, H. X. (2008). Dynamic rheology and microstructure of polypropylene/clay nanocomposites prepared under scCO_2 by melt compounding, *Polym. Testing,* **27**, pp. 129–134.

58. Liu, J., Thompson, M. R., Balogh, M. P., Speer, R. L., Fasulo, P. D., and Rodgers, W. R. (2011). Influence of supercritical CO_2 on the interactions between maleated polypropylene and alkyl-ammonium organoclay, *J. Appl. Polym. Sci.,* **119**, pp. 2223–2234.

59. Su, F. H., Huang, H. X., and Zhao, Y. (2011). Microstructure and mechanical properties of polypropylene/poly (ethylene-co-octene copolymer)/clay ternary nanocomposites prepared by melt blending using supercritical carbon dioxide as a processing aid, *Comp. Part B: Eng.,* **42**, pp. 421–428.

60. Yu, Y., and Yang, X. (2011). Molecular simulation of swelling and interlayer structure for organoclay in supercritical CO_2, *Phys. Chem. Chem. Phys.,* **13**, pp. 282–290.

61. Yuvaraj, H., Woo, M. H., Park, E. J., Jeong, Y. T., and Lim, K. T. (2008). Polypyrrole/γ-Fe_2O_3 magnetic nanocomposites synthesized in supercritical fluid, *Eur. Polym. J.,* **44**, pp. 637–644.

62. Yuvaraj, H., Park, E. J., Gal, Y. S., and Lim, K. T. (2008). Synthesis and characterization of polypyrrole–TiO_2 nanocomposites in supercritical CO_2, *Colloids Surf., A,* **313**, pp. 300–303.

63. Matsuyama, K., and Mishima, K. (2009). Preparation of poly (methyl methacrylate)–TiO_2 nanoparticle composites by pseudo-dispersion polymerization of methyl methacrylate in supercritical CO_2, *J. Supercrit. Fluids,* **49**, pp. 256–264.

64. Hasell, T., Wood, C. D., Clowes, R., Jones, J. T., Khimyak, Y. Z., Adams, D. J., and Cooper, A. I. (2009). Palladium nanoparticle incorporation in conjugated microporous polymers by supercritical fluid processing, *Chem. Mater.,* **22**, pp. 557–564.

65. Nguyen, V. H., Haldorai, Y., Pham, Q. L., and Shim, J. J. (2011). Supercritical fluid mediated synthesis of poly (2-hydroxyethyl methacrylate)/Fe_3O_4 hybrid nanocomposite, *Mater. Sci. Eng.: B,* **176**, pp. 773–778.

66. Goren, K., Okan, O. B., Chen, L., Schadler, L. S., and Ozisik, R. (2012). Supercritical carbon dioxide assisted dispersion and distribution of silica nanoparticles in polymers, *J. Supercrit. Fluids,* **67**, pp. 108–113.

67. Zhang, F., Xu, Q., Zhang, H., and Zhang, Z. (2009). Polymer supermolecular structures built on carbon nanotubes via a supercritical carbon dioxide-assisted route, *J. Phys. Chem. C,* **113**, pp. 18531–18535.

68. Chen, C., Bortner, M., Quigley, J. P., and Baird, D. G. (2012). Using supercritical carbon dioxide in preparing carbon nanotube nanocomposite: improved dispersion and mechanical properties, *Polym. Comp.*, **33**, pp. 1033–1043.

69. Jin, H., Li, S., Hu, D., and Zhao, Y. (2012). Preparation of PLA-PEG nanoparticles by the solution enhanced dispersion with enhanced mass transfer using ultrasound in supercritical CO_2, *Power Technol.*, **227**, pp. 17–23.

70. Haldorai, Y., Shim, J. J., and Lim, K. T. (2012). Synthesis of polymer-inorganic filler nanocomposites in supercritical CO_2, *J. Supercrit. Fluids*, **71**, pp. 45–63.

71. Lee, Y. H., Park, C. B., Wang, K. H., and Lee, M. H. (2005). HDPE-clay nanocomposite foams blown with supercritical CO_2, *J. Cell. Plast.*, **41**, pp. 487–502.

72. Lee, Y. H., Wang, K. H., Park, C. B., and Sain, M. (2007). Effects of clay dispersion on the foam morphology of LDPE/clay nanocomposites, *J. Appl. Polym. Sci.*, **103**, pp. 2129–2134.

73. Liao, W., Wu, B. Z., Nian, H., Chen, H. Y., Yu, J. J., and Chiu, K. (2012). Fabrication of a form-and size-variable microcellular-polymer-stabilized metal nanocomposite using supercritical foaming and impregnation for catalytic hydrogenation, *Nanoscale Res. Lett.*, **7**, pp. 1–7.

74. Chang, Y. W., Lee, D., and Bae, S. Y. (2006). Preparation of polyethylene-octene elastomer/clay nanocomposite and microcellular foam processed in supercritical carbon dioxide, *Polym. Int.*, **55**, pp. 184–189.

75. Ding, J., Ma, W., and Zhong, Q. (2013). Foaming of Homogeneous polypropylene and ethylene-polypropylene block copolymer using supercritical carbon dioxide, *Polym. Plast. Technol. Eng.*, **52**, pp. 592–598.

76. Oh, K., Seo, Y. P., Hong, S. M., Takahara, A., Lee, K. H., and Seo, Y. (2013). Dispersion and reaggregation of nanoparticles in the polypropylene copolymer foamed by supercritical carbon dioxide. *Phys. Chem. Chem. Phys.*, **15**, pp. 11061–11069.

77. Bhattacharya, S., Gupta, R. K., Jollands, M., and Bhattacharya, S. N. (2009). Foaming behavior of high-melt strength polypropylene/clay nanocomposites, *Polym. Eng. Sci.*, **49**, pp. 2070–2084.

78. Jiang, X. L., Liu, T., Xu, Z. M., Zhao, L., Hu, G. H., and Yuan, W. K. (2009). Effects of crystal structure on the foaming of isotactic polypropylene using supercritical carbon dioxide as a foaming agent, *J. Supercrit. Fluids*, **48**, pp. 167–175.

79. Zhai, W., Kuboki, T., Wang, L., Park, C. B., Lee, E. K., and Naguib, H. E. (2010). Cell structure evolution and the crystallization behavior of polypropylene/clay nanocomposites foams blown in continuous extrusion, *Ind. Eng. Chem. Res.,* **49**, pp. 9834–9845.

80. Jiang, X. L., Bao, J. B., Liu, T., Zhao, L., Xu, Z. M., and Yuan, W. K. (2009). Microcellular foaming of polypropylene/clay nanocomposites with supercritical carbon dioxide, *J. Cell. Plast.,* **45**, pp. 515–538.

81. Rende, D., Schadler, L. S., and Ozisik, R. (2013). Controlling Foam morphology of poly (methyl methacrylate) via surface chemistry and concentration of silica nanoparticles and supercritical carbon dioxide process parameters, *J. Chem.*, pp. 1–13.

82. Zhu, B., Zha, W., Yang, J., Zhang, C., and Lee, L. J. (2010). Layered-silicate based polystyrene nanocomposite microcellular foam using supercritical carbon dioxide as blowing agent, *Polymer,* **51**, pp. 2177–2184.

83. Yang, J., Wu, M., Chen, F., Fei, Z., and Zhong, M. (2011). Preparation, characterization, and supercritical carbon dioxide foaming of polystyrene/graphene oxide composites, *J. Supercrit. Fluids,* **56**, pp. 201–207.

84. Zhang, C., Zhu, B., and Lee, L. J. (2011). Extrusion foaming of polystyrene/carbon particles using carbon dioxide and water as co-blowing agents, *Polymer,* **52**, pp. 1847–1855.

85. Saraeian, P., Tavakoli, H. R., and Ghassemi, A. (2013). Production of polystyrene-nanoclay nanocomposite foam and effect of nanoclay particles on foam cell size, *J. Comp. Mater.,* **47**, pp. 2211–2217.

86. Yeh, S. K., Huang, C. H., Su, C. C., Cheng, K. C., Chuang, T. H., Guo, W. J., and Wang, S. F. (2013). Effect of dispersion method and process variables on the properties of supercritical CO_2 foamed polystyrene/graphite nanocomposite foam, *Polym. Eng. Sci.,* **53**, pp. 2061–2072.

87. Yang, J., Sang, Y., Chen, F., Fei, Z., and Zhong, M. (2012). Synthesis of silica particles grafted with poly (ionic liquid) and their nucleation effect on microcellular foaming of polystyrene using supercritical carbon dioxide, *J. Supercrit. Fluids,* **62**, pp. 197–203.

88. Michaeli, W., and Heinz, R. (2000). Foam extrusion of thermoplastic polyurethanes (TPU) using CO_2 as a blowing agent, *Macromol. Mater. Eng.,* **284**, pp. 35–39.

89. Prasad, A., Fotou, G., and Li, S. (2012). The effect of polymer hardness, pore size, and porosity on the performance of thermoplastic polyurethane-based chemical mechanical polishing pads, *J. Mater. Res.,* **28**, pp. 2380–2393.

90. Yeh, S. K., Liu, Y. C., Wu, W. Z., Chang, K. C., Guo, W. J., and Wang, S. F. (2013). Thermoplastic polyurethane/clay nanocomposite foam made by batch foaming, *J. Cell. Plast.*, **49**, pp. 119–130.
91. Wu, H. B., Haugen, H. J., and Wintermantel, E. (2012). Supercritical CO_2 in injection molding can produce open porous polyurethane scaffolds: a parameter study, *J. Cell. Plast.*, **48**, pp. 141–159.
92. Dai, C., Zhang, C., Huang, W., Chang, K. C., and Lee, L. J. (2013). Thermoplastic polyurethane microcellular fibers via supercritical carbon dioxide based extrusion foaming, *Polym. Eng. Sci.*, **53**, pp. 2360–2369.
93. Bernal, M., Martin-Gallego, M., Romasanta, L. J., Mortamet, A. C., López-Manchado, M. A., Ryan, A. J., and Verdejo, R. (2012). Effect of hard segment content and carbon-based nanostructures on the kinetics of flexible polyurethane nanocomposite foams, *Polymer,* **53**, pp. 4025–4032.
94. Sarier, N., and Onder, E. (2010). Organic modification of montmorillonite with low molecular weight polyethylene glycols and its use in polyurethane nanocomposite foams, *Thermochim. Acta,* **510**, pp. 113–121.
95. Madaleno, L., Pyrz, R., Crosky, A., Jensen, L. R., Rauhe, J. C. M., Dolomanova, V., Viegas de Barros Timmons A. M., Pinto J. J., and Norman, J. (2012). Processing and characterization of polyurethane nanocomposite foam reinforced with montmorillonite-carbon nanotube hybrids, *Comp. Part A: Appl. Sci. Manufact.*, **44**, pp. 1–7.
96. Lorenzetti, A., Hrelja, D., Besco, S., Roso, M., and Modesti, M. (2010). Improvement of nanoclays dispersion through microwave processing in polyurethane rigid nanocomposite foams, *J. Appl. Polym. Sci.,* **115**, pp. 3667–3674.
97. Tsimpliaraki, A., Tsivintzelis, I., Marras, S. I., Zuburtikudis, I., and Panayiotou, C. (2013). Foaming of PCL/clay nanocomposites with supercritical CO_2 mixtures: the effect of nanocomposite fabrication route on the clay dispersion and the final porous structure, *J. Supercrit. Fluids,* **81**, pp. 86–91.
98. Salerno, A., Zeppetelli, S., Di Maio, E., Iannace, S., and Netti, P. A. (2011). Design of bimodal PCL and PCL-HA nanocomposite scaffolds by two step depressurization during solid-state supercritical CO_2 foaming, *Macromol. Rapid Commun.*, **32**, pp. 1150–1156.
99. Salerno, A., Zeppetelli, S., Di Maio, E., Iannace, S., and Netti, P. A. (2011). Processing/structure/property relationship of multi-scaled PCL

and PCL–HA composite scaffolds prepared via gas foaming and NaCl reverse templating, *Biotechnol. Bioeng.,* **108**, pp. 963–976.

100. Salerno, A., Di Maio, E., Iannace, S., and Netti, P. A. (2011). Solid-state supercritical CO_2 foaming of PCL and PCL-HA nano-composite: effect of composition, thermal history and foaming process on foam pore structure, *J. Supercrit. Fluids,* **58**, pp. 158–167.

101. Kumar, V., and Schirmer, H. G. (1995). Semi-continuous production of solid state PET foams, *ANTEC'95,* **2**, pp. 2189–2192.

102. Leicher, S., Will, J., Haugen, H., and Wintermantel, E. (2005). MuCell® technology for injection molding: a processing method for polyether-urethane scaffolds, *J. Mater. Sci.,* **40**, pp. 4613–4618.

103. Hwang, S. S., Liu, S. P., Hsu, P. P., Yeh, J. M., Chang, K. C., and Lai, Y. Z. (2010). Effect of organoclay on the mechanical/thermal properties of microcellular injection molded PBT–clay nanocomposites, *Int. Commun. Heat Mass Transfer,* **37**, pp. 1036–1043.

104. Minaei-Zaim, M., Ghasemi, I., Karrabi, M., and Azizi, H. (2012). Effect of injection molding parameters on properties of cross-linked low-density polyethylene/ethylene vinyl acetate/organoclay nanocomposite foams, *Iranian Polym. J.,* **21**, pp. 537–546.

105. Hwang, S. S., Liu, S. P., Hsu, P. P., Yeh, J. M., Yang, J. P., Chang, K. C., and Chu, S. N. (2011). Morphology, mechanical, thermal and rheological behavior of microcellular injection molded TPO-clay nanocomposites prepared by kneader, *Int. Commun. Heat Mass Transfer,* **38**, pp. 597–606.

106. Sauceau, M., Fages, J., Common, A., Nikitine, C., and Rodier, E. (2011). New challenges in polymer foaming: a review of extrusion processes assisted by supercritical carbon dioxide, *Prog. Polym. Sci.,* **36**, pp. 749–766.

107. Tomasko, D. L., Burley, A., Feng, L., Yeh, S. K., Miyazono, K., Nirmal-Kumar, S., Kusaka I., and Koelling, K. (2009). Development of CO_2 for polymer foam applications, *J. Supercrit. Fluids,* **47**, pp. 493–499.

Chapter 16

Coating and Impregnation Processes Using Dense-Phase CO_2

Concepción Domingo, Carlos A. García-González, and Pedro López-Aranguren
Materials Science Institute of Barcelona (CSIC), Campus de la UAB, 08193 Bellaterra, Spain
conchi@icmab.es

scCO_2 technology has been proposed for the production of a great range of knowledge-based multifunctional materials with better control of surface properties and purity, and for the industrial high-rate, high-volume fabrication of surface hybridized or composite materials. This chapter shows the development of the innovative sc-CO_2-based surface technology, applicable to existing and novel high-performance functional products, which had led to the development of procedures that enable the design of complex surface structures and products with unique characteristics in relation to composition, purity, physicochemical properties, and effectiveness.

Supercritical Fluid Nanotechnology: Advances and Applications in Composites and Hybrid Nanomaterials
Edited by Concepción Domingo and Pascale Subra-Paternault

ISBN 978-981-4613-40-8 (Hardcover), 978-981-4613-41-5 (eBook)
www.panstanford.com

16.1 Surface and Interphase Modification

In advanced nanomaterials, the option of modifying the surface for stabilization, control of the chemical behavior, functionalization, biocompatibilization, (super)hydrophobization, etc., as well as the interphase to shape composite nanostructures, is essential to achieve end products with the desired functionality. Nanometric entities, involving both nanoparticles and nanoporous matter, have extremely large surface-to-volume ratios and very energetic surfaces, which makes it difficult the control of the surface activity or reactivity. An important limitation in the application of traditional solution-phase surface modification processes to nanoentities is the restriction of the kinetics of mass transport of the modifier agent to nooks in intricate geometries or into small pores. Supercritical carbon dioxide ($scCO_2$) routes seem a promising alternative for the individual surface coating of micro- and nanoparticles, even agglomerated, as well as for the deep impregnation of meso- and microporous materials. This is due to the low viscosity and absence of surface tension in the supercritical fluid that allows the complete wetting of complex-shaped nanostructured substrates [1]. For this purpose, $scCO_2$ has been used as a solvent in a large number of processes in which the supercritical solution gets in contact with a dry or wet solid surface upon which components may adsorb (e.g., high-value nonvolatile organics separation, impregnation for drug delivery, protective coatings, and surface functionalization) or from which components may desorb (e.g., cleaning and drying, regeneration of sorbents, supercritical fluid chromatography, and supercritical fluid extraction) [2]. Besides being used as a solvent, $scCO_2$ can also play other primary roles as antisolvent, solute, or reaction medium, which offer a unique flexibility as a surface engineering technology.

The choice of supercritical processes for surface design is related with the substrate and the solute behaviors in $scCO_2$, following the scheme in Fig. 16.1. Two different types of matrixes are considered, those that interact with $scCO_2$, like polymers that swell and those that do not interact with $scCO_2$, mainly inorganic substrates in the particulate or porous forms. The use of the so-called supercritical micronization techniques to surface modification depends primarily on the solubility of the coating material in the supercritical fluid.

For solutes soluble in $scCO_2$, the fluid is used as a transport solvent, either pure or with the addition of cosolvents, chelating agents, etc. For solutes insoluble in $scCO_2$, antisolvent and particles from gas-saturated solutions (PGSS®) processes are mainly used. Surface modification protocols can be applied to either the external surface of particles, normally denoted as *coating processes*, or the internal surface of porous materials, labeled as *impregnation processes*.

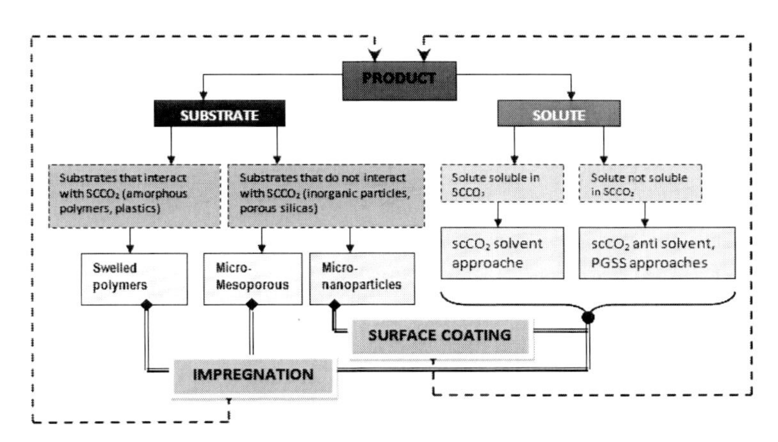

Figure 16.1 Process design as a function of substrate and solute behavior in $scCO_2$.

$scCO_2$ dyeing of textile is a clear example of surface modification technology developed at an industrial scale. In conventional textile dyeing large quantities of wastewater are produced. This environmental and economical burden is avoided when $scCO_2$ is used as the dyeing medium instead of water [3, 4]. Separating residual dye from the CO_2 and recycling of CO_2 are relatively easy processes. Energy is saved because textiles do not need to be dried after the dyeing process. An additional advantage of $scCO_2$ is the high diffusivity and low viscosity that allow the dye to diffuse faster toward and into the textile fibers. This results in a faster dyeing process. Textiles can be classified into nonpolar, synthetic polymers (e.g., polyester) and polar, natural textiles. The second category can be divided into polymers built from amino acids (e.g., silk and wool) or cellulose (e.g., cotton). In polyester dyeing, $scCO_2$ penetrates and swells the fibers, thereby making them accessible for dye molecules [5]. Upon depressurization, the dye molecules are trapped inside the shrinking polyester fibers. In natural textiles, the dye molecules can be fixed by either physical (e.g., van der Waals) or chemical (e.g., covalent) bonds [6]. Since the dyes used in a $scCO_2$-dyeing process are nonpolar and natural fibers are polar the affinity between dyes and textiles is low so physical bonds are weak. Therefore, dyeing processes are being developed for dyeing natural textiles in $scCO_2$ with reactive dyes that create covalent dye–textile bonds [7]. This technology is fully described in Chapter 17.

16.2 Supercritical CO_2 Coating of Nanoparticles

The high-quality dispersion of discrete inorganic fillers in continuous organic phases is essential for the production of homogeneous hybrid composite materials used as paints, plastics, adhesives, restorative biomaterials, electronic, packages, etc. [8–10]. In these systems, the interfacial interactions between the inorganic particles and the organic matrix must be enhanced by particle surface modification on a controlled basis. The dispersibility, reactivity, bioefficacy, or flowability of micro- or nanoparticles is also often modified by surface treatment in applications such as cosmetics, pharmaceuticals, food and, in general, materials industries [11–13]. For nanomaterials applied in the single particle form, with important applications in electronics, optics, and optoelectronics, the energy needed for particle deagglomeration is reduced after coating. Surface coating has also been applied for metals protection against corrosive environments [14], water repellency [15], protection of labile biomolecules against denaturization [16], metal recovery [17], and the design of chromatographic stationary phases [18].

16.2.1 Coating Agents and Methods

Organic compounds (surfactants, fats, lipids, and waxes) and natural or synthetic polymers are among the most commonly used coating materials [9–24]. In addition, bifunctional silanes, a group of nontoxic and environmentally compliant chemicals, are widely used to form spontaneous self-assembled monolayer coatings on inorganic surfaces [25]. Numerous examples of functionalized fillers prepared through organic surface modification can be found in the literature [26, 27]. The coating of nanoparticles surface can be performed either by in situ deposition during synthesis (e.g., dispersion or microemulsion processes [28]) or by postsynthesis grafting (e.g., Langmuir–Blodgett, self-assembly, emulsion, spray drying, vapor deposition, or polymerization methods [29–31]). However, few publications have described efficient methods for the coating of fine powders, due to the intrinsic necessity of having the nanoparticles in a completely dispersed state [32–36]. The high surface energy of nanoparticles causes them to agglomerate in the synthesis or postsynthesis process (Fig. 16.2a). Upon contact, primary particles form bonds of different nature. As a baseline of bond energy, physical van der Waals forces are always present, leading to agglomerates.

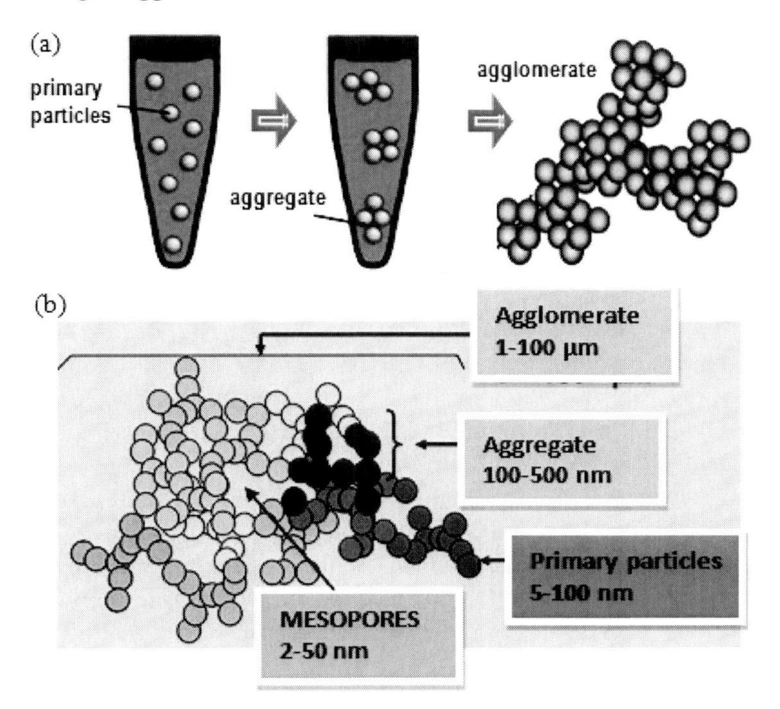

Figure 16.2 Agglomeration of nanometric particles: (a) synthesis and postsynthesis scheme and (b) mesopore formation.

Most commercially available nanoparticulated systems are large agglomerates of about 1–100 μm composed of nanoparticles with sizes ranging from 5 to 100 nm and/or strongly interacting aggregates of 100–500 nm. The strength of these interparticle bonds determines to a large extent the physical properties and applicability in materials processing [37]. The agglomeration of primary particles in the nanometric range leads to the formation of mesoporous networks with intricate internal geometries (Fig. 16.2b). As a consequence, the coating mechanism of agglomerated nanoparticles has strong similarities to the impregnation of the internal surface of intrinsic mesoporous matter. $scCO_2$ solutions can readily permeate and penetrate into the interparticle voids [38]. The superior kinetics of mass transfer from the supercritical fluid to the interior of the mesopores induces the coating agent to cover preferentially primary particles.

16.2.2 Supercritical CO_2 Polymer Coating

The technologies currently used in the industry for polymer coating of particulate substrates, based on either mechanical movement (mixing by agitation) or solid fluidization (Würster process), involve organic solution chemistry and the use of high temperatures to eliminate the solvent [39, 40]. $scCO_2$ has a limited solvent strength for many coating polymers. Thus, most of the work performed in polymer coating using $scCO_2$ technology has been carried out by antisolvent methods [41, 42] and only a little by straight polymer precipitation from the solvent, as in the rapid expansion of supercritical solutions (RESS) process [43–46]. In a further evolution, the RESS and fluidized bed technologies have been combined using $scCO_2$ as the fluidizing gas and as a solvent [47–51]. The PGSS® process and the co-injection route, in which the $scCO_2$ is dissolved in the polymer to reduce its melting temperature, are used to coat thermally sensitive compounds [52]. However, individual coating is difficult to achieve using these processes, especially for small-size particles. In a different approach, particles have been coated by in situ dispersion polymerization in $scCO_2$ [53, 54]. This coating method has shown excellent results for the encapsulation of fine particles. However, it is severely limited for pharmaceutical applications due to the possibility of significant residual monomer content. Surface-initiated polymerization has been used for the coating of aerogel discs with a hydrogel [55]. Antisolvent approaches have been successfully applied to the coating of micrometric organic and inorganic particles [56–58], although only in some particular cases they are devoted to the coating of nanometric entities [59–61]. In Fig. 16.3, uncoated and $scCO_2$ antisolvent coated silica micro- (Fig. 16.3a) and nanoparticles (Fig. 16.3b) are shown, demonstrating the high ability of the supercritical antisolvent approach to perform a continuous and homogeneous coating of almost individual particles [62].

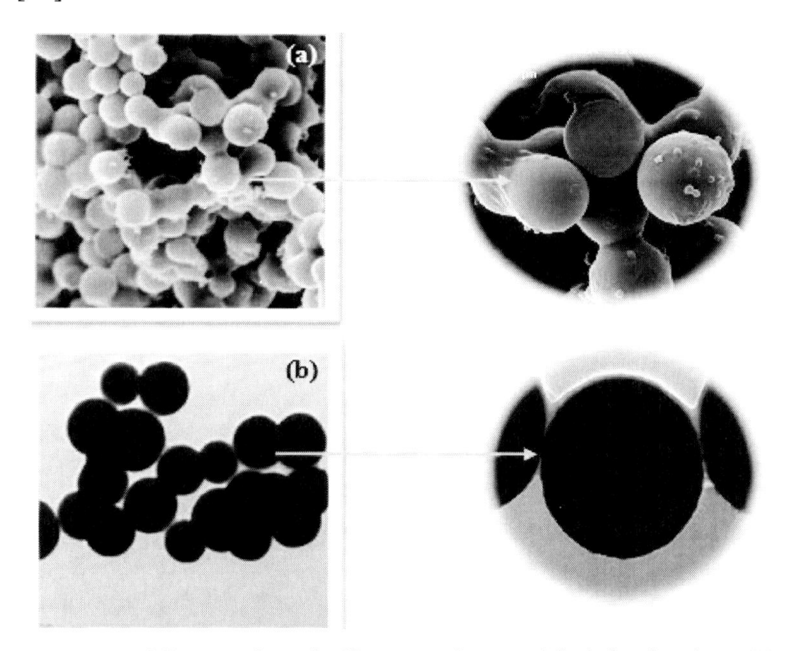

Figure 16.3 Micrographs of silica samples modified by batch $scCO_2$ antisolvent processing of slurries of SiO_2/polymer/acetone (10 MPa and 309 K): (a) SEM of silica microparticles (2–3 μm) modified with polymethylmethacrylate [62] and (b) TEM of silica nanoparticles (200–250 nm) coated with Eudragit RL100 [63].

16.2.3 Supercritical CO_2 Anhydrous Silane Coating

The need for novel approaches for the fabrication of surface-functionalized materials arises from the exponential increase of the domains of technological applications of small particles and nanoporous products. In this view, a simple and cost-effective way of regularly organize functional entities on surfaces is represented by their self-assembly. One of the most successful self-assembly approach is the chemical grafting of long hydrocarbon chains (R) on hydrated surfaces via trifunctional silanes ($RSiX_3$), forming a well-ordered monolayer [63]. Organofunctional silanes have been used in the filler industry for years to provide a stable bond between two otherwise poorly bonding surfaces (Fig. 16.4a). Silane coupling agents are silicon-based chemicals that contain two types of reactivity in the same molecule. The organic functional group R interacts to the organic phase or renders hydrophobicity to the surface. The X group hydrolyzes to silanol either through the addition of water or from residual water present on the inorganic surface (Fig. 16.4b). The formed silanols can condensate in the bulk and coordinate with metal hydroxyl groups on the inorganic surface, either by hydrogen bond (#-OH···HO-Si) or siloxane linkage (#-O-Si). Hydrogen bridges are described as the predominant binding mode [64]. Bonds formed with neighboring silane molecules (Si-O-Si) give a crosslinked horizontal structure. It has been reported that the deposition of trialkoxysilane from solutions can give rise to a number of possible surface structures, schematized in Fig. 16.4c: covalent, self-assembled, and polycondensate [65, 66]. The final configuration mainly depends on the curing temperature and water concentration in the bulk and adsorbed on the surface [67, 68].

Deposition of a self-assembled monolayer of a great variety of silane molecules on flat, polished metallic and metal oxide surfaces has been extensively studied using either gaseous- or liquid-phase reactions [69–71] or even supercritical methods [72, 73]. The gas-phase treatment is limited to few monodentate organosilanes possessing a significant vapor pressure and thermal stability. Deposition from aqueous-alcohol solutions is the most facile and common method used for bidentate and tridentate silanes deposition. Given that most silanes have a relatively high solubility in $scCO_2$, silane coating can be also performed through diffusion from a supercritical solution [74–76]. The high solubility in $scCO_2$ is due to the interaction between CO_2, a weak Lewis acid, and the strong electron-donating capacity of the siloxane group. In Fig. 16.5a, the curves obtained as a function of pressure and temperature for the octyltriethoxysilane–$scCO_2$ system is shown as an example.

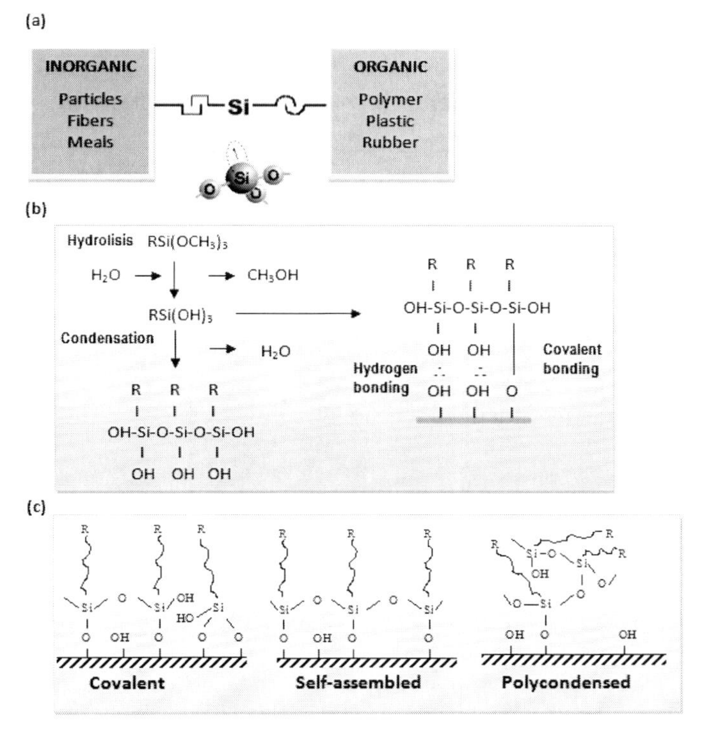

Figure 16.4 Silane chemistry: (a) coupling behavior, (b) self-assembly in the bulk or on hydroxilated surfaces, (c) feasible surface configurations.

Solubility data was determined under the pressure range of 8–18 MPa and at the operating temperatures of 318 and 348 K. The data can be correlated using the density-based equation proposed by Chrastil, eq. (16.1) [77]:

$$S_{C8} = \rho_{CO2}{}^{k} \cdot \exp(a/T + b) \quad (16.1)$$

where S_{C8} is the measured silane solubility ($g \cdot L^{-1}$), ρ_{CO2} is the supercritical fluid density ($g \cdot L^{-1}$), T is the temperature (K), and a, b, and k are empirical fitting parameters. The density of compressed CO_2 was calculated using a cubic equation of state. Data correlation is shown in Fig. 16.5b. For the octylsilane, the solubility increased with increasing pressure for both studied temperatures [78]. A crossover phenomenon [79] was observed with the crossover pressure occurring at ca. 16 MPa. Below the crossover pressure, an isobaric increase in temperature from 318 to 348 K decreased the silane solubility. Conversely, above the crossover pressure, a similar isobaric increase in temperature increased the silane solubility.

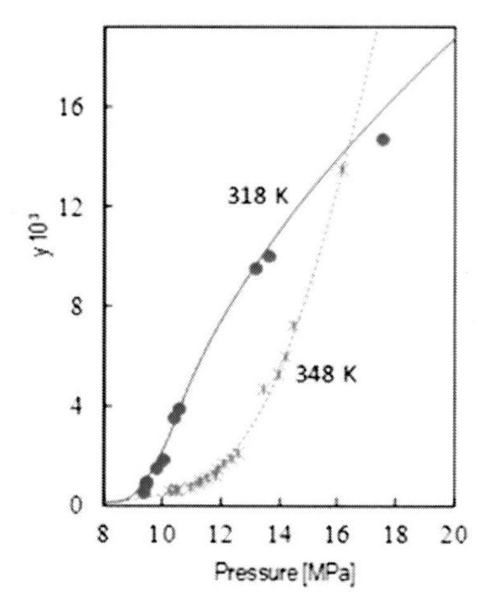

Figure 16.5 Experimental solubility data and correlation lines obtained with the Chrastil equation. The natural logarithmic form of experimental data in the Chrastil equation is adjusted by straight lines with a R^2 of 0.994 and 0.993 for 348 and 318 K, respectively.

The alternative and generic $scCO_2$ method described to efficiently coat the surface of nanoparticulate is placed between the conventional gas deposition and liquid-phase adsorption methods. The alcohol conventional liquid-phase method is a wet method in which water molecules are added to the solution bulk (Fig. 16.6a). Instead, the supercritical procedure is an anhydrous method and only moisture adsorbed on the inorganic substrate surface is present in the medium (Fig. 16.6b). Since the solubility of water in $scCO_2$ is low [80], water molecules are expected to be associated with the surface and not free in solution. Following this assumption, the hydrolysis step that initiate the silanization reaction (Fig. 16.4b) is only favored near the substrate surface and not in the solution bulk, promoting the formation of a self-assembled monolayer through horizontal polymerization. The kinetics of hydrolysis and condensation of alkoxysilanes in water has been reported to be strongly dependent on pH, being the optimum value for fast hydrolysis around 3–4 [81].

Water in contact with carbon dioxide becomes acidic (pH ≈ 3) due to the formation and dissociation of carbonic acid. Therefore, the process at supercritical conditions is performed under acid catalysis. In the wet methods, a high-quality self-assembled monolayer is not easily obtained, mainly because of the difficulty in controlling the amount of water in the system. Alkoxysilane polycondensation prior to deposition onto the surface is promoted by the presence of excess of water in the medium, and therefore is favored in the alcoholic solution.

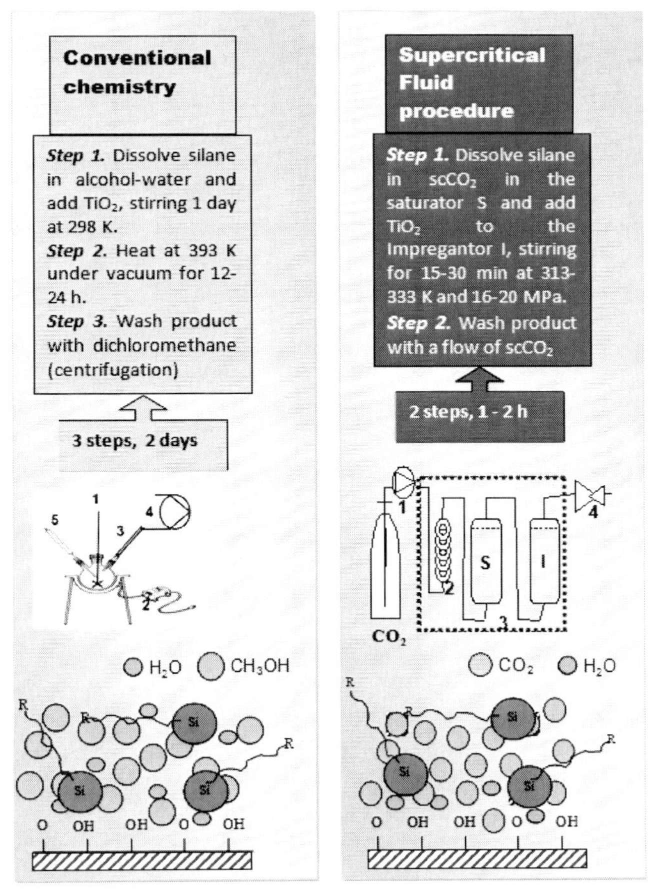

Figure 16.6 Schematic representation of (a) conventional alcoholic and (b) supercritical procedures used for silane deposition and grafting.

An additional difficulty in using the traditional solution-phase silanization process for the coating of fine particles is that the solution containing the dispersed inorganic powder and the reactive silane turn out to be very viscous during reaction and solvent evaporation step. Therefore, kinetics became restricted by the mass transport of the silane to the inner mesopore surface of the agglomerates and, eventually, transport is hindered by channels blockage [82]. Covalent bonds (Fig. 16.4c) are promoted by increasing the curing temperature. Typical conventional thermal curing of silanes is performed at temperatures of ~373–393 K, whereas the complete $scCO_2$ process is carried out at 313–333 K (Fig. 16.6). In consequence, silane covalent binding is expected to be favored in the conventional method, while hydrogen bonding is likely the predominant conformation formed by supercritical silanization.

Nanometric TiO_2 (Fig. 16.7a) has been coated following the conventional aqueous-alcohol solution [83] and the supercritical fluid methods [84]. Modified TiO_2 is widely employed in the industry of photocatalysis and as a pigment or filler in the paper, plastic and cosmetic industries [85–87]. Nanoparticles are coated with octyltriethoxysilane ($C_8Si(OEt)_3$, Fig. 16.7b), possessing a nonreactive organic functionality (hydrophobic alkyl chain). In Fig. 16.7c, the most important steps of the silanization process of TiO_2 nanoparticles are schematized. In the conventional alcoholic process, the final amount of silane deposited on the surface depends on the initial amount of water and silane added to the medium that should be calculated very carefully during the experimental design. Excess of silane results in polymerization in the bulk and eventual deposition of polysiloxanes, resulting in poor coating control as well as incomplete and nonuniform individual particle coverage. TiO_2 powder silanized by the supercritical approach reveals agglomerates similar to those corresponding to nontreated powder (Fig. 16.8a,d), in which the primary particles are still evident (Fig. 16.8b,e). The mesoporous character of the aggregates is maintained. Conversely, alcoholic coating is performed mainly in the external surface of the agglomerates, thus losing the mesoporous character (Fig. 16.8c). Transmission electron microscopy (TEM) characterization is also performed in the supercritically silanized particles, employing a negative staining solution deposited around the electron transparent silane layer, with the objective of visualizing the silane-deposited layer. On the TEM images in Fig. 16.8f, the staining of the sample caused the silane layer to appear pale against the dark background and TiO_2 nanoparticles.

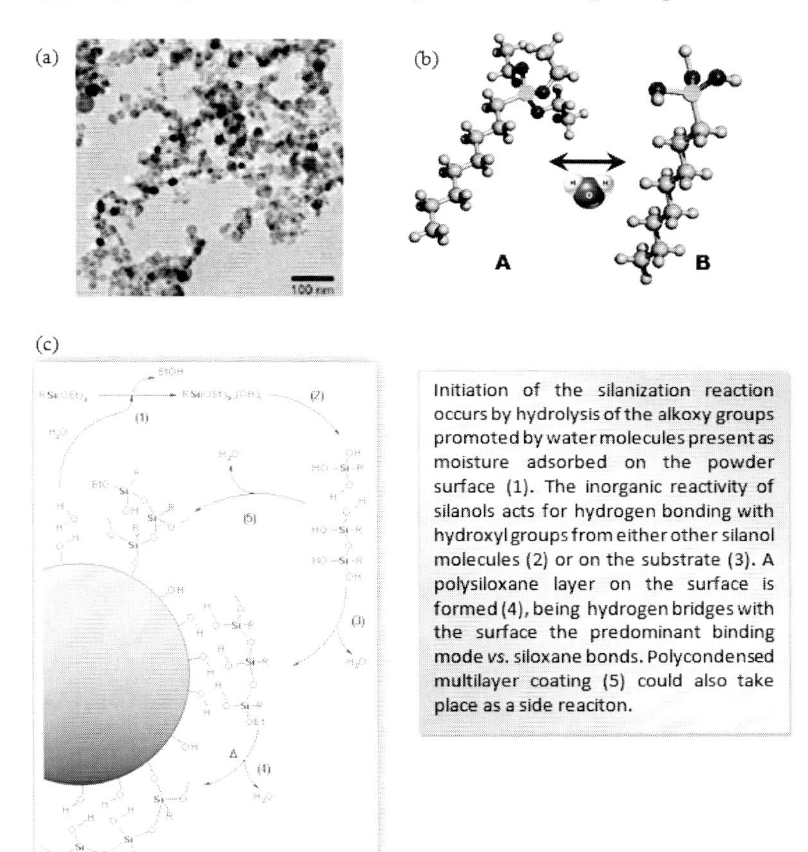

Figure 16.7 Sianization reagents: (a) TiO_2 20 nm from Degussa P25, (b) silane reagent (A octyltriethoxysilane) and hydrolized molecule (B octysilicilic acid), and (c) schematic representation of the different steps (1–5) of the silanization reaction occurring on the surface of TiO_2 nanoparticles with triethoxysilanes.

Figure 16.8 Microscopy images of raw and silanized TiO_2 nanoparticles: SEM (Hitachi S-570) images of (a) pristine powder, (b) supercritically silanized, and (c) conventionally silanized powder and TEM images of (d) pristine powder, (e) supercritically silanized, and (f) supercritically silanized powder treated with an electron opaque heavy metal salt (uranyl acetate).

A value of grafting densities of ~3 molecules/nm^2 is found for the supercritically silanized mesoporous aggregates of TiO_2 nanoparticles, suggesting the formation of a closely packed silane monolayer. A theoretical limit of 5 molecules/nm^2 has been estimated as the maximum surface density for a close-packed silane monolayer deposited on a flat surface [88, 89]. Despite of the low curing temperature used in the supercritical experiments (313–333 K vs. 293 K in the conventional method), the main portion of mass decay in the thermogravimetric analysis takes place at relatively high temperature (648–873 K), suggesting siloxane chemisorption on the surface and a high degree of horizontal crosslinking for coatings prepared under supercritical conditions (Fig. 16.9a,b) [90]. Static contact angles have been measured on a compressed disk of particles to provide qualitative information related to the surface properties of the powder before and after surface modification (Fig. 16.9c).

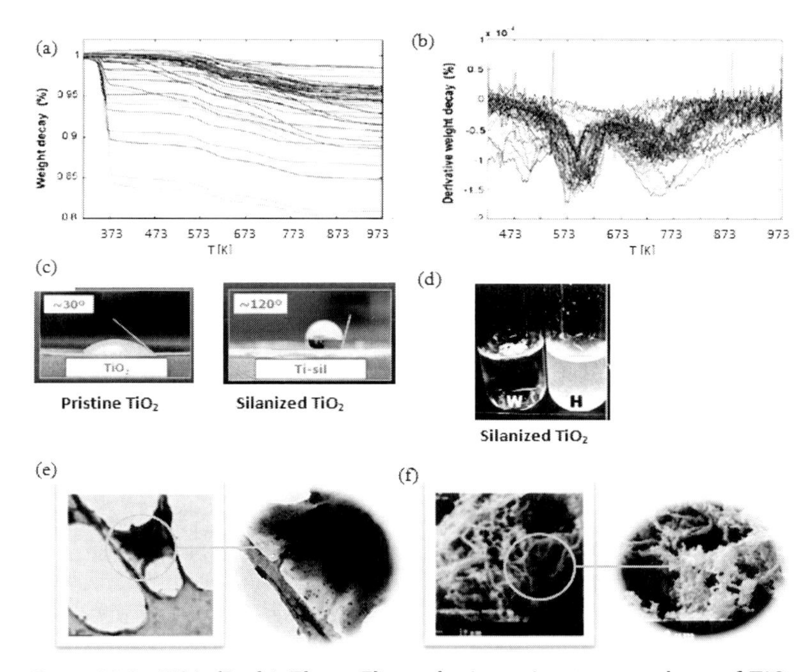

Figure 16.9 TGA (PerkinElmer 7) graphs in an inert atmosphere of TiO_2 nanoparticles reacted with silane under $scCO_2$ and having different degrees of silane deposited: (a) weight loss vs. temperature and (b) derivative of the rescaled profiles to eliminate the effect of the weight decay in the 313–423 K (water) temperature range; hydrophobic character of the supercritically silanized particles verify in (c) compacted pellets in contact with a drop of water and (d) dispersed in polar water (W) and in apolar hexane (H); (e) TEM micrographs of $scCO_2$ PGSS-precipitated composite particles involving solid lipid particles and dispersed silanized TiO_2 (small black dots); and (f) SEM pictures of a composite involving polymethylmethacrylate/poly(ε-caprolactone) fibers loaded with silanized TiO_2 nanoparticles, precipitated using a $scCO_2$ antisolvent procedure.

Using water as the probe fluid, the disc constituted by untreated TiO_2 nanoparticles exhibited a contact angle of ~30°, standard value observed for water deposited on hydrophilic surfaces [91]. On the contrary, for discs formed by compacting $scCO_2$ silanized TiO_2 nanoparticles, the contact angle increased to a value of ~115–120°. This increase indicates a pronounced hydrophobic character of the surface after supercritical silanization. The changes in powder wettability are also estimated by dispersibility analysis in polar water and apolar hexane, with dielectric constants of 80 and 2, respectively [92, 93]. The dispersion of coated nanoparticles gives a transparent suspension when added to the apolar solvent, while they are not dispersible at all in water (Fig. 16.9d).

Finally, surface hydrophobization allows the dispersion of the silanized TiO_2 powder into viscous solids. In a first example, TiO_2 is dispersed into solid lipid particles (Fig. 16.9e) using the PGSS® supercritical procedure [94–96]. For both pharmaceutical and cosmetic applications, solid lipid particles have advantages over other carriers, such as liposomes or emulsions, in terms of stability and protection of the incorporated active compounds. The production of particulate hybrid solid lipid particles containing silanized TiO_2 is investigated with the aim of producing sunscreens with UV radiation protection properties. In a second example, a $scCO_2$ antisolvent technique is used to precipitate networks of fibers of the blend polymethylmethacrylate/poly(ε-caprolactone) (PMMA/PCL) loaded with supercritically silanized TiO_2 nanopowder (Fig. 16.9f) [97]. Loadings of 20 wt% of TiO_2 were achieved. Introduction of small quantities of nanosized inorganic particles can significantly improve mechanical and physical properties of the polymer matrix. These materials found applications in the area of tissue engineering as scaffolds.

16.3 Impregnation of Porous Matter

The deposition of materials in intrinsic porous substrates and swelled nonporous polymers (cellular materials or foams) using supercritical fluids was first proposed by McHugh and Krukonis [98]. Later, Debenedetti et al. [99] fabricated polymeric microspheres loaded with pharmaceuticals for drug delivery applications using the technique of rapid expansion of supercritical solutions. In general, technology based on supercritical fluids is considered as an alternative to overcome some of the problems associated with the use of traditional organic solvents for the modification of materials involving polymers, due to $scCO_2$ plasticization capacity [100, 101]. $scCO_2$ can be used either as a medium for polymer synthesis

[102, 103] or for the postmodification of polymer morphology and/or composition [104, 105]. For nonporous polymers, impregnation must be preceded of polymer swelling, which is responsible for the enhanced diffusion of solute molecules in such matrixes.

16.3.1 Intrinsic Porous Matter

Over the last two decades, significant interest has been devoted to the synthesis and applications of both nanoporous materials (2–100 nm) and hierarchically structural porous materials, which are defined as materials that contain a porous structure consisting of interconnected pores of different length scales from micro- (<2 nm) to meso- (2–50 nm) to macropores (>50 nm) [106]. Micro- and mesopores provide size selectivity for guest molecules enhancing host–guest interactions. Besides, synthesis processes performed in restricted space, that is, inside micro- and small mesopores, in which the dimensions of the reaction vessel are comparable to those of solutes or reactants, provide a new access for the design of novel functional materials [107, 108].

Some of the most important types of highly porous materials with diverse structures are those involving silicon, including zeolites and silicas. Zeolites are the hydrated aluminosilicate members of the family of the crystalline microporous solids known as molecular sieves. Aluminosilicate minerals are composed of aluminum, silicon, and oxygen, plus countercations. Molecular sieves with completely new framework compositions have been synthesized on the basis of aluminophosphates. Those materials are commonly used as commercial adsorbents and in separation processes, since they can easily incorporate guest molecules or ions in their open framework [109, 110], acting also as molecular sieves. The encapsulation of suitable molecular species in zeolite cavities by adsorption or reaction led to materials with specific functionalities and applications in chromotropism, size/shape selective catalysis, gas separation/purification, photocatalysis, or even drug delivery [111–114].

Silica is one of the most complex and most abundant families of materials, existing both as several minerals and being produced synthetically. Notable examples include nonporous quartz, crystal or glass, and amorphous porous silica gel, xerogels, cryogels, and aerogels. Mesoporous silica supports have a random pore size distribution and extremely high pore interconnectivity. Hydrophilic mesoporous silica gels are important commercial substances, while organically modified silica gel (ormosil) is a class of novel material designed to host a large variety of organic and inorganic guests for applications in chromatography, catalysis, photochemistry, etc. [115]. The choice of such amorphous silica materials relied on their low cost and adsorbent properties that are suitable for a large amount of bulk applications. Mesoporous amorphous silica gels are obtained by random aggregation of primary dense silica nanospheres dispersed in a colloidal suspension, following the sol–gel process schematized in Fig. 16.10a.

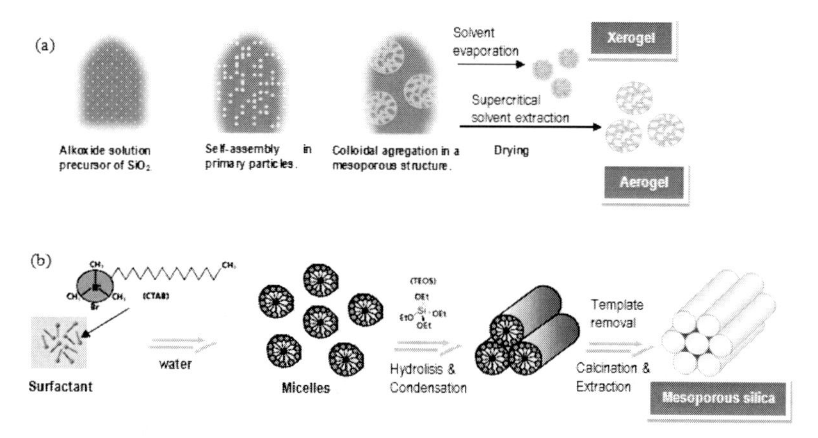

Figure 16.10 Schematic representation of the synthesis procedures for (a) amorphous silica with a random distribution of pores and (b) ordered mesoporous silica.

The sol–gel polymerization of silica alkoxides produces primary sol nanoparticles, which are further linked in a 3D solid network of gel, forming an alcogel with the pores filled with solvent (methanol/acetone and water). A silica aerogel is obtained if a suitable drying technique able to eliminate the liquid solvent from the gel, while avoiding the collapse of the already existing nanoporous structure is picked up. To reach these processing requirements, supercritical drying of alcogels assisted by the use of supercritical fluids (usually $scCO_2$) emerges as the ideal approach to get aerogels [116]. The supercritical drying process leads to the presence of supercritical fluid mixtures in the gel pores without remnants of the liquid phase. The presence of any intermediate vapor–liquid transition and surface tensions in the gel pores is thus avoided, preventing the gel structure from the pore collapse phenomenon during solvent removal [117]. In contrast, traditional drying procedures, for example, air-drying, are not able to preserve the gel structure, leading to pore collapse and massive shrinkage (xerogels).

In a different approach, a new family of large-pore molecular sieves generically called M41S consisting on ordered mesoporous silica materials were discovered in the early 1990s [118]. The most well-known and common mesoporous silica materials include the MCM-41, MCM-48, and SBA-15 families, which have pore sizes in the range of 2–10 nm and different structural characteristics from 2D hexagonal to 3D cubic [119–121]. These mesoporous silica materials are synthesized in the presence of cationic surfactants, which serve as structure-directing agents for polymerizing silica units by electrostatic interaction, as illustrated in Fig. 16.10b. The obtained mesoporous structure has a large surface area with an outstanding percentage of silanol groups, high pore volume, uniform and tunable pore size (easily realized by varying the surfactant used), low bulk density, nontoxic nature, easily modified surface properties, and good biocompatibility. Ordered mesoporous materials have been shown to be important candidates in sensors, catalysts, biomedicine, and environmental applications.

Porous loaded matter is obtained via crystallization inclusion that involves the addition of the guest component to batches for the hydrothermal synthesis of the porous material. An alternative procedure involves impregnation after the matrix has been synthesized, for example, by ion exchange or diffusion of the guest compound from a liquid solution or a gas phase [122]. However, a deep homogeneous impregnation in micro- and mesopores is difficult to achieve using conventional solvent methods. When the loading of nanoporous materials is carried out from a liquid solution, the possibility of competition between solvent and solute molecules for the substrate adsorption sites often leads to the incorporation of both components into the cavities. $scCO_2$ has important advantages as a solvent to carry out the impregnation process. First, the adsorption behavior of pure supercritical fluids has been reported to be different from that of gases at temperatures lower than the critical, since condensation is impossible for a supercritical fluid [123]. Physical van der Waals adsorption is essentially only valid for vapors. Nonspecific physical adsorption of CO_2 molecules on meso- and macroporous substrates is known to be rather weak at temperatures higher than the critical [124]. Only microporous materials are slightly effective on adsorbing $scCO_2$, as physical adsorption is enhanced by the overlapping of the molecule–surface interaction potentials from opposite pore walls. Hence, under supercritical conditions, competition between the solvent and the solute for the substrate adsorption sites is basically eliminated. Second, enhanced values of transport coefficients are found in $scCO_2$ [125]. Third, supercritical fluid gas-like viscosities allow rapid penetration of the fluid molecules into the cavities of the nanoporous material, while this process is highly hindered in normal liquids.

As an example, the adsorption of benzoic acid into the aluminophosphate VPI-5 channels by diffusion from saturated $scCO_2$ solutions has been studied [126, 127]. The VPI-5 contains 1D circular channels with a free diameter of 1.2 nm and a surface area of 300–400 m^2g^{-1} (Fig. 16.11a) [128].

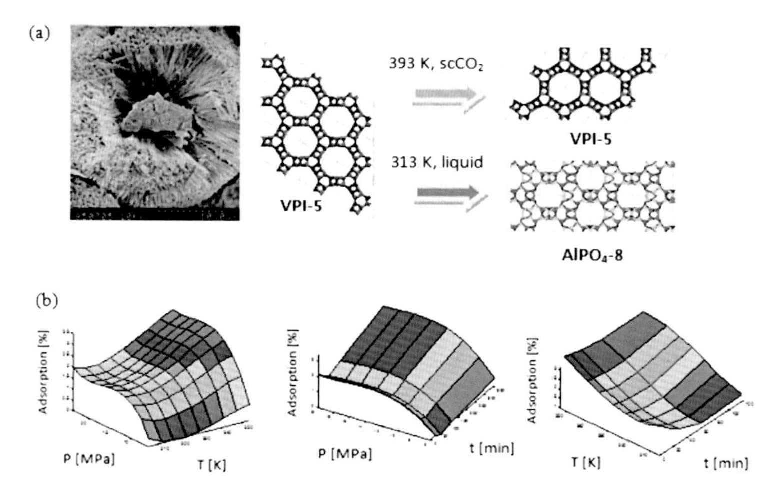

Figure 16.11 (a) SEM of a VPI-5 crystal and a CERIUS-2 structure and (b) chemometric approach based on nonlinear partial least-squares modeling applied to characterize the effect of the experimental variables on the supercritical impregnation of VPI-5 with benzoic acid.

VPI-5 is frequently chosen as adsorbent because its pore opening is significantly larger than that of other zeolites, thus offering the possibility of incorporating larger molecules in its open framework. Obtained loadings under supercritical conditions are in the order of 1.5–3 wt% [129]. Solute uptake and adsorption kinetics can be controlled through the modification of different process parameters, such as time, pressure, and/or temperature of adsorption (Fig. 16.11b) [130]. Most importantly, the structural transformation of VPI-5 to $AlPO_4$-8, with only 0.8 nm pore size, described at temperatures higher than 333 K is avoided by using the supercritical procedure up to 393 K (Fig. 16.11a).

The supercritical silanization process previously described for nanoparticles coating has been extended to the modification of the internal surface of porous silica substrates (i.e., impregnation) of diverse pore size and architectures (Fig. 16.12) [131, 132]. The organic functionalization allows tuning surface properties, such as hydrophobicity and reactivity. The resulting class of hybrid solid porous sorbents has new uses in organic matter separation, with applications ranging from sorption of oil and hydrocarbon contaminants to CO_2 capture. In the disordered silica gel substrates, loading of octytriethoxysilane (0.5–1.5 molecules/nm^2) as a probe molecule, the hydrophobicity of the matrix, determined by Karl Fischer, increased considerably.

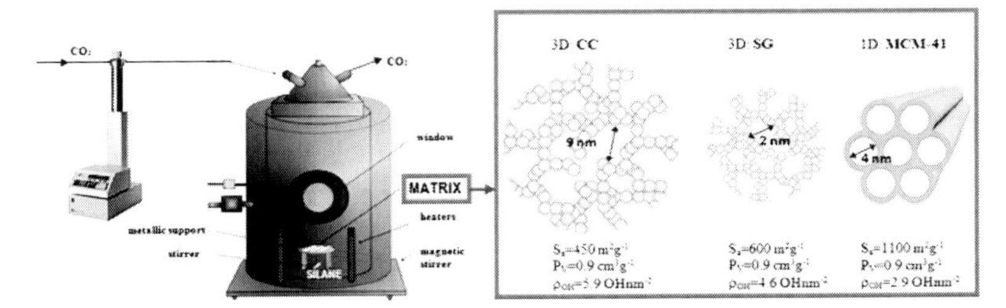

Figure 16.12 Schematic representation of the experimental setup used for the $scCO_2$ silanization (P = 90–20 MPa, T = 318–408 K, t = 30–90 min, octyltriethoxysilane) of several silica matrices with different structures.

The water uptake of the mesoporous samples was reduced to ca. 50 wt% after silanization (Fig. 16.13a). In the SG sample, the modification of the water adsorption capacity was evidenced qualitatively from the color change of an added moisture indicator from blue (dry) to pink (moisturized) (Fig. 16.13b).

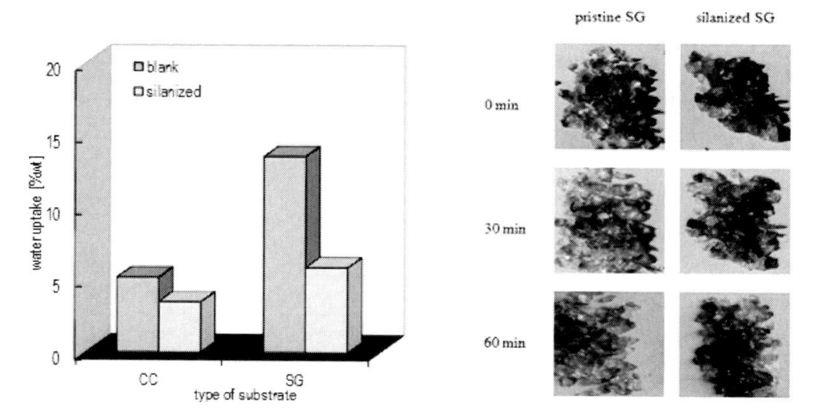

Figure 16.13 Water uptake for two different mesoporous silica gels (CC and SG), pristine and supercritically silanized, after one week of hydration under ambient conditions (293–394 K and 60%–65% relative humidity) and (b) optical pictures of the modification of the color (from blue to pink) of silica blue (SG) sample with a moisture indicator (0.5–1 wt% cobalt chloride) exposed to ambient conditions during 0 (dry samples at 393 K), 30 and 60 min.

Chemical absorption using aqueous alkanolamines is considered a benchmark mature technology in CO_2 capture for the separation of dilute combustion flue gases [133]. The use of solid porous sorbents modified with aminopolymers [134] or aminosilanes [135] is now being widely considered as an alternative separation technology to alkanolamines, potentially less energy intensive [136–138]. In this respect, the immobilization of aminosilanes on the internal surface of silica porous supports (aminosilica) has several advantages as reduced evaporation and degradation of amine. The preparation of the aminosilica can be carried out either by co-condensation of the aminosilane with the silica source during matrix synthesis or by grafting aminoalkoxysilane molecules to the previously synthesized silica support [139, 140]. Grafted compounds have an exceptional high thermal stability and are used for a large variety of applications ranging from catalysis and adsorption to intermediates in advanced synthesis procedures [141]. Supercritical anhydrous silanization using $scCO_2$ as a solvent has result in one of the most effective, simple and reproducible methods for producing homogeneous, covalently bonded, high-density silane films on the internal surface of porous materials. Then, the functionalization of similar silica supports via supercritical silanization with aminosilanes has also been prospected [142]. The process carried out for grafting aminosilanes to silica supports is complex, since it involves both the silane and the amine chemistry. In addition, in the presence of compressed CO_2, primary and secondary alkylamines react with the solvent to give nonsoluble carbamates. Hence, an extensive study of process optimization in regard of working supercritical conditions is necessary prior to hybrid materials preparation. The use of the supercritical method for silanization requires an appreciable solubility of the solute molecules in $scCO_2$ to get them into the pores of the substrate. For alkylsilanes, the operating pressure and temperature determined the solute mole fraction in the fluid phase. Similar behavior is expected for aminosilanes, although for primary and secondary aminoalkylsilanes, the formation of carbamate species with CO_2 take places giving place to an insoluble waxy solid (Fig. 16.14a) [143, 144]. The reaction of CO_2 with primary and secondary amines is usually described by the zwitterion mechanism (reaction A). For the monoamine in an anhydrous environment, the zwitterion undergoes deprotonation by another amine molecule (reaction B), thereby resulting in carbamate formation. For mono and diamines, initially, the addition of pressurized CO_2 causes the formation of a white waxy solid corresponding to carbamate formation (Fig. 16.14b, species II at 333 K), though liquid amine is again visible when the temperature reached a value higher than 373 K (Fig. 16.14b, species I at 393 K). Hence the supercritical procedure should be applied at a minimum temperature of 373 K. $scCO_2$ 3-(methylamino)propyltrimethoxysilane (MPA)-loaded MCM-41 (Fig. 16.14c) and CC (Fig. 16.14d) supports (3–5 mmol·N·g^{-1}) have high thermal stability.

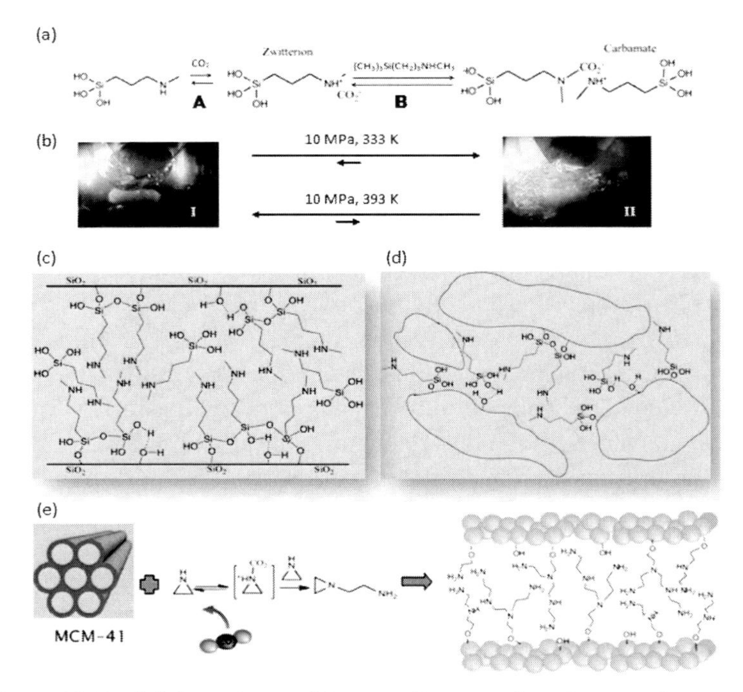

Figure 16.14 (a) Reactions of intermediates and carbamate formation between MAP and CO_2; (b) images captured from the high-pressure reactor window of a MAP and CO_2 mixture as a function of temperature: I at 393 K and II at 333 K; mesoporous supports loaded with MAP (c) MCM-41 and (d) CC; and (e) aziridine ring-opening polymerization in CO_2 and formation of the MCM-41/PEI hybrid product.

The CO_2 adsorption capacity of prepared products, together with the ability in the selective capture of CO_2 from a mixture with other light gases, is an important factor to be considered for applications development. The MAP-grafted mesoporous supports produced by the supercritical procedure showed enhanced CO_2 adsorption selectivity in mixtures of gases, owing to the affinity of the functional group -NH-CH_3 toward CO_2. Adsorption runs are carried out at atmospheric pressure with a mixture of CO_2 (10 vol%) and N_2 at 298 and 318 K to examine these parameters in the supercritically prepared hybrid products. The concentration of adsorbed CO_2 for MCM-41 and CC aminosilanized samples is represented in Fig. 16.15a as a function of the amine loading In the figure, the CO_2 adsorption capacity of supercritically prepared monoamine materials is compared to adsorption literature data (best-performing material) of monoamine-modified silica products synthesized using the toluene approach [145–160].

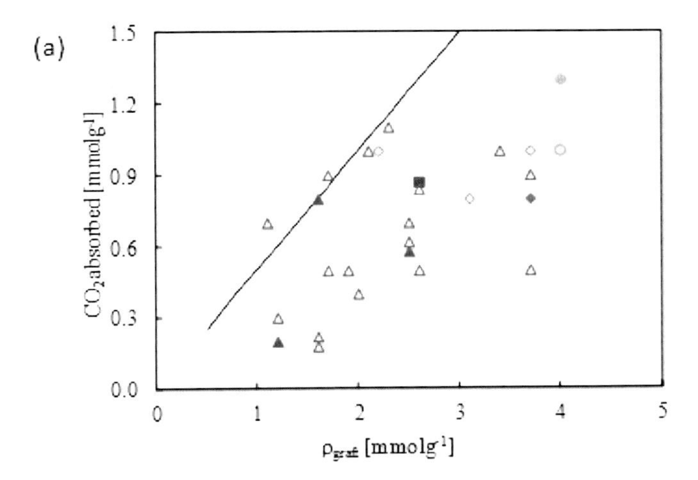

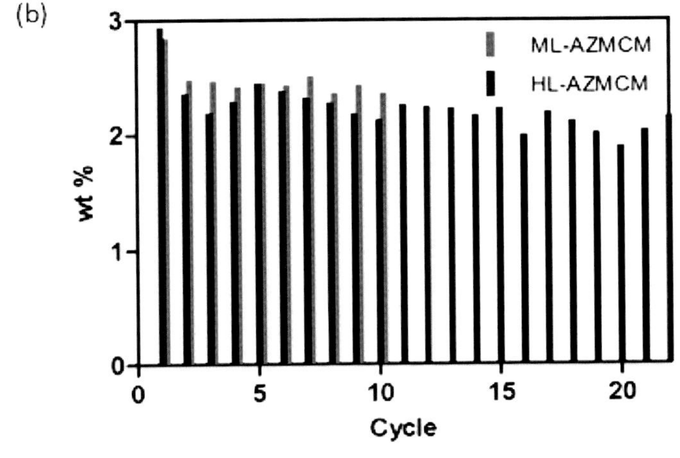

Figure 16.15 Microbalance (IMS HP HT based on a magnetically coupled Rubotherm GmbH unit) results of CO_2 adsorption: (a) first adsorption cycle of MAP aminosilane-loaded MCM-41 and CC hybrid products [142] compared to literature data [145–160] and (b) 10–20 adsorption cycles of PEI-loaded MCM-41 at two different concentrations ML (6 mmol·N·g^{-1}) and HL (8 mmol·N·g^{-1}).

Most of the literature reported amine uptake values are within the range 1–4 mmol·g^{-1}, corresponding to CO_2 adsorption values between ca. 0.4 and 1 mmol·g^{-1}. Under the assumption of carbamate formation at dry conditions, the CO_2/monoamine loading ratio has a theoretical maximum value of 0.5 and it is represented by a solid line in Fig. 16.15a. It is clear from the comparison to literature data that high loading values can be attained by using the supercritical approach without sacrificing the CO_2 capture capabilities. The CO_2 adsorption value for the CC products is in line with the values obtained in the literature, while the MCM-41 samples show an extraordinary high adsorption capacity. The performance of the supercritically prepared hybrid products is fairly stable after 20 adsorption/desorption cycles.

Grafted hyperbranched aminopolymers are often preferred as modifiers instead of aminosilanes, as they have a higher amine density, which is considered advantageous in CO_2 adsorption applications [161]. Current developed methods to prepare hyperbranched polyamine grafted into mesoporous silica need the use of organic solvents, catalysts, temperature, and/or long reaction times [12–164]. A further development of the supercritical technology applied to the surface modification of porous materials has focused on the design of a new eco-efficient method, based in the use of compressed CO_2 as a solvent, reaction media and catalyst, for the in situ polymerization of aminomonomers inside mesoporous silica, obtaining a highly loaded material at low temperatures and short reaction times [165]. The three-membered ring ethyleneimine, also referred to as aziridine, is used to impregnate mesoporous MCM-41 (Fig. 16.14e). The aziridine gives polyethyleneimine (PEI) by ring-opening polymerization, containing primary, secondary, and tertiary amines, and forms covalent bonds with the silica [166]. Hyperbranched PEI loading efficiency in MCM-41 using the supercritical method was in the order of 5–9 mmol·N·g^{-1} at pressures of 2–10 MPa and temperatures of 318 K in only 10 min. Reported data of amine loading for similar materials is in the same range regardless of the used synthetic method. Temperature swing in CO_2 adsorption/desorption cycles test using a microbalance under a flow of CO_2 (10 vol%) and N_2 gave CO_2 adsorption values in the order of 1 mmol·g^{-1} at 298 K. The CO_2 capture performance of the prepared materials was comparable with data for similar products. The CO_2 adsorption was easily reversed with temperature and no appreciable loss in efficiency was observed after 20 cycles (Fig. 16.15b), indicating high thermal stability, as required to any industrial solid adsorbent.

16.3.2 Polymer Bulk Modification by $scCO_2$ Impregnation

16.3.2.1 Semicrystalline polymers

$ScCO_2$ induces depression in the glass transition (T_g), crystallization (T_c), and melting (T_m) temperatures of semicrystalline polymers, which has been reported to affect the crystallization kinetics [167, 168] and the resulting mechanical properties (Fig. 16.16). For semicrystalline polymers, three different behaviors are described as a function of the working temperature (Fig. 16.16a) [169]: (i) by processing the polymer at temperatures near its glass transition temperature, a solvent-induced crystal growth process on the existing crystals is facilitated by the presence of $scCO_2$ (Fig. 16.16c); (ii) by handling the polymer at temperatures between the glass transition and the cold crystallization temperatures, isothermal annealing leading to a heterogeneous nucleation process of the polymer chains in the amorphous region is more likely to occur (Fig. 16.16d); and (iii) polymer treatment at temperatures between the cold crystallization and the melt of the raw polymer provokes nucleation and crystallization from the melt during system depressurization and simultaneous cooling (Fig. 16.16e). For semicrystalline polymers, the $scCO_2$-induced nucleation and/or crystallization effect in the amorphous fraction prevented the molecular dispersion of organic compounds in the matrix, which crystallizes segregated from the polymer (Fig. 16.16b).

16.3.2.2 Amorphous polymers

$ScCO_2$ can reversibly swell glassy and rubbery polymers and reduce the viscosity of the polymer melts up to an order of magnitude. For nonporous polymers, the impregnation process is based on polymer plasticization and swelling caused by CO_2, which facilitates the diffusion of impurities (residual monomer or solvent, initiator, etc.) to the fluid phase, and active agent infusion or impregnation into the polymeric phase. Impregnation processes can be classified on the criteria of solute affinity by the polymer matrix [170, 171]. A first scenario involves solutes with low affinity by the matrix but high solubility in $scCO_2$, in which the deposition of the solute into the polymer matrix occurs upon depressurization.

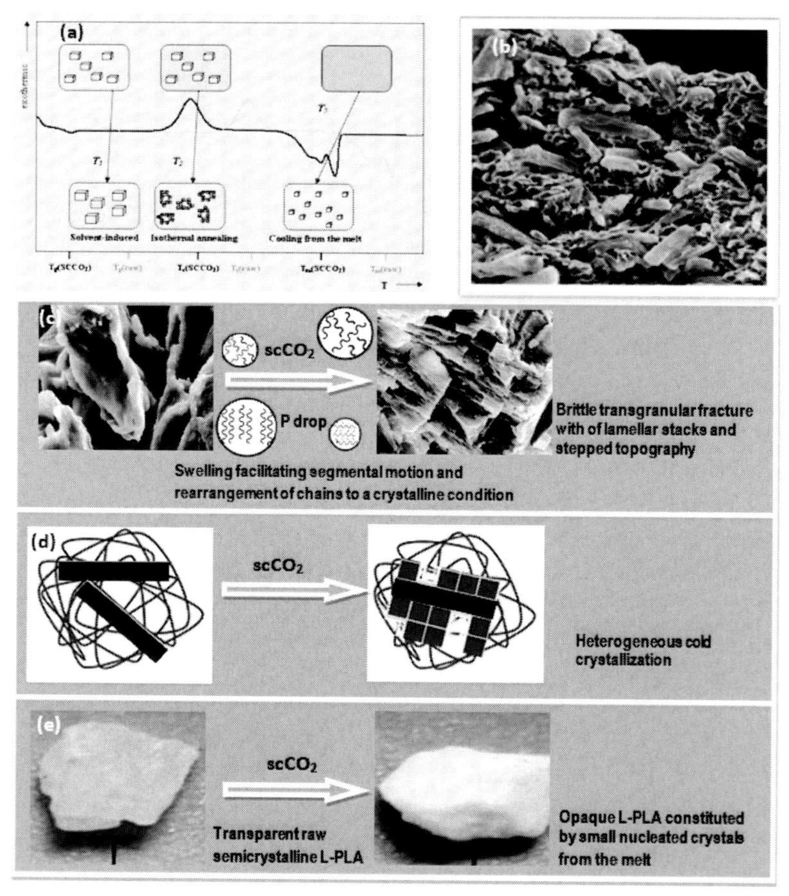

Figure 16.16 (a) Schematic description of the possible L-PLA annealing processes occurring under $scCO_2$ as a function of temperature, (b) segregated drug crystals from semicrystalline L-PLA, (c) SEM micrographs of the morphological modifications produced by $scCO_2$ solvent-induced crystallization, (d) schematic representation of the heterogeneous cold crystallization process, and (e) optical pictures of the transparent/opaque transition caused by fast melt crystallization [169].

The solute is trapped within the polymer matrix by a reprecipitation process, without a molecularly dispersed formulation. Crystallization of the solute in the polymer matrix could lead to a noncontrolled impact on the dissolution and diffusion rate in the polymer. A different scenario considers a high partition coefficient of solute between the polymer and fluid phases due to a high affinity of the solute for the polymer matrix [172, 173]. This mechanism has tremendous potential for the $scCO_2$ molecular impregnation of active agents into polymers for pharmaceutical applications. For active agents with low water solubility, the objective is to enhance drug dissolution rate via dispersion of the drug within water-soluble polymer matrices. On the contrary, for water-soluble and short-half-life drugs, sustained release systems involving an insoluble hydrophobic polymer should be formulated. One of the main limitations that pharmaceutical industry encounters in the preparation of this kind of systems involves the use of organic solvents, since solvent residues get trapped in the polymer together with the drug and may be responsible of some toxic effects.

PMMA is an acrylic hydrophobic biostable polymer that is widely used in the biomedical field as an implant carrier for sustained local delivery of anti-inflammatory or antibiotic drugs. Generally, the polymeric matrixes used for drug delivery are in the form of beads of micro- or nanometric size. However, in some cases monolithic pieces are also loaded to be used as implants or in tissue engineering. Strong interactions between the PMMA and CO_2 molecules have been described, which lead to a dramatic T_g depression (60 K or more at the relatively low pressures of 4–6 MPa) [174, 175]. Therefore, at standard supercritical working conditions (20 MPa and 308 K), the PMMA adsorbed enough gas to lower its glass transition temperature below the processing temperature. The release of CO_2 leads to bubble nucleation and pore growth. Depressurization causes polymer vitrification when the glass transition temperature is reached. At this point, the structure could not further grow and is locked in with a specific morphology. A uniform distribution of macropores throughout the polymer matrix has been described to be formed [176]. The microstructure can be controlled via the rate of pressure release. In general, a rapid depressurization leads to a rapid formation of bubble nuclei and, therefore, a small pore size. A drawback noted with this processing strategy is the formation of a nonporous skin which results from rapid diffusion of the gas away from the surface during depressurization. For the PMMA polymer, scanning electron microscopy (SEM) pictures show reduced apparent external porosity in either beads (Fig. 16.17a) or monolithic bars (Fig. 16.17b) after depressurization [177]. For the PMMA beads, no significant modification in either the shape or the size of the beads is observed. On the contrary, for the monolithic pieces after swelling the rod during 24 h in $scCO_2$, the diameter increases in ca. 25%. The exposure time was not enough to swell the entire rod homogeneously, as it was evident in the images of the rod cross section. An interface between the nonswelled (inner) and plasticized (exterior) regions is observed. The $scCO_2$ foaming process leads to an integral foam structure with a microcellular core encased by a nonporous skin, which resulted from rapid diffusion of the gas away from the surface. The microstructure can be controlled via the rate of pressure release. In general, a rapid depressurization leads to a rapid formation of bubble nuclei and, therefore, a small pore size.

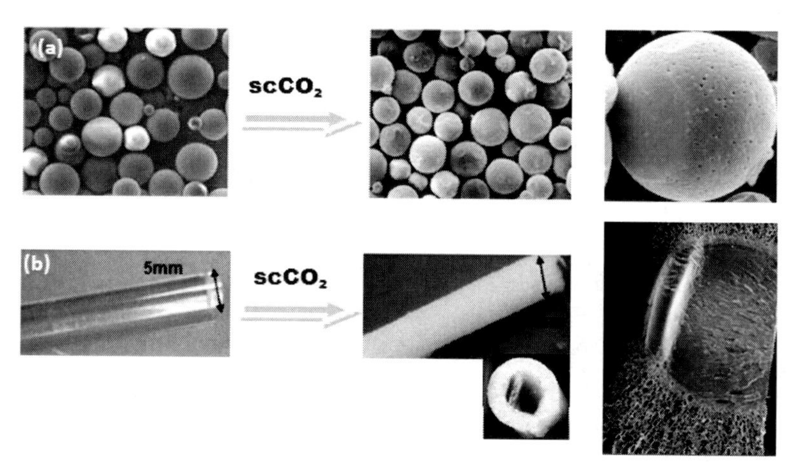

Figure 16.17 PMMA samples before and after treating with $scCO_2$ (20 MPa, 318 K): (a) SEM micrograph of beads with a diameter of 50–100 μm and (b) optical and SEM pictures of monolithic pieces consisting of rods of 5 mm diameter [177].

Acknowledgments

The financial support of the Spanish government under project BIOREG MAT2012-35161 is gratefully acknowledged.

References

1. Beckman, E. J. (2004). Supercritical and near-critical CO_2 in green chemical synthesis and processing, *J. Supercrit. Fluids*, **28**, pp. 121–191.
2. Arai, Y., Sako, T., and Takebayashi, Y. (2002). *Supercritical Fluids, Molecular Interactions, Physical Properties and New Applications* (Springer Series in Materials Processing, USA).
3. Montero, G. A., Smith, C. B., Hendrix, W. A., and Butcher, D. L. (2000). Supercritical fluid technology in textile processing: an overview, *Ind. Eng. Chem. Res.*, **39**, pp. 4806–4812.
4. Banchero, M. (2013). Supercritical fluid dyeing of synthetic and natural textiles – a review, *Color. Technol.*, **129**, pp. 2–17.
5. Draper, S. L., Montero, G. A., Smith, B., and Beck, K. (2000). Solubility relationships for disperse dyes in supercritical carbon dioxide, *Dyes Pigments*, **45**, pp. 177–183.
6. Schmidt, A., Bach, E., and Schollmeyer, E. (2003). The dyeing of natural fibers with reactive disperse dyes in supercritical carbon dioxide, *Dyes Pigments*, **56**, pp. 27–35.
7. van der Kraan, M., Fernandez-Cid, M. V., Woerlee, G. F., Veugelers. W. J. T., and Witkamp, G. J. (2007). Dyeing of natural and synthetic textiles in supercritical carbon dioxide with disperse reactive dyes, *J. Supercrit. Fluids*, **40**, pp. 470–476.
8. Mark, J. E., Lee, C. C.-Y., and Bianconi, P. A. (1995). *Hybrid Organic-Inorganic Composites* **585** (ACS Symposium Series, USA).
9. Jana, S. C., and Jain, S. (2001). Dispersion of nanofillers in high performance polymers using reactive solvents as processing aids, *Polymer*, **42**, pp. 6897–6905.
10. Hakim, L. F., King, D. M., Zhou, Y., Gump, C. J., George, S. M., and Weimer, A. W. Nanoparticle coating for advanced optical, mechanical and rheological properties, *Adv. Funct. Mater.*, **17**, pp. 3175–3181.
11. Feng, W., Patel, S. H., Young, M.-Y., Zunino III, J. L., and Xanthos, M. (2007). Smart polymeric coatings-recent advances, *Adv. Polym. Technol.*, **26**, pp. 1–13.

12. Aguirre, M., Paulis, M., and Leiza, J. R. (2013). UV screening clear coats based on encapsulated CeO_2 hybrid latexes, *J. Mater. Chem. A*, **1**, pp. 3155–3162.

13. Nam, J., Won, N., Bang, J., Jin, H., Park, J., Jung, S., Jung, S., Park, Y., and Kim, S. (2013). Surface engineering of inorganic nanoparticles for imaging and therapy, *Addv. Drug Delivery Rev.*, **65**, pp. 622–648.

14. Palanivel, V., Zhu, D., and van Ooij, W. J. (2003). Nanoparticle-filled silane films as chromate replacements for aluminum alloys, *Prog. Org. Coat.*, **47**, pp. 384–392.

15. Kulkarni, S. A., Ogale, S. B., and Vijayamohanan, K. P. (2008). Tuning the hydrophobic properties of silica particles by surface silanization using mixed self-assembled monolayers, *J. Colloid Interface Sci.*, **318**, pp. 372–379.

16. Gill, I. (2001). Bio-doped nanocomposite polymers: sol-gel bioencapsulates, *Chem. Mater.*, 13, pp. 3404–3421.

17. Wu, P., and Xu, Z. (2005). Silanation of nanostructured mesoporous magnetic particles for heavy metal recovery, *Ind. Eng. Chem. Res.*, **44**, pp. 816–824.

18. Albert, K. (2008). Effect of silane reagent functionality for fluorinated alkyl and phenyl silica bonded stationary phases prepared in supercritical carbon dioxide, *J. Chromatogr. A*, **1191**, pp. 99–107.

19. Yan, F. Y., Gross, K. A., Simon, G. P., and Berndt, C. C. (2003). Peel-strength behavior of bilayer thermal-sprayed polymer coatings, *J. Appl. Polym. Sci.*, **88**, pp. 214–226.

20. Skirtach, A. G., Volodkin, D. V., and Möhwald, H. (2010). Bio-interfaces-interaction of PLL/HA thick films with nanoparticles and microcapsules, *Chem. Phys. Chem.*, **11**, pp. 822–829.

21. Leyton, P., Domingo, C., Sanchez-Cortes, S., Campos-Vallette, M., and Garcia-Ramos, J. V. (2005). Surface enhanced vibrational (IR and Raman) spectroscopy in the design of chemosensors based on ester functionalized *p-tert*-butylcalix arene hosts, *Langmuir*, **21**, pp. 11814–11820.

22. Gomes, J. F. P. S., Rocha, S., do Carmo Pereira, M., Peres, I., Moreno, S., Toca-Herrera, J., and Coelho, M. A. N. (2010). Lipid/particle assemblies based on maltodextrin-gum arabic core as bio-carriers, *Colloids Surf., B*, **76**, pp. 449–455.

23. Sarkar, R., Pal, P., Mahato, M., Kamilya, T., Chaudhuri, A., and Talapatra, G. B. (2010). On the origin of iron-oxide nanoparticle formation using phospholipid membrane template, *Colloids Surf., B*, **79**, pp. 384–389.

24. Vezz, K., Campolmi, C., and Bertucco, A. (2009). Production of lipid microparticles magnetically active by a supercritical fluid-based process, *Int. J. Chem. Eng.*, pp. 1–9.

25. Plueddemann, E. P. (1991). *Silane Coupling Agents*, 2nd Ed. (Kluwer Academic, Plenum Press, New York, USA).

26. Nebhani, L., and Barner-Kowollik, C. (2009). Orthogonal transformations on solid substrates: efficient avenues to surface modification, *Adv. Mater.*, **21**, pp. 3442–3468.

27. Caruso, F. (2001). Nanoengineering of particle surfaces, *Adv. Mater.*, **13**, pp. 11–22.

28. López-Quintera, M. A. (2003). Synthesis of nanomaterials in microemulsions: formation mechanism and growth control, *Curr. Opin. Colloid Interface Sci.*, **8**, pp. 137–144.

29. Ulman, A. (1991). *An Introduction to Ultrathin Organic Film from Langmuir-Blodgett to Self-Assembly* (Academic Press, Boston, MA, USA).

30. Mendel, J. (1999). Dispersions and coating, in *Nanostructure Science and Technology. A Worldwide Study*, eds. Siegel, R. W., Hu, E., and Roco, M. C. (WTEC, Loyola College, MD, USA), pp. 35–48.

31. Ruys, A. J., and Mai, Y.-W. (1999). The nanoparticle-coating process: a potential sol-gel route to homogeneous nanocomposites, *Mater. Sci. Eng.*, **A265**, pp. 202–207.

32. Das, S., Jain, T. K., and Maitra, A. (2002). Inorganic-organic hybrid nanoparticles from n-octyl triethoxy silane, *J. Colloid Interface Sci.*, **252**, pp. 82–88.

33. Ma, M., Zhang, Y., Yu, W., Shen, H.-Y., Zhang, H.-Q., and Gu, N. (2003). Preparation and characterization of magnetite nanoparticles coated by amino silane, *Colloids Surf., A*, **212**, pp. 219–226.

34. Zhang, M., and Singh, R. P. (2004). Mechanical reinforcement of unsaturated polyester by $A1_20_3$ nanoparticles, *Mater. Lett.*, **58**, pp. 408–412.

35. Selvin, T. P., Kuruvilla, J., and Sabu, T. (2004). Mechanical properties of titanium dioxide-filled polystyrene microcomposites, *Mater. Lett.*, **58**, pp. 281–289.

36. Borum, L., and Wilson Jr., O.C. (2003). Surface modification of hydroxyapatite. Part II. Silica, *Biomaterials*, **24**, pp. 3681–3688.

37. Rotello, V. M. (2004). *Nanoparticles: Building Blocks for Nanotechnology* (Springer Series Nanostructure Science and Technology, USA).

38. Uchida, H., Iwai, Y., Amiya, M., and Arai, Y. (1997). Adsorption behaviors of 2,6- and 2,7-dimethylnaphthalenes in supercritical carbon dioxide using NaY-type zeolite, *Ind. Eng. Chem. Res.*, **36**, pp. 424–429.
39. Fan, J., Chen, S., and Gao, Y. (2013). Coating of gold nanoaprticles with peptide molecules via a peptide elongation approach, *Colloids Surf., B*, **28**, pp. 199–207.
40. Ebrahiminezhad, A., Ghasemi, Y., Rasoul-Amini, S., Barar, J., and Davaran, S. (2013). Preparation of novel magnetic fluorescent nanoparticles using amino acids, *Colloids Surf., B*, **102**, pp. 534–539.
41. Cansell, F., and Aymonier, C. (2009). Design of functional nanostructured materials using supercritical fluids, *J. Supercrit. Fluids*, **47**, pp. 508–516.
42. Seremeta, K. P., Chiappetta, D. A., and Sosnik, A. (2013). Poly(ε-caprolactone). Eudragit® RS 100 and poly(ε-caprolactone)/Eudragit® RS 100 blend submicron particles for the sustained release of the antiretroviral efavirenz, *Colloids Surf., B*, **102**, pp. 441–449.
43. Kim, J. H., Paxton, T. E., and Tomasko, D. L. (1996). Microencapsulation of naproxen using rapid expansion of supercritical solutions, *Biotechnol. Prog.*, **12**, pp. 650–661.
44. Mishima, K., Matsuyama, K., Tanabe, D., Yamauchi, S., Young, T. J., and Johnston, K. P. (2000). Microencapsulation of proteins by rapid expansion of supercritical solution with a nonsolvent, *AIChE J.*, **46**, pp. 857–865.
45. Glebov, E. M., Yuan, L., Krishtopa, L. G., Usov, O. M., and Krasnoperov, L. N. (2001). Coating of metal powders with polymers in supercritical carbon dioxide, *Ind. Eng. Chem. Res.*, **40**, pp. 4058–4068.
46. Ovaskainen, L., Rodriguez-Meizoso, I., Birkin, N. A., Howdle, S. M. Gedde, U., Wågberg, L., and Turner, C. (2013). Towards superhydrophobic coatings made by non-fluorinated polymers sprayed from a supercritical solution, *J. Supercrit. Fluids*, **77**, pp. 134–141.
47. Vogt, C., Schreiber, R., Werther, J., and Brunner, G. (2004). Coating of particles in a fluidized bed operated at supercritical fluid conditions, *Chem. Eng. Technol.*, **27**, pp. 943–945.
48. Tsutsumi, A., Nakamoto, S., Mineo, T., and Yoshida, K. (1995). A novel fluidized-bed coating of fine particles by rapid expansion of supercritical fluid solutions, *Powder Technol.*, **85**, pp. 275–278.
49. Wang, T. J., Tsutsumi, A., Hasegawa, H., and Mineo, T. (2001). Mechanism of particle coating granulation with RESS process in a fluidized bed, *Powder Technol.*, **118**, pp. 229–235.

50. Kröber, H., and Teipel, U. (2005). Microencapsulation of particles using supercritical carbon dioxide, *Chem. Eng. Proc.*, **44**, pp. 215–219.
51. Rodríguez-Rojo, S., Marienfeld, J., and Cocero, M. J. (2008). RESS process in coating applications in a high pressure fluidized bed environment: bottom and top spray experiments, *Chem. Eng. J.*, **144**, pp. 531–539.
52. Calderone, M., Rodier, E., Lochard, H., Marciacq, F., and Fages, J. (2008). A new supercritical coinjection process to coat microparticles, *Chem. Eng. Proc.*, **47**, pp. 2228–2237.
53. Yue, B., Yang, J., Wang, Y., Huang, C.-Y., Dave, R., and Pfeffer, R. (2004). Particle encapsulation with polymers *via* in situ polymerization in supercritical CO_2, *Powder Technol.*, **146**, pp. 32–45.
54. Lock, E. H., Merchan-Merchan, W., D'Arcy, J., Saveliev, A. V., and Kennedy, L. A. (2007). Coating of inner and outer carbon nanotube surfaces with polymers in supercritical CO_2, *J. Phys. Chem. C.*, **111**, pp. 13655–13658.
55. Giray, S., Bal, T., Kartal, A. M., Kizilel, S., and Erkey, C. (2012). Controlled drug delivery through a novel PEGn hydrogel encapsulated silica aerogel system, *J. Biomed. Mater. Res.*, **100A**, pp. 1307–1315.
56. Wang, Y., Dave, R. N., and Pfeffer, R. (2004). Polymer coating/encapsulation of nanoparticles using a supercritical anti-solvent process, *J. Supercrit. Fluids*, **28**, pp. 85–99.
57. Wang, Y., Pfeffer, R., and Dave, R. (2005). Polymer encapsulation of fine particles by a supercritical antisolvent process, *AIChE J.*, **51**, pp. 440–455.
58. Marre, S., Cansell, F., and Aymonier, C. (2008). Tailor-made surface properties of particles with a hydrophilic or hydrophobic polymer shell mediated by supercritical CO_2, *Langmuir*, **24**, pp. 252–258.
59. Chen, A. Z., Kang, Y.-Q., Pu, X.-M., Yin, G.-F., Li, Y., and Hu, J.-Y. (2009). Development of Fe_3O_4-poly(L-lactide) magnetic microparticles in supercritical CO_2, *J. Colloid Interface Sci.*, **330**, pp. 317–322.
60. Adami, R., and Reverchon, E. (2012). Composite polymer-Fe3O4 microparticles for biomedical applications, produced by supercritical assisted atomization, *Powder Technol.*, **218**, pp. 102–108.
61. Barbé, C., Bartlett, J., Kong, L., Finnie, K., Lin, H. Q., Larkin, M., Calleja, S., Bush, A., and Calleja, G. (2004). Silica particles: a novel drug delivery system, *Adv. Mater.*, **16**, pp. 1959–1966.
62. Roy, C., Vega-González, A., García-González, C. A., Tassaing, T., Domingo, C., and Subra-Paternault, P. (2010). Assessment of $scCO_2$ techniques for surface modification of micro- and nanoparticles: process design methodology based on solubility, *J. Supercrit Fluids*, **54**, pp. 362–368.

63. Mittal, K. L. (Ed.). (2009). *Silanes and Other Coupling Agents* **5** (CRC Press, Taylor and Francis, USA).
64. Fadeev, A. Y., and McCarthy, T. J. (2000). Self-assembly is not the only reaction possible between alkyltrichlorosilanes and surfaces: monomolecular and oligomeric covalently attached layers of dichloro- and trichloroalkylsilanes on silicon, *Langmuir,* **16**, pp. 7268–7274.
65. Brzoska, J. B., Shahidzzdeh, N., and Rondelez, F. (1992). Evidence of a transition temperature for the optimum deposition of grafted monolayer coatings, *Nature,* **360**, pp. 719–721.
66. Le Grange, J. D., Markham, J. L., and Kurkjian, C. R. (1993). Effects of surface hydration on the deposition of silane monolayers on silica, *Langmuir,* **9**, pp. 1749–1753.
67. Loste, E., Fraile, J., Fanovich, A., Woerlee, G. F., and Domingo, C. (2004). Anhydrous supercritical carbon dioxide method for the controlled silanization of inorganic nanoparticles, *Adv. Mater.*, **16**, pp. 739–744.
68. Domingo, C., Loste, E., and Fraile, J. (2006). Grafting of trialkoxysilane on the surface of nanoparticles by conventional wet alcoholic and supercritical carbon dioxide deposition methods, *J. Supercrit. Fluids,* **37**, pp. 72–86.
69. van der Woort, P., and Vansant, E. F. (1996). Silylation of the silica surface: a review, *J. Liq. Chromatogr. Relat. Technol.*, **19**, pp. 2723–2731.
70. Schreiber, F. (2000). Structure and growth of self-assembling monolayers, *Prog. Surf. Sci.*, **65**, pp. 151–256.
71. Yoshida, W., Castro, R. P., Jou, J.-D., and Cohen, Y. (2001). Multilayer alkoxysilane silylation of oxide surfaces, *Langmuir*, **17**, pp. 5882–5888.
72. Tripp, C. P., and Combes, J. R. (1998). Chemical modification of metal oxide surfaces in supercritical CO_2: the interaction of supercritical CO_2 with the adsorbed water layer and the surface hydroxyl groups of a silica surface, *Langmuir*, **14**, pp. 7348–7352.
73. Cao, C., Fadeev, A. Y., and McCarthy, T. J. (2001). Reactions of organosilanes with silica surfaces in carbon dioxide, *Langmuir*, **17**, pp. 757–761.
74. Xiong, Y., and Kiran, E. (1995). Miscibility, density and viscosity of poly(dimethylsiloxane) in supercritical carbon dioxide, *Polymer*, **36**, pp. 4817–4826.
75. Fink, R., and Beckman, E. J. (2000). Phase behavior of siloxane-based amphiphiles in supercritical carbon dioxide, *J. Supercrit. Fluids*, **18**, pp. 101–110.

76. Kazarian, S. G., Vincent, M. F., Bright, F. V., Liotta C. L., and Eckert, C. A. (1996). Specific intermolecular interaction of carbon dioxide with polymers, *J. Am. Chem. Soc.*, **118**, pp. 1729–1736.

77. Chrastil, J. (1982). Solubility of solids and liquids in supercritical gases, *J. Phys. Chem.*, **86**, pp. 3016–3021.

78. García-González, C. A., Fraile, J., López-Periago, A. M., Saurina, J., and Domingo, C. (2009). Measurements and correlation of octltriethoxysilane solubility in supercritical CO_2 and assembly of functional silane monolayers on the surface of nanometric particles, *Ind. Eng. Chem. Res.*, **48**, pp. 9952–9960.

79. Chimowitz, E. H. (2005). *Introduction to Critical Phenomena in Fluids* (Topics in Chemical Engineering Series, Oxford University Press, UK).

80. Toews, K. L., Shroll, R. M., Wai, C. M., and Samrt, N. G. (1995). pH-Defining equilibrium between water and supercritical CO_2. Influence on SFE of organics and metal chelates, *Anal. Chem.*, **67**, pp. 4040–4043.

81. Pope, E. J. A., and Mackenzie, J. D. (1986). Sol-gel processing of silica, *J. Non-Cryst. Solids*, **87**, pp. 185–198.

82. Wöhrle, D., and Schulz-Ekloff, G. (1994). Molecular sieve encapsulated organic dyes and metal chelates, *Adv. Mater.*, **11**, pp. 875–880.

83. Santos, C., Clarke, R. L., Braden, M., Guitian, F., and Davy, K. W. (2002). Water absorption characteristics of dental composites incorporating hydroxyapatite filler, *Biomaterials*, **23**, pp. 1897–1904.

84. García-González, C. A., Saurina, J., Ayllón, J. A., and Domingo, C. (2009). Preparation and characterization of surface silanized TiO2 nanoparticles under compressed CO_2: reaction kinetics, *J. Phys. Chem. C*, **113**, pp. 13780–13786.

85. Lowe, N. J., Shaath, N. A., and Pathak, M. A. (1997). *Sunscreens: Development, Evaluation, and Regulatory Aspects*, 2nd ed. (Marcel Dekker, New York, USA).

86. McConnell, R. D. (2002). Assessment of the dye-sensitized solar cell, *Renew. Sust. Energ. Rev.*, **6**, pp. 271–293.

87. Herrmann, J. M. (2005). Photocatalysis: state of the art and present applications, *Top. Catal.*, **34**, pp. 49–65.

88. Fadeev, A. Y., Helmy, R., and Marcinko, S. (2002). Self-assembled monolayers of organosilicon hydrides supported on titanium, zirconium and hafnium dioxides, *Langmuir*, **18**, pp. 7521–7529.

89. Helmy, R., and Fadeev, A. Y. (2002). Self-assembled monolayers supported on TiO_2: comparison of $C_{18}H_{37}SiX_3$ (X = H, Cl, OCH_3), $C_{18}H_{37}Si(CH_3)_2Cl$, and $C_{18}H_{37}PO(OH)_2$, *Langmuir*, **18**, pp. 8924–8928.

90. García-González, C. A., Andanson, J. M., Kazarian, S. G., Domingo, C., and Saurina, J. (2009). Application of principal component analysis to the thermal characterization of silanized nanoparticles obtained at supercritical carbon dioxide conditions, *Anal. Chim. Acta*, **635**, pp. 227–234.
91. García-González, C. A., Fraile, J., López-Periago, A. M., and Domingo, C. (2009). Preparation of silane-coated TiO2 nanoparticles in supercritical CO_2, *J. Colloid Interface Sci.*, **338**, pp. 491–499.
92. Luo, K., Zhou, S. Wu, L., and Gu, G. (2008). Dispersion and functionalization of nonaqueous synthesized zirconia nanocrystals via attachment of silane coupling agents, *Langmuir*, **24**, pp. 11497–11505.
93. Iijima, M., Kobayakawa, M., and Kamiya, H. (2009). Tuning the stability of TiO_2 nanoparticles in various solvents by mixed silane alkoxides, *J. Colloid Interface Sci.*, **337**, pp. 61–65.
94. Argemí, A., Domingo, C., Sampaio de Sousa, A. R., Duarte, C. M. M., García-González, C. A., and Saurina, J. (2011). Characterization of new topical ketoprofen formulations prepared by drug entrapment in solid lipid matrices, *J. Pharm. Sci.*, **100**, pp. 4783–4789.
95. García-González, C. A., Sampaio da Sousa, A. A. R., Argemí, A., López-Periago, A., Saurina, J., Duarte, C. M. M., and Domingo, C. (2009). Production of hybrid lipid-based particles loaded with inorganic nanoparticles and active compounds for prolonged topical release, *Int. J. Pharm.*, **382**, pp. 296–304.
96. García-González, C. A., Argemí, A., Sampaio da Sousa, A. R., Duarte, C. M. M., Saurina, J., and Domingo, C. (2010). Encapsulation efficiency of solid lipid hybrid particles prepared using the PGSS® technique and loaded with different polarity active agents, *J. Supercrit. Fluids*, **54**, pp. 342–347
97. García-González, C. A., Vega-González, A., López-Periago, A. M., Subra-Paternault, P., and Domingo, C. (2009). Composite fibrous biomaterials for tissue engineering obtained using a supercritical CO_2 antisolvent process, *Acta Biomater.*, **5**, pp. 1094–1103.
98. McHugh, M., and Krukonis, V. (1986). *Supercritical Fluid Extraction: Principles and Practice* (Butterworths, Stoneham, USA)
99. Debenedetti, P. G., Tom, J. W., Yeo, S.-D., and Lim, G.-B. (1993). Application of supercritical fluids for the production of sustained delivery devices, *J. Controlled Release*, **24**, pp. 27–44.
100. Fleming, O. S., and Kazarian S. G. (2005). Supercritical fluids polymer processing, in *Supercritical Carbon Dioxide in Polymer Reaction Engineering*, eds. Kemmere, M. F., and Meyer, T. (Wiley-VCH, Germany), pp. 205–238.

101. Bahrami, M., and Ranjbarian, S. (2007). Production of micro- and nano-composite particles by supercritical carbon dioxide, *J. Supercrit. Fluids*, **40**, pp. 263–283.

102. Ganapathy, H. S., Hwang, H. S., Jeong, Y. T., Lee, W. K., and Lim, K. T. (2007). Ring-opening polymerization of L-lactide in supercritical carbon dioxide using PDMS based stabilizers, *Eur. Polym. J.*, **43**, pp. 119–126.

103. Bratton, D., Brown, M., and Howdle, S. M. (2003). Suspension polymerization of L-lactide in supercritical carbon dioxide in the presence of a triblock copolymer stabilizer, *Macromolecules*, **36**, pp. 5908–5911.

104. Khan, F., Czechura, K., and Sundararajan, P. R. (2006). Modification of morphology of polycarbonate/ thermoplastic elastomer blends by supercritical CO_2, *Eur. Polym. J.*, **42**, pp. 2899–2904.

105. Jenkins, M. J., Harrison, K. L., Silva, M. M. C. G., Whitaker, M. J., Shakesheff, K. M., and Howdle, S. M. (2006). Characterisation of microcellular foams produced from semi-crystalline PCL using supercritical carbon dioxide, *Eur. Polym. J.*, **42**, pp. 3145–3151.

106. Su, B.-L., Sanchez, C., and Yang, X.-Y. (2012). *Hierarchically Structured Porous Materials: From Nanoscience to Catalysis, Separation, Optics, Energy, and Life Science* (Wiley-VCH Weinheim).

107. van Bekkum, H., Flanigen, E. M., and Jansen J. C. (1991). *Introduction to Zeolite Science and Practice* (Elsevier, Amsterdam, The Netherlands).

108. Corma, A., Garcia, H., Sastre, G., and Viruela, P. M. (1997). Activation of molecules in confined spaces: an approach to zeolite-guest supramolecular systems, *J. Phys. Chem. B*, **101**, pp. 4575–4582.

109. Davis, M. E., and Lobo, R. F. (1992). Zeolite and molecular sieve synthesis, *Chem. Mater.*, **4**, pp. 756–768.

110. Ozin, G. A., Kuperman, A., and Stein, A. (1989). Advanced zeolite materials science, *Angew. Chem., Int. Ed. Engl.*, **28**, pp. 359–376.

111. Clennan, E. L. (2008). Mechanisms of oxygenations in zeolites, *Adv. Phys. Org. Chem.*, **42**, pp. 225–261.

112. Dutta, P. K., and Kim, Y. (2003). Photochemical processes in zeolites: new developments, *Curr. Opin. Solid State Mater. Sci.*, **7**, pp. 483–490.

113. Corma, A., and Garcia, H. (2004). Zeolite-based photocatalysts, *Chem. Commun.*, **13**, pp. 1443–1459.

114. Weitkamp, J., and Puppe, L. (2000). *Catalysis and Zeolites* (Springer, Berlin, Germany).

115. Dash, S., Mishra, S., Patel, S., and Mishra, B. K. (2008). Organically modified silica: synthesis and applications due to its surface interaction with organic molecules, *Adv. Colloid Interface Sci.*, **140**, 77–94.

116. Hüsing, N., and Schubert U. (Eds.) (1998). Aerogels-airy materials: chemistry, structure, and properties, *Angew. Chem., Int. Ed. Engl.*, **37**, pp. 22–45.

117. García-González, C. A., and Smirnova I. (2013). Use of supercritical fluid technology for the production of tailor-made aerogel particles for delivery systems, *J. Supercrit. Fluids*, **79**, pp. 152–158.

118. Degnan, T. F. (2009). Mesoporous materials (M41S): from discovery to application, in *Dekker Encyclopedia of Nanoscience and Nanotechnology*, 2nd ed., Vol. 6 (Taylor and Francis).

119. Takei, T., Houshito, O., Yonesaki, Y., Kumada, N., and Kinomura, N. (2007). Porous properties of silylated mesoporous silica and its hydrogen adsorption, *J. Solid State Chem.*, **180**, pp. 1180–1187.

120. Fryxell, G. E., Mattigod, S. V., Lin, Y., Wu, H., Fiskum, S., Parker, K., Zheng, F., Yantasee, W., Zemanian, T. S., Addleman, R. S., Liu, J., Kemner, K., Kelly, S., and Feng, X. (2007). Design and synthesis of self-assembled monolayers on mesoporous supports (SAMMS): the importance of ligand posture in functional nanomaterials, *J. Mater. Chem.*, **17**, pp. 2863–2874.

121. Angloher, S., and Bein, T. (2005). Organic functionalisation of mesoporous silica, *Stud. Surf. Sci. Catal.*, **158**, pp. 2017–2026.

122. Karge, H. G., and Weitkamp J. (Eds.), (2002). *Molecular Sieves-Science and Technology Book, Post-Synthesis Modification I*, Vol. 3 (Springer, Berlin).

123. Kaneko, K. (1996). Micropore filling mechanism in inorganic sorbents, in *Adsorption on New and Modified Inorganic Sorbents*, eds. Dabrowski, A., and Terykh V. A. (Elsevier Science BV, The Netherlands), pp. 573–598, *Studies Surf. Sci. Catal.*, **99**.

124. Aranovich, G., and Donohue, M. (1997). Determining surface areas from linear adsorption isotherms at supercritical conditions, *J. Colloid Interface Sci.*, **194**, pp. 392–397.

125. Paulatis, M. E., Krukonis, V. J., Kurnik, R. T., and Reid, R. C. (1983). Supercritical fluid extraction, *Rev. Chem. Eng.*, **1**, pp. 179–250.

126. Domingo, C., García-Carmona, J., Llibre, J., and Rodríguez-Clemente, R. (1998). Organic-guest/microporous-host composite materials obtained by diffusion from a supercritical solution, *Adv. Mater.*, **10**, pp. 672–676

127. Domingo, C., García-Carmona, J., Fanovich, M. A., Llibre, J., and Rodríguez-Clemente, R. (2001). Single or two-solute adsorption processes at supercritical conditions: an experimental study, *J. Supercrit. Fluids*, **21**, pp. 147–157.

128. Meier, W. M., and Olson, D. H. (1992). *Atlas of Zeolite Structure Types*, 3rd ed. (Butterworth-Heinemann, London, UK).

129. García-Carmona, J., Fanovich, M. A., Llibre, J., Rodríguez-Clemente, R., and Domingo, C. (2002). Processing of microporous VPI-5 molecular sieve by using supercritical CO_2: stability and adsorption properties, *Microp. Mesop. Mater.*, **54**, pp. 127–137.

130. Domingo, C., García-Carmona, J., Fanovich, M. A., and Saurina, J. (2001). Application of chemometric techniques to the characterization of impregnated materials obtained following supercritical fluid technology, *Analyst*, **126**, pp. 1792–1796.

131. López-Aranguren, P., Saurina, J., Vega, L. F., and Domingo, C. (2012). Sorption of tryalkoxysilane in low-cost porous silicates using a supercritical CO_2 method, *Microp. Mesop. Mater.*, **148**, pp. 15–24.

132. Builes, S., López-Aranguren, P., Fraile, J., Vega, L. F., and Domingo, C. (2012). Alkylsilane-functionalized microporous and mesoporous materials: molecular simulation and experimental analysis of gas adsorption, J. *Phys. Chem. C,* **116**, pp. 10150–10161.

133. Yu, C. H., Huang, C.-H., and Tan, C.-S. (2012). A review of CO_2 capture by absorption and adsorption, *Aerosol Air Qual. Res.*, **12**, pp. 745–769.

134. Xu, X., Song, C., Andrésen, J. M., Miller, B. G., and Scaroni, A. W. (2003). Preparation and characterization of novel CO_2 molecular basket adsorbents based on polymer-modified mesoporous molecular sieve MCM-41, *Microp. Mesop. Mater.*, **62**, pp. 29–45.

135. Samanta, A., Zhao, A., Shimizu, G. K. H., Sarkar, P., and Gupta, R. (2011). Post-combustion CO_2 capture using solid sorbents: a review, *Ind. Eng. Chem. Res.*, **51**, pp. 1438–1463.

136. Vaidya, P. D., and Kenig, E. Y. (2007). CO_2-alkanolamine reaction kinetics: a review of recent studies, *Chem. Eng. Technol.*, **30**, pp. 1467–1474.

137. Rao, A. B., and Rubin, E. S. (2002). A technical, economic, and environmental assessment ofamine-based CO_2 capture technology for power plant greenhouse gas control, *Environ. Sci. Technol.*, **36**, pp. 4467–4475.

138. Choi, S., Drese, J. H., and Jones, C. W. (2009). Adsorbent materials for carbon dioxide capture from large anthropogenic point sources, *Chem. Sus. Chem.*, **2**, pp. 796–854.

139. Hoffmann, F. Cornelius, M., Morell, J., and Fröba, M. Silica-based mesoporous organic–inorganic hybrid materials, *Angew. Chem., Int. Ed.*, **45**, pp. 3216–3251.
140. Fryxell, G. E. (2006). The synthesis of functional mesoporous materials, *Inorg. Chem. Commun.*, **9**, pp. 1141–1150.
141. McKittrick, M. W., and Jones, C. W. (2003). Toward single-site functional materials preparation of amine-functionalized surfaces exhibiting site-isolated behavior, *Chem. Mater.*, **15**, pp. 1132–1139.
142. López-Aranguren, P., Fraile, J., Vega, L. F., and Domingo, C. (2014). Regenerable solid CO_2 sorbents prepared by supercritical grafting of aminoalkoxysilane into low-cost mesoporous silica, *J. Supercrit. Fluids*, **85**, pp. 68–80.
143. Dijkstra, Z. J., Doornbos, A. R., Weyten, H., Ernsting, J. M., Elsevier, C. J., and Keurentjes, J. T. F. (2007). Formation of carbamic acid in organic solvents and in supercritical carbondioxide, J. *Supercrit. Fluids*, **41**, pp. 109–114.
144. Mahajani, V. V., and Joshi, J. B. (1988). Kinetics of reactions between carbon dioxide and alka-nolamines, *Gas Sep. Purif.*, **2**, pp. 50–64.
145. Aziz, B., Zhao, G., and Hedin, N. (2011). Carbon dioxide sorbents with propylamine groups–silica functionalized with a fractional factorial design approach, *Langmuir*, **27**, pp. 3822–3834.
146. Bacsik, Z. N., Atluri, R., Garcia-Bennett, A. E., and Hedin, N. (2010). Temperature-induced uptake of CO_2 and formation of carbamates in mesocaged silica modified with n-propylamines, *Langmuir*, **26**, pp. 10013–10024.
147. Chang, F. Y., Chao, K. J., Cheng, H. H., and Tan, C. S. (2009). Adsorption of CO_2 onto amine-grafted mesoporous silicas, *Sep. Purif. Technol.*, **70**, pp. 87–95.
148. Hiyoshi, N., Yogo, K., and Yashima, T. (2004). Adsorption of carbon dioxide on amine modified SBA-15 in the presence of water vapor, *Chem. Lett.*, **33**, pp. 510–511.
149. Huang, H. Y., Yang, R. T., Chinn, D., and Munson, C. L. (2002). Amine-grafted MCM-48 and silica xerogel as superior sorbents for acidic gas removal from natural gas, *Ind. Eng. Chem. Res.*, **42**, pp. 2427–2433.
150. Kim, S., Ida, J., Guliants, V. V., and Lin, Y. S. (2005). Tailoring pore properties of MCM-48 silica for selective adsorption of CO_2, *J. Phys. Chem. B*, **109**, pp. 6287–6293.
151. Knowles, G. P., Graham, J. V., Delaney, S. W., and Chaffee, A. L. (2005). Aminopropyl-functionalized mesoporous silicas as CO_2 adsorbents, *Fuel Process. Technol.*, **86**, pp. 1435–1448.

152. Ko, Y. G., Lee, H. J., Oh, H. C., and Choi, U. S. (2013). Amines immobilized double-walled silica nanotubes for CO_2 capture, *J. Hazard. Mater.*, **250–251**, pp. 53–60.

153. Ko, Y. G. Shin, S. S., and Choi, U. S. (2011). Primary, secondary, and tertiary amines for CO_2 capture: designing for mesoporous CO_2 adsorbents, *J. Colloid Interface Sci.*, **361**, pp. 594–602.

154. Leal, O., Bolívar, C., Ovalles, C., García, J. J., and Espidel, Y. (1995). Reversible adsorption of carbon dioxide on amine surface-bonded silica gel, *Inorg. Chim. Acta*, **240**, pp. 183–189.

155. Mello, M. R., Phanon, D., Silveira, G. Q., Llewellyn, P. L., and Ronconi, C. M. (2011). Amine-modified MCM-41 mesoporous silica for carbon dioxide capture, *Microp. Mesop. Mater.*, **143**, pp. 174–179.

156. Sanz-Pérez, E. S., Olivares-Marín, M., Arencibia, A., Sanz, R., Calleja, G., and Maroto-Valer, M. M. (2013). CO_2 adsorption performance of amino-functionalized SBA-15 under post-combustion conditions, *Int. J. Greenhouse Gas Control*, **17**, pp. 366–375.

157. Selva, M., Tundo, P., and Perosa, A. (2002). The synthesis of alkyl carbamates from primary aliphatic amines and dialkyl carbonates in supercritical carbon dioxide, *Tetrahedron Lett.*, **43**, pp. 1217–1219.

158. Serna-Guerrero, R., Da'na, E., and Sayari, A. (2008). New insights into the interactions of CO_2 with amine-functionalized silica, *Ind. Eng. Chem. Res.*, **47**, pp. 9406–9412.

159. Wörmeyer, K., Alnaief, M., and Smirnova, I. (2012). Amino functionalised silica-aerogels for CO_2-adsorption at low partial pressure, *Adsorption*, **18**, pp. 163–171.

160. Zelenak, V., Halamova, D., Gaberova, L., Bloch, E., and Llewellyn, P. (2008). Amine-modified SBA-12 mesoporous silica for carbon dioxide capture: effect of amine basicity on sorption properties, *Microp. Mesop. Mater.,* **116**, pp. 358–364.

161. Yu, C. H., Huang C. H., and Tan, C. S. (2012). A review of CO_2 capture by absorption and adsorption, *Aerosol Air Qual. Res.,* **12**, pp. 745–769.

162. Hicks, J. C., Drese, J. H., Fauth, D. J., Gray, M. L., Qi, G., and Jones, C. W. (2008). Designing adsorbents for CO_2 capture from flue gas-hyperbranched aminosilicas capable of capturing CO_2 reversibly, *J. Am. Chem. Soc.*, **130**, pp. 2902–2903.

163. Drese, J. H., Choi, S., Lively, R. P., Koros, W. J., Fauth, D. J., Gray M. L., and Jones, C. W. (2009). Synthesis–structure–property relationships for hyperbranched aminosilica CO_2 adsorbents, *Adv. Funct. Mater.*, **19**, pp. 3821–3832.

164. Chaikittisilp, W., Didas, S. A., Kim H. J., and Jones, C. W. (2013). Vapor-phase transport as a novel route to hyperbranched polyamine-oxide hybrid materials, *Chem. Mater.*, **25**, pp. 613–622.

165. Lopez-Aranguren, P., Vega, L. F., and Domingo, C. (2013). A new method using compressed CO_2 for the in situ functionalization of mesoporous silica with hyperbranched polymers, *Chem. Commun.*, **49**, pp. 11776–11778.

166. Rosenholm, J. M., Penninkangas A., and Linden, M. (2006). Amino-functionalization of large-pore mesoscopically ordered silica by a one-step hyperbranching polymerization of a surface-grown polyethyleneimine, *Chem. Commun.*, **37**, pp. 3909–3911.

167. Gupper, A., Chan, K. L. A., and Kazarian, S. G. (2004). FT-IR imaging of solvent induced crystallization in polymers, *Macromolecules*, **37**, pp. 6498–6503.

168. Takada, M., Hasegawa, S., and Ohshima, M. (2004). Crystallization kinetics of poly(L-lactide) in contact with pressurized CO_2, *Polym. Eng. Sci.*, **44**, pp. 186–196.

169. López-Periago, A. M., García-González, C. A., and Domingo, C. (2009). Solvent- and thermal-induced crystallization of poly-L-lactic acid in supercritical CO_2 medium, *J. Appl. Polym. Sci.*, **111**, pp. 291–300.

170. Kikic, I., and Vecchione, F. (2003). Supercritical impregnation of polymers, *Curr. Opin. Solid State Mater. Sci.*, **7**, pp. 399–405.

171. Kazarian, S. G.(2000). Polymer Processing with Supercritical Fluids, *Polym. Sci., Ser. C*, **42**, pp. 78–101.

172. Elvira, C., Fanovich, A., Fernández, M., Fraile, J., San Román, J., and Domingo, C. (2004). Evaluation of drug deliver characteristics of microspheres of PMMA/PCL-cholesterol obtained by supercritical-CO_2 impregnation and by dissolution-evaporation techniques, *J. Controlled Release*, **99**, pp. 231–240

173. Andanson, J. M., López-Periago, A., García-González, C. A., Domingo, C., and Kazarian, S. G. (2009). Spectroscopic analysis of triflusal impregnated into PMMA from supercritical CO_2 solution, *Vib. Spectrosc.*, **49**, pp. 183–189.

174. Condo, P. D., and Johnston, K. P. (2006). In situ measurement of the class transition temperature of polymers with compressed fluid diluents, *J. Polym. Sci. B: Polym. Phys.*, **32**, pp. 523–533.

175. Pantoula, M., and Panayiotou, C. (2006). Sorption and swelling in glassy polymer/carbon dioxide systems Part I. Sorption, *J. Supercrit. Fluids*, **37**, pp. 254–262.

176. Üzerm, S., Akman, U., and Hortaçsu, Ö. (2006). Polymer swelling and impregnation using supercritical CO_2: a model-component study towards producing controlled-release drugs, *J. Supercrit. Fluids*, **38**, pp. 119–128.

177. López-Periago, A., Argemí, A., Andanson, J. M., Fernández, V., García-González, C. A., Kazarian, S. G., Saurina, J., and Domingo, C. (2009). Impregnation of a biocompatible polymer aided by supercritical CO_2: evaluation of drug stability and drug-matrix interactions, *J. Supercrit. Fluids*, **48**, pp. 56–63.

Chapter 17

Supercritical Dyeing

M. Vanesa Fernandez Cid[a,b] and Geert F. Woerlee[b]
[a]*Echo Pharmaceuticals, B.V., GMP Production and Pharmaceutical Development Center, Rijnkade 16A, 1382 GS Weesp, The Netherlands*
[b]*FeyeCon, B.V., Rijnkade 17a, 1382 GS Weesp, The Netherlands*
vanesa.fernandez@echo-pharma.com

Dyeing textiles by traditional water-based methods demands large amounts of water. Roughly to dye 1 kg of textile 100 L of clean water are used; annually this requires 9 trillion liters of water or four times the water requirement of a country like Taiwan. At the end of the dyeing process, the same equivalent amount of wastewater is generated that needs to be treated before discharge. The dye bath effluents contain salts, bases, and hydrolyzed dye molecules, which make water purification expensive. For the last 25 years, $scCO_2$ has been developed as an alternative to water as a dyeing medium to eliminate and reduce the usage of water and wastewater production in the conventional textile dyeing industry. The physical properties of $scCO_2$ with regard to high diffusion rates and low mass transfer resistance facilitate the penetration of the dye into the textile,

Supercritical Fluid Nanotechnology: Advances and Applications in Composites and Hybrid Nanomaterials
Edited by Concepción Domingo and Pascale Subra-Paternault

ISBN 978-981-4613-40-8 (Hardcover), 978-981-4613-41-5 (eBook)
www.panstanford.com

decreasing dyeing times. Since no water is use, the drying of the textile is not necessary, saving a large amount of energy. Dyes are dissolved in the $scCO_2$ and can be reused as is the CO_2. Nowadays, supercritical dyeing of polyester is commercially available. Several years of equipment engineering and process development have been crystallized in industrial supercritical dyeing machines. This development has opened up a sustainable way of textile dyeing that will be followed soon by natural textiles such as wool and cotton on a commercial scale. In this chapter the state of the art and applications of supercritical textile dyeing are described.

17.1 Introduction

17.1.1 The Conventional Textile-Dyeing Process

In 2012 the textile industry produces 82 million tons of fibers, of which 61% are synthetic fibers and 39% are natural fibers. Approximately 70% of the synthetic fiber is polyester, while cotton dominates the natural fibers with approximately 80% (Fig. 17.1).

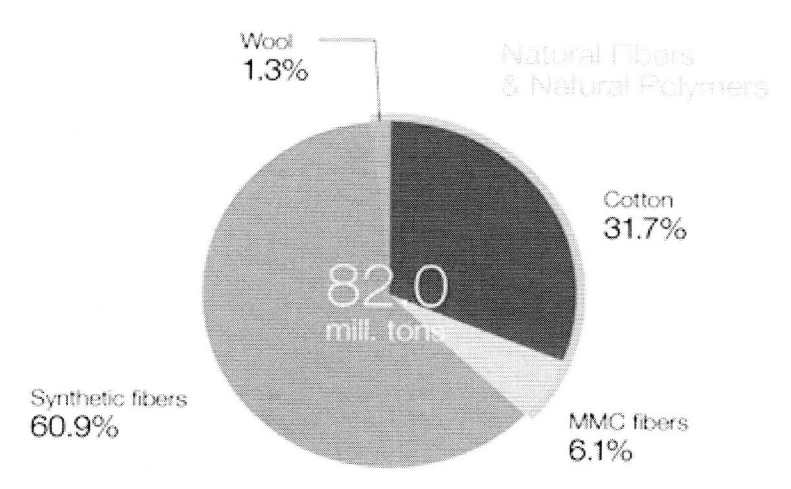

Figure 17.1 Global fiber market.

In the conventional textile dyeing process, it is estimated that 100 to 150 L of water is required for dyeing 1 kg of textile, giving a total amount of up to 9 billion cubic meters of wastewater annually. Approximately 5% to 10% of the nonreactive dyes are lost with the wastewater; for reactive dyes this is even 50% that are commonly used in cotton dyeing. The excellent wet fastness, due to the fixation of the dye via covalent bonding with the substrate, and the brilliance and variety of hues are responsible for the extended used of reactive dyes to dye cotton [1]. Besides dyes, large amounts of salts, alkalis, and dispersing agents are demanded in the dyeing process. An electrolyte is required for overcoming the anionic dye–fiber repulsion, so adsorption can take place. An alkali is needed for the fixation because cellulosic anions are generated, which easily react with the reactive dye. During the dyeing process up to 40% of the dye is hydrolyzed. This hydrolyzed dye is highly substantive for cotton, which needs to be removed via a wash-off step in order to achieve the characteristic wet fastness of reactive dyes. The wash-off step is a laborious operation, which is often longer than the dyeing step itself and requires large volumes of water. Dispersing agents are required to enable the solubilization of the disperse dyes, which are hydrophobic, in water for dyeing polyester. Nowadays, in the textile industry over 50% of the dyeing of cotton in water is carried out with reactive dyes [2]. At the end of the dyeing process high levels of polluted water are produced. The effluent contains up to 0.1 million tons of hydrolyzed dye (when a dye concentration of 2% on weight of the fiber is applied), 16 million tons of salt and 5 million tons of alkali per year [3]. The treatment of the dyeing effluents is complicated and constitutes a major economic and environmental issue. As much as 50% of the total cost of the process is attributed to the washing-off stages and wastewater treatment of the effluent [4, 5]. Therefore, the conventional water-based process of dyeing textiles is an expensive, time-consuming process and a hazard for the environment. The development of a more ecological dyeing process without compromising the desirable attributes of the reactive dyes is greatly demanded.

17.1.2 Supercritical Dyeing: An Alternative Dyeing Process

Dyeing textiles using supercritical carbon dioxide ($scCO_2$) as a dye solvent instead of water has been developed during the last two decades as an alternative process to eliminate the water usage and water pollution [6–8]. From an environmental and safety point of view, $scCO_2$ is the best solvent to replace water in textile dyeing. It is a waste product, is inexpensive, nontoxic, nonflammable, and environmentally friendly, and is chemically inert under many conditions. Simple flow schemes of the alternative supercritical dyeing and the current process are shown in Fig. 17.2.

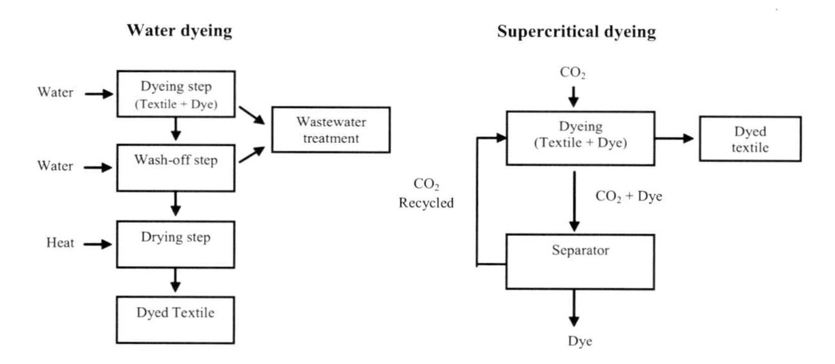

Figure 17.2 Flow schemes of dyeing steps involved in the water dyeing process and in supercritical dyeing process.

17.1.3 Advantages of Supercritical CO_2 for Textile Dyeing

The textile industry can greatly benefit from the use of $scCO_2$ as dye solvent. The high diffusion rates and low mass transfer resistance observed in $scCO_2$ compared to water facilitate dye penetration into the fibers, which allows a reduction on dyeing times. For polyester, the glass transition of the fibers is lowered by the CO_2 that acts as a swelling agent, thus enhancing the diffusion of the dye into the fibers and accelerating the process by a factor of two. Since water is eliminated from the process, the textiles do not need to be dried, saving a great deal of energy. Furthermore, the reactive dye cannot be hydrolyzed; therefore essentially all dye molecules are available for reaction with the fiber. The CO_2 loaded with dyestuff penetrates deep into the pore and capillary structure of fibers, providing effective coloration of the textiles. Unlike water dyeing, the dye can be easily removed from the $scCO_2$ by simply lowering the pressure. The reduction of chemicals and dyes, the lowers amount of energy and the speed of the dyeing process, makes the process economically feasible on top of the environmentally advantages [9–11]. A comparison of the conventional water dyeing and the supercritical dyeing processes is given in Table 17.1.

Table 17.1 Comparison of the conventional water dyeing and the supercritical dyeing process

Conventional water textile dyeing	Alternative textile dyeing with $scCO_2$
Large usage of water	Elimination of water usage
High levels of salt, alkali, and disperse agents	No additives
Hydrolysis of dye molecules	No hydrolysis of dye molecules
Costly water purification	No production of polluted water
Drying step of textile	No drying step (energy saving)
	Shorter process time due to high diffusion coefficients and low mass transfer resistance
	Easy separation of the dye from $scCO_2$ with dye recovery
	Carbon dioxide and dyes reusable

17.2 Challenges of Textile Dyeing in Supercritical CO_2

$scCO_2$ for dyeing synthetic fibers has been shown excellent results on a laboratory scale [12] and the dyeing mechanism has been fully unraveled [13]. Basically, for nonpolar textiles, such as polyester, nonpolar, nonreactive dyes are dissolved into the supercritical phase, transported to the fiber and adsorbed on the surface. Finally, the dye molecules diffuse into the CO_2-swollen polymer matrix, where they are bound to the polyester molecules by physical attraction, mainly dispersion forces. Upon depressurization, the CO_2 molecules exit the shrinking fibers and the dye molecules are retained. Polyester is currently dyed on an industrial scale that allows the dyeing of 180 kg of polyester per batch. The equipment developed by DyeCoo Textile Systems B.V. (the Netherlands) is commercially available and is already in use at different dye houses in Asia [14].

Natural fibers, and especially cotton, have been more challenged to be dyed in $scCO_2$. The dyeing mechanism is not as straight forward as with polyester or other nonpolar fibers, implying a deeper investigation and an understanding of the chemistry involved in the dyeing process in $scCO_2$. Considering that 32% of the world market share is represented by cotton [15], the development of a method for dyeing cotton is essential. To understand the limitations of the dyeing process with cotton, it is necessary to examine the structure of cotton, the properties of CO_2 as a dyeing medium and the dye itself. Cotton consists of an assembly of cellulose chains connected via interchain hydrogen bonding. The chemical structure of cellulose can be described as a condensation polymer of β-D-glucopyranose with 1,4-glycosidic bonds. The intermediate units possess one primary and two secondary alcohols groups each, which are capable of reaction. However, it is generally accepted that the primary alcohol group is more reactive than the secondary groups. Figure 17.3 illustrates the cellulose structure.

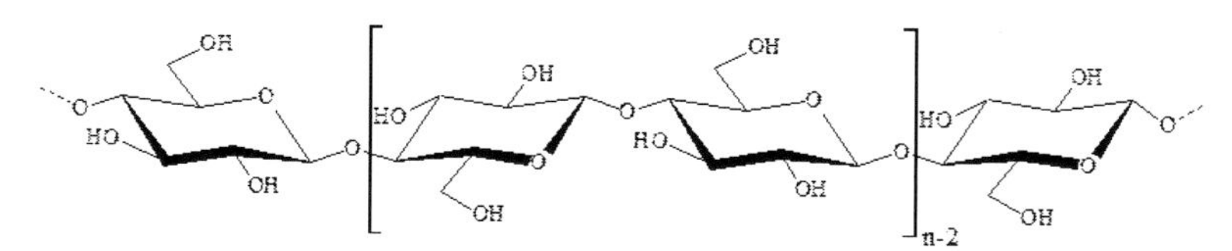

Figure 17.3 Cellulose structure.

The dyeing properties of cotton are determined by its structure, which contains crystalline as well as amorphous domains. The dye can only access the reactive site of cotton in the amorphous region, which represents only some 30% of the whole cotton structure. The dyeability of cotton can be improved, however, by swelling the fibers. The cotton structure is, then, partially disrupted, increasing the accessibility of the reactive sites of cotton. The major limitation of carbon dioxide as a dye medium is its inability to swell the cotton [16], which is a completely opposite effect for polyester or other hydrophobic fibers where $scCO_2$ is an excellent swelling agent. $scCO_2$ is unable to break hydrogen bonds between adjacent molecule chains to partially disrupt the cotton structure and open ways for the dye. Consequently, the diffusion of the dye into the fiber is significantly reduced. In addition, $scCO_2$ can extract the natural moisture present in the cotton fibers. Hence, the cotton loses its flexibility and its glass transition temperature (T_g) increases up to 220°C [17], which is a prohibited operating temperature because it causes cotton damage.

Considering the dyes, the commercial reactive dyes for cotton are salts, generally sulfonate groups introduced for water solubility, and, therefore, they are insoluble in carbon dioxide. The existing nonpolar disperse dyes applied in polyester dyeing are soluble in $scCO_2$; however, they have very low affinity for cotton resulting in very low effective coloration and extremely low dye fixation. To overcome these limitations, disperse dyes are modified by attaching a reactive group that can react with cotton [18]. Reactive disperse dyes are up to now only developed for investigation purposes. Currently, and as part of the up-scaling of the dyeing process of cotton in $scCO_2$ toward a commercial process as polyester, a palette of reactive disperse dyes is under development. An example of a reactive dye is given in Fig. 17.4.

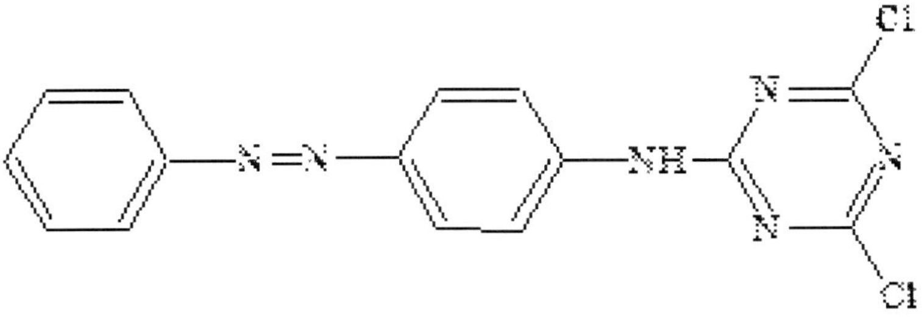

Figure 17.4 Structure of a reactive disperse dye.

17.3 Advances in Textile Dyeing in Supercritical CO_2

17.3.1 The Dyeing Process

As mentioned before, dyeing of hydrophobic fibers, such as polyester, is a well-known and developed process that has found more the challenges in the engineering of the equipment than in the process itself. Equipment is discussed in the next section. On the contrary, dyeing of natural fibers, and especially cotton, in $scCO_2$ have required a more thorough study of the dyeing process and the chemistry involved. In this process development, many aspects are involved, such as the transportation of the dye dissolved in CO_2 to the fiber, the overcoming of the limited swelling properties of CO_2 for natural fibers, the improvement of the affinity of the reactive disperse dyes for cotton, and the understanding of the kinetics of the reaction between the reactive disperse dye and the natural fiber once the dye has reached the reactive sites of cotton. The reaction kinetics of reactive dyes has been studied by different authors [19–22], as an essential step for the development of the dyeing process of natural fibers, especially cotton. In these kinetic studies, the heterogeneous dye–textile system is replaced by the homogeneous reactive dye–alcohol system. Different alcohols have been proposed as models for cotton dyeing [23]; from all of them methanol has been demonstrated to be a satisfactory model [24]. The same approach has been used to investigate the kinetics of the reaction of a reactive disperse dye and methanol in $sscCO_2$ as a solvent [25, 26] as part of the process development of cotton dyeing. From the kinetic studies in $scCO_2$, it was possible to establish the most suitable dye structure and dyeing conditions for reaction with cotton, approaching the dyeing rates observed in the conventional dyeing process of cotton [27].

The major limitation is the fact that $scCO_2$ is a hydrophobic solvent, which does not swell the hydrophilic cotton fibers. It is unable to break hydrogen bonds between adjacent molecule chains to partially disrupt the cotton structure and open ways for the dye. Consequently, the diffusion of the dye into the fiber is significantly reduced. Several processes have been experimentally tested in the literature to improve the dyeability of cotton and other natural fibers in $scCO_2$. A number of studies have been performed to chemically modify the cotton with hydrophobic groups in order to increase its affinity for the hydrophobic disperse dyes and the $scCO_2$ medium as well. Özcan et al. modified the fiber using benzoylation [28]; Schmidt et al. [29] modified the fiber by reaction with small amounts of trichlorotriazine. Disperse dyes with an amine group are necessary to dye the trichlorotriazine-pretreated fiber, essentially forming a reactive dye on the surface of the cotton. These pretreatments have the disadvantage that a structural modification due to a chemical reaction occurs and the properties of the cotton are altered. Other publications mentioned the use of auxiliary agents to facilitate the dissolution and transportation of the dye to the fiber. By adding this auxiliaries or bulking agent, the hydrogen bonds between the cellulosic fibers break, improving the accessibility of the cotton to the dye. The most common agent is polyethylene glycol [30]. Glyezin CD has been also experimented with, and thiodiglycols are mentioned as an option. Several problems arise using these auxiliary agents, however. When disperse dyes are used, low substantivity is found, the dye can be rinsed off the fabric after dyeing. A rinsing step to remove the swelling agent after dyeing is also necessary. It is suggested that dyes with high substantivity, such as reactive dyes, would be more effective with this kind of pretreatment [31].

Another approach is to use polar dyes with a proven affinity for the cellulose fibers, and simply to adapt the dyeing medium for their solubility. Sicardi et al. [32] found a process in which the simple addition of alcohols or water will allow fibers to be sufficiently dyed with standard cellulosic dyes. However, no reports of exact dyeing results or what is understood under "good" dye strength is given. According to another article [33], water-soluble dyes are insufficiently soluble in the medium for it to be useable for a dyeing process. Sawada et al. [34, 35] and Jun et al. [36] enhanced the dyeability of cotton with direct dyes in CO_2 by forming a reverse micellar system with water and a number of different surfactants, most of them fluorinated and not commercially available. Moreover, salts and other additives are necessary for a good dyeing. None of the aforementioned methods have been shown to improve the dyeability of cotton to commercially acceptable values. Often, the improvements made use of chemicals or extra process steps that made the process less commercially viable as well.

Fernandez et al. [37] developed a method to enhance the dyeability of cotton in $scCO_2$ without chemical modification of the fabric, but through physical interactions with the cotton. Presoaking in methanol and addition of extra cosolvent were both found to be necessary to improve coloration and fixation. Reactive disperse dyes were used in this study, where coloration of the cotton was determined by K/S measurements. In this novel method, a rinse step of the fabric was not required, after either the pretreatment or the dyeing for removing any trace of solvents. Hence, the pretreatment and dyeing of cotton can be performed in the same equipment in one batch process, which makes the pretreatment process economically attractive. Dyeing cotton with $scCO_2$ was achieved by Fernandez et al. [38], yielding dye fixations of nearly 100% and evenly distribution of the dye. In this works, reactive disperse dyes with different reactive groups were tested in several reactor sizes, up to a 4 L dyeing vessel. These results have provided the basis for further development of dye palette for cotton and other natural fibers in $scCO_2$, and are a valuable step forward toward the commercialization of a water-free dyeing process for natural fibers. Currently, the dyeing process for natural fibers, such as cotton, wool, silk, etc., in $scCO_2$ is being scaled up to become commercially available, as polyester.

17.3.2 Equipment

A supercritical dyeing process should be operated, as is shown in Fig. 17.5. During the dyeing, the CO_2 is circulated in the vessel in which the beam of fabric is placed. The dye is dissolved in the CO_2 and brought into the vessel, where the dye is delivered to the textile. When the desired coloration is attained, dye is still left in the CO_2, which is removed by passing the CO_2 through a pressure-reducing valve into a separator vessel. In the separator, the CO_2 is gasified, so that the dye precipitates and the clean CO_2 can be recycled by pumping it back to the dyeing vessel.

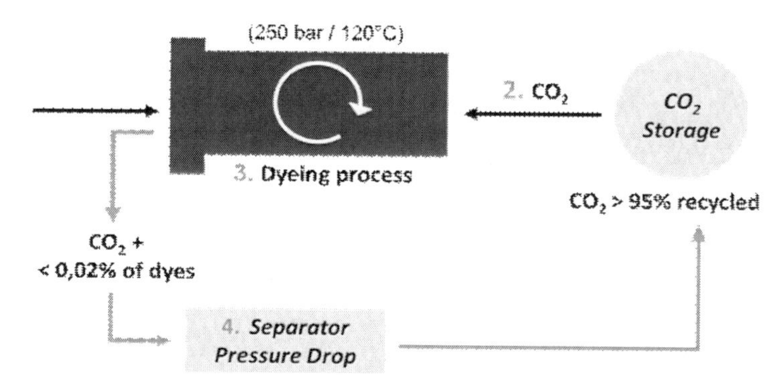

Figure 17.5 Simplified flow diagram of a process to dye textiles in supercritical CO_2.

Pilot plants have been constructed, but it was not until 2013 that a commercial-size machine was built. Currently, a commercial-size dyeing machine for polyester is a reality and in operation at different dye houses in Asia. This dye machine has been developed by DyeCoo Textile Systems B.V. (the Netherlands) and has a capacity of 180 kg of textile per batch. Design and construction of a commercial size dyeing machine has, of course, not been a straightforward process. Years of development have been required to overcome several engineering issues. Pilot-scale machines were designed and constructed first, which have provided excellent information to make the final step toward the commercial size. Engineering solutions have been performed in many features of the dyeing machine; the pressure vessel containing the textile has been an important factor, next to the circulation of the CO_2 to create evenness. Considering the operating pressure of 250 bar required in supercritical dyeing in order to dissolve the dyes, and that industrial dye baths have volumes in the order of 1000 L, the thick-walled supercritical dyeing vessels could consist of a large amount of steel compared to aqueous dyeing vessels. The impact on investment and operating costs could be enormous. Van der Kraan [39] discussed in his research the engineering advances performed with respect to the pressure vessel and other features as the circulation pump for the $scCO_2$ and the heat exchanger for high-pressure CO_2. A technical-scale, 100 L machine was designed and built to test polyester beam dyeing in $scCO_2$, with respect to the performance of the equipment and the process. This technical-scale machine is a polyester beam-dyeing machine also used in aqueous dyeing; the knitted or woven polyester cloth is wound around a perforated pipe. The pipe is closed at one end and the water or, in this case, the $scCO_2$, is pumped into the other end of the pipe. The flow is then forced through the polyester layers; the dye molecules simultaneously diffuse into the fibers. Figure 17.6 gives a schematic impression of the dyeing principle. Both the process and the equipment were designed in close cooperation with FeyeCon D&I B.V. (the Netherlands). Thirteen kilograms of polyester were dyed in this technical-scale machine at 120°C and 300 bar in a dyeing cycle of two hours, taking into account loading, dyeing, and unloading.

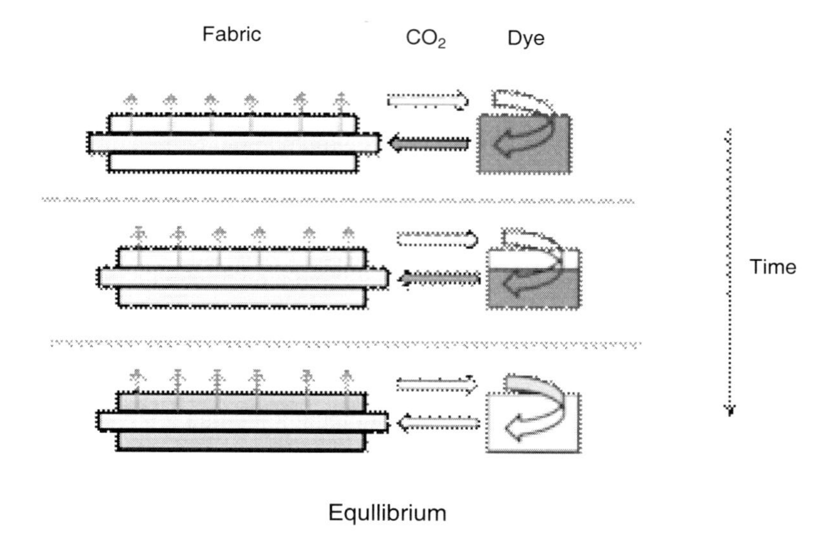

Figure 17.6 Schematic representation of a beam-dyeing process.

The technical-scale beam-dyeing machine has been the fundament for the industrial-scale dyeing machine commercialized by DyeCoo Textile Systems B.V. (the Netherlands). The beam dye concept has been selected because it has the lowest pressure volume requirements. Over a period of 15 years the focus of the development by FeyeCon and later DyeCoo has been on the reduction of investment costs to decrease the hurdle of the larger investment costs (Fig. 17.7). Essential improvements on the engineering are related to the evenness of the dyeing process the dyestuff supply, pumps, cleaning, and utilities. Even more important has been the quality development of the fabrics. Today, the quality of supercritical dyed polyester fabrics is equal or better than for water dyeing. All industrial standards related on light fastness, rubbing, washing, and staining have been addressed. Another, quality advantages is related to the superior reproducibility of the dyeing results. This is due to the fact that the CO_2 quality is far more consistent. It does not vary in pH or salt or mineral content. This enables a smooth transition from one site to another.

Figure 17.7 Commercial supercritical dyeing machine DyeCoo Textile Systems B.V.

17.4 Economical Evaluation

Although the machines are substantially more expensive than water dyeing machines, the significant reduction of operational costs of the machine enables dyeing mills a return of investment of less than three years. Especially, the cost reduction on energy, water, dyes, and chemicals is substantial. Today, the economic evaluations do not yet include benefits related to the better quality. Particularly the reproducibility everywhere in the world will be a big advantage for the larger dye houses. The ability to extract dyes from the fabric will enable an important decrease of nonapproved quality grades. Probably, the largest economic benefit of supercritical dyeing is not even possible to evaluate and will become apparent at a later stage of the cycles. This concerns the use of green field sites. It will be possible have to dye fabrics independent of water, meaning less expensive land, reduction of utility, logistic advantages, and above all a shorter time to market. These advantages will become more apparent in the years to come. It is therefore not surprising that the rivals Nike and Adidas have embraced supercritical dyeing and are introducing marketing campaigns around the technology (Nike: NikeColorDye; Adidas: DryDye).

References

1. Hunter, A., and Renfrew, M. (1999). Reactive dyes for textile fibres, *J. Soc. Dyers Colour.*, pp. 168–177.
2. Phillips, D. A. S. (1996). Environmentally friendly, productive and reliable: priorities for cotton dyes and dyeing processes, *J. Soc. Dyers Colour.*, **112**, pp. 183–199.
3. Blackburn, R. S., and Burkinshaw, S. M. (2002). A greener approach to cotton dyeings with excellent wash fastness, *Green Chem.*, **4**, pp. 47–61.
4. Allegre, C., Maisseu, M., Charbit, F., and Moulin, P. (2004). Coagulation-flocculation-decantation of dye house effluents: concentrated effluents, *J. Hazard. Mater.*, **B116**, pp. 57–64.
5. Shore, J. (1995). Cellulosic dyeing, *J. Soc. Dyers Colour.*, pp. 191–196.
6. Knittel, D., and Schollmeyer, E. (1995). Environmentally friendly dyeing of synthetic fibres and textile accessories, *Int. J. Clothing Sci. Technol.*, **7**, pp. 36–45.
7. Montero, G. A., Smith, C. B., Hendrix W. A., and Butcher, D. L. (2000). Supercritical fluid technology in textile processing: an overview, *Ind. Eng. Chem. Res.*, **39**, pp. 4806–4812.
8. Saus, W., Knittel, D., and Schollmeyer, E. (1993). Dyeing of textiles in supercritical carbon dioxide, *Text. Res. J.*, **63**, pp. 135–142.
9. Bach, E., Cleve, E., and Schollmeyer, E. (2002). Past, present and future of supercritical dyeing technology: an overview, *Rev. Prog. Color*, **32**, pp. 88–101.
10. Fernandez Cid, M. V. (2005). *Cotton Dyeing in Supercritical Carbon Dioxide*, PhD Dissertation, Delft University of Technology, Delft, the Netherlands.
11. Hendrix, W. A. (2001). Progress in supercritical CO_2 dyeing, *J. Ind. Text.*, **31**, pp. 43–56.
12. Bach, E., Cleve E., Schollmeyer E., Bork M., and Koerner P. (1998). Experience with the Uhde CO_2-dyeing plant on technical scale. Part 1: optimization steps of the pilot plant and first dyeing results, *Melliand Int.*, **3**, pp. 192–199.
13. Tabata, I., Lyu, J., Cho, S., Tominaga, T., and Hori, T. (2001). Relationship between the solubility of disperse dyes and the equilibrium dye adsorption in supercritical fluid dyeing, *Color. Technol.*, **117**, pp. 346–351.
14. www.dyecoo.com.
15. Johnson, T. (1999). Outlook for man-made fibers to 2005/2010, *Chem. Fibers Intern.*, **49**, pp. 455–459.

16. Saus, W., Hoger, S., Knittel, D., and Schollmeyer, E. (1993). Fäber aus überkritischem kohlendioxid: dispersionsfarbstoffe und baumwollgewebe, *Textilveredlung*, **28**, pp. 38–40.

17. Beltrame, P. L., Castelli, A., Selli, E., Mossa, A., Testa, G., Bonfatti, A. M., and Seves, A. (1998). Dyeing of cotton in supercritical carbon dioxide, *Dyes Pigm.*, **39**, pp. 335–340.

18. Schmidt, A., Bach, E., and Schollmeyer, E. (2003). The dyeing of natural fibres with reactive disperse dyes in supercritical carbon dioxide, *Dyes Pigm.*, **56**, pp. 27–35.

19. Klančnik, M. (2000). The influence of temperature on the kinetics of concurrent hydrolysis and methanolysis reactions of a monochlorotriazine reactive dye, *Dyes Pigm.*, **46**, pp. 9–15.

20. Klančnik, M., and Gorenšek, M. (1997). Kinetics of hydrolysis of monofunctional and bifunctional monochloro-s-triazine reactive dye, *Dyes Pigm.*, **33**, pp. 337–350.

21. Renfrey, A. H. M., and Taylor, J. A. (1989). Reactive dyes for cellulose. Concurrent methoxide-hydroxide reactions of triazinyl reactive systems: a model system for assessment of potential fixation efficiency, *J. Soc. Dyers Colour.*, **105**, pp. 441–445.

22. Xiao-Tu, L., Zheng-Hua, Z., and Kong-Chang, C. (1989). The kinetics of the hydrolysis and alcoholysis of some model monofluorotriazinyl reactive dyes, *Dyes Pigm.*, **11**, pp. 123–136.

23. Zheng-Hua, Z., Kongchang, C., and Ronggeng, Y. (1990). Study of competitive alcoholysis and hydrolysis of vinylsulfonyl reactive dyes, *Dyes Pigm.*, **14**, pp. 129–142.

24. Bentley, T. W., Ratcliff, J., Renfrew, A. H. M., and Taylor, J. A. (1995). Homogeneous models for the chemical selectivity of reactive dyes on cotton-development of procedures and choice of model, *J. Soc. Dyers Colour.*, **11**, pp. 288–301.

25. Fernandez Cid, M. V., van Spronsen, J., van der Kraan, M., Veugelers, W. J. T., Woerlee, G. F., and Witkamp, G. J. (2007). Acid-catalysed methanolysis reaction of non-polar triazinyl reactive dyes in supercritical carbon dioxide, *J. Supercrit. Fluids*, **39**, pp. 389–398.

26. Fernandez Cid, M. V., van der Kraan, M., Veugelers, W. J. T., Woerlee, G. F., and Witkamp, G. J. (2004). Kinetics study of a dichlorotriazine reactive dye in super-critical carbon dioxide, *J. Supercrit. Fluids*, **32**, pp. 147–152.

27. Özcan, A. S., Clifford, A. A., Bartle, K. D., and Lewis, D. M. (1998). Dyeing of cotton fibres with disperse dyes in supercritical carbon dioxide, *J. Supercrit. Fluids*, **32**, pp. 147–152.

28. Özcan, A. S., Clifford, A. A., Bartle, K. D., Broadbent, P. J., and Lewis, D. M. (1998). Dyeing of modified cotton fibres with disperse dyes from supercritical carbon dioxide, *J. Soc. Dyers Colour.*, **114**, pp. 169–173.
29. Schmidt, A., Bach, E., and Schollmeyer, E. (2003). Supercritical fluid dyeing of cotton modified with 2,4,6-trichloro-1,3,5-triazine, *Color. Technol.*, **119**, pp. 31–40.
30. Schlenker, W., Liechti, P., Werthemann, D., and Della Casa, R. (1993). *Verfahren zum Färben von Textilmaterial aus Cellulose mit Dispersionsfarbstoffen*, German Patent DE 42 30 325 A1.
31. Gebert, B., Saus, W., Knittel, D., Buschmann, H. J., and Schollmeyer, E. (1994). Dyeing natural fibres with disperse dyes in supercritical carbon dioxide, *Text. Res. J.*, **64**, pp. 371–374.
32. Sicardi, S., and Frigerio, M. (2001). *A Method of Dyeing Natural Textile Fibres with a Dyeing Medium Comprising Supercritical Carbon Dioxide*, Italian Patent WO 01/04410 A1.
33. Sawada, K., Takagi, T., Jun, J. H., Ueda, M., and Lewis, D. M. (2002). Dyeing natural fibres in supercritical carbon dioxide using a nonionic surfactant re-verse micellar system, *Color. Technol.*, **118**, pp. 233–237.
34. Sawada, K., and Ueda, M. (2003). Adsorption and fixation of a reactive dye on cotton in non-aqueous systems, *Color. Technol.*, **119**, pp. 182–186.
35. Sawada, K., Takagi, T., and Ueda, M. (2004). Solubilization of ionic dyes in super-critical carbon dioxide: a basic study for dyeing fibre in non-aqueous media, *Dyes Pigm.*, **60**, pp. 129–135.
36. Jun, J. H., Sawada, K., and Ueda, M. (2004). Application of perfluoropolyether reverse micelles in supercritical CO_2 on dyeing process, *Dyes Pigm.*, **61**, pp. 17–22.
37. Fernandez Cid, M. V., Gerstner, K. N., van Spronsen, J., van der Kraan, M., Veugelers, W. J. T., Woerlee, G. F., and Witkamp, G. J. (2007). Novel process to enhance the dyeability of cotton in supercritical carbon dioxide, *Text. Res. J.*, **77**, pp. 38–43.
38. Fernandez Cid, M. V., van Spronsen, J., van der Kraan, M., Veugelers, W. J. T., Woerlee, G. F., and Witkamp, G. J. (2007). A significant approach to dye cotton in supercritical carbon dioxide with fluorotriazine reactive dyes, *J. Supercrit. Fluids*, **40**, pp. 477–483.
39. Van der Kraan, M. (2005). *Process and Equipment Development for Textile Dyeing in Supercritical Carbon Dioxide*, PhD Dissertation, Delft University of Technology, Delft, the Netherlands.

Chapter 18

Introduction to the Analytical Characterization of Materials: Application of Chemometrics to Process Optimization and Data Analysis

Javier Saurina
Department of Analytical Chemistry, University of Barcelona, Martí I Franquès 1-11, 08028 Barcelona, Spain
xavi.saurina@ub.edu

The analytical characterization of different types of (nano)materials prepared by means of supercritical fluid technology is discussed in this section. The assessment of the performance of such materials may result in a great challenge that requires a thorough evaluation of physicochemical, morphological, and textural properties. The application of chemometrics to supercritical fluid technologies for both process optimization and data analysis is an important subject of this chapter. Hence, strategies for systematic design of experiments are applied to investigate the key factors affecting a given process. Besides, principal component analysis and related methods are frequently used to fish into the complex multivariate

Supercritical Fluid Nanotechnology: Advances and Applications in Composites and Hybrid Nanomaterials
Edited by Concepción Domingo and Pascale Subra-Paternault

ISBN 978-981-4613-40-8 (Hardcover), 978-981-4613-41-5 (eBook)
www.panstanford.com

nature of supercritical fluid processes. Representative examples dealing with factor evaluation, optimization of interrelated variables, classification of samples, and prediction of product properties are described. Specific examples are summarized in Chapters 19 and 20. In Chapter 19, vibrational spectroscopy is used to reveal valuable information about polymeric samples treated with $scCO_2$, while methods based on the axisymmetric drop shape analysis for the evaluation of the interfacial and surface tension of fluid interfaces are reviewed in Chapter 20.

18.1 Introduction

Complex composites and hybrid materials and nanomaterials have a broad variety of applications in industrial, technological, chemical, pharmaceutical, and medical fields. In many cases, materials are produced using organic solvents for manufacturing and postprocess media (e.g., methanol, toluene, xylene, methyl ethyl ketone, or dichloromethane), with well-known unwanted properties like cost, toxicity, flammability, and environmental concerns [1]. Furthermore, the manipulation of micro- and nanoentities in organic solvents is extremely difficult due to undesired phenomena such as agglomeration, degradation, or contamination that may damage the delicate surfaces of materials. Although other conventional solvent-free approaches (e.g., high-pressure homogenization or freeze drying [2, 3]) have been proposed to try to minimize these drawbacks, additional problems for dealing with labile compounds have been encountered. As a result, new strategies for elaborating complex (nano)materials with high added value are welcome.

A highly promising manufacturing alternative to the conventional approaches mentioned above relies on supercritical fluids as a type of benign solvent. Our society is increasingly interested in green production to avoid the use and generation of hazardous substances, while preventing environmental and health impacts [4]. Nowadays, the weight of supercritical technologies is growing dramatically in the context of industrial production to develop products, especially for human consumption (e.g., pharmaceuticals and cosmetics).

Dimensions of the materials are important features to be taken into account when planning their evaluation. The study of nanostructured materials certainly entails a higher degree of complexity in comparison with the macro- and microcounterparts. In general, phenomena at the nanosize scale seem to be more severe. The effects of nanomaterials on biological systems are related to some scale-dependent properties, such as capillary forces, magnetism, surface energy, and reactivity [5]. Indeed, nanostructured surfaces are extremely reactive in front of catalytic and oxidative processes, and thus, they can result in a source of unexpected phenomena. The continuous advances on nanoprodutcs are bringing an increasing interest in their potential harmful concerns [6].

This chapter is organized in various sections as follows. First, chemometric methods, focused on characterization of materials, are briefly revised. Subsequently, the principal strategies for structural, physicochemical and textural assays are described. In the last section, examples illustrating representative applications are given.

18.2 Chemometrics for the Study of Supercritical Fluid Processes and Materials

Chemometrics can be defined as the branch of chemistry that uses mathematics, statistics and logic to design and select optimum experimental conditions, and to extract useful chemical information from a given system under study [7, 8]. It is an interdisciplinary approach that can be used in all experimental scientific fields including chemistry, biochemistry, and biology. The application of chemometrics to the study of supercritical processes and materials is gaining momentum, as demonstrated by the large number of papers recently published on this topic [9–11].

18.2.1 Design of Experiments for Screening and Optimization

In this section, design of experiments (DOE) is introduced to provide a better knowledge of the key variables (also referred to as factors) affecting a given process, as well as a more efficient optimization of operational conditions. The optimization of chemical processes is sometimes tackled by trial and error in which runs are carried out without a pre-established plan of experiments. Then, results obtained from some previous assays are interpreted and used to perform further experiments. This approach is poorly systematic, expensive, and time consuming and does not reveal possible interdependences among factors. As a result, the experimental conditions finally chosen as optimal may be wrong.

The strategy towards a more feasible and comprehensive optimization of supercritical processes for manufacturing (nano) materials is schematized in Fig. 18.1.

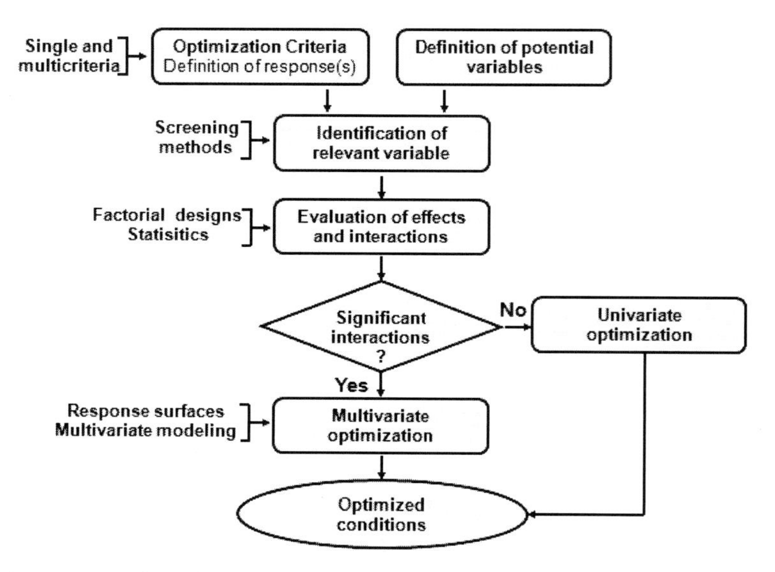

Figure 18.1 Flowchart of the strategy for the study and optimization of supercritical processes.

Exhaustive revisions of optimization methods are given in Refs. [10, 11]. In the figure, the starting point is the definition of the optimization criterion and the potential variables affecting the process. A first concept to be considered is the response, in optimization also referred to as objective response. Response is a (numerical) result provided by a given run, and it can be a single value or a more complex multivariate data (see Section 18.2.2). For instance, the yield of product, the mean particle size, the surface area, the infrared (IR) or X-ray diffraction (XRD) spectra, etc., are typical responses commonly involved in the characterization of materials. Depending on our interests, one or several objective responses can be utilized to establish a suitable criterion to try to find the best conditions. In supercritical processes, a single objective response may be insufficient to express the optimal situation. For instance, in drug delivery systems, the amount of drug loaded in the vehicle is important, of course, but other characteristics have to be taken into account as well. It should be noted that if the process is too fast or to slow the material will be deficient regardless of the loaded amount. Hence, the best product does not correspond necessarily to that with the higher percentage of the active agent, but release kinetics, particle homogeneity, and absence of degradation compounds may be equally important.

The following step consists of screening all potential variables that may influence the process in order to identify those actually important. In supercritical technologies, these variables can be, for instance, pressure, temperature, time, reagent concentration, etc. Each factor can take one, two, or several levels, that is, values that we have previously defined. Hence, only those factors found relevant are considered for a more comprehensive optimization, while the irrelevant ones can be obviated. As the number of variables involved in supercritical processes is quite high, one general, preliminary screening is recommended. Screening methods like fractional or Plackett–Burman designs provide simple models with information about variables with noticeable influence on the behavior of the system. Besides, only a few experiments are required to be run.

Screening designs are usually a prelude to further evaluation of main effects and interactions, typically, by full factorial design. Two-level factorial designs are the most common ones for an exploratory evaluation of variables. Signs + and – are used as the nomenclature to refer the high and low level, respectively. As shown in Fig. 18.2a, experiment ++ corresponds to that assayed at the high level of A and B. When more levels are used (Fig. 18.2b,c), the behavior of such a factor is better defined, although, obviously, the experimental cost is higher. The nature of the variable is important in the definition of levels. For instance, temperature is continuous so any value compatible with the technological equipment can be adopted. Conversely, the nozzle type is intrinsically discrete since devices available are limited to a few shapes (e.g., conical, spherical, etc.). Focusing our attention on the number variables that can be simultaneously explored, experimental designs of two, three, four, and more factors can be planned (see Fig. 18.2). The number of experiments to be performed is L^f, where L is the number of levels and f the number of variables. Hence, experimental efforts required increase dramatically with the number of factors under study.

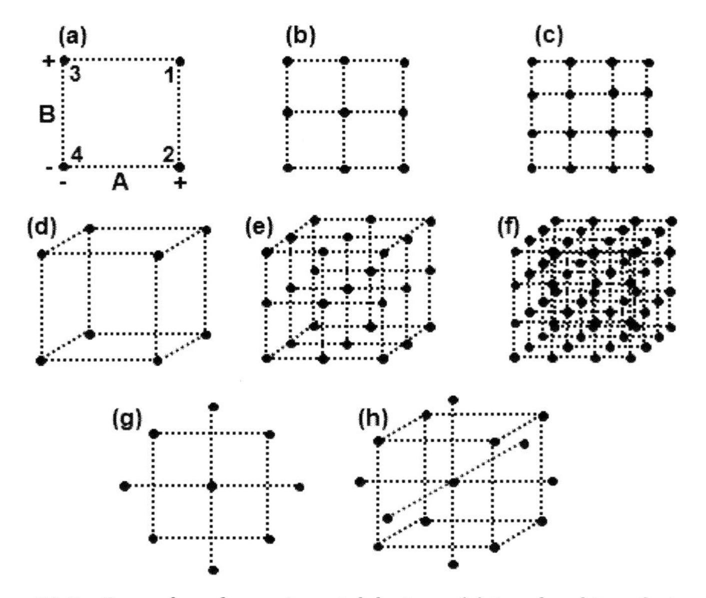

Figure 18.2 Examples of experimental designs: (a) two-level two-factor, (b) three-level two-factor, (c) four-level two-factor, (d) two-level three-factor, (e) three-level three-factor, (f) four-level three-factor, (g) two-factor central composite, and (h) three-factor central composite.

The intensity of a main effect and the degree of interaction among variables can be estimated mathematically. The main effect is accounted from the difference of the responses of assays carried out at the high and low levels (keeping constant the rest of variables). In the example of Fig 18.2a, the main effect of A is estimated from the average of subtracted responses of runs 1 minus 3 and 2 minus 4. The concept of interaction between two or more variables indicates that the behavior of one of them is different depending on the level of the others. In our example, the intensity of interactions between A and B can be scored from the average of subtracted responses of runs 1 minus 4 and 2 minus 3. Statistic tests can be applied to identify factors and interactions that are actually important.

Then, contributions recognized as statistically significant will be investigated more deeply to find optimal conditions.

18.2.1.1 Types of objective functions in optimization

When two or more objective responses are considered to be relevant in a given process, the simplest optimization criterion consists of an independent evaluation of each of these objectives. In many cases, the information is processed separately and, further, a suitable compromise satisfying all predefined objectives is attained. However, the way of reaching such a compromise is often arbitrary and some prejudices may occur. An excellent way to deal with multiple objectives while minimizing the arbitrariness of the previous approach is based on multicriteria decision making. It consists of the definition of a response function that measures the overall suitability or quality of the experimental results. Hence, single objectives such as yield, cost, size, homogeneity, etc., are combined in an objective response function. Multicriteria response functions have been implemented as mathematical expressions involving, for instance, the sum of weighted contributions of objectives according to the following generic equation: $F = \Sigma w_i \times r_i$, where F is the overall response, r_i represents each individual response, and w_i the corresponding weighting coefficient. In this case, certain arbitrariness still persists as w_i values are assigned by the analyst. Objective functions can also be based on product expressions such as $F = \Pi r_i^{wi}$, where F, r_i, and w_i are as above. Again, if an r_i term is considered to be more important than the others the exponent will take a higher value.

A particular case of product functions is the Derringer desirability, one which has been widely used in diverse scientific fields. The overall desirability D is calculated from the individual desirability d_j of each objective response as in a geometric mean: $D = (d_1 \times d_2 \times d_3 \ldots d_k)^{1/k}$, where k corresponds to the number of objectives considered. The desirability values d_j range between 1 and 0, with 1 indicating the best results and 0 the unacceptable ones. The maximum of D corresponds to the optimal experimental conditions reached from the combined objectives. It should be noted that if one of the individual objectives scores 0, D will be 0 independently on the suitability of the rest of objectives (for a practical example see Section 18.4 and Fig. 18.6).

18.2.1.2 Univariate vs. multivariate optimization

According to the scheme shown in Fig. 18.1, the presence or absence of interactions conditions marks the strategy to be followed. For instance, those variables independent from the rest can be studied in a univariate manner. In this case, a given variable is examined at two or several levels by measuring the corresponding objective responses, while the rest of variables are maintained constant. Once that variable has been tested we proceed with another one in a similar way. The procedure is finish when all variables have been explored. In contrast, the simultaneous optimization of interacting variables requires the use of methods such as central composite and grid designs. The resulting data can be fitted to a response surface. When dealing with three or more variables, 3D structures and higher-order structures can be built. Obviously, representing more than two dimensions is difficult, so slices of pairs of variables can be plotted at different levels.

18.2.2 Chemometric Methods for Data Analysis

The characterization of composites and hybrid (nano)materials prepared by supercritical fluid technologies entails several physicochemical, morphological and textural assays. The resulting data can be treated by statistic and mathematical tools to try to extract the maximum amount of relevant chemical information. Using the arsenal of analytical techniques available in the laboratories, data of different dimensionality, ranging from scalars to tensors, can be generated (see Fig. 18.3) [12, 13]. For instance, microscopy performs pore and particle size measurements, so the mean diameter is a discrete data to be used as a characteristic of materials. Similarly, the percentage of loaded drug, the residual solvent content or the concentration of an impurity can also be expressed numerically as a single data value. Data of higher dimensionality is easily obtained from other analytical measurements. For instance, in delivery studies, the drug dissolution can be monitored as a function of time resulting in concentration values taken over the release process. Similarly, spectroscopic measurements from X-ray, Raman, or IR techniques record data in the form of intensity values versus wavelength. Hence, kinetic or spectral profiles consist of lists of numbers arranged as in a data vector. Imaging techniques, 2D spectroscopies (2D nuclear magnetic resonance [NMR], tandem mass spectrometry [MS/MS], excitation/emission fluorescence, . . .) and spectrokinetic measurements generate data of higher dimensionality. In sample imaging, for example, specimens are scanned regularly over two spatial dimensions by recording the property of interest in small square portions or pixels, thus resulting in a table of values, the so-called data matrix (see Fig. 18.3a). Furthermore, when a full spectrum is registered for each pixel the data structure corresponds to a cube (3D) matrix, a data tensor in mathematical jargon.

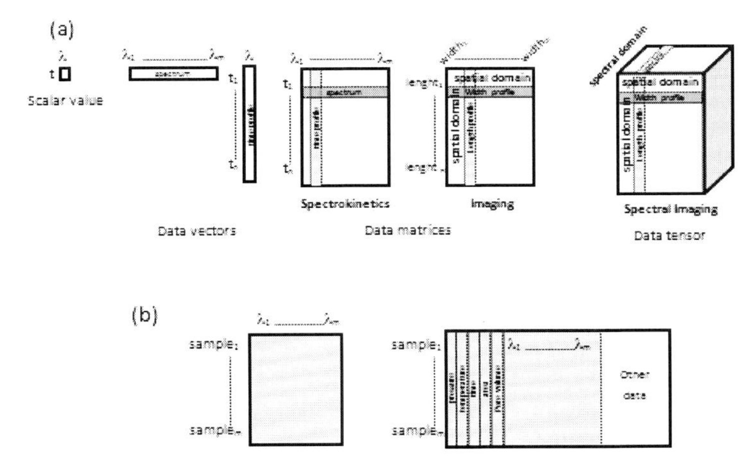

Figure 18.3 Types of data in the study of materials. (a) Data from a given sample according to dimensionality and (b) example of a data matrix to be used in principal component analysis.

To summarize, following to the chemometric nomenclature, data is classified according to its dimensionality or domains of measurement as follows [13]: zero-way or scalar data, one-way data or data vector, two-way data or data matrix, and three-way data or data tensor (although unusual, data of higher dimensionality would be possible as well). The informative content and the analytical potentiality increase from zero- to multiway systems.

18.2.2.1 Exploratory methods

Principal component analysis (PCA) is a powerful multivariate modeling method widely used for exploratory studies, including sample clustering and classification, evaluation of descriptive variables, study of correlations, etc. [14]. In PCA, experimental data obtained from the analysis of materials is arranged in a matrix (**X**) in which each row corresponds to a sample/process and each column to a variable (experimental, estimated or bibliographic data). All columns in the matrix can be of the same nature (e.g., spectral or kinetic profiles) or different (from different techniques or measurements) as represented in the scheme of Fig. 18.3b.

PCA concentrates the information (often referred to as variance) contained in the original variables into a few number of new and more informative variables called principal components (PCs). PCs are lineal combinations of the original variables obtained by decomposition of **X** into scores (**T**) and loadings (**P**) in the following manner: $\mathbf{X} = \mathbf{T} \times \mathbf{P}^{\mathrm{T}} + \mathbf{E}$, where **E** corresponds to the matrix of nonmodeled residuals and the superindex ^{T} stands for the transposed matrix. **T** represents the coordinates of the samples in the space of PCs and **P** the eigenvectors that give the direction of PCs in the space of the original variables. Mathematically, PC1 is built to retain the maximum amount of information. PC2 captures the maximum amount of residual variance and is orthogonal to PC1. The rest of PCs are extracted under the same criteria so vectors are orthogonal to each other and maximize variance captured. In this way, representations of the first PCs display highly concentrated information from both samples and variables.

The plot of scores shows the sample distribution map that sometimes reveals clusters and patterns that can be related to sample or process characteristics. Hence, samples with similar scores, that is, lying in the same zone, display analogous behaviors. In contrast, materials appearing far away are considered different. In general, the 2D scatter plot of PC1 versus PC2 describes quite efficiently the sample behavior as a high percentage of the variance is captured by these PCs. Other 2D and 3D scatter plots may deserve attention if further PCs are found to be relevant. The plot of loadings explains the behavior of variables. Typically, the most descriptive ones have the highest loadings, so they are located in the extreme positions (left, right, top, and bottom) in the corresponding scatter plot. Conversely, variables appearing in the central area are often scarcely important to describe behavior of the system. The correlation between variables is easily investigated by PCA from the similarities of loadings. Visually, variables occurring in close positions are expected to be correlated so they contain the same kind of information. Variables found in opposite positions (e.g., left/right or top/bottom) are negatively correlated, so high values of one correspond to low values of the other and vice versa. In these cases, we talk about redundancy since variables provide equivalent (nondiscriminative) information. On the contrary, uncorrelated variables contain complementary information of great interest for description and interpretation purposes. As in the case of scores, scatter plots of PC1 versus PC2 are typically studied, although other graphs can also be interpreted if necessary. The simultaneous analysis of scores and loadings from the so-called biplot allows materials and features to be related. From the comparison of the relative positions of samples and variables in overlapping spaces the characteristics of materials can be deduced (see examples 2 and 3 in Section 18.4).

18.2.2.2 Multivariate calibration methods

In this part, the principal multivariate calibration methods applied to characterize products and processes from supercritical fluid technologies are introduced. Due to the complex nature of supercritical processes, the final product properties are not dependent on a single experimental factor but they may be affected by multiple instrumental and experimental conditions. As a result, making predictions on product properties from univariate calibration methods may not be very reliable. Under these circumstances, multivariate calibration methods seem to be much more recommendable. The concept of calibration means the establishment of a relationship between the experimental response(s) and another sample property of interest such as the product yield, the analyte concentration, or others. The objective of the calibration is to assess a mathematical model to predict the value of the desired property in unknown samples. Hence, calibration comprises two stages. First, the relationship between instrumental response and the property of the interest is modeled using standard information, that is, data from well-known samples. Second, the property is estimated for the new test (unknown) samples using the corresponding experimental responses and the calibration model previously established.

Since the 1980s, principal component regression (PCR) and partial least-squares (PLS) regression are the most extensively used methods for multivariate calibration. PCR and PLS have essentially been developed for linear modeling of relationships between responses (independent variables, **X** matrix) and sample characteristics (dependent variables, **Y** matrix) [14, 15].

Data contained in matrix **X** is analogous to that described above for PCA (see some matrix possibilities in Fig. 18.3b). For each sample, for instance, the spectrum recorded by a given instrumental technique, the chromatographic or kinetic profiles, or data collected from various sources including operational conditions (time, pressure, temperature) and product characteristics (e.g., surface area, pore volume, particle diameter, . . .) can be integrated in the **X** matrix. The dependent variable (or variables) to be estimated consists of one or various columns of data such as the product yield of the corresponding samples. As in PCA, PCR first decomposes the original variables **X** into scores (matrix **T**) and loadings (matrix **P**). Using appropriate standard samples, in the calibration step, multilinear regression is applied to model the relationship of **T** and **Y**, thus giving the matrix of the regression coefficients **B** follows: $\mathbf{Y} = \mathbf{T} \times \mathbf{B}$. In the prediction, the score matrix of the unknown sample $\mathbf{T}_{test}$ is obtained and the property of interest ($\mathbf{Y}_{test}$) is calculated according to the expression $\mathbf{Y}_{test} = \mathbf{T}_{test} \times \mathbf{B}$. PLS determines latent variables, that are linear combinations of the original variables, but maximal covariance between responses and properties to predict is searched, that is, **X** and **Y** matrices. Thus, eigenvectors and scores calculated using PLS are not orthogonal so they are quite different from those of PCR. The algorithm is iterative and complex mathematically so its description is not pertinent here (for additional details see Refs. [14, 15]).

18.3 Analytical Techniques for the Characterization of Materials

Complex products, such as composites and hybrid (nano)materials, need to be evaluated according to a multidisciplinary approach involving physicochemical morphological, textural, and thermal assays. The composition and structure of such materials is often characterized by XRD, vibrational, and NMR spectroscopies on the solid state. Chromatography is used for more accurate quantification of active compounds loaded, and detection of impurities and solvent residues. The morphology and size of particles can be checked by microscopy, sometimes combining spectroscopic measurements to gain complementary chemical information. In this case, the homogeneity and spatial distribution of active components throughout the sample can be studied by imaging techniques. The thermal transitions and stability of material can be assessed by calorimetric and thermogravimetric analysis (TGA). In the following subsections, a more detailed introduction to different analytical possibilities is given.

18.3.1 Chemical Characterization

18.3.1.1 Solid-state assays

X-ray spectroscopy is used routinely in structural studies to determine the atomic arrangement of sample crystals and to identify unknown substances by comparing diffraction data against databases. Scattering and structure are related through the Bragg's law, where the scattering angle θ is inversely proportional to the interplanar distance [16, 17]. X-ray powder diffraction (XRPD) is the most used technique to obtain the crystallographic structure, crystallite size, and preferred orientation in polycrystalline or powdered solid samples.

Fourier transform infrared (FTIR) spectroscopy in the midrange (4000–400 cm^{-1}) is mainly focused on the identification of the molecules, since the pattern of absorption bands is often considered as a fingerprint of the composition. Attenuated total reflectance (ATR) accessories are of special interest for obtaining spectra of samples that cannot be examined by the normal transmission method [18]. IR imaging in ATR mode allows the spatial distribution of different components to be visualized, thus providing a 2D picture of chemicals. If full spectral resolution is not accomplished for the different species of interest, chemometric methods can be used to increase the resolution by mathematical means.

Raman spectroscopy allows simple sample manipulation, since there is no need to dissolve solids, prepare pellets, or use any other physical or chemical treatment prior analysis. Raman spectra offer acceptable sensitivities for molecules containing aromatic rings and carbon double bonds [19]. Confocal Raman represents a powerful tool to analyze morphological gradients, polymorphism of drugs, and diffusion in polymer systems in a nondestructive manner. This may be of great importance when mechanical sectioning would result in changes in the polymer morphology and/or distribution of components in the matrix. Applications concerning Raman spectroscopy are still scarce in comparison with the FTIR counterpart, especially for the assessment of *dds* prepared by green technologies.

Recent advances on NMR instrumentation have allowed measurements on the solid state to evaluate materials. NMR has the unique ability to simultaneously detect the chemical structure together with the lateral and rotational molecular mobility of every single component, thus differentiating between solids and liquids. Most of the studies involve ^{13}C-NMR, since ^{1}H-NMR measurements remain extremely difficult in the solid state due the small isotropic chemical shift of ^{1}H. Despite this drawback, high resolution ^{1}H-NMR spectra of solids can be obtained with the magic-angle spinning (MAS) technique [20].

TGA is widely used to study the thermal stability of materials. Besides, the technique permits a rough estimation of contents of drugs and other active organic chemicals especially in inorganic matrices. In the case of hybrid or polymeric systems, the progressive decomposition of the organic matrix components may hinder the measurement of the drug decay and, thus, an accurate evaluation of the drug percentage is difficult. Quantification by TGA is based on the percentage of weight lost due to the volatilization of the sample components. Specimens are heated in the range of 40°C to 700°C at a constant rate (e.g., 10°C/min). The weight decays underwent by the product as a consequence of the thermal treatment can be associated to various processes. For example, water removal takes place at temperatures below 100°C and analyte decomposition/volatilization up to 350°C. Other processes associated with the matrix decomposition may occur at higher temperatures. Additionally, the combination of TGA with gas-phase IR analysis provides complementary information about the identification of the volatile component associated to each weight loss process.

18.3.1.2 Wet assays

Assays in dissolution are fundamental for the quantification of chemical compounds of interest including drugs, impurities, and decomposition species. In the case of organic polymeric matrices, for example, poly(lactic acid), polymethylmethacrylate, poly(ε-caprolactone-co-lactide), etc., organic solvents such as acetonitrile, methylene chloride, or acetone are often used to dissolve completely the polymer with the consequent release of the active compound [21]. Subsequently, the sample suspension can be filtered and/or the organic solvent can be evaporated to dryness. The drug is then redissolved in a proper aqueous or hydro-organic solution to be further assessed with the proper analytical method. When dealing with inorganic matrices, the destruction of the carrier structure commonly entails a more aggressive treatment that might also damage the active agent [22]. In these circumstances, alternative approaches such as lixiviation seems to be a good choice when the release is fast and complete. The drug content in the solution recovered from either organic or inorganic matrices is often quantified by ultraviolet-visible (UV-Vis) spectroscopy or high-performance liquid chromatography (HPLC) [23]. UV-Vis methods are simple but they are, in general, inefficient to detect the occurrence of side components. In contrast, chromatographic methods are appropriate to monitor side processes such as drug degradation or release of unwanted components and impurities from the matrix. Despite the claimed null or reduced use of organic solvents in the supercritical manufacturing, small amounts of solvents are sometimes added to enhance the solubility of drugs. Gas chromatography (GC) equipped with a flame ionization detector (FID) is an excellent technique to deal with the detection and quantification of residual organic solvents.

The elaboration of complex (nano)pharmaceuticals focused on controlled drug delivery results in one of the most important applications of supercritical fluid technologies. The performance of these products and their capacity to deliver the desired active agents in a controlled manner is monitored under appropriate experimental conditions. For intravenous and oral forms, for instance, delivery kinetics is often obtained from bulk experiments using devices such as those depicted in Fig. 18.4a. In general, the release solution medium tries to simulate the conditions of the physiological environments. For instance, 0.01 M HCl (pH 2, 37°C) is used to reproduce the gastric conditions or 0.01 M $H_2PO_4^-$ / HPO_4^{2-} (pH 7.4, 37°C) plasmatic ones. Suitable volumes of solution can be withdrawn with a micropipette at the desired (preselected) times over the entire period of study to be further analyzed by chromatography or spectroscopy. The drug release from topic and transdermal pharmaceuticals (e.g., creams, lotions, patches, etc.) is often evaluated by using diffusion cells. Figure 18.4b shows a scheme of the so-called Franz cell consisting of a receptor chamber of 10–100 mL volume and a donor chamber, which are separated by a skin-like membrane supported on a glass frit.

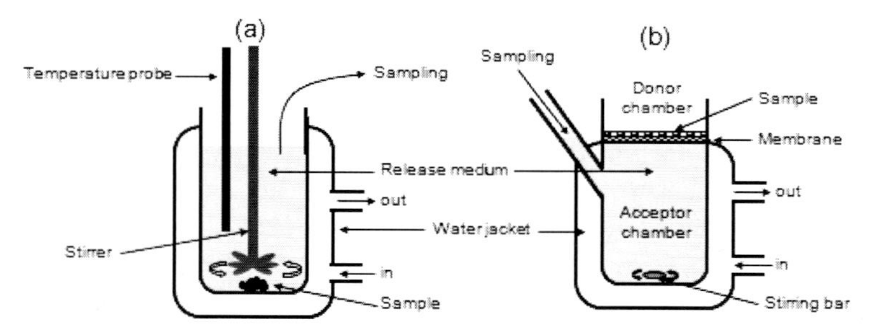

Figure 18.4 Experimental setups for the study of drug release: (a) off-line batch release vessel and (b) Franz cell for diffusion and permeation processes through membranes of topical and transdermal formulations.

As a representative model of skin conditions, the receptor chamber is filled with 0.1 M phosphate buffer solution at pH 7.4 and the temperature of the device is maintained at 32°C ± 0.5°C. Aliquots of 300 μL are withdrawn at preselected times to follow the process by the corresponding analytical processes.

18.3.2 Particle Characterization

Particle characteristics affect the physical and thermal stability, redispersability, the in vivo distribution and, in general, performance of (nano)particulate products. Advanced microscopic techniques including scanning electron microscopy (SEM), transmission electron microscopy (TEM), and atomic force microscopy (AFM) are currently used to evaluate the size, size distribution, morphology, and surface topography of (nano)particles [24, 25].

SEM gives 3D morphological and surface information with a direct visualization of the product. Samples need to be converted first into a dry powder and then they are gold-coated before measurements in vacuum conditions. Hence, in general, SEM is not recommendable for labile samples. SEM instruments are sometimes equipped with an energy-dispersive X-ray spectrometer (EDX). An EDX records X-rays emitted from the sample when bombarded by the electron beam. A combination SEM-EDX performs an elemental analysis on different microscopic sections of the sample. In other cases, microscopes can be furnished with fluorescence probes to monitor the spatial distribution of fluorescent compounds. TEM gives a 2D image of the sample when a beam of electrons is transmitted through a specimen. The sample preparation results in a complex procedure as the thickness required for TEM specimens is in the order of magnitude of 100 nm. AFM offers 3D surface profiles with ultrahigh resolution of 1 nm, approximately. The microscope consists of a cantilever with a sharp tip (probe) at its end that is used to scan the specimen surface [25]. AFM records images of nonconducting samples so none special treatment such as coating is required. As a result, this microscopy provides real pictures of delicate biological and polymeric nanostructures.

Dynamic light scattering (DLS) is a well-established noninvasive technique for measuring the size distribution of (nano)particles, emulsions, lipids and proteins. DLS relies on the Doppler shift effect occurring when a monochromatic light (laser) hits particles with Brownian motion in a colloidal suspension, changing the wavelength of the incoming light and generating a wide spectrum of different frequencies.

Sample stability, melting, sublimation, glass and polymorphic transitions, or chemical degradation of (nano)materials can be studied by differential scanning calorimetry (DSC) [26]. In DSC, sample and reference are kept at the same temperature and the heat flow required to maintain the equality in temperature is measured so thermal events can be thus detected. For polymeric materials, DSC is applied to the study of properties as melting behavior, glass transitions, curing processes, and polymerizations. For crystalline materials, the melting point and its associated enthalpy are used to estimate purity, degree of crystallinity, and particle size.

Surface and textural features of materials are often evaluated by physical gas adsorption. In this method, an inert gas, mostly nitrogen, is adsorbed on the surface of a solid material at low temperature and then desorbed at room temperature. Sorption occurs on the outer surface and, in case of porous materials, also on the surface of pores. The adsorption/desorption isotherms can be interpreted using the equations of the Brunauer–Emmett–Teller (BET) method. For a more accurate evaluation of materials with small micropores argon adsorption assays are more efficient. As an alternative, mercury porosimetry is very successful technique for the evaluation of materials showing broad distributions of pore sizes. The pore size distribution can be obtained as a function of pressure conditions, starting from 4 nm (400 MPa) up to 800 μm (vacuum).

18.4 Examples of Application of Chemometrics to Product Characterization

This section describes representative examples of chemometric studies of evaluation of effects and interactions, assessment of multicriteria response functions, fitting data to response surfaces, comparison of sample/process characteristics, and prediction of product properties.

18.4.1 Example 1: Screening of Factors Influencing the Supercritical Silanization of TiO_2

The surface modification of nanometric titanium dioxide (TiO_2) with octyltriethoxysilane using supercritical carbon dioxide ($scCO_2$) as a solvent was investigated by experimental design [27]. Pressure (P), temperature (T), and processing time (t) have been chosen as variables of the process. Their effects on the properties of the silanized TiO_2 nanoparticles are preliminarily evaluated by means of a three-factor two-level factorial design involving 2^3 experiments (see Table 18.1). The specific surface area (a_s) determined by the BET method is used as an objective response. The magnitude of effects and interactions is estimated according to the Yates's method [28]. Results depicted in Fig. 18.5 show that P is an influencing variables with a positive sign, so that increasing pressure leads to increasing a_s.

The interaction *PT* results in the most significant contribution to the process. Such interaction is attributed to changes in compressibility of $scCO_2$ at the different pressures and temperatures used during silanization.

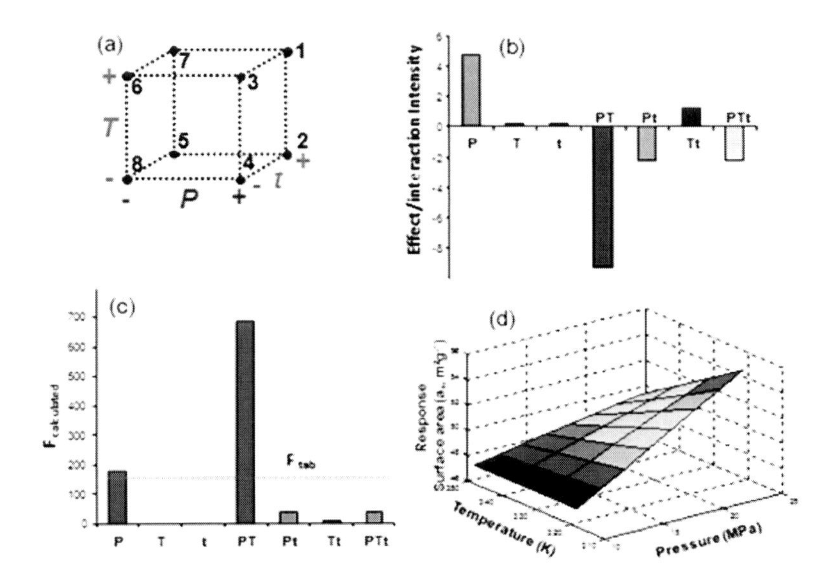

Figure 18.5 Factorial design for the study of the influence of experimental variables on the surface area. (a) Scheme of the design, (b) estimation of the magnitude of effects and interactions, (c) statistic evaluation of significant factors, and (d) response surface of the simultaneous influence of *P* and *T* on the surface area.

In this design, process time is not important, since within the experimental domain of the study (from 15 to 50 min) the kinetics of the silanization reached a steady state. Figure 18.5c shows the statistic evaluation of significant contributions to the response. *P* and *PT* exceed the threshold of F_{cal} and they are relevant statistically. From the point of view of process optimization, this finding indicates that pressure and temperature should be assessed more deeply and simultaneously to find the best operational conditions.

Data of Table 18.1 is exploited to construct a multilinear model for describing the relationship of response a_s as a function of the experimental factors *P*, *T*, and *t*. A general equation to be fitted can be written as: Response(a_s) = $b_0 + b_1\boldsymbol{P} + b_2\boldsymbol{T} + b_3\boldsymbol{t} + b_{12}\boldsymbol{PT} + b_{13}\boldsymbol{Pt} + \ldots + b_{123}\boldsymbol{PTt}$, $b_0, b_1, b_2, \ldots$ being the corresponding coefficients. In this case, however, the model can be simplified neglecting irrelevant contributions so the expression is $R(a_s) = b_1\boldsymbol{P} + b_2\boldsymbol{T} + b_{12}\boldsymbol{PT}$. For this model, coefficients b_1, b_2, and b_{12} are 4.68, 0.131, and -0.013, respectively. The expression is used to estimate the a_s behavior, as depicted in Fig. 18.5d. It can be seen that the effect of *P* is important in the whole experimental domain, while *T* influences the process only at high levels. The interdependence of *P* and *T* is also evidenced.

Table 18.1 Factorial design (2^3) for the study of silanized TiO_2 nanoparticles. High (+) and low (-) levels of variables: pressure (MPa), 22.5 and 10.0; temperature (K), 348 and 318; processing time (min); 50 and 15

Sample	Pressure	Temperature	Time	a_s, m^2g^{-1}
Ti1	+	+	+	46
Ti2	+	–	+	56
Ti3	+	+	–	49
Ti4	+	–	–	57
Ti5	–	–	+	42
Ti6	–	+	+	55
Ti7	–	+	–	49
Ti8	–	–	–	43

18.4.1.1 Multiobjective responses

For a more comprehensive evaluation of silanized materials, apart from a_s, other responses are considered to be of interest. In the example reported by García-González et al. [27], the optimum is reached from the combination of the following objectives: *A*, surface area a_s; *B*, percentile of 95 of the particle size distribution ($d_{p(0.95)}$) giving information on particle homogeneity; *C*, silane weight loss in the temperature range 525–850 K (Δm) by TGA, as an index of overall silane impregnation; and *D*, the ratio between the weight losses in the ranges of 650–850 K and 525–650 K ($\Delta m_2/\Delta m_1$) as a way of expressing the coating stability. The experimental measurements corresponding to each sample are shown in Table 18.2. Since the nature and magnitude of these objective responses are different, normalization is recommendable. From normalized values a summatory function can be built as follows: $F = w_1A + w_2B + w_3C + w_4D$, w_1 to w_4 being the weighting coefficients, taking values between 1 and 5 to rank the importance of each response depending on the intended applications. For instance, in the field of cosmetics, for manufacturing sun creams, wi values assigned to terms A, B, C, and D are 5, 5,1 and 2, respectively. The criterion for choosing these values is as follows: $w_1 = 5$, because the effectiveness of micrometerized TiO_2 as a UV filter mainly depends on having a high surface area; $w_2 = 5$, since the consumer acceptance of sun creams is higher if its whitening effect on the skin caused by TiO_2 agglomerates is minimized; $w_3 = 1$, as silane consumption in this application is not crucial due to the high-added value of the resulting sun cream; $w_4 = 2$, because coating stability is not critical in cosmetic products supposedly not exposed to high temperature or to aggressive chemical conditions. Conversely, for the plastics industry the assigned values of w_i are 3, 5, 4, and 4. In this application, the priority is focused on the homogeneity ($w_2 = 5$), coating efficiency ($w_3 = 4$) and stability ($w_4 = 4$). As shown in the table, sample Ti2 is the best option for cosmetics, as this product displayed high surface area and homogeneity. In contrast, in plastic paints the best options seems to be Ti5, because of high coating percentage and stability.

a_s, surface area; $d_{p(0.95)}$, percentile of 95 of the particle size distribution; Δm, silane weight loss in the temperature range 525–850 K; $\Delta m_2/\Delta m_1$, ratio between the weight losses in the ranges of 650–850 K and 525–650 K; Norm., normalized data corresponding to each objective; OoC: overall objective cosmetics; OoP: overall objective painting.

A similar study can be carried out using desirability functions. In this case, original responses given in Table 18.2 are transformed into desirability values, according to considerations of optimal or unacceptable (see Fig. 18.6). As an example, in the case of silane impregnation by $\Delta m > 2\%$ is considered as a poor coverage ($d = 0$), while $<5\%$ is considered excellent ($d = 1$). Between 2% and 5%, the desirability function $d_{\Delta m}$ take values according to a linear model. Analogous methods are used for the other desirability objectives. Once individual desirabilities are calculated (see Table 18.3) the overall function is as follows: $D = ({d_{as}}^{w1} \times {d_{dp}}^{w2} \times {d_{\Delta m}}^{w3} \times {d_{\Delta m2/\Delta m1}}^{w4})^{1/\Sigma wi}$, where d_i corresponds to desirabilities and w_i to the weighting coefficients adjustable depending on the application

Table 18.2 Experimental data considered as objective responses for multicriteria optimization. Bold numbers represent the best choice of cosmetics and plastic painting. a_s, surface area; $d_{p(0.95)}$, percentile of 95 of the particle size distribution; Δm, silane weight loss in the temperature range 525–850 K; $\Delta m_2/\Delta m_1$, ratio between the weight losses in the ranges 650–850 and 525–650 K; Norm., normalized data corresponding to each objective; OoC: Overall objective Cosmetics; OoP: Overall objective Painting

Sample	a_s (m^2g^{-1})	$d_{p(0.95)}$ (μm)	Δm (%)	$\Delta m_2/\Delta m_1$ (%/%)	Norm. a_s	Norm. d_p	Norm. Δm	Norm. $\Delta m_2/\Delta m_1$	OoC	OoP
Ti1	43	7.8	3.0	0.76	0.07	0.71	0.35	0.34	4.90	6.50
Ti2	57	6.7	2.8	0.75	1.00	0.92	0.27	0.32	**10.51**	9.96
Ti3	49	6.4	3.4	0.79	0.47	0.97	0.50	0.41	8.53	9.92
Ti4	49	7.9	3.2	0.68	0.47	0.69	0.43	0.15	6.52	7.17
Ti5	42	7.9	4.7	1.04	0.00	0.68	1.00	1.00	6.40	**11.40**
Ti6	56	9.5	3.0	0.76	0.93	0.38	0.35	0.34	7.61	7.48
Ti7	55	6.2	2.3	0.77	0.87	1.00	0.08	0.36	10.14	9.36
Ti8	46	11.5	2.1	0.62	0.27	0.00	0.00	0.00	1.33	0.80

(see above). According to this function, best samples for cosmetics and plastic painting applications are Ti2 and Ti5, respectively. In this case, conclusions reached are the same as those extracted from summatory functions.

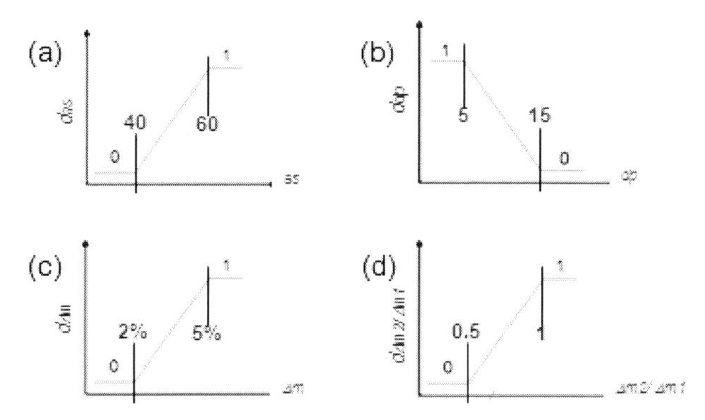

Figure 18.6 Estimation of individual desirability: (a) d_{as}, desirability of surface area; (b) d_{dp}, desirability of the particle size distribution; (c) $d_{\Delta m}$, desirability of coating coverage; and (d) $d_{\Delta m2/\Delta m1}$, desirability of coating.

Table 18.3 Desirability values from experimental data given in Table 18.2 and overall desirability results for cosmetics and plastic painting applications. Bold numbers represent the best choice of cosmetics and painting samples. d_{as}, desirability of surface area; d_{dp}, desirability of the particle size distribution; $d_{\Delta m}$, desirability of coating coverage; $d_{\Delta m2/\Delta m1}$, desirability of coating; OdC: Overall desirability Cosmetics; OdP: Overall desirability Painting

Sample	d_{as}	$d_{\text{dp}(0.95)}$	$d_{\Delta m}$	$d_{\Delta m2/\Delta m1}$	OdC	OdP
Ti1	0.15	0.72	0.34	0.52	0.35	0.41
Ti2	0.85	0.83	0.27	0.50	**0.71**	0.55
Ti3	0.45	0.86	0.47	0.58	0.60	0.59
Ti4	0.45	0.71	0.40	0.36	0.51	0.48
Ti5	0.10	0.71	0.90	1.08	0.36	**0.58**
Ti6	0.80	0.55	0.34	0.52	0.61	0.52
Ti7	0.75	0.88	0.10	0.54	0.65	0.44
Ti8	0.30	0.35	0.03	0.23	0.26	0.17

18.4.2 Example 2: Principal Component Analysis Applied to the Study of TiO_2 Silanization

This example describes the application of PCA to characterize the thermal stability and decomposition processes regarding silanized TiO_2 materials obtained by supercritical technology [29]. Different reaction conditions were used to prepare the silanized samples in order to study the influence of experimental factors on their properties. TGA profiles consisting of weight loss values taken as a function of temperature in the range of 40°C to 700°C are recorded for each sample. The corresponding data are treated by PCA to gain information on the linkage between silane and TiO_2, as well as on the thermal stability of these materials.

The quality of data can be improved using some pretreatment procedures. For instance, mean centering and autoscaling are commonly used to equalize the influence of all variables on the model. In this particular example, as shown in Fig. 18.7a, weight decays corresponding to the different processes are quite overlapped, so sample discrimination based on this type of measurements might be difficult. Here, instead of working with the original data sets, derivative ones are used to enhance the discrimination capacity by providing more resolved peaks of each different process (Fig. 18.7b). The resulting derivative TGA profiles from 150°C to 700°C showed two main peak ranges corresponding to weight decays at the temperature ranges of 250°C–350°C and 350°C–550°C, approximately.

PCA results using derivative TGA profiles from 150°C to 700°C indicated that PC1 retained 88% of variance, while PC2 and PC3 captured 8.5% and 2% of the remaining information, respectively. The representation of scores on PC1 and PC2 is given in Fig. 18.7c. Most of the samples are in the cluster of the center of the graph. Other samples with more different TGA profiles are segregated from this main group. The percentage of silanization of TiO_2 increased from right to left. For instance, samples 1 (raw TiO_2), 58, 59, and 60 (processed under subcritical conditions) result in negligible silanization. PC2 explained the differences among materials as a function of loss percentage in the ranges of 150°C–350°C and 350°C–600°C separately. For instance, samples with high PC2 scores underwent a low decay in the range of 150°C–350°C. On the contrary, samples displaying high weight losses in the range of 350°C–600°C were mainly located to the bottom of Fig. 18.7c. The variance retained by PC3 cannot be underestimated as it modeled subtle differences in stability of the silanization. It is concluded that samples lying on the bottom part present superior thermal stability than those at the top. From these results, the corresponding chemical interpretation is as follows. The weight loss in the 250°C–350°C region is related to silane adsorbed on the TiO_2 surface via hydrogen bonding. Above 350°C and up to 550°C the mass decay is related to the decomposition of the chemically bonded octyl phase.

18.4.2.1 Comparison of prepared products with a commercial material

This study aimed at finding a potential substitute of a commercial TiO_2 silanized sample from the set of materials prepared by supercritical fluid technologies. TGA profiles of homemade and purchased samples are analyzed by PCA for a more comprehensive comparison. From the map of scores, samples appearing closely to the position of the commercial material present similar properties (see Fig. 18.7c). On the contrary, samples located on distant positions with respect to the reference display markedly different properties either regarding the amount of silane immobilized or the thermal stability.

18.4.3 Example 3: Multivariate Calibration Applied to the Study of Drug Impregnation on Absorbing Matrices

This example describes the characterization of the supercritical impregnation of model drugs (benzoic, salicylic, and acetylsalicylic acid) in various adsorbing matrices such as florisil, zeolite, silica gel, and amberlite [30]. Although a thorough evaluation of impregnation processes according to a conventional exploratory approach would require a large number of runs covering all possible sources of variance, here, assays were carried out without a suitable experimental design. In these circumstances, extracting information from such a poorly structured set of assays became much more difficult. This example is focused on showing how chemometics contributed to gain knowledge on the system under study in a very attractive way, even when dealing with a deficient plan of experiments [31].

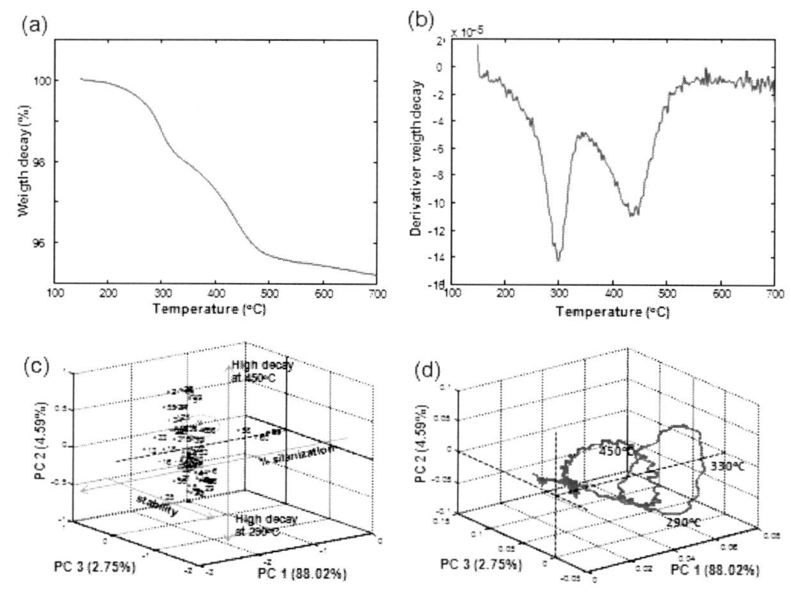

Figure 18.7 Application of principal component analysis to study the thermogravimetric data of TiO_2 silanization. (a) Example of thermogravimetric prolife, (b) example of derivative thermogravimetric profile, (c) scores plot, and (d) loadings plot. Specimen 9 (in red) indicates the commercial sample.

18.4.3.1 Study of impregnation processes by PCA

Pressure (P), temperature (T), and time of adsorption (t) are expected to have a significant effect on adsorption. As an example, in the case of salicylic acid adsorbed on zeolite, a set of 17 experiments was run at different working conditions as depicted in Fig. 18.8a. Despite the fact that assays are not distributed according to a systematic design, PCA could still be used to model the behavior of drug adsorption as a function of the experimental variables. Multivariate data to be treated consisted of impregnation percentages measured for six absorbing matrices materials obtained for each triad of P, T, and t. Dimensions of the corresponding data matrix were 17 objects by 6 variables (see Fig. 18.8b). Variance captured by PC1 is 98%, by PC2 is 1%, and less than 1% is retained by further PCs. The scatter plot of loading vectors of PC1 versus PC2 (Fig. 18.8c) indicated that PC1 is correlated to the adsorption capacity. Sorbents presenting the higher adsorption of solute are, then, to the right, while matrices resulting in a lower adsorption capacity are to the left. Besides, within the experimental domain of the study, the processing time is the most relevant descriptor of the absorption yield. Hence, impregnation increased with the increase of *t*. Regarding loadings, matrices are clearly discriminated depending on their chemical nature. The hydrophilic macroporous inorganic materials are at the top, while the organic materials are at the bottom (Fig. 18.8d).

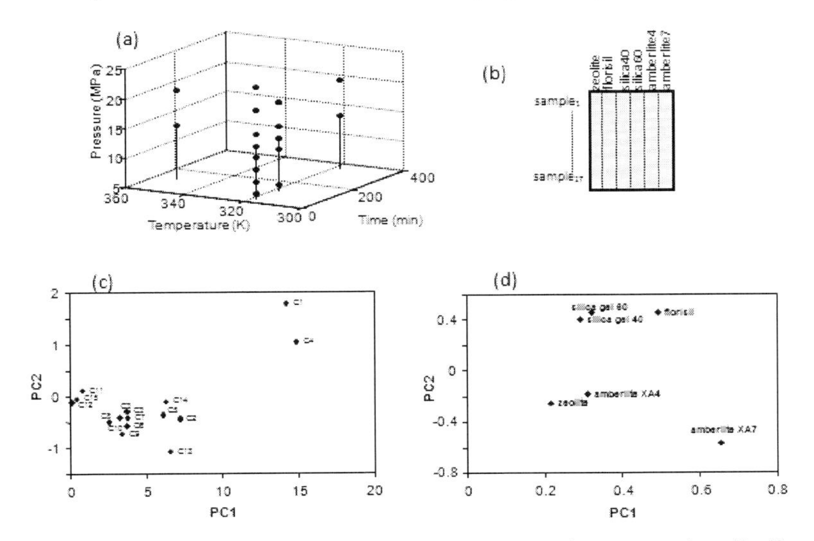

Figure 18.8 Application of principal component analysis to study salicylic acid adsorption. (a) Set of experiments, (b) scheme of the data matrix, (c) scores plot, and (d) loadings plot.

18.4.3.2 Prediction of drug impregnation by using multivariate calibration

Impregnation values obtained experimentally for several of adsorbing materials can be used to estimate the adsorption properties in other materials using multivariate calibration methods. PLS is applied to predict benzoic, salicylic, and acetylsalicylic acid amounts impregnated supercritically for a series of common adsorbent materials (silica gel, florisil, amberlite and zeolite). As an example, Fig. 18.9a shows the 3D plot of the benzoic acid adsorption on zelolite versus pressure and temperature for a constant running time of 30 min. From this figure, a remarkable increase in the adsorption with pressure is observed in the range of 6 to 12 MPa, and remained almost constant from 12 to 24 MPa. The influence of temperature is less noticeable. In addition, a slightly interaction between P and T is found for this particular sorbent material. In a similar way, other 3D plots of P and t, and T and t can be built from PLS results (Fig. 18.9b,c). The influence of P is rather similar to that evidenced in Fig. 18.9a. With respect to the time domain, increasing this variable produced a slight increase in the amount of solute adsorbed. A certain degree of interaction between T and t is observed as adsorption decreases with time at low temperatures but increases at high temperatures.

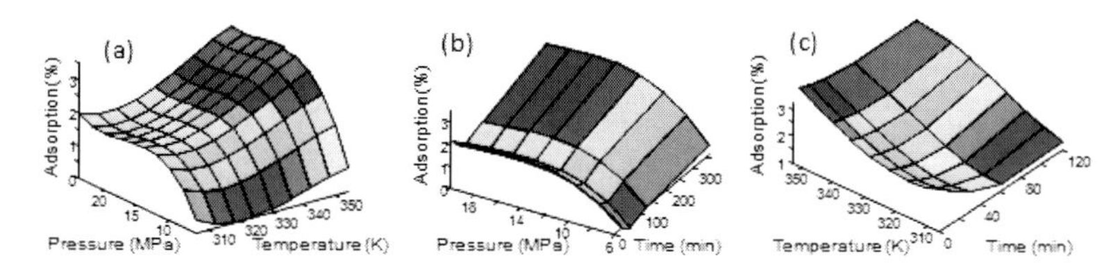

Figure 18.9 Estimation of benzoic acid adsorption on zeolite by PLS: (a) effect of P and T, (b) effect of P and t, and (c) effect of T and t.

18.5 Conclusions

This chapter introduces the application of chemometric methods to facilitate the optimization of supercritical processes and the comprehensive evaluation of product features. The complexity of analytical characterizations dealing with a broad variety of assays often requires multifactorial interpretations of experimental results. Hence, strategies based on experimental design may be exploited to perform these studies in a more elegant and efficient way. Otherwise, some subtle but important details and information regarding samples and processes may be missed. In this way, approaches based on multicriteria responses, multifactor optimization, and multivariate calibration may be of great interest. Once the key factors affecting a given process have been found out, the performance of samples produced can thoroughly be assessed. PCA and related methods have demonstrated excellent possibilities for a better knowledge of global characteristics, and analogies and differences among samples. As a final consideration, it is expected that the impact of chemometrics to carry out the analytical characterization of materials will increase in the next years.

References

1. Chan, H. K., and Kwok, P. C. L. (2011). Production methods for nanodrug particles using the bottom-up approach, *Adv. Drug Delivery Rev.* **63**, pp. 406–416.
2. Mathiowitz, E. (1999). *Encyclopaedia of Controlled Drug delivery* (Wiley-Interscience, New York, USA).
3. Kydonieus, A. (1992). *Treatise on Controlled Drug Delivery* (Marcel Dekker, New York, USA).
4. Anastas, P., and Warner J. (1998). *Green Chemistry: Theory and Practice* (Oxford University Press, New York, USA).
5. Hagens, W. I., Oomen, A. G., de Jong, W. H., Cassee, F. R., and Sips, A. J. A. M. (2007). What do we (need to) know about the kinetic properties of nanoparticles in the body?, *Regul. Toxicol. Pharmacol.*, **49**, pp. 217–229.
6. Fischer, H. C., and Chan, W. C. W. (2007). Nanotoxicity: the growing need for in vivo study *Curr. Opin. Biotechnol.*, **18**, pp. 565–571.
7. Massart, D. L., Vandeginste, B. G. M., Buydens, L. M. C., de Jong, S., Lewi, P. J., and Smeyers-Verbeke, J. (1997). *Handbook of Chemometrics and Qualimetrics* (Elsevier, Amsterdam).
8. Vandeginste, B. G. M., Massart, D. L., Buydens, L. M. C., de Jong, S., Lewi, P. J., and Smeyers-Verbeke, J. (1998). *Handbook of Chemometrics and Qualimetrics: Part B* (Elsevier, Amsterdam).
9. Domingo, C., and Saurina, J. (2012). An overview of the analytical characterization of nanostructured drug delivery systems: towards green and sustainable pharmaceuticals: a review, *Anal. Chim. Acta*, **744**, pp. 8–22.
10. West, C., and Lesellier, E. (2012). Chemometric methods to classify stationary phases for achiral packed column supercritical fluid chromatography, *J. Chemometr.*, **26**, pp. 52–65.

11. Eriksson, L., Johansson, E., Kettaneh, N., Wikstrom, C., and Wold, S. (2008). *Design of Experiments. Principles and Applications* (Umetrics AB, Umea, Sweden).

12. Sentellas, S., and Saurina, J. (2003). Chemometrics in capillary electrophoresis. Part B: methods for data analysis, *J. Sep. Sci.*, **26**, pp. 1395–1402.

13. Booksh K. S., and Kowalski B. R. (1994). Theory of analytical chemistry, *Anal. Chem.*, **66**, pp. A782–A791.

14. Brown, S. D., Tauler, R., and Walczak, B. (2009). *Comprehensive Chemometrics. Chemical and Biochemical Data Analysis*, Vol. 2 (Elsevier, Amsterdam).

15. Martens, H., and Naes, T. (1989). *Multivariate Calibration* (John Wiley and Sons, New York, USA).

16. Brown, S. D., Tauler, R., and Walczak, B. (2009). *Comprehensive Chemometrics. Chemical and Biochemical Data Analysis*, Vol. 3 (Elsevier, Amsterdam).

17. Dong, Y. D., and Boyd, B. J. (2011). Applications of X-ray scattering in pharmaceutical science, *Int. J. Pharm.*, **417**, pp. 101–111.

18. Panella, B., Vargas, A., Ferri, D., and Baiker, A. (2009). Chemical availability and reactivity of functional groups grafted to magnetic nanoparticles monitored in situ by ATR-IR spectroscopy, *Chem. Mater.*, **21**, pp. 4316–4322.

19. Mansour, H. M., and Hickey, A. J. (2007). Raman characterization and chemical imaging of biocolloidal self- assemblies, drug delivery systems, and pulmonary inhalation aerosols: a review, *AAPS PharmSciTech*, **8**, pp. 99–102.

20. Ansari, K. A., Torne, S. J., Vavia, P. R., Trotta, F., and Cavalli, R. (2011). Paclitaxel loaded nanosponges: in-vitro characterization and cytotoxicity study on MCF-7 cell line culture, *Curr. Drug Delivery*, **8**, pp. 194–202.

21. Argemi, A., Domingo, C., de Sousa, A. R. S., Duarte, C. M. M., Garcia-Gonzalez, C. A., and Saurina, J. (2011). Characterization of new topical ketoprofen formulations prepared by drug entrapment in solid lipid matrices, *J. Pharm. Sci.*, **100**, pp. 4783–4789.

22. Murillo-Cremaes, N., Lopez-Periago, A. M., Saurina, J., Roig, A., and Domingo, C. (2010). A clean and effective supercritical carbon dioxide method for the host-guest synthesis and encapsulation of photoactive molecules in nanoporous matrices, *Green Chem.*, **12**, pp. 2196–2204.

23. Argemi, A., Lopez-Periago, A., Domingo, C., and Saurina, J. (2008). Spectroscopic and chromatographic characterization of triflusal delivery systems prepared by using supercritical impregnation Technologiews, *J. Pharm. Biomed. Anal.*, **46**, pp. 456–463.

24. Tumer, Y. T. A., Roberts, C. J., and Davies, M. C. (2007). Scanning probe microscopy in the field of drug delivery, *Adv. Drug Delivery Rev.*, **59**, pp. 1453–1473.

25. Sitterberg, J., Ozcetin, A., Ehrhardt, C., and Bakowsky, U. (2010). Utilising atomic force microscopy for the characterisation of nanoscale drug delivery Systems, *Eur. J. Pharm. Biopharm.*, **74**, pp. 2–13.

26. Giron, D. (2002). Applications of thermal analysis and coupled techniques in pharmaceutical industry, *J. Therm. Anal. Calorim.*, **68**, pp. 335–357.

27. García-González, C. A., Fraile, J., López-Periago, A., Saurina, J., and Domingo, C. (2009). Measurements and correlation of octyltriethoxysilane solubility in supercritical CO_2 and assembly of functional silane monolayers on the surface of nanometric particles, *Ind. Eng. Chem. Res.*, **48**, pp. 9952–9960.

28. Morgan, E. (1991). *Chemometrics: Experimental Design* (ACOL, John Wiley and Sons, London, UK).

29. García-Gonzalez, C. A., Andanson, J. M., Kazarian, S. G., Domingo, C., and Saurina, J. (2009). Application of principal component analysis to the thermal characterization of silanized nanoparticles obtained at supercritical carbon dioxide conditions, *Anal. Chim. Acta*, **635**, pp. 227–234.

30. Domingo, C., García-Carmona, J., Fanovich, M. A., and Saurina, J. (2001). Application of chemometric techniques to the characterisation of impregnated materials obtained following supercritical fluid technology, *Analyst*, **126**, pp. 1792–1796.

31. Domingo, C., García-Carmona, J., Fanovich, M. A., and Saurina, J. (2002). Study of adsorption processes of model drugs at supercritical conditions using partial least squares regression, *Anal. Chim. Acta*, **452**, pp. 311–319.

Chapter 19

Interaction of Supercritical Carbon Dioxide with Polymers Studied by Vibrational Spectroscopy

Andrew V. Ewing and Sergei G. Kazarian
Department of Chemical Engineering, Imperial College London, South Kensington Campus, London SW7 2AZ, UK
s.kazarian@imperial.ac.uk

High-pressure and supercritical CO_2 ($scCO_2$) have the potential to be employed as viable alternatives to the harmful volatile organic solvents (VOCs) currently used for industrial processes. Developments using this technology are driven by our society continuing to move towards environmentally friendly processes. As well as conforming to the modern industrial environmental requirements, high-pressure and $scCO_2$ offer fundamental advantages for polymer processing that include the reduction of polymer melting temperatures (T_m) and glass transition temperatures (T_g) and can induce polymer swelling. These effects are crucial to the modification of polymeric samples. Thus, the importance of processing polymeric materials with CO_2 can lead to the production of novel resources that have

Supercritical Fluid Nanotechnology: Advances and Applications in Composites and Hybrid Nanomaterials
Edited by Concepción Domingo and Pascale Subra-Paternault

ISBN 978-981-4613-40-8 (Hardcover), 978-981-4613-41-5 (eBook)
www.panstanford.com

unique characteristics and properties. This book chapter will review developments discovered by the use of vibrational spectroscopic approaches, specifically infrared and Raman spectroscopy. Vibrational spectroscopy can reveal valuable information about the polymeric samples, as well as highlighting interactions between CO_2 and the functional groups present within the polymeric materials. Understanding the behaviour of these systems is integral as a means to develop more efficient processing routes. Finally, the most recent advancements in this field will be summarized with an outlook to future opportunities using this "green" processing technology.

19.1 Introduction

In this chapter we review some of the most important discoveries associated with the use of supercritical fluids (SCFs) for polymer processing. The unique physical properties that SCFs possess give great potential to study a range of different systems, not only in polymer science, but in areas such as organic synthesis, catalysis and coordination chemistry [1, 2]. Applications using SCFs, specifically supercritical carbon dioxide ($scCO_2$), is a growing area for industrial processes, which has been motivated by the need for "greener" solvent alternatives arising from environmental concerns. The most widely employed SCF is $scCO_2$ because it is readily available, relatively inexpensive and has low toxicity, flammability and zero ozone depletion levels. One of the key industrially significant properties that $scCO_2$ possesses is that it is a gas under ambient temperature and pressure conditions. This means that removal from a sample after processing is facile, saving both cost and time by preventing the need for solvent removal and drying stages postprocessing. As a result, $scCO_2$ has much potential to be an alternative for the harmful volatile organic solvents (VOCs) currently used in such industrial processes.

The first section of this chapter will summarize some of the effects that high-pressure and $scCO_2$ have on polymeric materials. The ability of CO_2 to interact with different functional groups in polymers has profound effects on the mechanical and physical properties of the polymers. The sorption of CO_2 into polymers can lead to swelling and plasticization. An understanding of the phase behavior of such systems and the interactions between the polymer and CO_2 is integral as a means to develop more efficient processing routes. This section will focus on $scCO_2$-induced plasticization, diffusion, crystallization and foaming of polymers. These are important factors determined by the behavior and interactions of CO_2 with polymers, which is a key factor in the outcome of materials produced with this method. Next, the principles of vibrational spectroscopy, namely Fourier transform infrared (FTIR) and Raman spectroscopy, will be described. This section will include a description about a powerful and versatile FTIR spectroscopic imaging approach which has been used to study high-pressure and $scCO_2$ processing with polymers, developed in the author's laboratory over last decade. Finally, examples and applications where vibrational spectroscopy has significantly advanced the understanding of high-pressure and $scCO_2$ systems have been reviewed. This includes both pioneering developments using spectroscopy, as well as recent applications where supercritical technology is being utilized today.

19.2 Effect of High-Pressure and Supercritical CO_2 on Polymers

19.2.1 Solubility of CO_2 in Polymers

The physical and chemical properties of CO_2 mean that it is a desirable solvent to use on a range of industrially relevant systems as an alternative to VOCs. High-pressure and $scCO_2$ have proven their applicability for polymer processing because they can act as a "temporary" plasticizer [3]. The solubility of high-pressure CO_2 into polymeric species is integral to the subsequent processing procedures that can be applied and is fundamental to all the processing techniques that will be described throughout this chapter. Solubility of a polymer occurs based on the weak interactions between CO_2 and basic sites in the sample (Fig. 19.1). Due to the interactions affecting the solubility of CO_2 into the polymer, many trends in the behavior of different polymers under CO_2 environments have been recognized, for example, looking at the equilibrium fraction of CO_2 in different polymers. Higher-molecular-weight polymers, such as poly(dimethylsiloxane) (PDMS), reach higher levels, 21% at 50°C. Whereas for polymers such as poly(methylmethacrylate) (PMMA), that have lower molecular weights, the maximum equilibrium fraction is ~10% at 35°C [4].

Much research has been carried out to broaden the understanding of the significance of polar functional groups within the polymeric materials which are subjected to high-pressure CO_2 [5]. Substantial attention has been paid to the positions of the polar groups, number of electron-donating groups and morphology of the bulk matrix within a given polymer [6–8]. For example, the steric hindrance of the main carbonyl functional group in the polymer has an influence on the solubility of CO_2 because the access and interaction with these functionalities is difficult. As a result, CO_2 solubility is greater in PMMA than in poly(ethylene terephthalate) glycol (PETG) which contains an ester side chain [9, 10]. Another example is polymers that are glassy in nature exhibit higher CO_2 sorption than those that are in a crystalline or semicrystalline state [9].

Although many experimental [6, 9–15] and theoretical [16–19] methods have been applied to such systems, one of the most well established spectroscopic methods for determining the solubility of CO_2 into polymer is using FTIR spectroscopy [20, 21]. The general principals of this technique will be described in detail further in this chapter, but initial research that has developed and established this approach will be briefly summarized.

The interaction between CO_2 and functional groups in polymers can be identified using this approach. This has impacted on the modification of polymers and consequently had implications on the synthesis, membrane technology and the processing of such samples. Kazarian et al. applied FTIR spectroscopy to study PMMA and the intermolecular interactions with CO_2 [20]. A splitting phenomenon was observed of the two degenerate CO_2 (ν_2) bending modes at ~660 cm^{-1} (Fig. 19.1). It has been proposed that this is caused by a Lewis acid-base interaction between the polar groups in the polymer, i.e., carbonyl groups, and CO_2. This work was further complimented by Nelson and Borkman where molecular orbital simulations were calculated to study the interaction of poly(ethylene terephthalate) (PET) with CO_2 [16].

Another condition to be discussed that has an effect on the solubility of CO_2 is temperature [22]. Adjusting and controlling the temperature has significant influence on the state of the bulk polymer matrix. CO_2-induced foaming is an example of a physical process that occurs to a polymer when there is a substantial temperature increase in a saturated CO_2 environment. At lower temperatures the solubility of CO_2 into the matrix is higher and hence as the temperature is increased the solubility becomes lower. The significance of this for polymeric processing is that even slight deviations from the desired temperature can affect the solubility of CO_2 into polymer. A comprehensive understanding of the solubility of CO_2 into polymers is essential to improve applications and optimize such processes. There are many reviews summarizing CO_2 solubility data and experimental approaches which explain this process in greater detail [23–26].

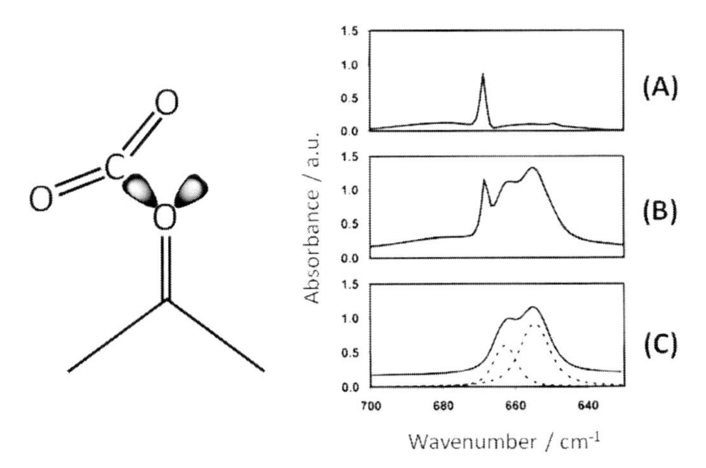

Figure 19.1 Right: Schematic representation of CO_2 with a polymer containing carbonyl functional groups. Left: Infrared spectra of the CO_2 (ν_2) bending mode region of (A) pure gaseous CO_2, (B) in a PMMA film immediately after CO_2 decompression, and (C) in the PMMA after removal of gaseous CO_2. Adapted with permission from Kazarian et al. Copyright (1996) American Chemical Society [20].

19.2.2 CO_2-Induced Plasticization of Polymers

The sorption of CO_2 into polymers can lead to swelling of the materials (Fig. 19.2), as well as initiating other physical and mechanical changes in the polymers. The pressure of CO_2 subjected to the system determines the extent of which these properties will change, where the most important effect to the systems is the reduction of the glass transition temperature (T_g), also known as plasticization. The phenomenon of plasticization occurs as a result of the Lewis acid–base-type interactions between the acidic sites from the CO_2 molecules and basic functional groups within the molecules [20]. As a result, both the mobility and the interchain distance between two polymeric chains will increase in the sample. This has been shown experimentally by Kazarian et al. in 1997, where increased mobility of carbonyl bands from ester functional groups in PMMA films was observed at 40°C and 100 bar in a CO_2 environment [27]. It was concluded that the CO_2 conditions give comparable polymer mobilities to those observed when the polymer is heated above its T_g indicating that the high-pressure conditions can mimic the effect of heating.

FTIR spectroscopy has been widely applied to confirm the presence of the intricate interactions between CO_2 and polymeric functional groups. The advantage of using spectroscopic approaches is that it can support previous suggestions about CO_2-induced plasticization with evidence on the molecular level and thus can broaden overall knowledge in this field.

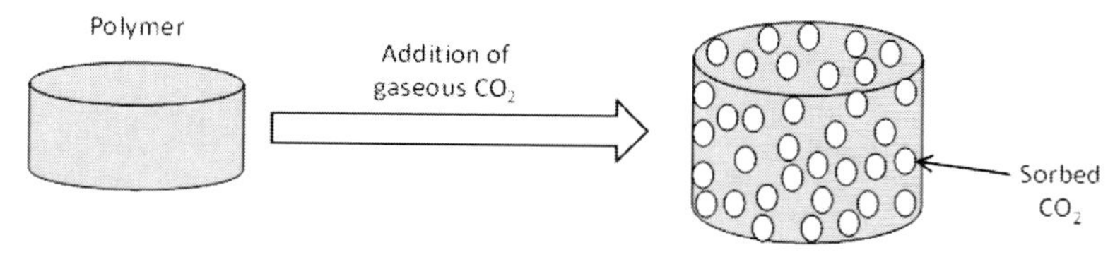

Figure 19.2 Diagram showing the effect of a high-pressure CO_2 atmosphere to induce swelling of a polymer.

19.2.3 Crystallization Induced by CO_2

Crystallization of certain materials can be induced as a result of the plasticization properties that CO_2 has on polymers. The greater mobility of amorphous and semicrystalline polymeric chains means that structural arrangement can occur in such a way that favors the kinetics and crystallites can form. A well-reported example of $scCO_2$-induced crystallization is for PET which has industrial significance because of its use for synthetic fibers and drinks bottles, accounting for >60% and ~30% of all of the globally synthesized PET, respectively [28]. Gas permeability through PET materials is important when considering the properties needed for bottles that will contain carbonated soft drinks since zero loss of CO_2 is desired [23]. Consideration also has to be taken when designing synthetic fibers for materials such as clothing. PET is a colorless polymer and thus dyeing is required to produce color in the materials. Hence, the degree of crystallinity in the PET fibers can affect the ability of the materials to be dyed. Many analytical approaches have been used to understand the effect of polymer morphology on PET materials processed with $scCO_2$, which includes X-ray scattering [28, 29], differential scanning calorimetry (DSC) [29], mid-infrared (IR) [28, 30–32] and Raman spectroscopy [32, 33].

CO_2-induced crystallization of polymers such as syndiotactic polystyrene (sPS) [34], polypropylene [35], *tert*-butyl poly(ether ether ketone) [36], and bisphenol A polycarbonate [37] have also been studied. The enhanced crystallization rate of polymers by CO_2 at increasing pressures can be explained by the relationship between the T_g and melting temperature (T_m). When under high (or increasing) pressures of CO_2 the depression of the T_g is far greater than the depression of the T_m. However, for some polymers, such as poly(ʟ-lactide) [38] and isotactic propylene [35], CO_2 is actually seen to suppress the rate of crystallization. This is due to prevention of nucleation sites since the T_g and T_m have similar rates of depression.

The interest in polymers for processing with CO_2 largely stems from their industrial relevance. sPS, for example, is typically affected by three factors: temperature; choice of solvent and physical strain. Handa et al. used DSC and X-ray diffraction to investigate the effects of CO_2 on the morphology of this polymer [34]. The authors report that the crystallization of the polymer was too fast to record any kinetic measurements at a temperature of 122°C. The DSC stabilization time is too long to measure the polymer before maximum crystallinity is achieved because at high temperatures and pressures in CO_2 the rate of crystallinity is increased. On the other hand, FTIR spectroscopy allows one to monitor very fast changes in polymer morphology. Kazarian and co-workers have been at the forefront of using IR and Raman spectroscopy to study these processes for many years. One of their reported articles reports an in-situ study of sPS using conventional FTIR spectroscopy [39]. Morphological changes were shown in the recorded infrared spectrum of sPS that was subjected to $scCO_2$. This demonstrated the power of FTIR spectroscopy to reveal information about polymer/CO_2 systems. A dynamic study of these systems, i.e., taking measurements as a function of time, allows one to learn about the kinetics of crystallization.

CO_2-induced crystallization is another example where FTIR spectroscopy can be used as a unique approach for the processing of polymeric materials. The ability to control the amount of crystallization, and the diffusion of CO_2 into these materials, can benefit a vast range of industrially relevant processes. Understanding and predicting the degree of crystallization has implications on the synthesis of novel polymers and offers the ability to tune such materials for specific use.

19.2.4 CO_2-Induced Extraction and Separation

For many years now the use of high-pressure and $scCO_2$ has been reported as an extremely useful method for the extraction and separation of polymers [23, 40]. One of the most well-known extraction applications used industrially is the $scCO_2$ extraction of caffeine from coffee beans. More than one hundred thousand tonnes of decaffeinated coffee are consumed every year globally. Although caffeine is not a polymer, it is an example of how powerful SCF technology can be used from an industrial point of view.

Applying such an approach to polymeric processing has advantages, for example by extracting residual solvents, unreacted species and side-products from the matrices. Plasticization of polymers under high-pressure CO_2 atmospheres can considerably enhance the extraction process of these materials. There are a number of different reviews [41–45] and book chapters [46] that have been published describing the role of SCFs for extraction applications. Typical examples of where this processing technique has been applied includes: degassing in the leather industry [47], developing the quantitative analysis of additives in polypropylene [48], and the extraction of copper ion into chelating agents [49].

19.2.5 Foaming

Polymeric foams are materials that have superior properties to unfoamed samples of the same polymer. They are defined as closed-shell structures with a diameter of ~10 μm and cell densities ranging between that of 10^9 to 10^{15} cells per cm^3 [50]. The preparation of foamed materials using high-pressure CO_2 can be done so using continuous [51–53] and discontinuous methods [54–56]. The approach works in three stages: (1) the high-pressure chamber containing the polymer sample is saturated with CO_2, at a temperature below that of the T_g, which plasticizes the bulk polymer matrix; (2) is a transfer stage and (3) is to rapidly increase the temperature above the T_g and/or quickly depressurize the chamber which means CO_2 leaves the matrix. As the gaseous CO_2 escapes the T_g rises and the new foamed structure is "frozen" in this form.

Modeling approaches have been applied to predict the conditions that are required to induce open-microcellular foamed polymers in batch processes [57]. Learning and understanding the relationship between the depression of the T_g and pressure is integral in the application of foaming as it dictates the degree to which the sample will nucleate and grow.

As mentioned above there are two methods of preparing the polymer foams. A characteristic of the discontinuous method is that there is a dense "skin" of unfoamed polymer on the outermost layer of the bulk matrix. This "skin" forms as a result of desorption of CO_2 during the transfer stage of processing. Controlling the thickness of this layer can be done so by varying the transfer time of the polymer and the morphology of the starting material. Krause et al. have studied the discontinuous foaming of thin polymer films including polysulfone (PSU), poly(ether sulfone) and cyclic olefin copolymer using scanning electron microscopy (SEM) [55]. The size of the foamed cell was controlled and influenced by the pressure of saturated CO_2 on the system. It was observed in this study that higher pressures of CO_2 resulted in a reduction of cell size diameter but an increase in cell distribution. Further to this, it was seen that a maximum cell density was observed within a finite-temperature window. The relevance of temperature influences can be understood when considering two competing processes in the polymer, cell growth and nucleation and the sorption of CO_2 into the matrix. SEM was also employed to image the foaming of PSU as a function of transfer time (Fig. 19.3). It is visible in the SEM micrographs shown in Fig. 19.3 that an increase in transfer time led to an increase in cell size. It was proposed that this was caused by a reduction of the CO_2 content in the film.

Finally, foaming architecture can be manipulated by the amount of CO_2 sorbed into the polymer as shown in PSU/polyimide bends [54]. Closed-microcellular materials were produced when low concentrations of CO_2 were in the film whereas at higher concentrations open-porous foams were formed with smaller cell sizes. Krause et al. confirmed this where gas permeation measurements were used to study such changes to conditions [54].

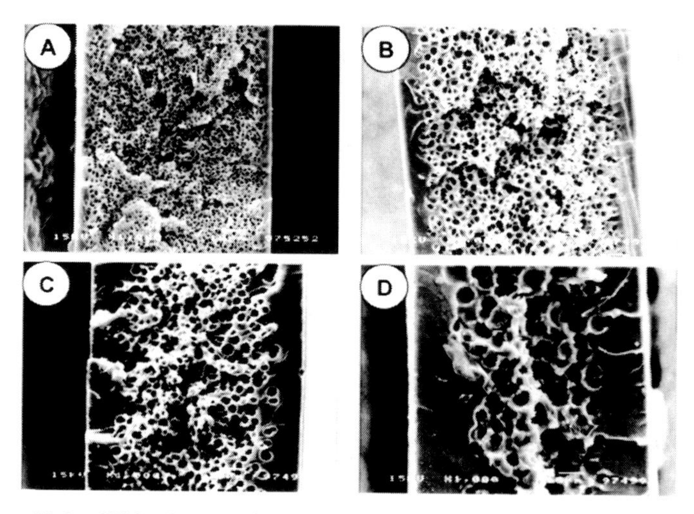

Figure 19.3 SEM micrographs (magnification 1000x) of a foamed PSU film as a function of transfer time. (A) 1 min;, (B) 5 min, (C) 10 min, and (D) 20 min. Reprinted with permission from Krause et al. Copyright (2001) American Chemical Society [55].

19.2.6 Rheology of Polymers

The rheology, or the flow, of viscous polymers under high-pressure CO_2 can be divided into four categories: pressure driven; falling body; rotational devices; and vibrating wire. These have implications for other processing techniques, such as extrusion which have been described in a previous book chapter [26] and thus will not be discussed in any more detail in this review.

To briefly summarize, pressure driven devices include capillary and slit-dye viscometers, where the flow is governed by pressure. In such cases, monitoring the flow rate and the pressure changes is used to determine the viscosity. It should be noted that problems may arise in phase separation. Capillary viscometers have a simple flow pathway, give reproducible data and only require a small amount of sample for measurements. Drawbacks for using these devices can be encountered where a pressure drop across the capillary can affect the viscosity. Nevertheless, these devices have been widely applied to study polymer/CO_2 systems [58–61]. Falling body viscometers share a common feature where the known volume of sample passes through a fluid under the influence of gravity. The viscosity can be determined by the time required for this to occur. To date these viscometers have been limited to the use of low-viscosity polymer systems [62, 63], including that of a PDMS/CO_2 (10 bar to 30 bar) system [64].

However, it is rotational viscometers that are most widely employed for determining the rheology of polymeric materials. Continuous measurements are made at a given shear rate or shear stress to perform subsequent measurements on the sample under differing conditions. One of the main advantages to using rotational viscometers is the fact that the shear rate can be distributed in a uniform manner across the sample allowing one to have more control. Yet, there were limitations with some rheometer designs when it came to use with high-pressure systems. Flichy et al. developed an approach that was capable of measuring pressures up to 150 bar and temperatures of 300°C [65]. This solved specific problems that had been a shortcoming of alternative rheometers such as a seal around the spinning shaft as well as a way to determine the torque. Since then, other operational devices for measuring the viscosity of polymers have been developed such as the magnetically levitated sphere rheometer where measurements are conducted under constant pressures by Royer et al. [66].

19.3 Vibrational Spectroscopy

The motivation to use SCFs for polymer processing does not just originate from the environmental impetus for their use as benign solvents but the fact that CO_2 has a number of effects on the polymer being processed. Vibrational spectroscopy, specifically FTIR and

Raman spectroscopy, are versatile and powerful approaches to study such interactions. The following section will summarize the fundamental principles of these analytical methods as a means to learn about the interactions of $scCO_2$ with polymeric materials [3].

The ability to probe characteristic vibrations of specific functional groups is possible because of the quantized nature of the vibrational energy levels. FTIR and Raman spectroscopic methods measure transitions between vibrational energy levels induced when the sample is interacting with incident light. Despite being complementary techniques, FTIR and Raman spectroscopy differ both in terms of the instrumentation used, and by the mechanisms in which the population of vibrational energy levels change.

The differences between the two spectroscopic approaches can be described by considering the mid-IR and the Stokes Raman transitions represented in Fig. 19.4. The overall result is to excite a molecular bond from the ground vibrational state (V = 0) to the first excited state (V = 1), governed by the selection rule of $\Delta V \pm 1$. However, mechanistic differences occur about the excitation to the V = 1 vibrational level. The FTIR transition is directly excited from V = 0 to V = 1. In contrast, this excitation occurs via a transitional (or "virtual") state in Raman spectroscopy (Fig. 19.4). The energy required for the IR transition is the same as the difference in energy that occurs between the incident and scattered light in Raman spectroscopy. It follows that FTIR spectroscopy measures the absorption of a specific quantized amount of energy while Raman measure the "shift" in energy between the incident and scattered light to give the same difference between the quantized energy levels.

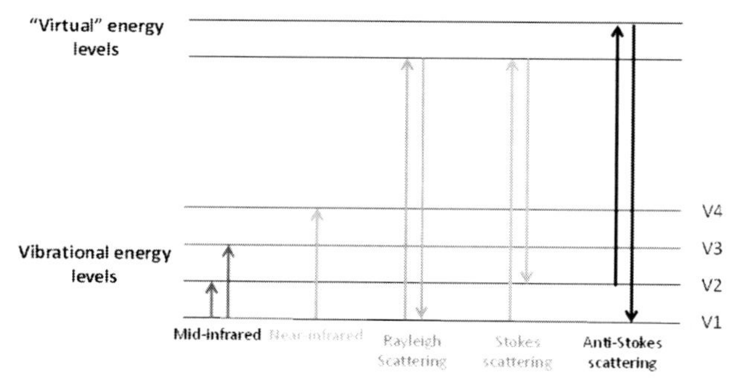

Figure 19.4 Schematic representation of the potential energy of a molecular bond showing typical infrared and Raman transitions.

19.3.1 FTIR Spectroscopy

19.3.1.1 Sampling methodologies

Transmission mode is one of the most commonly used methods to record an IR spectrum. This sampling mode works by passing infrared light directly through the whole sample measuring the frequencies which are absorbed. The popularity of transmission mode stems from the fact that the apparatus used is relatively inexpensive and straightforward to operate. However, particularly for high-pressure CO_2 research, this mode can prove challenging. Typically, the sample has to be prepared so that it is thin enough for the IR radiation to pass through (between 5–50 μm). Solid samples are commonly cast as thin films onto a non-mid-IR absorbing material such as silicon or potassium bromide. However, transmission measurements of polymeric samples can be performed as demonstrated by Fleming et al. where this sampling mode was used to study the cross-sections of PET film under exposure to $scCO_2$ [31].

Alternative sampling methodologies are those of reflection modes, such as attenuated total reflection (ATR). These differ from that of transmission mode because it is the reflected IR light from the surface layer of the sample that is measured. ATR mode is based on the phenomenon that total internal reflection will occur at the interface between the ATR crystal (high-refractive-index material) and the sample (low-refractive-index material). At this interface an evanescent wave of IR light penetrates into the surface layer of the sample. The distance by which this wave enters into the sample is assigned as the depth of penetration (d_p), which can be defined as the distance the electric field of the evanescent wave travels into the sample before its amplitude falls to e^{-1} of its original surface value (Eq. 19.1), where λ is the wavelength of radiation, n_1 and n_2 are the refractive indices of the ATR crystal and sample respectively and θ is the angle of incidence.

$$d_\mathrm{p} = \frac{\lambda}{2\pi n_1\left(\sin^2\theta - \left(\frac{n_2}{n_1}\right)^2\right)^{1/2}} \tag{19.1}$$

The d_p is wavelength dependent, where radiation of shorter wavelengths will have a shorter penetration into the sample.

Furthermore, Eq. 19.1 states that the angle of incidence must be greater than that of the critical angle in order for internal reflection to occur.

Several different materials have been used as suitable ATR crystals, each with differing physical properties, these commonly include diamond, germanium (Ge) and zinc selenide (ZnSe). Since high-pressure experiments can cause damage to fragile and brittle materials, ZnSe has not been used in such studies but diamond and Ge ATR crystals have been reported for these applications [21]. One of the main advantages of studying the interactions of CO_2 with polymeric samples in ATR mode is that little sample preparation is required and the measurement is independent of the sample thickness. This means the polymer can be placed to cover the ATR crystal within a high-pressure cell, allowing different pressures to be applied in-situ (Fig. 19.5) [21, 67].

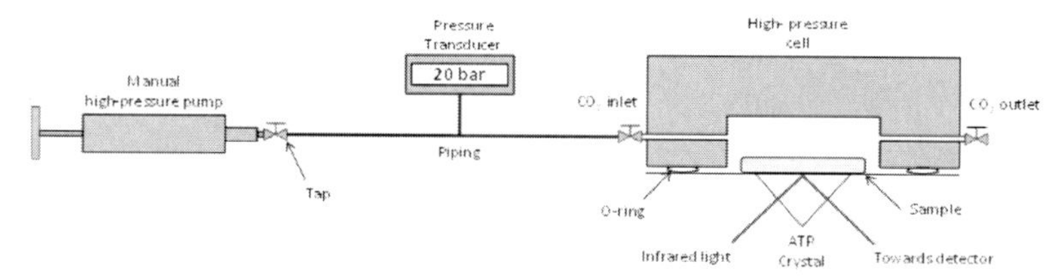

Figure 19.5 Diagram showing a typical high-pressure CO_2 setup for use with a diamond or Ge ATR sampling accessory for FTIR measurements.

19.3.1.2 FTIR spectroscopic imaging

FTIR spectroscopic imaging has recently emerged as an important tool to study a range of different systems such as polymers [31, 68, 69], pharmaceutical tablets [70–72], proteins [73, 74] and biological materials [75]. This approach combines an FTIR spectrometer with a focal plane array (FPA) detector enabling thousands of spectra to be measured from different regions within the imaging area simultaneously. Spectroscopic images can be obtained by selecting unique absorption bands for the material of interest and integrating the area under this band as a function of all pixels. A typical example of an imaging setup has been demonstrated by Kazarian and Chan where an array size of 64 × 64 pixels was used, thus 4096 individual spectra can be obtained from the sample simultaneously [68]. This approach means that both chemical and spatial information can be obtained from the polymer blend and this was the first application using FTIR spectroscopic imaging to study the behavior of polymers under high-pressure and $scCO_2$. The in-situ imaging approach studied the phase separation of a lower critical solution temperature (LCST) blend of polystyrene (PS) and poly(vinyl methyl ether) (PVME). The polymer blend was cast onto the diamond accessory and after subjecting the blend to $scCO_2$ for different periods of time, spectroscopic images showing the spatial distribution of both components were recorded [68].

19.3.2 Raman Spectroscopy

The interaction of electromagnetic radiation with a molecule can be viewed as the electric field of the incident energy perturbing the molecular system. The oscillating electric field in the incident light induces an oscillating dipole moment in the molecule and hence the electron distribution in a specific chemical bond is distorted from the original equilibrium position. The ease by which this distortion occurs, producing an induced electric dipole moment under the influence of an external electric field, is determined by the molecule's polarisability, which is modulated by molecular vibrations. When a molecule is subjected to an external electric field, polarisation occurs in a time-dependent manner which raises the energy of the system into a "virtual" state. This "virtual" state induces a dipole and upon relaxation emits a photon resulting in scattering. Rayleigh scattering, or elastically scattered light, occurs when there is zero energy exchange between the incident radiation and the molecule. On the other hand, the Raman effect describes situations when there is a difference in energy between the incident and emitted radiation, producing inelastically scattered light. The inelastically scattered radiation can be defined as Stokes or anti-Stokes, depending on the mechanism of excitation. These lines are displayed in the spectrum as pairs that are symmetrically distributed around the Rayleigh line. Photons that lose energy appear at a higher wavenumber (lower frequency) than the Rayleigh line and are known as Stokes lines. Whereas, anti-Stokes lines are cause by photons that gain energy and thus appear at a lower wavenumber (higher frequency) to the Rayleigh line.

The relative intensity of the Stokes peak is greater than that of the anti-Stokes peak, which can be understood by considering the population distributions of the first excited and ground vibrational state (Boltzmann distribution). In Raman spectroscopy it is standard practice to study the Stokes response because of the higher intensity of the corresponding Raman bands.

19.4 Advancing CO_2 Technologies Using Vibrational Spectroscopy

19.4.1 Diffusion of Materials into Polymeric Species

Firstly, the diffusion of materials into polymeric matrices will be described. The use of supercritical technologies for dyeing PET is a commercially viable alternative to the traditional aqueous methods. Tuning the polymer's properties by adjusting the pressures and temperatures of the system allows the mass transfer within the polymer to be controlled. Subsequently, polymers can be doped with materials in a "green" and essentially solvent-free manner. In circumstances where the intended doping materials, such as a fabric dye, have low solubility in CO_2, phase separation can occur thus impregnation into the polymer is possible (Fig. 19.6).

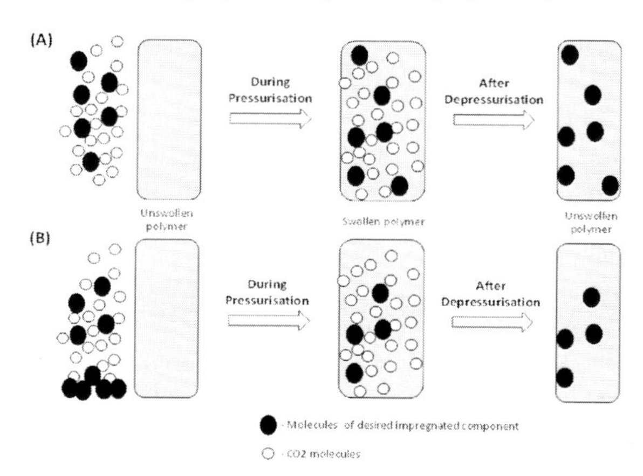

Figure 19.6 Schematic representation of the impregnation of a molecule into a polymer before, during and after depressurization. (A) Represents a highly-CO_2-soluble molecule and (B) a poorly-CO_2-soluble molecule. Adapted from Kazarian (2004) [76].

Optimizing this dyeing process requires further knowledge of the mass transport of the materials when dissolved in CO_2 through the polymer. In 2005, Fleming et al. used confocal Raman microscopy (CRM) to probe the concentration and distribution of the dyes within polymeric fibers in a nondestructive manner [33]. Depth profiling through the sample was measured which allowed the researchers to collect spectral information from different areas within the material; hence the diffusion of the dye was evaluated (Fig. 19.7). An obvious concentration gradient was detected through the sample and shown to be coincident with the exposure time. Based on the ability to obtain diffusion coefficients directly from the CRM data of CO_2-induced dyeing of PET, the study was extended to include mathematical modelling of the diffusion process [77].

Prior to this work the feasibility of CRM to investigate polymeric dyeing was reported by Chan and Kazarian in 2003 where the distribution of dye in a CO_2 processed PET film was studied [78]. This approach has also been combined with IR spectroscopic imaging where a gradient of differing morphology was induced by exposure to high-pressure CO_2 [31].

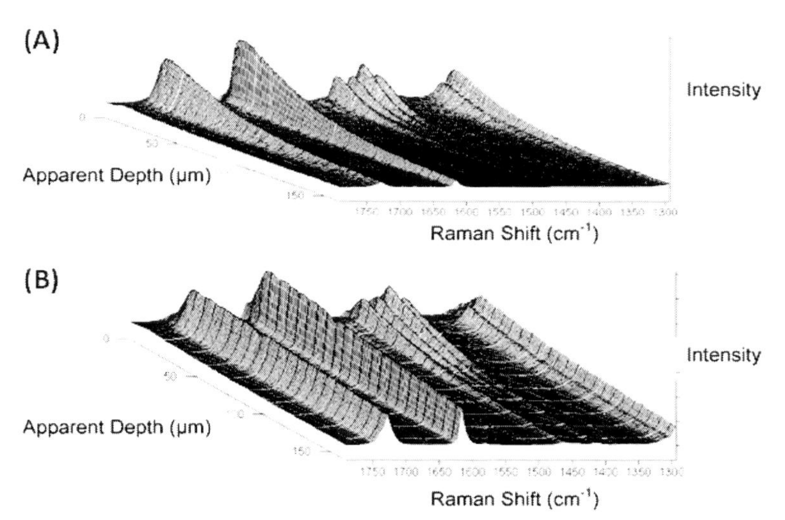

Figure 19.7 Confocal Raman depth profiles of PET fibers dyed for 3 hours (A) collected using a 50x dry objective and (B) collected using a 50x oil objective. All Raman spectra were recorded at 1 μm increments along the line normal to the fiber surface. Reprinted with permission from Fleming et al. Copyright (2005) Elsevier [33].

19.4.2 Polymeric Blending

Pioneering research by Watkins and McCarthy described in-situ polymerization to prepare immiscible blends [79]. Since then, much research has been applied based upon this methodology to prepare novel polymer blends useful for a range of applications [80–82], including conducting electricity [83].

The diffusion of the monomer and initiator within a swollen polymer matrix at high pressures determines the extent to which polymerization will occur. This can be related to the blending of polymers, since, at very short exposure times only the surface of the swollen polymers will be modified, whereas the bulk will remain unchanged [26]. The fact that CO_2 has been shown to induce morphological changes in polymers means that more information about the behavior of these changes is needed as a way to produce more efficient polymer blends. A range of approaches have been applied to research the morphological changes for the blending of polymers: which includes FTIR spectroscopy and spectroscopic imaging [68], DSC [84, 85], and atomic force microscopy (AFM) [86, 87]. More information about these techniques can be found within the referenced articles but for the purpose and topic of this book chapter, research using FTIR spectroscopy and spectroscopic imaging will be focused on.

In 2003, Busby et al. reported a study that combined three analytical techniques, including FTIR spectroscopy, to characterize novel nanostructured polymer blends [88]. The research demonstrated that blends of ultrahigh-molecular-weight polyethylene (UHMWPE) and polymethacrylate (PMA) could be made using supercritical fluid technology. PMA was dissolved in CO_2 and the loading, or blending, of the polymers could be controlled by a number of factors i.e. by adjusting the temperature and pressure and the PMA concentration in the CO_2. It was reported here, that processing with traditional VOCs did not produce great incorporation of the two polymers due to the low penetration of the solvents into the bulk of the matrix. The swelling of UHMWPE under a high-pressure environment meant there was much better incorporation of the two species and thus the article demonstrates that this technology has much potential in the field of characterizing molecular blends. The benefit to using a combination of FTIR spectroscopy, DSC analysis and AFM means much more information about the morphology in the UHMWPE matrix can be collected. As a result, it helped to realize other factors, such as size of the polymer phases, which may need to be controlled to tune the properties of the blended polymeric materials.

Kazarian and Chan presented a novel application to study polymers subjected to high-pressure and $scCO_2$ environments using FTIR spectroscopic imaging in ATR mode [68]. In-situ experiments allowed the researchers to obtain spatially resolved images (~15 μm) of the LCST phase separation of PS/PVME polymer blends. The same analytical methodology was applied to study the sorption of CO_2 into PMMA and poly(ethylene glycol) (PEO), subjected to identical conditions. The two polymers were cast to cover half of the diamond ATR crystal, allowing the two species to be measured simultaneously and thus their behaviors compared. Quantitative information can be obtained about the system using FTIR spectroscopy by considering the relative absorbance of the polymeric bands. Kazarian and Chan calculated the extent of polymer swelling and CO_2 sorption within the first few micrometers of the surface of the sample under high-pressure CO_2 conditions [68]. The spectroscopic images showed that swelling at 50 bar was greater in PMMA, as shown by the decreasing absorbance of PMMA bands compared to PEO. However, the degree of swelling of PMMA was seen to decrease as temperature was increased (from 30 to 50 °C). PEO was affected by the CO_2 by the fact that the T_m was reduced. This was characterized by spectral features that were characteristic of molten PEO.

The above articles have demonstrated the versatility of FTIR spectroscopy and spectroscopic imaging to simultaneously measure different components under high-pressure conditions. The ability to reveal chemical information and spatial information about the components within the imaged area has opened up new areas for developing supercritical fluid technologies using these approaches.

19.4.3 CO_2-Enhanced Polymer Interdiffusion and Dissolution Studies

Plasticisation of polymeric materials can cause a reduction in both the viscosity and interfacial tension in the polymers thus enhancing the miscibility and blending [61, 89]. Therefore, subjecting polymeric systems to a CO_2 environment can be predicted to accelerate the interdiffusion of miscible polymers. The importance of understanding the interdiffusion of two or more polymers to produce homogenous blends and possible phase separation is important for optimizing the design of materials. It should be noted that there are many experimental challenges involved with studying polymer interdiffusion facilitated by CO_2.

Investigations utilizing $scCO_2$ in this area include the study of reactive coupling at polymer–polymer interfaces at low-temperatures [90], as well as investigating density fluctuations in thin polymer films [91, 92]. A thorough understanding of this novel processing route is important as the enhanced interdiffusion of polymer blends has the potential to open up new areas in this processing technology.

In 2006, Fleming et al. applied ATR-FTIR spectroscopic imaging to study interdiffusion and dissolution of miscible polymers under high pressure and supercritical CO_2 [69]. The spectroscopic imaging approach allowed a model system of polyvinylpyrrolidone (PVP) and PEO with different molecular weights to be studied in-situ. The significance of using these two polymers is due to the attention they receive from the pharmaceutical industry as polymer excipients used in tablet formulations [7]. One of the main advantages of using this approach is that all components in the ternary system could be simultaneously monitored as a function of time. Three different ATR-FTIR spectroscopic imaging results were presented in the article (shown in Fig. 19.8) that represent time-dependent interdiffusion of PVP and PEO 600 in the absence of CO_2 and when subjected to pressures of 40 and 80 bar [69]. The results show that when the system was under high-pressure conditions the interdiffusion of PEO into PVP was more rapid. As a result, controlling the pressure can be used as a way to control and "tune" the speed and mechanism of the interdiffusion process.

To explain further this phenomenon, Fleming et al. studied the role of CO_2 on the interdiffusion profiles. Spectroscopic imaging allowed for the spatial distribution of CO_2 to be observed within the system [69]. It was proposed that the presence of CO_2 molecules dissolved in the polymer system significantly enhances the interdiffusion process under isothermal conditions. This was studied both as a function of pressure and molecular weight of PEO. Similar studies have been carried out by Bairamov et al. where the effect of different molecular weight polymers on the diffusion characteristics during heating were studied using optical wedge microinterferometry [93].

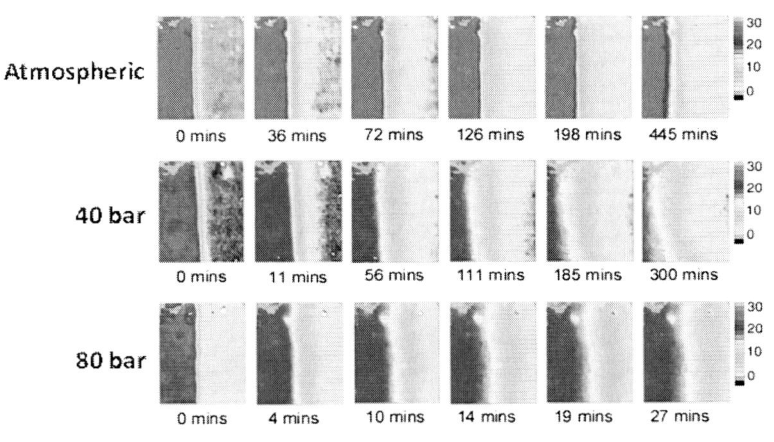

Figure 19.8 FTIR spectroscopic images, based on the distribution of the integrated absorbance of PVP carbonyl band, showing PVP (red regions) in contact with PEO (blue regions). The images show the interdiffusion of the two polymer species as a function of time. Reprinted with permission from Fleming et al. Copyright (2006) Elsevier [69].

Overall, such studies demonstrated that ATR-FTIR spectroscopic imaging can be a powerful approach to understand the interdiffusion and dissolution of miscible polymeric systems in-situ. The chemical specificity offered by FTIR spectroscopic imaging enables the behavior of the systems to be characterized and understood at a molecular level. Regulating the pressure allows one to control the diffusion characteristics. The combinations of all the parameters in this study meant that a comprehensive understanding of a dynamic system could be learned. Further work from the group probed H-bonding interactions between PVP and PEO using ATR-FTIR spectroscopy [94]. Overall, it was found that the extent of H-bonding was greatest in blends comprised of low-molecular-weight polymers and CO_2-induced shielding of the H-bonded interactions is reversible upon depressurization.

19.4.4 Drug-Loaded Polymers

Polymers that are soluble in biological fluids, give nontoxic products from degradation and are biodegradable are useful when it comes to drug delivery. Active pharmaceutical ingredients (APIs) are commonly dispersed within these polymeric materials as a means to increase their bioavailability and control the release rate. Molecularly dispersed formulations are important for APIs that have both high and low solubility in aqueous media. For example, APIs that are readily soluble can be dispersed within a polymer which will slow down their dissolution rate in a biological system and hence control their release. Dispersing a poorly water soluble drug within a polymer can prevent crystallization and thus improve the therapeutic value of the APIs [7, 95]. Solid dispersions are widely used in the pharmaceutical industry and common preparation techniques include solvent evaporation, hot melt extrusion and mechanical mixing [96].

Opportunities exist to utilize scCO_2 technologies which are desirable based on the fact that no VOCs are used when processing materials and thus the "green" process has an environmentally benign nature [76, 97]. The preparation of solid dispersions can be done so by supercritical impregnation into a polymeric material which works on the same principle as the incorporation of dye molecules. Impregnation by scCO_2 works by doping a solute, in this case an API, into a bulk polymeric material. One of the major concerns with API impregnation into the polymers is the effect that the processing stage has upon the bioactivity of the components i.e. degradation of the species. The ability of scCO_2 to mimic heat and lower the T_g of polymers means that thermally labile compounds can be processed. Additionally, the lower viscosity of polymers under the influence of a high-pressure CO_2 environment facilitates the mixing of the polymer with the API resulting in a homogenous material as the end product.

Kazarian and Martirosyan carried out a detailed vibrational spectroscopic investigation of the impregnation of ibuprofen into a PVP matrix [7]. The results concluded that molecular dispersion of the API was achieved within the polymer and no evidence of crystallization was observed by analysis of both the Raman and IR spectra. Although the results from this study were consistent with those obtained from conventionally prepared samples, there was no evidence of the crystalline peaks in the spectra, indicating molecular dispersion. Conversely, in the samples prepared by conventional methods the absorption bands associated with the crystalline form of ibuprofen were observed. Overall, this research demonstrates the importance of $scCO_2$ impregnation of drug into polymers and vibrational spectroscopy in terms of achieving more efficient processing and gaining a more thorough understand pharmaceutical systems. Nevertheless, despite the advantages of using this technology, $scCO_2$ processing still has room for further development in the future particularly for applications on industrial scales where it is still in the initial stages [45, 97–100].

19.4.5 CO_2 Adsorption into Porous Materials

Potential applications for gas storage and CO_2 capture mean that much can be learnt about studying the adsorption of high-pressure CO_2 into micropouros materials [101–104]. Hasell et al. have recently applied conventional ATR-FTIR spectroscopy to determine the concentration of sorbed CO_2 into porous polymers, porous crystalline frameworks and molecular organic cage materials in-situ [105]. The significance of applying this spectroscopic approach is that gas sorption isotherms can be measured at pressures up to 140 bar which is relevant for applications such as precombustion CO_2 capture. The concentration of the species can be quantified based on the absorbance of the chemically specific vibrational bands related to the molecule. Different organic imine cage materials were studied, where the CO_2 adsorption was measured at different pressure values. The ATR-FTIR spectral data was compared to manometric techniques and molecular simulations as a means to validate the feasibility of this spectroscopic approach for such applications.

The spectra showed that for imine samples subjected to differing pressures of CO_2 at a constant temperature (35°C), the band assigned for the asymmetric stretching mode of CO_2 (ν_3) at 2333 cm^{-1} increased in absorbance as the pressure increased (Fig. 19.9). In contrast, Fig. 19.9 also shows that the absorbance band assigned to the organic imine cage material (1647 cm^{-1}) remained relatively constant throughout the experiment. The uptake of CO_2 into the imine cage complexes was quantified from the absorbance of the bands in the spectra and was compared to manometric crystallographic data from the same complexes. It was shown that for pressures up to 1 bar there was very close correlation between the two approaches. Thus, using ATR-FTIR spectroscopy to measure CO_2 adsorption into these complexes at higher pressures, up to 120 bar, was valid as an approach to learn about such systems.

The results presented by Hasell et al. have shown an application where vibrational spectroscopy is validated as a means to measure the CO_2 adsorption into complex organic cage molecules [105]. The advantage of using ATR-FTIR spectroscopy for this study is significant because it allows measurements to be recorded at pressures that are above the critical temperature of CO_2, which can be challenging to achieve using other techniques. Validation of the vibrational spectroscopic approach has been shown by comparing the results with both manometric data and molecular simulations. This gives ATR-FTIR spectroscopy scope to be used to study CO_2 sorption, not only into organic cage molecules, but into a range of different microporous systems.

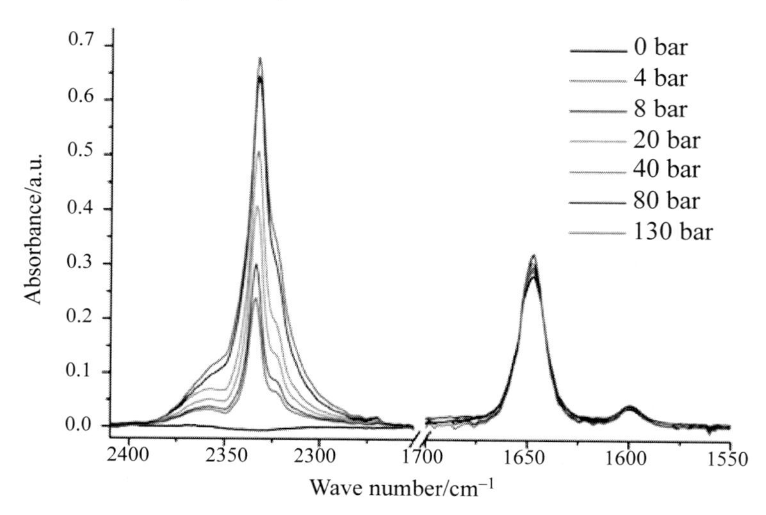

Figure 19.9 ATR-FTIR spectra of the organic imine film subjected to a range increasing high-pressure CO_2 conditions. The spectra show the ν_3 band of CO_2 (2335 cm^{-1}) and the imine stretching vibration (1647 cm^{-1}). Reproduced from Hasell et al. (2013) with permission from The Royal Society of Chemistry [105].

19.4.6 CO_2 Functionalization of Natural Biomaterials

The final research area to be discussed in this book chapter where vibrational spectroscopy has aided the development of supercritical technologies is for the functionalization of natural biomatrices. In this field, the use of SCFs as a cleaner alternative for particle design and drug impregnation is one emerging example of improving the properties of biomatrices [17, 106, 107]. Most biocompatible polymeric materials are swellable under high-pressure CO_2 conditions, which allows both the extraction and impregnation of other components into the systems.

Tissue engineering is an area where processing natural polymers using CO_2 has been advanced in recent years [108, 109]. The swelling of biomaterials with CO_2 is a proven method of producing 3D scaffolds. Desired scaffolds require an environment that facilitates cell attachment and growth, allows transport of the materials in and out of the scaffolds and can be mechanically tuned for specific use. Applications of these materials have been demonstrated for the delivery of chondrocytes [110], providing a local environment able to promote osteogenesis [111] and the ability to deliver a range of growth factors [112]. It should be noted that one of the main drawbacks using CO_2 to swell such natural polymers can result in closed 3D scaffolds which is not favorable for the transport of materials. However, this limitation can be overcome with the addition of particle leaching step as reported by Sheradin et al. in 2000 [113].

A recent research paper by Silva et al. has reported the modification of chitin (found in shells of crustaceans) using *green chemistry principles* [114]. FTIR spectroscopy was applied to study the behavior of dexamethasone impregnated into chitin aggregated 3D scaffolds using SCF conditions. The release of the drug incorporated into the scaffolds was investigated with the aim to control the release of the systems. The researchers used SCF technology to improve the flexibility of these biomaterials for use as drug delivery systems. The shape of the scaffold and osteo-inductive behavior was also modified due to the presence of certain functional groups in the system. The topic of using SCFs to impregnate and process natural biomaterials (such as chitosan derivatives) have been discussed in more detail in a number of reviews [106, 107, 115, 116].

19.5 Conclusions and Future Outlook

The main aim of this chapter was not only to provide a summary of polymeric processing using high-pressure and $scCO_2$ but to give an overview of how vibrational spectroscopy can be used to aid development of this emerging technology. Considering the costs associated with high-pressure processing, the benefits for using this method must be significant. As mentioned in the above sections the key, and thus major, advantage to using high-pressure CO_2 is its ability to act as a temporary plasticizer due to the weak interactions it has with certain functional groups in the polymers. Hence, the following physical changes occur within the materials: lowering of T_g, swelling or foaming, reduction in T_m and more facile solute mass transfer in the bulk for the sample.

The applications for CO_2 processing is by no means exhausted and, as shown in the chapter, there are many more emerging research areas and fields where this technology can be utilized. It is intended that this chapter will summarize some of the key research which has been carried out over the last few decades and demonstrate where vibrational spectroscopy can help as a development tool in both academia and industry. There has been much advancement in this field over recent years, the extent by which has even exceeded expectations that were predicted in an earlier review paper [23]. One of the most powerful and innovative approaches used for such applications is that of FTIR spectroscopic imaging. This allows one to collect spatially resolved "chemical images" of high-pressure CO_2 interacting with polymeric materials. The technique has shown great potential to contribute and add valuable information about systems that was not previously possible and offers improved knowledge about the reliability of analysis and data. Emerging developments include the functionalization of natural biomaterials and the impregnation of materials into porous substrates, including drug loaded polymers. It comes as no surprise that this processing technology has extended to both biomedical and pharmaceutical applications because of the drive towards "greener" processing routes in industry. Finally, the authors hope this review further stimulates research ideas for polymer processing with supercritical fluids by highlighting some of the most recent developments of this technology.

References

1. Cooper, A. I. (2000). Polymer synthesis and processing using supercritical carbon dioxide, *J. Mater. Chem.*, **10**, pp. 207–234.
2. Wood, C. D., Cooper, A. I., and DeSimone, J. M. (2004). Green synthesis of polymers using supercritical carbon dioxide, *Curr. Opin. Solid State Mater. Sci.*, **8**, pp. 325–331.
3. Kazarian, S. G. (2002). Polymers and supercritical fluids: Opportunities for vibrational spectroscopy, *Macromol. Symp.*, **184**, pp. 215–228.
4. Gerhardt, L. J., Manke, C. W., and Gulari, E. (1997). Rheology of polydimethylsiloxane swollen with supercritical carbon dioxide, *J. Polym. Sci., Part B: Polym. Phys.*, **35**, pp. 523–534.
5. Koros, W. J. (1985). Simplified analysis of gas/polymer selective solubility behavior, *J. Polym. Sci., Part B: Polym. Phys.*, **23**, pp. 1611–1628.
6. Aubert, J. H. (1998). Solubility of carbon dioxide in polymers by the quartz crystal microbalance technique, *J. Supercrit. Fluids*, **11**, pp. 163–172.
7. Kazarian, S. G., and Martirosyan, G. G. (2002). Spectroscopy of polymer/drug formulations processed with supercritical fluids: in situ ATR-IR and Raman study of impregnation of ibuprofen into PVP, *Int. J. Pharm.*, **232**, pp. 81–90.
8. Shen, Z., McHugh, M., Xu, J., Belardi, J., Kilic, S., Mesiano, A., Bane, S., Karnikas, C., Beckman, E., and Enick, R. (2003). CO_2-solubility of oligomers and polymers that contain the carbonyl group, *Polymer*, **44**, pp. 1491–1498.
9. Shieh, Y. T., Su, J. H., Manivannan, G., Lee, P. H., Sawan, S. P., and Dale Spall, W. (1996). Interaction of supercritical carbon dioxide with polymers. I. Crystalline polymers, *J. Appl. Polym. Sci.*, **59**, pp. 695–705.
10. Shieh, Y. T., Su, J. H., Manivannan, G., Lee, P. H., Sawan, S. P., and Dale Spall, W. (1996). Interaction of supercritical carbon dioxide with polymers. II. Amorphous polymers, *J. Appl. Polym. Sci.*, **59**, pp. 707–717.
11. Zhang, Y., Gangwani, K. K., and Lemert, R. M. (1997). Sorption and swelling of block copolymers in the presence of supercritical fluid carbon dioxide, *J. Supercrit. Fluids*, **11**, pp. 115–134.
12. Webb, K. F., and Teja, A. S. (1999). Solubility and diffusion of carbon dioxide in polymers, *Fluid Phase Equilib.*, **158**, pp. 1029–1034.
13. Wong, B., Zhang, Z., and Handa, Y. P. (1998). High-precision gravimetric technique for determining the solubility and diffusivity of gases in polymers, *J. Polym. Sci., Part B: Polym. Phys.*, **36**, pp. 2025–2032.

14. Zhang, C., Cappleman, B., Defibaugh-Chavez, M., and Weinkauf, D. (2003). Glassy polymer-sorption phenomena measured with a quartz crystal microbalance technique, *J. Polym. Sci., Part B: Polym. Phys.*, **41**, pp. 2109–2118.

15. Von Solms, N., Nielsen, J. K., Hassager, O., Rubin, A., Dandekar, A., Andersen, S. I., and Stenby, E. H. (2004). Direct measurement of gas solubilities in polymers with a high-pressure microbalance, *J. Appl. Polym. Sci.*, **91**, pp. 1476–1488.

16. Nelson, M. R., and Borkman, R. F. (1998). Ab initio calculations on CO_2 binding to carbonyl groups, *J. Phys. Chem. A*, **102**, pp. 7860–7863.

17. Kikic, I., Lora, M., Cortesi, A., and Sist, P. (1999). Sorption of CO_2 in biocompatible polymers: experimental data and qualitative interpretation, *Fluid Phase Equilib.*, **158**, pp. 913–921.

18. Fried, J., and Hu, N. (2003). The molecular basis of CO_2 interaction with polymers containing fluorinated groups: computational chemistry of model compounds and molecular simulation of poly [bis (2, 2, 2-trifluoroethoxy) phosphazene], *Polymer*, **44**, pp. 4363–4372.

19. Hamedi, M., Muralidharan, V., Lee, B., and Danner, R. P. (2003). Prediction of carbon dioxide solubility in polymers based on a group-contribution equation of state, *Fluid Phase Equilib.*, **204**, pp. 41–53.

20. Kazarian, S. G., Vincent, M. F., Bright, F. V., Liotta, C. L., and Eckert, C. A. (1996). Specific intermolecular interaction of carbon dioxide with polymers, *J. Am. Chem. Soc.*, **118**, pp. 1729–1736.

21. Flichy, N. M. B., Kazarian, S. G., Lawrence, C. J., and Briscoe, B. J. (2001). An ATR–IR study of poly (dimethylsiloxane) under high-pressure carbon dioxide: simultaneous measurement of sorption and swelling, *J. Phys. Chem. B*, **106**, pp. 754–759.

22. Brantley, N. H., Kazarian, S. G., and Eckert, C. A. (2000). In situ FTIR measurement of carbon dioxide sorption into poly (ethylene terephthalate) at elevated pressures, *J. Appl. Polym. Sci.*, **77**, pp. 764–775.

23. Kazarian, S. G. (2000). Polymer processing with supercritical fluids, *Polymer Science Series C*, **42**, pp. 78–101.

24. Tomasko, D. L., Li, H. B., Liu, D. H., Han, X. M., Wingert, M. J., Lee, L. J., and Koelling, K. W. (2003). A review of CO_2 applications in the processing of polymers, *Ind. Eng. Chem. Res.*, **42**, pp. 6431–6456.

25. Cansell, F., Aymonier, C., and Loppinet-Serani, A. (2003). Review on materials science and supercritical fluids, *Curr. Opin. Solid State Mater. Sci.*, **7**, pp. 331–340.

26. Fleming, O. S., and Kazarian, S. G. (2005) *Supercritical Carbon Dioxide: in Polymer Reaction Engineering,* eds. Kemmere, M. F., and Meyer, T., Chapter 10 "Polymer Processing with Supercritical Fluids", (WILEY-VCH Verlag GmbH & Co. KGaA: Germany). pp. 206–238.

27. Kazarian, S. G., Brantley, N. H., West, B. L., Vincent, M. F., and Eckert, C. A. (1997). In situ spectroscopy of polymers subjected to supercritical CO_2: plasticization and dye impregnation, *Appl. Spectrosc.*, **51**, pp. 491–494.

28. Mizoguchi, K., Hirose, T., Naito, Y., and Kamiya, Y. (1987). CO_2-induced crystallization of poly (ethylene terephthalate), *Polymer*, **28**, pp. 1298–1302.

29. Zhong, Z., Zheng, S., and Mi, Y. (1999). High-pressure DSC study of thermal transitions of a poly (ethylene terephthalate)/carbon dioxide system, *Polymer*, **40**, pp. 3829–3834.

30. Kazarian, S. G., Brantley, N. H., and Eckert, C. A. (1999). Applications of vibrational spectroscopy to characterize poly (ethylene terephthalate) processed with supercritical CO_2, *Vib. Spectrosc.*, **19**, pp. 277–283.

31. Fleming, O. S., Chan, K. L. A., and Kazarian, S. G. (2004). FT-IR imaging and Raman microscopic study of poly(ethylene terephthalate) film processed with supercritical CO_2, *Vib. Spectrosc.*, **35**, pp. 3–7.

32. Fleming, O. S., and Kazarian, S. G. (2004). Confocal Raman microscopy of morphological changes in poly(ethylene terephthalate) film induced by supercritical CO_2, *Appl. Spectrosc.*, **58**, pp. 390–394.

33. Fleming, O. S., Kazarian, S. G., Bach, E., and Schollmeyer, E. (2005). Confocal Raman study of poly(ethylene terephthalate) fibres dyed in supercritical carbon dioxide: dye diffusion and polymer morphology, *Polymer*, **46**, pp. 2943–2949.

34. Handa, Y. P., Zhang, Z., and Wong, B. (1997). Effect of compressed CO_2 on phase transitions and polymorphism in syndiotactic polystyrene, *Macromolecules*, **30**, pp. 8499–8504.

35. Takada, M., Tanigaki, M., and Ohshima, M. (2001). Effects of CO_2 on crystallization kinetics of polypropylene, *Polym. Eng. Sci.*, **41**, pp. 1938–1946.

36. Handa, Y. P., Zhang, Z., and Roovers, J. (2001). Compressed-gas-induced crystallization in tert-butyl poly (ether ether ketone), *J. Polym. Sci., Part B: Polym. Phys.*, **39**, pp. 1505–1512.

37. Beckman, E., and Porter, R. S. (1987). Crystallization of bisphenol a polycarbonate induced by supercritical carbon dioxide, *J. Polym. Sci., Part B: Polym. Phys.*, **25**, pp. 1511–1517.

38. Takada, M., Hasegawa, S., and Ohshima, M. (2004). Crystallization kinetics of poly (L-lactide) in contact with pressurized CO_2, *Polym. Eng. Sci.*, **44**, pp. 186–196.
39. Kazarian, S. G., Lawrence, C. J., and Briscoe, B. J. (1999). In-situ spectroscopy of polymers processed with supercritical carbon dioxide, *Proc. SPIE*, **4060**, pp. 210–216.
40. Cooper, A. I., Kazarian, S. G., and Poliakoff, M. (1993). Supercritical fluid impregnation of polyethylene films, a new approach to studying equilibria in matrices; the hydrogen bonding of fluoroalcohols to (η^5-C_5Me_5)$Ir(CO)_2$ and the effect on C-H activation, *Chem. Phys. Lett.*, **206**, pp. 175–180.
41. McNally, M. E. P. (1995). Advances in environmental SFE, *Anal. Chem.*, **67**, pp. 308A–315A.
42. Vandenburg, H. J., Clifford, A. A., Bartle, K. D., Carroll, J., Newton, I., Garden, L. M., Dean, J. R., and Costley, C. T. (1997). Critical review: Analytical extraction of additives from polymers, *Analyst*, **122**, pp. 101R–116R.
43. Martinez, J., and de Aguiar, A. C. (2014). Extraction of triacylglycerols and fatty acids using supercritical fluids - Review, *Curr. Anal. Chem.*, **10**, pp. 67–77.
44. Reverchon, E. (1997). Supercritical fluid extraction and fractionation of essential oils and related products, *J. Supercrit. Fluids*, **10**, pp. 1–37.
45. Kazarian, S. G. (1997). Applications of FTIR spectroscopy to supercritical fluid drying, extraction and impregnation, *Appl. Spectrosc. Rev.*, **32**, pp. 301–348.
46. McHugh, M., and Krukonis, V. (1994) *Supercritical Fluid Extraction*. 2nd Ed. (Butterworth-Heinemann, USA).
47. Marsal, A., Celma, P., Cot, J., and Cequier, M. (2000). Supercritical CO_2 extraction as a clean degreasing process in the leather industry, *J. Supercrit. Fluids*, **16**, pp. 217–223.
48. Martial, F., Huguet, J., and Bunel, C. (1999). Development of a quantitative analysis method for polypropylene additives using on-line SFE/SFC, *Polym. Int.*, **48**, pp. 299–306.
49. Zacharia, R., Simon, S., Beckman, E., and Enick, R. (1999). Improving the thermal stability of a polymer through liquid carbon dioxide extraction of a metal compound, *Polym. Degrad. Stab.*, **63**, pp. 85–88.
50. Rodriguez, F., Claude, C., Ober, C., and A., A. L. (2005) *Principles of Polymer Systems*. 5th Ed. (Taylor & Francis, USA).

51. Han, X., Koelling, K. W., Tomasko, D. L., and Lee, L. J. (2002). Continuous microcellular polystyrene foam extrusion with supercritical CO_2, *Polym. Eng. Sci.*, **42**, pp. 2094–2106.

52. Han, X., Shen, J., Huang, H., Tomasko, D. L., and Lee, L. J. (2007). CO_2 foaming based on polystyrene/poly (methyl methacrylate) blend and nanoclay, *Polym. Eng. Sci.*, **47**, pp. 103–111.

53. Han, X., Koelling, K. W., Tomasko, D. L., and Lee, L. J. (2003). Effect of die temperature on the morphology of microcellular foams, *Polym. Eng. Sci.*, **43**, pp. 1206–1220.

54. Krause, B., Diekmann, K., Van der Vegt, N., and Wessling, M. (2002). Open nanoporous morphologies from polymeric blends by carbon dioxide foaming, *Macromolecules*, **35**, pp. 1738–1745.

55. Krause, B., Mettinkhof, R., Van der Vegt, N., and Wessling, M. (2001). Microcellular foaming of amorphous high-T g polymers using carbon dioxide, *Macromolecules*, **34**, pp. 874–884.

56. Krause, B., Sijbesma, H., Münüklü, P., Van der Vegt, N., and Wessling, M. (2001). Bicontinuous nanoporous polymers by carbon dioxide foaming, *Macromolecules*, **34**, pp. 8792–8801.

57. Rodeheaver, B. A., and Colton, J. (2001). Open-celled microcellular thermoplastic foam, *Polym. Eng. Sci.*, **41**, pp. 380–400.

58. Han, C. D., and Ma, C. Y. (1983). Rheological properties of mixtures of molten polymer and fluorocarbon blowing agent. II. Mixtures of polystyrene and fluorocarbon blowing agent, *J. Appl. Polym. Sci.*, **28**, pp. 851–860.

59. Lee, M., Park, C. B., and Tzoganakis, C. (1999). Measurements and modeling of PS/supercritical CO_2 solution viscosities, *Polym. Eng. Sci.*, **39**, pp. 99–109.

60. Royer, J. R., Gay, Y. J., Desimone, J. M., and Khan, S. A. (2000). High-pressure rheology of polystyrene melts plasticized with CO_2: Experimental measurement and predictive scaling relationships, *J. Polym. Sci., Part B: Polym. Phys.*, **38**, pp. 3168–3180.

61. Royer, J. R., DeSimone, J. M., and Khan, S. A. (2001). High-pressure rheology and viscoelastic scaling predictions of polymer melts containing liquid and supercritical carbon dioxide, *J. Polym. Sci., Part B: Polym. Phys.*, **39**, pp. 3055–3066.

62. Xiong, Y., and Kiran, E. (1995). Miscibility, density and viscosity of poly (dimethylsiloxane) in supercritical carbon dioxide, *Polymer*, **36**, pp. 4817–4826.

63. Yeo, S.-D., and Kiran, E. (1999). Viscosity reduction of polystyrene solutions in toluene with supercritical carbon dioxide, *Macromolecules*, **32**, pp. 7325–7328.

64. Bae, Y. C., and Gulari, E. (1997). Viscosity reduction of polymeric liquid by dissolved carbon dioxide, *J. Appl. Polym. Sci.*, **63**, pp. 459–466.

65. Flichy, N. M. B., Lawrence, C. J., and Kazarian, S. G. (2003). Rheology of poly(propylene glycol) and suspensions of fumed silica in poly(propylene glycol) under high-pressure CO_2, *Ind. Eng. Chem. Res.*, **42**, pp. 6310–6319.

66. Royer, J. R., Gay, Y. J., Adam, M., DeSimone, J. M., and Khan, S. A. (2002). Polymer melt rheology with high-pressure CO_2 using a novel magnetically levitated sphere rheometer, *Polymer*, **43**, pp. 2375–2383.

67. Kazarian, S. G., Briscoe, B. J., and Welton, T. (2000). Combining ionic liquids and supercritical fluids: in situ ATR-IR study of CO_2 dissolved in two ionic liquids at high pressures, *Chem. Commun.*, 2047–2048.

68. Kazarian, S. G., and Chan, K. L. A. (2004). FTIR imaging of polymeric materials under high-pressure carbon dioxide, *Macromolecules*, **37**, pp. 579–584.

69. Fleming, O. S., Chan, K. L. A., and Kazarian, S. G. (2006). High-pressure CO_2-enhanced polymer interdiffusion and dissolution studied with in situ ATR-FTIR spectroscopic imaging, *Polymer*, **47**, pp. 4649–4658.

70. Kazarian, S. G., and Chan, K. L. A. (2003). "Chemical photography" of drug release, *Macromolecules*, **36**, pp. 9866–9872.

71. Kazarian, S. G., Kong, K. W. T., Bajomo, M., Van der Weerd, J., and Chan, K. L. A. (2005). Spectroscopic imaging applied to drug release, *Food and Bioprod. Processing*, **83**, pp. 127–135.

72. Kazarian, S. G., and Ewing, A. V. (2013). Applications of Fourier transform infrared spectroscopic imaging to tablet dissolution and drug release, *Expert Opin. Drug Deliv.*, **10**, pp. 1207–1221.

73. Glassford, S. E., Govada, L., Chayen, N. E., Byrne, B., and Kazarian, S. G. (2012). Micro ATR FTIR imaging of hanging drop protein crystallisation, *Vib. Spectrosc.*, **63**, pp. 492–498.

74. Glassford, S., Chan, K. L. A., Byrne, B., and Kazarian, S. G. (2012). Chemical imaging of protein adsorption and crystallization on a wettability gradient surface, *Langmuir*, **28**, pp. 3174–3179.

75. Kazarian, S. G., and Chan, K. L. A. (2013). ATR-FTIR spectroscopic imaging: recent advances and applications to biological systems, *Analyst*, **138**, pp. 1940–1951.

76. Kazarian, S. G. (2004) *Supercritical Fluid Technology for Drug Product Development,* eds. York, P., Kompella, U. B., and Y., S. B., Chapter 8 "Supercritical Fluid Impregnation of Polymers for Drug Delivery", (Marcel Dekker, Inc. : USA). pp. 343–366.

77. Fleming, O. S., Stepanek, F., and Kazarian, S. G. (2005). Dye diffusion in polymer films subjected to supercritical CO_2: Confocal Raman microscopy and modelling, *Macromol. Chem. Phys.*, **206**, pp. 1077–1083.

78. Kazarian, S. G., and Chan, K. L. A. (2003). Confocal Raman microscopy of supercritical fluid dyeing of polymers, *Analyst*, **128**, pp. 499–503.

79. Watkins, J. J., and McCarthy, T. J. (1995). Polymerization of styrene in supercritical CO_2-swollen poly (chlorotrifluoroethylene), *Macromolecules*, **28**, pp. 4067–4074.

80. Arora, K. A., Lesser, A. J., and McCarthy, T. J. (1999). Synthesis, characterization, and expansion of poly (tetrafluoroethylene-co-hexafluoropropylene)/polystyrene blends processed in supercritical carbon dioxide, *Macromolecules*, **32**, pp. 2562–2568.

81. Rajagopalan, P., and McCarthy, T. J. (1998). Two-step surface modification of chemically resistant polymers: blend formation and subsequent chemistry 1, *Macromolecules*, **31**, pp. 4791–4797.

82. Kung, E., Lesser, A. J., and McCarthy, T. J. (1998). Morphology and mechanical performance of polystyrene/polyethylene composites prepared in supercritical carbon dioxide, *Macromolecules*, **31**, pp. 4160–4169.

83. Shenoy, S. L., Kaya, I., Erkey, C., and Weiss, R. (2001). Synthesis of conductive elastomeric foams by an in situ polymerization of pyrrole using supercritical carbon dioxide and ethanol cosolvents, *Synth. Met.*, **123**, pp. 509–514.

84. Wang, J., Cheung, M. K., and Mi, Y. (2002). Miscibility and morphology in crystalline/amorphous blends of poly (caprolactone)/poly (4-vinylphenol) as studied by DSC, FTIR, and ^{13}C solid state NMR, *Polymer*, **43**, pp. 1357–1364.

85. Tsuji, H. (2002). Autocatalytic hydrolysis of amorphous-made polylactides: effects of L-lactide content, tacticity, and enantiomeric polymer blending, *Polymer*, **43**, pp. 1789–1796.

86. Böltau, M., Walheim, S., Mlynek, J., Krausch, G., and Steiner, U. (1998). Surface-induced structure formation of polymer blends on patterned substrates, *Nature*, **391**, pp. 877–879.

87. Ton-That, C., Shard, A., Teare, D., and Bradley, R. (2001). XPS and AFM surface studies of solvent-cast PS/PMMA blends, *Polymer*, **42**, pp. 1121–1129.

88. Busby, A. J., Zhang, J., Naylor, A., Roberts, C. J., Davies, M. C., Tendler, S. J., and Howdle, S. M. (2003). The preparation of novel nano-structured polymer blends of ultra high molecular weight polyethylene with polymethacrylates using supercritical carbon dioxide, *J. Mater. Chem.*, **13**, pp. 2838–2844.

89. Watkins, J. J., and McCarthy, T. J. (1994). Polymerization in supercritical fluid-swollen polymers: a new route to polymer blends, *Macromolecules*, **27**, pp. 4845–4847.

90. Harton, S. E., Stevie, F. A., Spontak, R. J., Koga, T., Rafailovich, M. H., Sokolov, J. C., and Ade, H. (2005). Low-temperature reactive coupling at polymer–polymer interfaces facilitated by supercritical CO_2, *Polymer*, **46**, pp. 10173–10179.

91. Koga, T., Seo, Y.-S., Hu, X., Shin, K., Zhang, Y., Rafailovich, M., Sokolov, J., Chu, B., and Satija, S. (2002). Dynamics of polymer thin films in supercritical carbon dioxide, *Europys. Lett*, **60**, pp. 559–565.

92. Koga, T., Seo, Y.-S., Zhang, Y., Shin, K., Kusano, K., Nishikawa, K., Rafailovich, M. H., Sokolov, J. C., Chu, B., and Peiffer, D. (2002). Density-fluctuation-induced swelling of polymer thin films in carbon dioxide, *Phys. Rev. Lett.*, **89**, pp. 1255061–1255064.

93. Bairamov, D. F., Chalykh, A. E., Feldstein, M. M., and Siegel, R. A. (2002). Impact of molecular weight on miscibility and interdiffusion between poly (n-vinyl pyrrolidone) and poly (ethylene glycol), *Macromol. Chem. Phys.*, **203**, pp. 2674–2685.

94. Labuschagne, P. W., Kazarian, S. G., and Sadiku, R. E. (2011). In situ FTIR spectroscopic study of the effect of CO_2 sorption on H-bonding in PEG-PVP mixtures, *Spectrochim. Acta A*, **78**, pp. 1500–1506.

95. Labuschagne, P. W., John, M. J., and Sadiku, R. E. (2010). Investigation of the degree of homogeneity and hydrogen bonding in PEG/PVP blends prepared in supercritical CO_2: Comparison with ethanol-cast blends and physical mixtures, *J. Supercrit. Fluids*, **54**, pp. 81–88.

96. Serajuddin, A. (1999). Solid dispersion of poorly water-soluble drugs: early promises, subsequent problems, and recent breakthroughs, *J. Pharm. Sci.*, **88**, pp. 1058–1066.

97. Girotra, P., Singh, S. K., and Nagpal, K. (2013). Supercritical fluid technology: a promising approach in pharmaceutical research, *Pharm. Dev. Technol.*, **18**, pp. 22–38.

98. Subramaniam, B., Rajewski, R. A., and Snavely, K. (1997). Pharmaceutical processing with supercritical carbon dioxide, *J. Pharm. Sci.*, **86**, pp. 885–890.

99. Naylor, A., Lewis, A. L., and Ilium, L. (2011). Supercritical fluid-mediated methods to encapsulate drugs: recent advances and new opportunities, *Ther. Deliv.*, **2**, pp. 1551–1565.

100. Dias, A. M. A., Rey-Rico, A., Oliveira, R. A., Marceneiro, S., Alvarez-Lorenzo, C., Concheiro, A., Junior, R. N. C., Braga, M. E. M., and de Sousa, H. C. (2013). Wound dressings loaded with an anti-inflammatory juca (Libidibia ferrea) extract using supercritical carbon dioxide technology, *J. Supercrit. Fluids*, **74**, pp. 34–45.

101. Cooper, A. I. (2009). Conjugated microporous polymers, *Adv. Mater.*, **21**, pp. 1291–1295.

102. Wood, C. D., Tan, B., Trewin, A., Su, F., Rosseinsky, M. J., Bradshaw, D., Sun, Y., Zhou, L., and Cooper, A. I. (2008). Microporous organic polymers for methane storage, *Adv. Mater.*, **20**, pp. 1916–1921.

103. Wood, C. D., Tan, B., Trewin, A., Niu, H., Bradshaw, D., Rosseinsky, M. J., Khimyak, Y. Z., Campbell, N. L., Kirk, R., and Stöckel, E. (2007). Hydrogen storage in microporous hypercrosslinked organic polymer networks, *Chem. Mater.*, **19**, pp. 2034–2048.

104. McKeown, N. B., and Budd, P. M. (2006). Polymers of intrinsic microporosity (PIMs): organic materials for membrane separations, heterogeneous catalysis and hydrogen storage, *Chem. Soc. Rev.*, **35**, pp. 675–683.

105. Hasell, T., Armstrong, J. A., Jelfs, K. E., Tay, F. H., Thomas, K. M., Kazarian, S. G., and Cooper, A. I. (2013). High-pressure carbon dioxide uptake for porous organic cages: comparison of spectroscopic and manometric measurement techniques, *Chem. Commun.*, **49**, pp. 9410–9412.

106. Knez, Z., Markocic, E., Novak, Z., and Hrncic, M. K. (2011). Processing Polymeric Biomaterials using Supercritical CO_2, *Chem. Ing. Tech.*, **83**, pp. 1371–1380.

107. Jeon, B., Kim, H. K., Cha, S. W., Lee, S. J., Han, M.-S., and Lee, K. S. (2013). Microcellular foam processing of biodegradable polymers - review, *Int. J. Prec. Eng. Manuf.*, **14**, pp. 679–690.

108. Sheridan, M. H., Shea, L. D., Peters, M. C., and Mooney, D. J. (2000). Bioadsorbable polymer scaffolds for tissue engineering capable of sustained growth factor delivery, *J. Controlled Release*, **64**, pp. 91–102.

109. Duarte, A. R. C., Santo, V. E., Alves, A., Silva, S. S., Moreira-Silva, J., Silva, T. H., Marques, A. P., Sousa, R. A., Gomes, M. E., Mano, J. F., and

Reis, R. L. (2013). Unleashing the potential of supercritical fluids for polymer processing in tissue engineering and regenerative medicine, *J. Supercrit. Fluids*, **79**, pp. 177-185.

110. Barry, J., Gidda, H., Scotchford, C., and Howdle, S. (2004). Porous methacrylate scaffolds: supercritical fluid fabrication and in vitro chondrocyte responses, *Biomater.*, **25**, pp. 3559–3568.
111. Yang, X., Roach, H., Clarke, N., Howdle, S., Quirk, R., Shakesheff, K., and Oreffo, R. (2001). Human osteoprogenitor growth and differentiation on synthetic biodegradable structures after surface modification, *Bone*, **29**, pp. 523–531.
112. Richardson, T. P., Peters, M. C., Ennett, A. B., and Mooney, D. J. (2001). Polymeric system for dual growth factor delivery, *Nat. Biotechnol.*, **19**, pp. 1029–1034.
113. Sheridan, M., Shea, L., Peters, M., and Mooney, D. (2000). Bioabsorbable polymer scaffolds for tissue engineering capable of sustained growth factor delivery, *J. Controll. Release*, **64**, pp. 91–102.
114. Silva, S. S., Duarte, A. R. C., Mano, J. F., and Reis, R. L. (2013). Design and functionalization of chitin-based microsphere scaffolds, *Green Chem.*, **15**, pp. 3252–3258
115. Woods, H. M., Silva, M. M., Nouvel, C., Shakesheff, K. M., and Howdle, S. M. (2004). Materials processing in supercritical carbon dioxide: surfactants, polymers and biomaterials, *J. Mater. Chem.*, **14**, pp. 1663–1678.
116. Moribe, K., Tozuka, Y., and Yamamoto, K. (2008). Supercritical carbon dioxide processing of active pharmaceutical ingredients for polymorphic control and for complex formation, *Adv. Drug Deliver. Rev.*, **60**, pp. 328–338.

Chapter 20

Online Analytical Methods: Axisymmetric Drop Shape Analysis

M. Giovanna Pastore Carbone and Ernesto Di Maio
Dipartimento di Ingegneria Chimica, dei Materiali e della Produzione Industriale, University of Naples Federico II, P.le V. Tecchio 80, 80125 Naples, Italy
edimaio@unina.it

The methods based on the axisymmetric drop shape analysis (ADSA) for the evaluation of the interfacial and surface tension of fluid interfaces are reviewed in this chapter. Interfacial and surface tensions are fundamental to a wide range of scientific disciplines and technologies, and several experimental methods have been developed for their measurement. Among all of these methods, those based on the ADSA are the most used due to their versatility. Actually, they consist in analyzing the profile of axisymmetric drops, such as pendant or sessile drop, and rely on the equilibrium between capillary and gravity forces. In particular, the use of the pendant drop method has gained enormous ground in the fields of polymer science and supercritical fluid nanotechnology. In this chapter, the theoretical background of the pendant drop method is firstly

Supercritical Fluid Nanotechnology: Advances and Applications in Composites and Hybrid Nanomaterials
Edited by Concepción Domingo and Pascale Subra-Paternault

ISBN 978-981-4613-40-8 (Hardcover), 978-981-4613-41-5 (eBook)
www.panstanford.com

discussed and details on the experimental apparatus are described. Furthermore, some examples of the use of pendant drop method in supercritical fluid technology are reviewed.

20.1 Introduction

The science of interfaces is fundamental to a broad range of scientific disciplines and technologies. Actually, the measurement of surface and interfacial tensions of fluid interfaces and the measurement of the contact angle represent an essential issue for a large number of applications embracing the fields of adhesion, coatings, dispersions, foaming, polymer blending and membranes [1]. Surface and interfacial tension are the most accessible experimental parameter describing the thermodynamic state and the structure of an interface: in particular, the surface tension is a measure of the free energy of an interface between a liquid and a gas or vapor, while interfacial tension is referred to a phase boundary between two incompletely miscible liquids [2]. Due to the scientific and technological importance of these parameters, a great deal of effort has been focused on the development of methods for their measurement [3–5]. Table 20.1 reports the classification of the techniques which have been developed in the last century and which are widely used in both surface chemistry laboratories and industrial R&D facilities [6]. The accuracy of most of these techniques for pure liquid–gas systems has been found to be about 0.1 mN/m; however, the suitability of the methods for a specific system depends upon the nature of the liquid under investigation, the conditions under which the surface tension has to be measured, and the stability of the surface under deformation [6].

Of all the methods which have been developed, those based on the axisymmetric drop shape analysis (ADSA) are the most used and versatile techniques. ADSA methods consist in analyzing the profile of axisymmetric drops, such as pendant or sessile drops (Fig. 20.1), and rely on the equilibrium between capillary and gravity forces [7–11], assuming that inertial forces are negligible. In the pendant drop configuration (Fig. 20.1a), a drop of liquid is suspended from the end of a support, which can be a cylinder or a syringe: surface (or interfacial) tension and gravity are the only forces shaping the drop.

Table 20.1 Classification of techniques for the measurement of interfacial/ surface tension

Direct measurement using a microbalance	Wilhelmy plate	The microbalance measures the forces exerted in lifting a vertical plate placed in a liquid surface.
	Du Noüya ring	The microbalance measures the forces exerted in lifting a ring from a liquid surface.
Measurement of capillary pressure	Maximum bubble pressure	A tensiometer measures the maximum pressure to force a gas bubble out of a capillary into the liquid sample.
	Growing drop	The pressure and the size of a drop growing and detaching from a capillary are monitored.
Analysis of capillary and gravity forces	Capillary rise	The height of the meniscus in a capillary semi-immersed into the liquid is measured.
	Drop volume	The weight or volume of a drop falling from a capillary is measured.
Analysis of drop profile	Pendant drop	The profile of a pendant drop hanging from a capillary or a rod is fitted to the force balance described by the Bashforth–Adams equation.
	Sessile drop	A drop is placed on a substrate. Profile is fitted and contact angle is measured.
Reinforced distortion of drop (ultralow interfacial tension measuring methods)	Spinning drop	The shape of a drop deformed in a horizontal rotating tube is used to derive surface tension.
	Micropipette	Droplets are captured at the tip of a micropipette and then sucked into it. Pressure is recorded and related to surface tension.

In this configuration, the surface (or interfacial) tension is calculated by fitting the shape of the drop as detected from the picture to the theoretical drop profile according to the force balance equation governing the mechanical equilibrium of the interface. In the sessile drop configuration, a small amount of liquid is disposed on a solid support and the shape of the drop is governed by the balance of surface tensions at the solid–liquid, liquid–gas, solid–gas interfaces. In the sessile drop configuration (Fig. 20.1b), the intersection of the liquid/solid interface and the liquid/air interface defines the contact angle: a high contact angle indicates a low solid surface energy or chemical affinity (also referred to as a low degree of wetting); a low contact angle indicates a high solid surface energy or chemical affinity, and a high or sometimes complete degree of wetting. Besides surface/interfacial tension, ADSA methods provide simultaneously contact angle (for sessile drop), surface area and drop volume. The pendant drop and the sessile drop methods are particularly suitable for researchers as they do not require large amount of the liquid sample, they are easy to handle and can be adopted at high pressure–temperature conditions, combined with a very simple instrumentation [12]. Furthermore, modern advances in image and computational analysis significantly improved the precision of the ADSA techniques and reduced the time of the measurement, providing a very reliable opportunity for online analyses. All of these advantages have made ADSA suitable for numerous systems, ranging from organic liquids to molten metals, and for different areas such as experimentation with biological systems and application to systems which show aging effects [12, 13]. In the following, a review of the pendant drop method will be developed as, in the last years, this configuration has been widely used in the fields of polymer science and supercritical fluid nanotechnology, revealing the high potential and the reliability of this method.

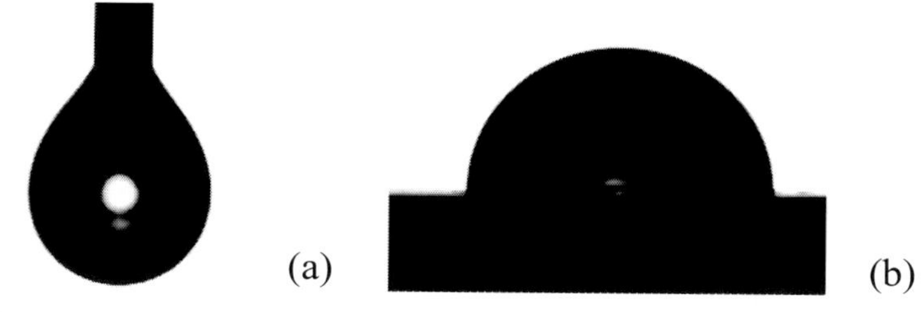

Figure 20.1 Pendant (a) and the sessile drops (b).

20.2 Theoretical Aspects of the ADSA–Pendant Drop Method

The shape of a drop is determined by a combination of surface and gravity effects. In fact, surface forces tends to make a drop spherical whereas gravity force tends to elongate a pendant drop or to flatten a sessile drop. Thus, in principle, surface tension can be measured by analyzing the shape of the drop, when gravitational and surface tension effects are comparable. The mathematical treatment consists in fitting the shape of an experimental drop to the theoretical drop profile according to the Young–Laplace equation of capillarity [14, 15]. In general, this equation describes the hydrodynamic and mechanical equilibrium condition for two homogeneous fluids separated by a curved interface, relating the pressure difference across an interface to the surface tension and the curvature of the interface:

$$\gamma\left(\frac{1}{R_1}+\frac{1}{R_2}\right)=\Delta P \tag{20.1}$$

where R_1 and R_2 are the two principal radii of curvature of the interface, and ΔP is the pressure difference across the interface. In case gravity is the only external force acting on the system, the pressure difference can be expressed as a linear function of the elevation:

$$\Delta P = \Delta P_0 + (\Delta\rho)\, gz \tag{20.2}$$

where ΔP_0 is the pressure difference at a reference plane and z is the vertical coordinate of the drop measured from the reference plane. The integration of the Young–Laplace equation is not straightforward for general irregular menisci. Fortunately, specific numerical procedures have been developed for the case of axisymmetric drops, as detailed reviewed in the work by Hoorfar and Neumann [12]. In fact, for axisymmetric interfaces, i.e., assuming that the interface is symmetric about the *z axis*, some geometrical considerations are used to solve the problem.

Referring to Fig. 20.2, the two principal radii of curvature, R_1 and R_2, can be related to the arc length, s, and to the angle of inclination of the interface to the horizontal, ϕ. Furthermore, the pressure difference at the apex is equal to:

$$\Delta P_0 = 2a\gamma \tag{20.3}$$

where a is the radius of curvature at the apex of the drop. Hence, the Young–Laplace equation can be written as the following system of ordinary differential equations:

$$\begin{aligned}\frac{dx}{ds} &= \cos\varphi \\ \frac{dz}{ds} &= \sin\varphi \\ \frac{d\varphi}{ds} &= 2a + cz - \frac{\sin\varphi}{x}\end{aligned} \tag{20.4}$$

where c is the capillary constant, defined as:

$$c = \frac{\Delta\varrho g}{\gamma} \tag{20.5}$$

The system Eq. 20.4 is solved with the following boundary conditions:

$$x(0) = z(0) = \varphi(0) = 0 \tag{20.6}$$

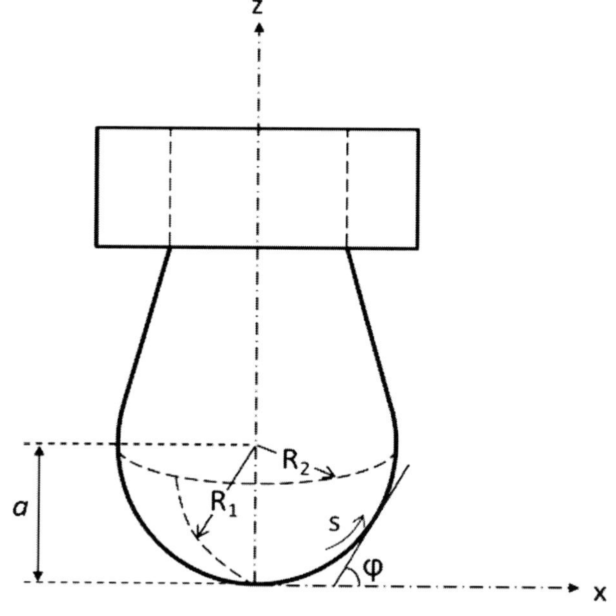

Figure 20.2 Geometry and notation of symbols of a pendant drop profile.

There is no analytical solution to the described problem and a numerical integration scheme is required. The first attempt was performed by Bashforth and Adams [16], before digital computers appeared. They generated sessile drop profiles for different values of surface tension and radius of curvature at the apex of the drop and the results were organized in tables. On the bases of Bashforth and Adams's tables, until the beginning of the 1980s, most of the methods were based on the determination of empirical relations between the capillary constant and dimensionless geometrical parameters of the pendant drop profile [9, 17–21]. In the last decades, several numerical methods—based on the Runge–Kutta, on the Burlisch–Stoer extrapolation and on predictor-corrector methods—have been proposed [22–25]. For a very detailed description of the numerical optimization methods, the reader is suggested to refer to the review by Hoorfar and Neumann [12].

20.3 Description of the Apparatus and of the Procedure

The general procedure of ADSA method is schematically reported in Fig. 20.3 and consists in: (1) drop image acquisition, (2) drop profile detection, and (3) numerical optimization [12].

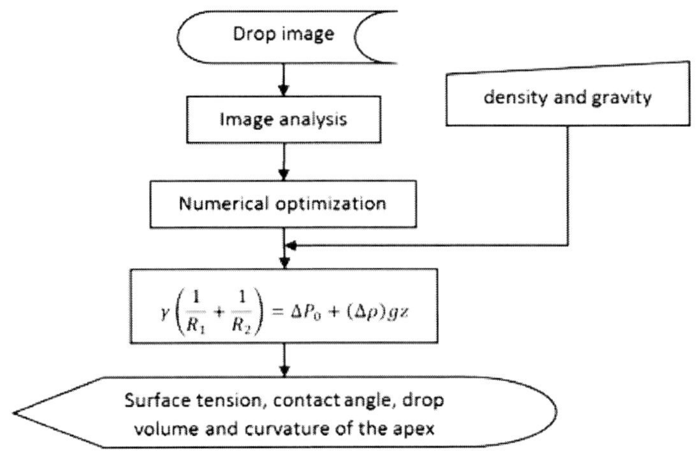

Figure 20.3 Scheme of the general procedure of axisymmetric drop shape analysis (ADSA).

The apparatus adopted for the whole process in made up of a device for forming and holding the drop (e.g., a syringe), a measuring chamber in which the pendant drop is set, optical components (i.e., light source, microscope lens, a CCD camera), a frame grabber used to generate a digital image of the sample into a PC and a commercial or custom-made software for drop profile fitting (Fig. 20.4). In the following, description of the components and details of image analysis will be given.

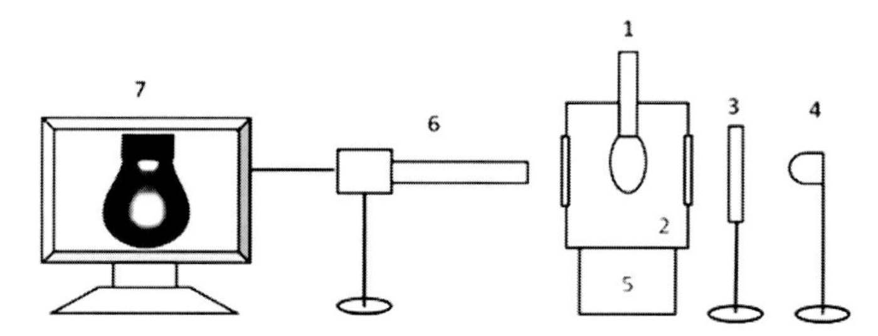

Figure 20.4 ADSA apparatus: (1) feeding/holding rod, (2) view cell, (3) light diffuser, (4) light source, (5) XYZ stage, (6) CCD with microscope, and (7) computer.

Measuring chamber and drop formation. A syringe is the most commonly used tool to form a pendant drop from low viscosity liquids. In order to automate the measurement and to create repeatable drops, the syringe can be actuated by a pump system. The formation of more viscous liquid drops (such as molten polymers) is not straightforward and several approaches have been proposed in order to ease the operation, without using heated syringes. For instance, in case of molten polymers, the proposed methods [26, 27] consist in fixing a small amount of solid polymer (such as a filament or a pellet) to the base of a special sample holders (such as capillaries or cylinder rods) and then heating in order to melt the polymer and to create the pendant drop. The liquid must not adhere the lateral surface of the holder, resulting in lack of the axisymmetry of the drop, thus compromising the reliability of the measurement (Fig. 20.5a). Also non perfectly vertical holders (rod or syringe) compromise the assumption of drop axisymmetry (Fig. 20.5b). Another critical aspect in drop preparation is the drop size (Fig. 20.5c–d) because it has been found that surface/interfacial tension value evaluated by using the pendant drop method might be drop volume dependent [28]. In particular, if the drop is larger than a critical value, it will neck and detach ("necking effect" [29]); conversely, if the drop is too small and close to spherical shape, numerical problems may arise in the fitting procedure [12]. In view of this limitation of ADSA for nearly spherical drop shapes, it is necessary to identify the range of drop sizes for which surface tension values are evaluated with a certain accuracy. As well described by Hoorfar et al. [12], the difference in shape between a given experimental drop and a spherical shape can be quantified by defining the shape parameter, which can be expressed as the difference between the drop volume and the volume of a sphere with radius a, i.e., the radius of curvature at the apex of the drop (Fig. 20.6a). Thus, a critical value of the shape parameter can be experimentally found to identify the range of drop sizes that guarantees a minimized error on the measurement. The drop shapes for which the shape parameter is larger than the critical value are thus acceptable for surface tension measurement. Another critical aspect in drop formation is sample contamination; in fact, impurities deposited on the surface of the drop compromise the results: it has been found that the presence of a dust layer on the drop act as a surfactant, thus reducing drastically the measured value of surface tension [30] (Fig. 20.5e). In order to avoid contamination, pure liquids, disposable needles and carefully cleaned holder should be adopted. Special measuring chamber equipped with view cells can be adopted for testing liquids at high temperatures, in special vapor atmospheres and at gas high pressures. The two optical quality windows mounted perpendicular to the axis of the cell allow for the acquisition of the drop image through the optical components. For the measurement of liquid–liquid interfacial tension, pendant drop of one liquid are formed at the tip of a stainless steel needle immersed in another liquid contained in a glass cuvette.

Optical components. The optical components are a light source, microscope lens and a charged couple device (CCD) camera. The light source provides a bright (preferably uniform) background, helping in the achievement of an optimal threshold for digitizing the drop image. A heavily frosted diffuser is commonly use to achieve the uniform bright background. A CCD camera equipped with microscope lens acquire the images or the video of the sample, with a large magnification. Spherical aberration and geometrical distortions are the most common problems of a lens system. The former can be minimized by using apochromatic lens; the latter can be avoided by correcting the digitalized image through the use of calibration grid pattern engraved on an optical glass slide.

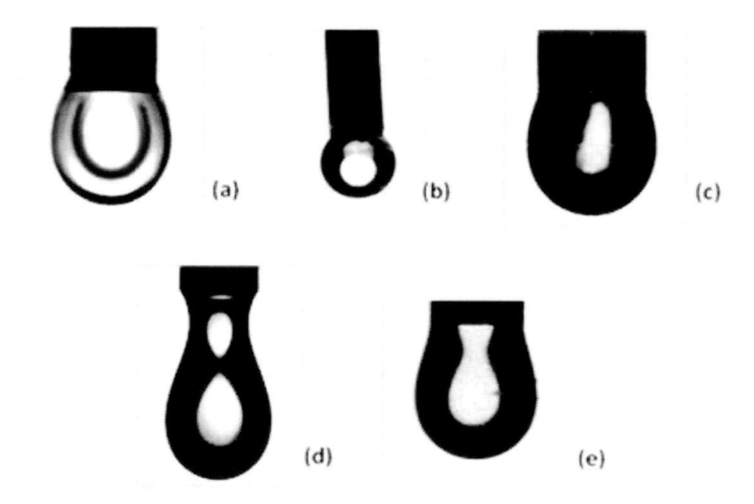

Figure 20.5 Common errors in pendant drop preparation: (a) drop sticking to the rod, (b) nonperfectly vertical rod, (c) drop smaller and (d) larger than a critical size, and (e) contaminated fluid.

Frame grabber. Image digitization consists in the acquisition of drop profile coordinates. Before digital image processing era, coordinates acquisition was performed manually: the 35 mm negatives were either projected onto a translucent (rear projection) manual digitization tablet or developed into photographic prints and digitized directly. The process was time consuming and the negatives or the photographic prints could shrink, warp, or the contrast could fade after a period of time [13]. Nowadays, thanks to considerable progresses in the field of digital image processing, image capture and digitization are automatic. In fact, the video source produces an analog video signal containing image data and this signal is transmitted to the image processor. The images are digitized by a frame grabber resident in the computer. This is a device that captures a single frame from the analog video signal and stores it as a digital image under computer control. The effect of frame grabber on the results of ADSA measurements have been investigated and has been found that, despite using a higher quality leads to more consistent and reliable results, the surface tension values obtained by using different frame grabbers differ by no more than 0.2 mJ/m^2 [12].

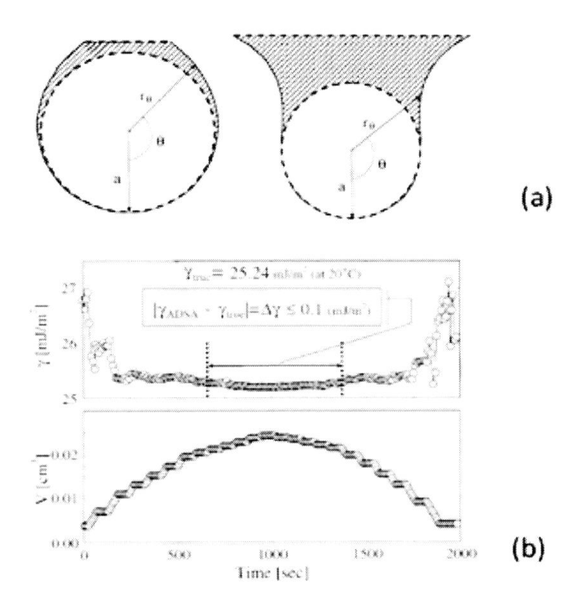

Figure 20.6 Effect of drop size: (a) sketches of nearly spherical drop and elongated drop (the hatched area between the drop profile and an inscribed circle with a radius of a corresponds represents the shape parameter), and (b) experimental results of a sequence of static experiments conducted for a pendant drop of cyclohexane: for a certain range of drop sizes, the difference between the surface tension values obtained from ADSA and the true surface tension of cyclohexane is less than ±0.1 mJ/m^2 [12]. This material is reproduced with permission of Elsevier.

ADSA software and image analysis. Having digitized the image in the computer, the image of the drop is analyzed in order to infer the value of surface/interfacial tension. The whole process of digitalization and analysis of the drop can be performed by using a commercial or a custom-made software and lasts less than 30 seconds. It consists of five steps:

1. Capture and digitalization of the image of the pendant drop (as described in the previous section).
2. Extraction of the drop profile: the Canny or the Sobel edge detector are adopted for searching for the pixels with maximum gradient along the drop profile.
3. Smoothing of the extracted contour of the drop using polynomial regression, to improve the precision of the detected edge [31, 32].
4. Correction and magnification: a calibration grid is adopted to convert the coordinates of the experimental profile in centimeters. The grid is also used to correct possible optical distortion due to the lenses [32].
5. Shape comparison between the theoretical and experimental drop: a numerical code fits a series of Laplacian, theoretical curves to the experimental profile. Details on the adopted numerical schemes can be found in Ref. 12.

20.4 Applications of Pendant Drop Method in Supercritical Fluid Technology

Supercritical fluid technology has made a giant step in the past decades in terms of fundamental understanding of solution behavior and commercial applications. In particular, a significant incentive has come from the environmental pressures on industry to move away from the use of volatile organic compounds (VOCs) and ozone-depleting substances (ODSs) as processing solvents. Supercritical fluids have shown their full potential as: (a) alternative solvents for classical separation processes (i.e., extraction, fractionation, adsorption, chromatography, and crystallization), (b) reaction media in polymerization or depolymerization processes, and (c) processing fluid for the production of particles, fibers, or foams. As summarized by Eggers and Jaeger [33], interfacial phenomena play a major role in a wide range of supercritical fluid-assisted processes. For instance, among the main factors determining the efficiency of the mass transfer in extraction processes is drop size which is strongly dependent on the interfacial tension value. Also, bubble nucleation rate is a strong function of interfacial tension thus playing a key role in polymer foaming processes. In the following, we present a review of papers focused on the application of the pendant drop method for the evaluation of the surface/interfacial tension in the fields of polymer science and particle formation. Table 20.2 lists some relevant examples of processes involving supercritical fluids and the role of interfacial properties of the polymer/supercritical fluids solutions.

Table 20.2 Interfacial phenomena in supercritical fluid processing [33]

Phenomenon	Process	Interfacial property to measure
Drop formation	Spray (RESS, GAS, etc.)	Interfacial tension, drop size, drop velocity
Bubble	Bubble column	Interfacial tension, bubble size/ shape
Wetting	Packed column	Contact angle, wetting
Foam	Bubble column	Foam stability, interfacial tension
Nucleation	Polymer processing	Critical pressure difference, interfacial tension
Emulsion	Reverse micelles	Phase behavior
Interfacial convection	Mass transfer operations	Schlieren method
Coalescence	Mixer-Settler	Coalescence rate

Polymer science. The surface tension of a molten polymer/gas solution or the interfacial tension between two immiscible polymer melts govern several processes in polymer technology, such as blending and foaming [34], extraction and fractionation [35, 36]. After the earliest works by Asakura [37] and Roe [38, 39], several researchers investigated the interfacial phenomena of molten polymers at high temperatures (polymer melts with ambient pressure gases [40], at high pressure (gases with oligomers or polymer solutions) [41, 42], and at simultaneous high-pressure and high-temperature (HP-HT) conditions [43, 44]. The pendant drop is the most commonly used method to measure the interfacial tension for polymer melts and its large potential has been revealed in the combined HP-HT applications, also in the supercritical regime. However, as previously stated, ADSA procedure requires the density of the molten polymer drop and the density of the fluid surrounding it. Very few experimental data on the density of polymer/gas solutions are available in literature [45]; hence, the application of ADSA at HP-HT conditions is generally based on a trial and error analysis of experimental data relying on theoretical prediction of the equilibrium mixture density obtained from solution theories grounded on statistical thermodynamics) [43, 44]. Some attempts to circumvent this hybrid approach are represented by the pioneering works by Dimitrov et al. [46], Jaeger et al. [26] and Pastore Carbone et al. [27]. In the approach proposed by Dimitrov [46], the observation of the polymer drop placed in CO_2 atmosphere is combined with a mass balance for the evaluation of the mass of dissolved CO_2 and, in turn, of the density of the gas-saturated polymer. Results by Dimitrov are reported in Fig. 20.7.

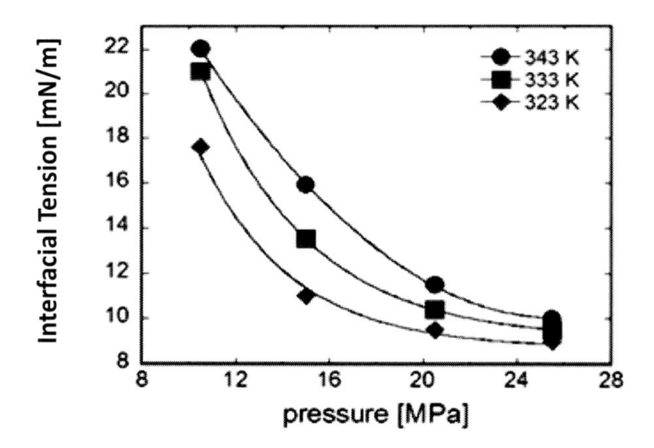

Figure 20.7 Effect of pressure on the interfacial tension in the system supercritical carbon dioxide/PEG–NPE. Data from Ref. [46].

The approaches proposed by Jaeger et al. [27] and Pastore Carbone et al. [28] are based on a simultaneous coupling of gravimetric gas sorption and ADSA measurements. The experimental setup consists of a magnetic suspension balance equipped with a high pressure and temperature view cell where both gravimetric and optical observation of the pendant drop are performed at the same time, at the same temperature and pressure (see Fig. 20.8). In the configuration proposed by Jaeger et al., sorption and ADSA tests are performed on the same sample as the pendant drop is directly mounted under the magnetic coupled balance. However, several practical reasons make gravimetric measurements difficult when carried out directly on the pendant drop: (1) pendant drops are very small in size and weight, thus reducing the accuracy of the gravimetric measurement, and (2) the positional lability of the rod-hook-balance coupling is responsible for continuous oscillation of the hanging sample, especially at high gas pressures, thus impeding a reliable optical monitoring of the drop profile and compromising the force balance due to the associated inertial forces. Furthermore, in the approach proposed by Jaeger et al., ADSA is adopted only for drop volume evaluation, which is needed for the appropriate correction of sorption data with the buoyancy effect; no surface tension data was provided in his work as it was focused on mass transport. In the light of the limits of Jaeger et al.'s approach, Pastore Carbone et al. have proposed the use of a double sample and extended it to the evaluation of surface tension of polymer–gas solutions. In this experimental configuration, while the magnetic suspension balance is measuring the weight change during sorption of the sample placed in a crucible, a high-resolution camera records the profile of a pendant drop hanging from a rod fixed to the typical metallic cage protecting the hook-balance coupling, thus minimizing oscillations and associated inertial forces. By adopting this approach, solubility, diffusivity, specific volume and surface tension of molten poly (ε-caprolactone)/CO_2 solutions were evaluated, in a single experiment, at several temperatures and CO_2 pressures (Fig. 20.9) [47].

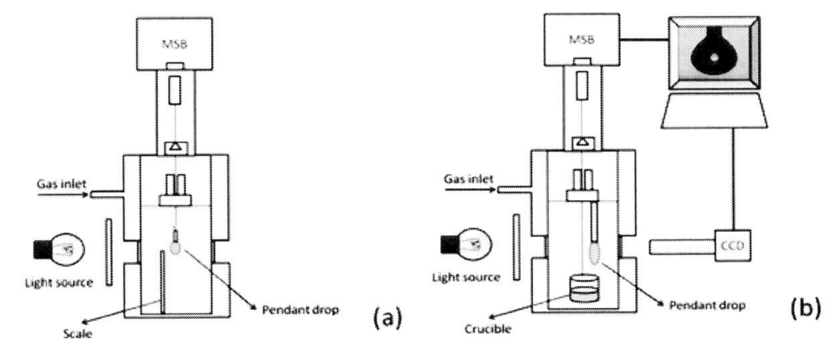

Figure 20.8 Experimental setup for simultaneous ADSA and gravimetric measurement as proposed by: (a) Jaeger et al. [27], and (b) Pastore Carbone et al. [28, 47].

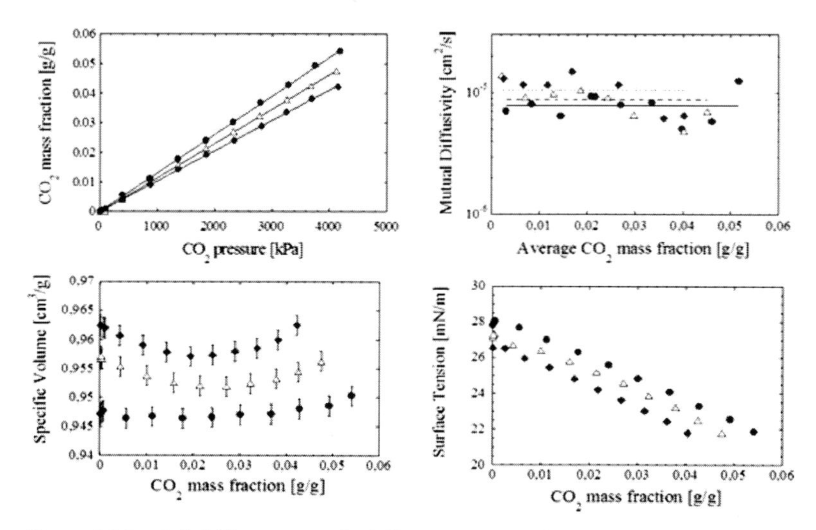

Figure 20.9 Solubility, mutual diffusivity, specific volume and surface tension of molten PCL/CO_2 solutions at different temperatures (80°C ●; 90°C Δ; 100°C ♦) as obtained by Pastore Carbone et al. by using a coupled gravimetric-ADSA technique [47].

Figure 20.10 exhibits the comparison between the surface tension as evaluated by semi-empirical approach and the one determined by using the fully-experimental method proposed by Pastore Carbone et al. [47]. This comparison highlights the differences among the fully experimental based calculations and the properties derived on the basis of theoretical assumptions, thus evidencing the importance of the adopted coupled measurement to obtain a reliable estimation of the surface tension of the solution.

Regarding the interfacial tension between immiscible polymers, very few data are reported in literature so far, despite they would be very useful for investigation of polymer blending. By using the pendant drop method, Arashiro and Demarquette [48, 49] studied the influence of temperature, molecular weight, and polydispersity of polystyrene on interfacial tension between low-density polyethylene (LDPE) and polystyrene (PS) (see Fig. 20.11). The effect of supercritical carbon dioxide ($scCO_2$) on the interfacial tension between molten polystyrene (PS) and low density polyethylene (LDPE) was studied by Xue et al. [50], by using the pendant drop method. By using a special high pressure syringe, the PS pendant drop was injected into the LDPE melt placed in high pressure optical cell. Values of interfacial tension of the PS/LDPE interface are shown as a function of CO_2 pressure in Fig. 20.12.

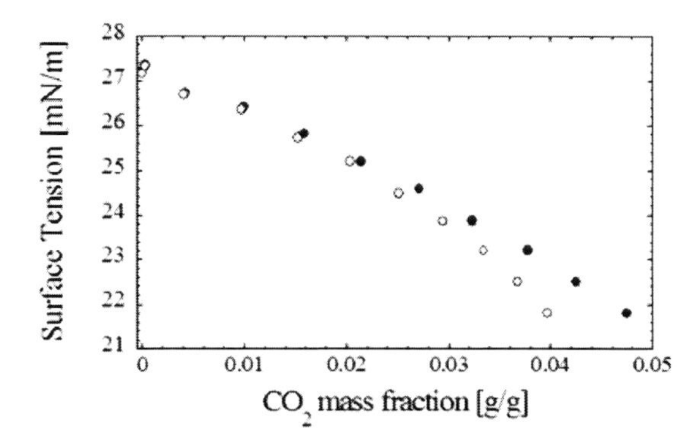

Figure 20.10 Comparison of the fully experimental approach (full symbols) with the semi-empirical one (open symbols): surface tension of the PCL/CO_2 solution as function of gas concentration at 90°C. Data from Ref. [47].

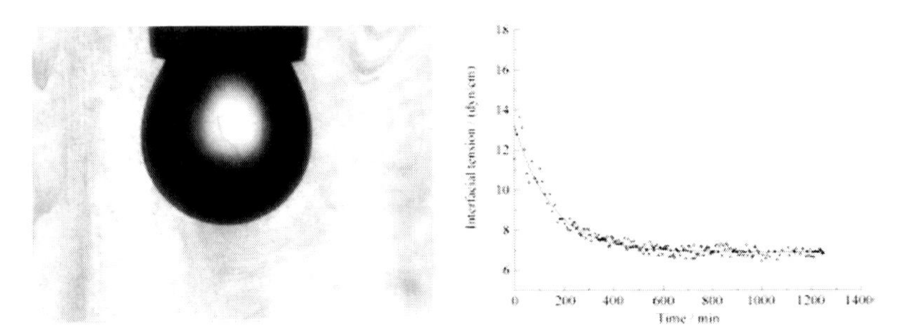

Figure 20.11 Pendant drop of molten polystyrene in low-density polyethylene at 202°C and interfacial tension values as a function of time [48]. This material is reproduced with permission of the publisher.

Particle formation. The use of supercritical fluids has gained ground in the production of particle with nanometer size, especially in pharmaceutical and food industries. The rapid expansion of supercritical solutions (RESS), the gas antisolvent process (GAS), the supercritical antisolvent process (SAS), and the particles from gas-saturated solution (PGSS) processes are the most used processes, which use different nucleation and growth mechanisms of precipitating particles. The supercritical fluid extraction of emulsions (SFEE) is a new promising technology that is based on the combination of nanoemulsions with supercritical fluids, allowing for the production of nanoparticles from natural substances [51]. It consists in extracting the organic solvent from the droplets of an oil-in-water emulsion by using supercritical carbon dioxide. SFEE offers the possibility of tailoring final particle size by changing the droplet size of the emulsion. Jaeger et al. [52] have investigated the phenomena occurring during the precipitation process of SFEE by using ADSA technique. Both pendant drop and sessile drop methods were adopted to elucidate what happens within a drop of dichloromethane in water in contact with CO_2 at high pressure dispersed phase while the precipitation process is being carried out. Experiments were performed with and without a solute (β-carotene) and a surfactant dissolved in the drop, and the evolution of the drop volume as well as of the interfacial tension between the drop and the aqueous phase was measured (Figs. 20.13 and 20.14). Interesting results were found: interfacial tension of the system was found to increases due to the diffusion of dichloromethane out of the drop, leading to a destabilizing effect on the emulsion that can have important implications for applications that involve long contact of the emulsion with the supercritical fluid.

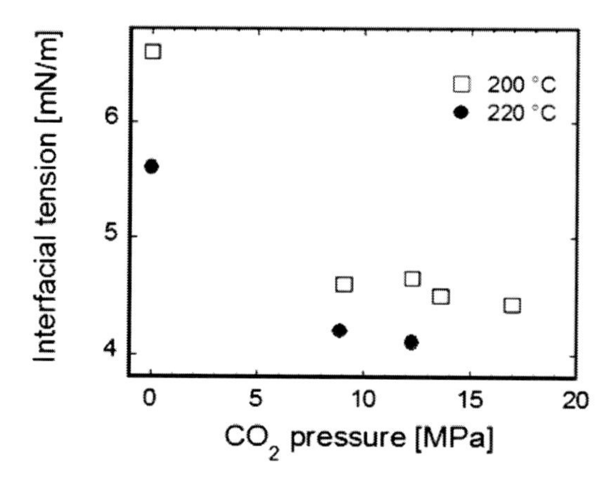

Figure 20.12 Effects of $scCO_2$ on the interfacial tension between PS and LDPE. Data from Ref. [50].

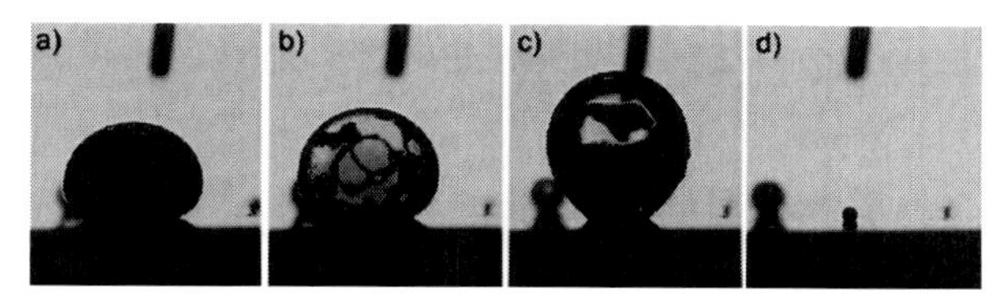

Figure 20.13 β-carotene particle precipitation during an experiment at pressure of 10 MPa and temperature of 308 K: (a) beginning of particle formation, (b, c) particle agglomeration, and (d) final condition after the drop detachment from the cell's surface [52]. This material is reproduced with permission of John Wiley & Sons, Inc.

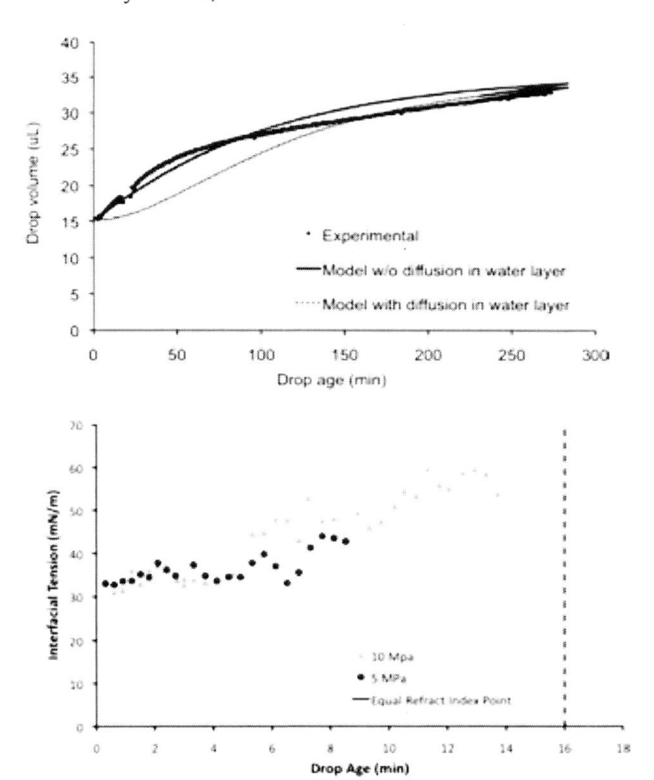

Figure 20.14 Drop volume evolution vs. drop age in an experiment at pressure of 10 MPa and temperature of 308 K, experimental and model results. Interfacial tension evolution curve of the system DCM–H_2O–CO_2 at 308 K at 5 and 10 MPa [52]. This material is reproduced with permission of John Wiley & Sons, Inc.

References

1. Anastasiadis, S. H., Chen, J.-K., Koberstein, J. T., Siegel, A. F., Sohn, J. E., and Emerson, J. A. (1987). The determination of interfacial tension by video image processing of pendant fluid drops, *J. Colloid Interface Sci.*, **119**, pp. 55–66.
2. Adamson, A. W., and Gast, A. P. (1997). *Physical Chemistry of Surfaces*, 6th Ed. (John Wiley & Sons Inc, New York, USA).
3. Padday, J. F. (1969). Surface tension. Part II. The measurement of surface yension, in *Surface and Colloid Science*, Matijevic, E. (Ed.), (John Wiley & Sons Inc, New York, USA), pp. 101–149.
4. Ambwani, D. S., and Fort, Jr., T. (1979). Pendent drop technique for measuring liquid boundary tensions, in *Surface and Colloid Science*, Good, R. J., and Stromberg, R. R. (Eds.), (Plenum, New York and London), pp. 93–119.
5. Neumann, A. W., and Good, R. J. (1979). Techniques of measuring contact angles, in *Surface and Colloid Science*, Good, R. J., and Stromberg, R. R. (Eds.), (Plenum, New York and London), pp. 31–91.
6. Drelich, J., Fang, C., and White, C. L. (2002). Measurement of interfacial tension in fluid-fluid systems, in *Encyclopedia of Surface and Colloid Science*, Hubbard, A. T. (Ed.), (Marcel Dekker, New York, USA), pp. 3152– 3166.
7. Gouy, G. (1889). On the osmotic balance and concentration of solutions by gravity, *C. R. Acad. Sci.*, **109**, pp. 102–110.
8. Gouy, G. (1888). Notes on brownian motion, *J. Phys.*, **7**, pp. 561–565.
9. Andreas, J. M., Hauser, E. A., and Tucker, W. B. (1938). Boundary tension by pendant drop, *J. Phys. Chem.*, **42**, pp. 1001–1019.
10. Fordham, S. (1948). On the calculation of surface tension from measurements of pendant drops, *Proc. R. Soc. Lond. Ser. A,* **194**, pp. 1–16.
11. Stauffer, C. J. (1965). Measurement of surface tension by pendant drop technique, *J. Phys. Chem.*, **69**, pp. 1933–1938.
12. Hoorfar, M., and Neumann A. W. (2006). Recent progress in axisymmetric drop shape analysis (ADSA), *Adv. Colloid Interf. Sci.*, **121**, pp. 25–49.
13. Cheng, P., Li, D., Boruvka, L., Rotemberg, Y., and Neumann A. W. (1990). Automation of axisymmetric drop shape analysis for measurements of interfacial tensions and contact angles, *Colloid Surf.*, **43**, pp. 151–167.

14. Young, T. (1805). An Essay on the cohesion of fluids, *Philos. Trans. R. Soc. Lond.*, **95**, pp. 65–87.

15. Laplace, P. S. (1805). Traité de Mécanique Céleste, Supplement to Book 10, (Gauthier–Villars, Paris).

16. Bashforth, F., and Adams, J. C. (1892). *An Attempt to Test the Theory of Capillary Action* (Cambridge).

17. Huh, C., and Reed, R. L. (1991). A method for estimating interfacial tensions and contact angles from sessile and pendant drop shapes, *J. Colloid Interface Sci.*, **91**, pp. 472–484.

18. Ramos, L., Redner, R. A., and Cerro, R. L. (1993). Surface tension from pendant drop curvature, *Langmuir*, **9**, pp. 3691–3694.

19. Stauffer, E. (1965). The measurement of surface tension by the pendant drop technique, *J. Phys. Chem.*, **69**, pp. 1933–1938.

20. Fordham, S. (1948). On the calculation of surface tension from measurements of pendant drops, *Proc. R. Soc. Lond. Ser. A*, **149**, pp. 1–16.

21. Roe, R. J. (1970). Surface tension of polymer liquids, *J. Phys. Chem.*, **72**, pp. 2013–2017.

22. Carlà, M., Cecchini, R., and Bordi, S. (1991). An automated apparatus for interfacial tension measurements by the sessile drop technique, *Rev. Sci. Instrum.*, **62**, pp. 1088–1092.

23. Girault, H. H., Schiffrin, D. J., and Smith, B. D. V. (1984). The measurement of interfacial tension of pendant drops using a video image profile digitizer, *J. Colloid Interface Sci.*, **101**, pp. 257–266.

24. Girault, H. H., Schiffrin, D. J., and Smith, B. D. V. (1982). Drop image processing for surface and interfacial tension by the axisymmetric drop technique, *J. Electroanal. Chem.*, **137**, pp. 207–217.

25. Hansen, F. K., and Rodsrud, G. (1991). Surface tension by pendant drop: I. A fast standard instrument using computer image analysis, *J. Colloid Interface Sci.*, **141**, pp. 1–9.

26. Jaeger, P. T., Schnitzler, J. V., and Eggers, R. (1996). Interfacial tension of fluid systems considering the nonstationary case with respect to mass transfer, *Chem. Eng. Tech.*, **19**, pp. 197–202.

27. Pastore Carbone, M. G., Di Maio, E., Iannace, S., and Mensitieri, G. (2011). Simultaneous experimental evaluation of solubility, diffusivity, interfacial tension and specific volume of polymer/gas solutions, *Polym. Test.*, **30**, pp. 303–309.

28. Morita, A. T., Carastan, D. J., and Demarquette, N. R. (2002). Influence of drop volume on surface tension evaluated using the pendant drop method, *Coll. Polym. Sci.*, **280**, pp. 857–864.
29. Park, H., Park, C. B., Tzoganakis, C., Tan, K. H., and Chen, P. (2006). Surface tension measurement of polystyrene melts in supercritical carbon dioxide, *Ind. Eng. Chem. Res.*, **45**, pp. 1650–1658.
30. Krishnan, A., Liu, Y. H., Cha, P., Woodward, R., Allara, D., and Vogler E. A. (2005). An evaluation of methods for contact angle measurement, *Colloid Surf. B*, **43**, pp. 95–98.
31. Lahooti, S., del Río, O. I., Cheng, P., and Neumann A. W. (1996). Axisymmetric drop shape analysis (ADSA), in *Applied Surface Thermodynamic*, Neumann, A. W., and Spelt, J. K. (Eds.), (Marcel Dekker Inc., New York, USA), pp. 441–507.
32. Cheng, P. (1990). *Automation of Axisymmetric Drop Shape Analysis Using Digital Image Processing*, PhD Thesis, University of Toronto, Toronto.
33. Eggers, R., and Jaeger P. (2004). Interfacial phenomena in countercurrent and spray processing using supercritical fluids, in *Supercritical Fluids as Solvents and Reaction Media*, Brunner, G. (Ed.), (Elsevier Science, Amsterdam), pp. 363–377.
34. Wu, S. (1987). Formation of dispersed phase in incompatible polymer blends: interfacial and rheological effects, *Polym. Eng. Sci.*, **27**, pp. 335–343.
35. McHugh, M. A., and Krukonis, V. J. (1994). *Supercritical Fluid Extraction: Principles and Practice*, 2nd ed., (Butterworths, Boston, MA, USA).
36. Kumar, S. K., Suter, U. V., and Reid, R. C. (1986). Fractionation of polymers with supercritical fluids, *Fluid Phase Eq.*, **29**, pp. 373–382.
37. Asakura, S., and Oosawa, F. (1954). On interaction between two bodies immersed in a solution of macromolecules, *J. Chem. Phys.*, **22**, pp. 1255–1256.
38. Roe, R. J. (1965). Parachor and surface tension of amorphous polymers, *J. Phys. Chem.*, **69**, pp. 2809–2816.
39. Roe, R. J. (1968). Surface tension of polymer liquids, *J. Phys. Chem.*, **72**, pp. 2013–2017.
40. Kwok, D. Y., Cheung, L. K., Park, C. B., and Neumann, A. W. (1998). Study on the surface tensions of polymer melts using axisymmetic drop shape analysis, *Polym. Eng. Sci.*, **38**, pp. 757–764.
41. Harrison, K. L., da Rocha, S. R. P., Yates, M. Z., and Johnston, K. P. (1998). Interfacial activity of polymeric surfactants at the polystyrene-carbon dioxide interface, *Langmuir*, **14**, pp. 6855–6863.

42. Li, H., Lee, J., and Tomasko, D. L. (2004). Effect of carbon dioxide on the interfacial tension of polymer melts, *Ind. Eng. Chem. Res.*, **43**, pp. 509–514.

43. Park, H., Park, C. B., Tzoganakis, C., Tan, K. H., and Chen, P. (2006). Surface tension measurement of polystyrene melts in supercritical carbon dioxide, *Ind. Eng. Chem. Res.*, **45**, pp. 1650–1658.

44. Liu, D., and Tomasko, J. (2007). Carbon dioxide sorption and dilation of poly(lactide-co-glycolide), *J. Supercrit. Fluids*, **39**, pp. 416–425.

45. Funami, E., Taki, K., and Ohshima, M. (2007). Density measurement of polymer/CO_2 single-phase solution at high temperature and pressure using a gravimetric method, *J. Appl. Polym. Sci.*, **105**, pp. 3060–3068.

46. Dimitrov, K., Boyadzhiev, L., and Tufeu, R. (1999). Properties of supercritical CO_2 saturated poly(ethylene glycol) nonylphenyl ether, *Macromol. Chem. Phys.*, **200**, pp. 1626–1629.

47. Pastore, M., Di Maio, E., Scherillo, G., Mensitieri, G., and Iannace, S. (2012). Solubility, mutual diffusivity, specific volume and interfacial tension of molten PCL/CO_2 solutions by a fully experimental procedure: effect of pressure and temperature, *J. Supercrit. Fluids*, **67**, pp. 131–138.

48. Arashiro, E. Y., and Demarquette, N. R. (1999). Use of the pendant drop method to measure interfacial tension between molten polymers, *Mater. Res.*, **2**, pp. 23–32.

49. Arashiro, E. Y., and Demarquette, N. R. (1999). Influence of temperature, molecular weight, and polydispersity of polystyrene on interfacial tension between low-density polyethylene and polystyrene, *J. App. Polym. Sci.*, **74**, pp. 2423–2431.

50. Xue, A., Tzoganakis, C., and Chen, P. (2004). Measurement of interfacial tension in PS/LDPE melts saturated with supercritical CO_2, *Polym. Eng. Sci.*, **44**, pp. 18–27

51. Shekunov, B. Y., Chattopadhyay, P., Seitzinger, J., and Huff, R. (2006). Nanoparticles of poorly water-soluble drugs prepared by supercritical fluid extraction of emulsions, *Pharm. Res.*, **23**, pp. 196–204.

52. Mattea, F., Martin, A., Schulz, C., Jaeger, P., Eggers, R., and Cocero M. J. (2010). Behavior of an organic solvent drop during the supercritical extraction of emulsions, *AIChE J.*, **56**, pp. 1184–1195.

Index